1401 Glueser

— 2022 —

THE FLOW OF
HOMOGENEOUS FLUIDS
THROUGH POROUS MEDIA

THE FLOW OF
HOMOGENEOUS FLUIDS
THROUGH POROUS MEDIA

BY

M. MUSKAT, Ph.D.

INTERNATIONAL HUMAN RESOURCES DEVELOPMENT CORPORATION
Boston

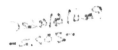

The Society of Petroleum Engineers, Inc., acknowledges permission granted by International Human Resources Development Corp. to reprint this book.

ISBN: 0-934634-16-5

Printed in the United States of America

DEDICATED TO MY PARENTS

PREFACE

In explaining the *raison d'être* of a technical book it is customary for the author to point out that the available texts or treatises do not cover or stress the particular features of the subject that he emphasizes or do not treat the subject from the same point of view. In the present case, however, the situation is quite reversed. The field has been a wide open one, for practical purposes untouched by the literary efforts of those writing books for either collegiate instruction or industrial consumption. Even as this work is completed, no comprehensive elementary treatments of the flow of homogeneous fluids through porous media are available to which one might refer as an introduction to the more advanced developments of the subject; nor are there any collected advanced discussions of any completeness requiring a more elementary introductory exposition as a background for the understanding of the more difficult problems of interest. In fact, the totality of *summarized* treatments of the flow of homogeneous fluids in porous media given heretofore in book form consists in single chapters devoted to the subject in "Hydraulik" by Ph. Forchheimer and "Wasserbauliche Stromungslehre" by P. Nemenyi, and the recently published book "Grundwasserstromung" (1936) by R. Dachler. In none of these, however, are problems of compressible liquids or gases discussed, while the treatments of incompressible liquids are brief and include only some of the cases presented in Chaps. IV and VI of this book.

In view of this situation, it has seemed best to make the present work as complete as possible. However, no attempt has been made simply to discuss every problem of fluid flow. Rather the purpose has been to treat all *typical* problems that are of some practical interest. Particular stress has been laid upon the illustration of the various analytical methods available for solving flow problems, and in some cases this has been carried out at the deliberate expense of brevity, which might have been attained by the use of a method already illustrated by another problem. In fact, from an analytical point of view, the whole of Part II might be summarized under the heading "Exercises in Potential Theory."

With respect to the physical significance of the problems treated in this work, they are admittedly in many cases of interest only in the oil or gas industries. However, it is believed that a considerable part of the discussion should be applicable to the study of hydrological questions, irrigation, and dam-construction problems, and to problems involving fluid flow through refractories and ceramic materials. Although no detailed engineering discussions of such applications are given in the text, and no attempt has been made to segregate the problems with respect to their interest in the fields of oil production, hydrology, irrigation, etc., references to these related fields are indicated wherever pertinent. On the other hand, no pretense is made of solving what is really the fundamental physical problem in most actual cases of oil production, namely, the nature of the flow of gas-oil *mixtures* through porous media. This problem is one in the flow of *heterogeneous* fluids through porous media, while the scope of this work is explicitly limited to that of the flow of *homogeneous* fluids. However, as explained in detail in the text, it is felt that there is nevertheless a considerable range of applicability to real oil-production problems of analytical results derived from a homogeneous-fluid theory, and indeed they may well find still wider application in the future as advanced production methods become more common and more efficient use is made of the possibilities of natural water drives.

This book has not been written simply as a college text, nor has it been prepared exclusively for reference purposes. A thorough knowledge of the calculus and some familiarity with the elements of differential equations are presupposed in the developments beyond Chap. II. Although no reasonably tractable problems have been excluded simply because of the complexity of the analysis, all those requiring special methods have been discussed in considerable detail. In fact, the treatments have been given with such detail as it was felt will clearly illustrate the analytical content of the method of solution as well as the physical significance of the problem. Indeed, it is hoped that the reader who goes through the analytical sections will find no difficulty in solving for himself other problems of similar nature that may occur to him. On the other hand, for those who are interested essentially in the results and do not care to follow the analytical derivations, extended nonmathematical summaries have been appended to each chapter, except the first two, giving the physical content of the various problems discussed and the salient features of the solutions derived theoretically. Moreover a complete list of the quantitative results derived in the text, in formula or graphical form, is given in the last appendix to the book.

The material treated in the text seemed to find a natural subdivision into four parts. In the first, essentially nonmathematical, are developed the foundations and background for the later detailed analytical treatments. The experimental basis for the fundamental laws of flow of homogeneous fluids through porous media is presented, and the technique is described for determining the basic dynamical constant characterizing a porous medium as the carrier of a homogeneous fluid, namely, its permeability. The empirical laws are then formulated generally into the three partial differential equations which form the starting points for the analysis of the other three parts. The next part, the most extended of all, treats problems of fluid flow in the steady state, the compressibility of the liquids being formally neglected in the sense that the fundamental dependent variable satisfying Laplace's equation is taken as the pressure rather than the density. Potential-theory methods are then applied successively for the solution of two-dimensional problems, those of three dimensions, gravity-flow systems, those in which the porous medium is of nonuniform permeability, systems involving two homogeneous fluids, and finally multiple-well systems. In several cases actual experiments with sand and electrical models, performed to give results for cases where the analysis becomes impractical, are described and discussed. Part III gives in a single chapter the treatment of typical cases of the non-steady-state flow of liquids, as based upon the Fourier heat-conduction equation in the fluid density. The closing Part IV, also a single chapter, is intended to give the solutions for typical cases of gas flow, though here the analysis for the unsteady state is only approximate owing to the difficulty of obtaining exact solutions to the basic differential equation. No pretense is made that the discussions of Parts III and IV exhaust these phases of the subject. Rather, as explained in detail in the text, the approximations of the steady state are well justified in most cases of practical interest, so that the non-steady-state solutions have been limited to those problems which at present definitely have some practical significance.

Although no claim is made for any originality in the *methods* of solution presented in this work—for they are all well-known developments of potential theory and the theory of heat conduction—a large part of the actual analysis has been taken from published papers of the author and his colleagues, and their reports written for the Gulf Research & Development Company during the last six years, or from derivations prepared for the purpose of this book. For, despite the fact that as long ago as 1899 C. S. Slichter in a classical paper pointed out that Darcy's fundamental law immediately led to the Laplace equation for the pressure distribution in a porous medium and gave the solutions for several specific examples, the literature, until very recently, has been practically barren of extensions or developments of this fundamental work. Perhaps the most notable exceptions are the significant papers of Hopf and Trefftz, Weaver, Hamel, and Wedernikow discussed in Secs. 6.2, 4.11, 6.3, 6.8, and 6.9, and several papers of Kozeny and Dachler. The published papers of most other earlier investigators of the problems of fluid flow through porous media have dealt with experimental studies, in particular the relation of the permeability to the grain structure of a porous medium.

It is a deep pleasure to acknowledge the help and encouragement of the many people who have made this work possible. First among these is R. D. Wyckoff of the Gulf Research & Development Company, with whom the author had the privilege of working from 1931 to 1936 while the major part of the manuscript was being written. In fact, this book was originally begun as a joint undertaking with Mr. Wyckoff, and its plan and much of the material were outlined jointly. However, the pressure of his other work forced him to withdraw from the original program, although he very generously did take the time to write the first chapter. In addition, Mr. Wyckoff actually collaborated with the author directly in the study of many of the problems treated in this book, and practically all of them were discussed with him during the $5\frac{1}{2}$ years he was Chief of the Physics Division of this Laboratory. For his first introduction to the subject of this book the author is indebted to Dr. A. E. Ruark, formerly of this Laboratory.

The other members of this Laboratory who have worked on the flow of fluids through porous media, Messrs. H. G. Botset, W. N. Arnquist, D. W. Reed, and M. W. Meres, have also been of invaluable aid to the author throughout the preparation of this manuscript in clarifying his ideas and making useful suggestions. Many of the discussions are the result of Mr. Botset's researches or of joint investigations of Mr. Botset with the author or other members of the Laboratory. And Mr. Meres has been most helpful in carrying through the numerical calculations for most of the quantitative solutions presented in this work, as well as in making all the drawings.

The whole of this work would, however, have been impossible without the constant assistance and encouragement of Dr. Paul D. Foote, Executive Vice-President of the Gulf Research & Development Company. Both his discussions of specific physical problems of fluid flow and criticisms of the plan of the book have contributed greatly to whatever merit it may have. And in addition to giving him the freedom for working on this book he has generously provided the author with complete stenographic service—excellently performed by Mrs. H. B. Crouse—and printing facilities, executed in expert manner by Messrs. C. G. Heim and W. S. Jones, all of this Laboratory.

For carrying through in a most painstaking and thorough manner the tedious task of proofreading, the author is grateful to Miss F. B. Metzger, Librarian of the Laboratory, who also was very helpful in securing the wide variety of reference volumes necessary for thoroughly covering the published literature of the subject of this book. Finally appreciation is extended to the many other members of this Laboratory whose kind interest and varied assistance on many occasions made the writing of this book a most pleasant experience.

MORRIS MUSKAT.

PITTSBURGH, PA.,
September, 1937.

CONTENTS

CHAPTER III

GENERAL HYDRODYNAMICAL EQUATIONS FOR THE FLOW OF FLUIDS THROUGH POROUS MEDIA

PART II. STEADY-STATE FLOW OF LIQUIDS

CHAPTER IV

TWO-DIMENSIONAL FLOW PROBLEMS AND POTENTIAL-THEORY METHODS 149

CONTENTS   xvii

PART III.   THE NONSTEADY-STATE FLOW OF LIQUIDS

CHAPTER X

PART IV.   THE FLOW OF GASES THROUGH POROUS
MEDIA

CHAPTER XI

ERRATUM

On page 549, line 10, read: *"has been seen"* for *"has seen"*

PART I
FOUNDATIONS

CHAPTER I

BY R. D. WYCKOFF

INTRODUCTION

1.1. Scope of Subject.—The subject matter of this treatise, the flow of homogeneous fluids through porous media, while distinctly limited by the qualifications "homogeneous" and "porous," is nevertheless sufficiently wide in scope to find application in many branches of applied science. In such fields as ground-water hydrology, encompassing the provision and maintenance of water supplies, irrigation, and drainage problems, and petroleum engineering, involving the production of gas and oil from their underground reservoirs, the applications are evidently of basic importance. Equally important are applications to specific problems encountered in civil engineering, agricultural engineering, and many industries. Thus the diffusion and flow of fluids through ceramic materials as bricks and porous earthenware has long been a problem of the ceramic industry, and the flow of gas through molding sand a problem of the foundry. The construction of filter beds for municipal water systems and the questions of water seepage through, around, and beneath dams, earthen reservoirs, and the like, have been important phases of civil engineering for which satisfactory discussions have been available only within the last few years. The scientific treatment of the problems of irrigation, soil erosion, and tile drainage, although still open to further development, can be considered already as a well-established branch of the general theory of the flow of fluids through porous media. The draining of artesian basins by deep wells or of rivers and canals by wells contiguous thereto is a problem in the flow of liquids through porous rocks or sands, as is the general subject of ground-water hydrology. And, of course, the whole physical problem of the production of oil and gas from underground sources is nothing more than that of fluid flow through porous media.

3

Such is the scope of this work. The specific problems that might be proposed and which would lie within its proper domain are almost unlimited in number, and the discussion of all of them would necessitate many volumes and unwarranted duplications. A selection has therefore been made to include typical problems of interest from a practical point of view, as well as to illustrate the various analytical methods that may be used in the mathematical treatment of the subject. It is hoped that these methods will suffice at least for the approximate solution of most of the related problems that may occur to the reader.

Before entering further into the consideration of the fundamental bases which will underlie the discussions to be presented in this work, it is necessary to define explicitly the limitations to be imposed upon its scope. As already suggested, these limitations may be divided into those which refer to the character of the fluid passing through the porous medium and those relating to the nature of the porous medium itself.

1.2. Homogeneous Fluids.—By the term "homogeneous fluid" is meant essentially a single-fluid phase. This may be either a gas or a liquid, but a mixture of the two resulting in gas-liquid interfaces such as characterize gas bubbles dispersed throughout a liquid are definitely to be excluded. Of course, if the fluid is a liquid, it may contain dissolved gas without invalidating the premises upon which the analysis will be based, provided, however, that the fluid pressures do not fall below the saturation pressure and thus release free gas within the system. Likewise, a gaseous fluid may be, or may contain, a condensible vapor and still fall within the scope of this work, provided such regions of the porous medium where the vapor is in equilibrium with its condensed phase at the temperature of the system are excluded from the discussion. Evidently a liquid system composed of immiscible components, as oil and water, cannot be considered as a homogeneous fluid if they are present as a dispersed mixture. In fact, as will become more apparent later, the requirement of homogeneity as referred to here is fulfilled only by mixtures of fluids which are completely miscible and remain so throughout the system.

Although the flow of heterogeneous fluids, in particular gas-liquid mixtures, is of fundamental importance for the understanding of many of the phases of oil production from underground "reservoirs," the interest in such heterogeneous fluid systems is largely confined to the science of oil production. While the foundations and specific basic facts regarding the behavior of such systems now appear to be definitely established empirically[1] and the analytical solutions for certain cases have been carried out from these formulations,[2] the graphical and numerical work involved in the development of such solutions is so lengthy that a great amount of time will be required for the accumulation of enough specific results to warrant practical generalizations and detailed discussion. In fact, until a much wider range of results from quantitative studies of heterogeneous fluid systems has been accumulated, it appears advisable to omit their discussion altogether.

On the other hand, it is not to be concluded that the applicability of the results of the present work to the subject of oil production is thereby completely nullified. On the contrary, most of the discussion given here, which inherently has any pertinence to flow systems of the type involved in oil production, such as that developed in Chap. V and in Chaps. VII to X, can, with careful and discriminating interpretation, be applied with reasonable certainty of qualitative validity to actual oil-production problems. For with respect to such parts as Chap. II in which is outlined the empirical technique for the determination of the significant constants of a porous medium, it is clear that while the porosity and homogeneous-fluid permeability will not completely define the sand as a carrier of a "heterogeneous fluid,"[3] the determination of these constants will at least be a necessary step in finding whatever the complete definition may require. As to the discussion of such problems as that of partially penetrating wells in Chap. V, it is evident that while the absolute values of the liquid flux from a sand bearing a gas-liquid mixture may be appreciably different from those found in Sec. 5.4, it is not unlikely that their variation with the well penetration will be closely approximated by the results given there.

Similarly, while problems of normal water encroachment and water flooding as applied in secondary recovery methods involve heterogeneous fluids as the intimate mixing of the water and oil

[1] Cf. WYCKOFF, R. D., and H. G. BOTSET, Physics, 7, 325, 1936.
[2] Cf. MUSKAT, M., and M. W. MERES, Physics, 7, 346, 1936.
[3] Cf. WYCKOFF and BOTSET, and MUSKAT and MERES, loc. cit.

within the porous medium develops, the primary features of such systems are adequately described by the theory to be presented here. In fact, from what is now known of the behavior of mixtures, it is not inconceivable that the rigorous solution of many of the systems involving heterogeneous fluids will always entail such laborious numerical work that the simplifications developed here, based on the assumption of the homogeneity of the fluids, may provide the only solution practically attainable.

It should be emphasized, however, that although many of the problems treated in the following chapters do deal directly with situations arising in the production of oil or gas, our main interest here will lie in the *flow* features of the oil or gas reservoir rather than in the production-engineering phases of the problem. In particular, our problems will be supposed to begin only at the sand face exposed by the well bore, the effect of the latter or of the equipment in it being of interest in our studies only as they affect the conditions of pressure or flux at the sand face. The actual operation of an oil well either while flowing or pumping, or of the technique of carrying through secondary recovery programs, such as water flooding or gas repressuring, are fully discussed in a number of petroleum-engineering texts to which the reader should turn for the strictly engineering aspects of the flow of fluids through porous media. In a similar manner, with regard to such problems as the seepage of water through or underneath dams, the discussions will again be limited strictly to the dynamics of the fluid movement, the stresses or forces on the porous structures themselves being explicitly considered as beyond the scope of this work. It is realized, of course, that the dynamical reactions experienced by the fluid-bearing media in such systems are in many cases of even greater practical significance than the motions of the fluids themselves. Indeed, the accumulated knowledge on "soil mechanics" may well be considered as an important science in itself, and for this very reason it has seemed better to omit altogether any discussion of it than to present a brief and sketchy treatment that would not do justice to the space and thoroughness which the subject merits.[1]

1.3. Porous Media.—A further qualification limiting the applicability of this work is contained in the definition of a

[1] The reader who is interested in this phase of the subject will find a thorough treatment in "Erdbaumechanik," by K. Terzaghi (1925).

porous medium as the carrier of the fluids. The ideal porous medium, fitting perfectly the definition required, may be most clearly comprehended by visualizing a body of ordinary unconsolidated sand. There are present in such a medium innumerable voids of varying sizes and shapes comprising "pore spaces" or interstices between the individual solid particles of sand. Moreover, each pore is connected by constricted channels to other pores, the whole forming a completely interconnected network of openings which form the channels through which the contained fluid may flow.

It is this complete multiple interconnection of the minute openings that characterizes the ideal "porous media" assumed in this work and definitely differentiates this subject from that of the usual hydrodynamics or hydraulics. Thus, instead of the open channels defined by impermeable boundaries characteristic of problems in hydraulics, we have here to deal with flow channels composed of multiply connected minute pores and limited in extent by impermeable boundaries or by the geometry of the flow system. Here indeed lies the fundamentally distinctive feature of a porous conducting medium. For, while the ordinary pipe or capillary considered in hydraulics is the equivalent of a single series of connected pores, it must be remembered that the whole channel composed of porous material involves not a single series of connected pores but a very large number of such elements with multiple *lateral* connections. In a capillary tube or pipe carrying a fluid in viscous flow, the velocity is not uniform across the section but has a parabolic distribution, with a maximum velocity at the center of the channel. And while in a linear system consisting of a porous medium the velocity distribution across a single pore opening will have similar characteristics, the velocity, when considered macroscopically across the whole channel, will be uniform. Thus the fluid flux carried by a linear conductor of this type is uniform across the section as contrasted with pipe flow wherein at the center of the fluid column the flux is a maximum and becomes vanishingly small at the walls of the conduit.

The porous materials which will concern us here will be the relatively fine-grained soils, unconsolidated sands, the consolidated sands which are the common sandstones, limestones, and other porous rocks. This may appear to provide serious limitations to the applicability of the results since many of the rocks containing fluids derive their fluid-carrying capacity largely from fractures. However, insofar as these fractures are numerous enough and randomly distributed, it is evident that on a large enough scale the multiple interconnection of the fractures will lead to a simulation of the characteristics of porous media outlined above.

1.4. General Qualifications Governing the Analytical Theory. The analytical work contained in this treatise is, as already outlined, based upon certain necessary assumptions and limitations regarding the type of fluids and the nature of the porous media. Moreover, it is evident that in those problems involving natural sedimentary materials or rock formations, one is faced with the uncertainty arising from the nonuniformity and unknown details characterizing the structure of such materials. Cursory consideration may, therefore, lead to the conclusion that the limitations are so drastic and the assumptions so idealistic as to preclude the application of the analytical results to problems of practical interest. Indeed, it is most probably to a reaction of this kind that we must attribute the apparent resistance maintained until very recently by hydrologists and engineers working in related fields against applying Darcy's law or the analytical formulations of Forchheimer or Slichter to practical field problems.

However, while it is true that the indeterminancy of certain conditions involved in problems of flow through porous media encountered in practice will prevent the attainment of exact quantitative results from any mathematical analysis, it is nevertheless certainly of value to analyze the problems as if they involved ideal systems. For it is only in this manner that the fundamental properties of the systems can be determined and their behavior under modified conditions ascertained. The fact that the actual system is not ideal as to conditions within the porous medium cannot nullify the results from a semiquantitative viewpoint.

As an illustration, one may consider a relatively simple problem of multiple-well systems discussed in Chap. IX. A municipal water supply is derived from deep artesian wells and it is desired to plan for future expansion by the drilling of additional wells. Because of interference effects between wells, doubling the density of the wells will not double the output, and the actual increase is dependent upon the well spacing and geometry of the multiple system used. By the analytical methods outlined in Chap. IX, the interference effects may be determined with assurance that even though the exact output of a given well may not be predictable, nevertheless statistically, the relative production rates of the various wells in the network will be quite accurately forecast, and the most economical development plan may be ascertained.

A more complex example lies in the problems of water flooding as applied to secondary oil recovery and discussed from the idealized viewpoint in Secs. 9.16 to 9.33. Here the actual field conditions, in which the injected water must move through a sand that contains both gas and oil, do depart markedly from those assumed for analytical convenience, and the physical factors responsible for this uncertainty are of such complexity as to preclude any hope of ascertaining accurately their quantitative effects. Nevertheless, the analysis does provide reliable information regarding the relative performance of various arrays of wells and the effects of modified spacing, which are questions that have been answered heretofore only by intuitive opinion.

Finally, it should be noted that the subject matter of this work does not directly encompass changes in the porous medium that may develop after an extended exposure to the fluid passing through it. Thus it is presupposed in the analytical developments that there is no plugging of the medium, owing either to changes in the medium itself or foreign matter brought in by the fluid, which grows as the time of flow continues. Such effects as a swelling or hydration or even direct plugging of the medium must be assumed to have reached their saturation magnitude and the corresponding modifications of the fluid-carrying capacity of the medium must be assigned from the beginning. Although such plugging effects could be easily taken into account if the history or law of their development be assumed or known a priori, the actual lack of such knowledge in practically all problems of physical importance has made it seem most advisable to explicitly exclude them as a general rule. When the pertinent information will be available, the reader should have no difficulty in appropriately modifying the formulas derived here to take into account the changing structure of the medium.

This discussion of the applicability of mathematical analyses to practical problems is intended to call attention to the need for very careful discrimination in properly qualifying the theoretical results before applying them to any given type of problem, and to remind the reader that while one phase of a problem may be answered only very qualitatively, the results for another phase of the same problem may represent a very close approximation to the true facts. Proper discrimination in these questions will develop as familiarity with the work progresses.

1.5. The Nature of Porous Media.—Brief mention has been made of the general nature of the porous media with which we are to deal. The significant properties which should be discussed in further detail are (1) porosity, which is a measure of the pore space and hence of the fluid capacity of the medium, and (2) permeability, which is a measure of the ease with which fluids may traverse the medium under the influence of a driving pressure. These properties will be considered now from the viewpoint of the structure of the medium, leaving other details to Chap. II where the technique of measurement will be considered.

Early work on flow through porous media was to a considerable extent confined to unconsolidated materials such as sand and gravels. The difficulty of measuring the porosity and permeability of such materials *in situ* or, to duplicate the natural packing of the particles for purposes of measurement in the laboratory, led naturally to the study of the geometry of such systems. Since it is a relatively simple matter to determine the grain sizes and their relative distribution by means of sieve analyses, attempts were made to calculate the porosity and permeability from these data. Evidently any analysis of this kind must be based upon ideal particles; hence the various modes of packing of uniform spheres were carefully investigated. A very extensive study was made by Slichter,[1] followed more recently by others,[2] of which latter the most comprehensive is that of Graton and Fraser.[3] While it would be quite impossible to consider the subject in detail as given by these investigators, the salient features are presented here with the sole purpose of clarifying conceptions of the structure of such assemblages in their role as an idealization of the clastic sediments which comprise the major portion of the porous media involved in our study.

1.6. Systematic Packing of Spheres.—Assuming for the purpose of analysis that the solid particles constituting an uncon-

---

[1] SLICHTER, C. S., U. S. Geol. Surv. 19th Ann. Rept. Part II, 295, 1897–1898.
[2] SMITH, W. O.,.P. D. FOOTE, and P. F. BUSANG, *Phys. Rev.*, **34**, 1271, 1929.
[3] GRATON, L. C., and H. J. FRASER, *Jour. Geol.*, **43**, 785, 1935.

solidated sand body are uniform spheres, it is possible to produce two[1] extremes in systematic arrangement which provide on the one hand a maximum possible pore space and on the other a minimum. Unit cells of such arrangements are shown in Fig. 1, case 1 being *cubic* and case 2 *rhombohedral* packing. Variations

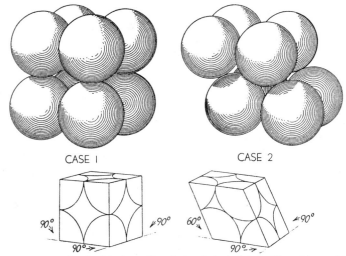

FIG. 1.—Unit cells of cubic (case 1) and rhombohedral (case 2) packing. (*After Graton and Fraser, J. Geol.*)

in the angles of the array will provide an infinite variety of intermediate arrangements. Figure 2 shows the unit pore for the cubic (case 1) and for the rhombohedral (case 2) array which represent respectively the "loosest" and "tightest" packing possible in a systematic assemblage of uniform spheres. From the geometry of these cases may be obtained the data of Table 1 wherein the porosities of these extremes of packing are given.

[1] GRATON and FRASER, *loc. cit.*, classify six cases. Their cases 1 and 3 are the above cases; case 2 is orthorhombic formed by skewing the cubic form parallel to one face so that four faces of the unit cell are squares of side $2R$ and two are rhombs of side $2R$ and acute angle of 60° (porosity 39.54 per cent); case 4 is identical with case 2 but with a different spatial orientation of the cell; case 5 is a "tetragonal-sphenoidal" packing formed by warping the cube cell into a very special array not briefly describable; case 6 is identical with case 3 but again differing only in spatial orientation. This classification has been used by these authors because of their special significance in determining the stability of the packing and their effect upon permeability. While undoubtedly useful in their comprehensive study, this more complex classification is not warranted here.

Intermediate arrays will evidently result in porosities lying between these limits.

It will be noted that since the porosity is the ratio of the volume of the unit pore to that of the unit cell (Table 1), it will be independent of the radius R of the uniform spheres comprising the

CASE 1   CASE 2

Fig. 2.—Pores of the unit cells of cubic (case 1) and rhombohedral (case 2) packing. (*After Graton and Fraser, J. Geol.*)

assemblage. However, the permeability of the array is dependent upon the actual dimensions of the pore openings and is proportional[1] to R^2. Thus the porosity of an assemblage cannot alone provide an accurate indication of its permeability. This lack of proportionality between porosity and permeability even in the ideal case is in itself sufficient to eliminate any possibility of deriving significant permeability estimates from porosity measurements.[2]

TABLE 1.—POROSITIES OF UNIFORM SPHERE PACKINGS

Packing	Cubic	Rhombohedral
Volume of unit cell............	$8.00R^3$	$5.66R^3$
Volume of unit pore...........	$3.81R^3$	$1.47R^3$
Porosity....................	47.64 %	25.95 %

Since between the limiting cases of systematic packing there is an infinitude of intermediate types, the question arises as to the type, and hence the porosity, of chance or natural packing. Examination of Fig. 1 shows clearly that the most stable array is rhombohedral, the spheres having sufficient points of contact to provide support from any direction, while the cubic array is stable only to forces normal to the cell faces and hence possesses a very critical stability in common with all intermediate forms. It might be expected, therefore, that by allowing the spheres to

[1] Slichter, *loc. cit.*

[2] Yet strangely enough, until recent years, the terms themselves were even used synonymously by some engineers, probably because the porous media with which they dealt showed in general that permeability and porosity varied in the same manner.

take their own positions under moderate agitation, they would assume the most stable array and hence the assemblage of lowest porosity. However, such an arrangement involving very large numbers of spheres requires such a high degree of perfection in the placement of individual members that it is very unlikely that it would be obtained by natural means. Rather it would appear that the required perfection in packing would be confined to some rather limited number of particles and beyond this point distortion of the array would begin. When such distortion of the array reaches sufficient magnitude, random or chaotic packing would result, for, unless the boundaries of a systematic array conform closely to the geometry of the unit cell, complete uniformity is impossible. Thus in natural assemblages, even when agitated to induce close packing, one should anticipate groups of spheres packed in orderly arrays separated by boundaries in which no orderly arrangements are present and where the porosity is even higher than that of cubic packing. Such zones can be maintained because of the "bridging" of groups of particles under pressures less than the crushing strength of the particles. That such is the actual condition is evident from simple experiments with single layers of spheres. Moreover, it is found experimentally that assemblages of spheres, or even sand particles, will have porosities averaging about 40 per cent in spite of careful efforts to induce closer packing, and even though the predominant array in the assemblage is rhombohedral with a porosity of only 26 per cent.

Theoretically, the actual size of the spheres has no influence on the porosity, but in assemblages of natural materials this does not prove true. Actual determinations[1] give the following porosity values: Coarse sand, 39 to 41 per cent; medium sand, 41 to 48 per cent; fine sand, 44 to 49 per cent; fine sandy loam, 50 to 54 per cent. And for small particles, the porosity may range from 50 per cent for particles larger than 0.02 mm. to about 95 per cent for material smaller than 0.002 mm.[2]

1.7. The Packing of Natural Materials.—Natural materials involve particles which may depart quite considerably from true spherical shape and, moreover, are far from uniform in size.

---

[1] ELLIS, A. J., and C. H. LEE, Geology and Ground Waters of the Western Part of San Diego Co., California, *U. S. Geol. Surv. Water-supply Paper* 446, 121–123, 1919.

[2] TERZAGHI, K., *Eng. News Record*, **95**, 914, 1925.

Nonuniformity in size in general permits smaller particles to fill the pores between the larger particles thus resulting in a material reduction in porosity. In fact, owing to the lack of sorting, glacial till, though containing large rocks and pebbles, has nevertheless a lower porosity and permeability than relatively fine-grained but better sorted sand. On the other hand, angularity of the grains tends to produce bridging, which results in random packing and higher porosities.

Data on porosities of beach sands, *in situ*, average 40 to 56 per cent,[1] agreeing well with laboratory experiments. However, freshly deposited clays and silts frequently exceed a porosity of 85 per cent, although when dried and compacted the same materials will have porosities of 40 to 50 per cent or less.

This wide range in the porosity of freshly deposited sediments of different grain size leads to important consequences after the sediment is buried. For under the weight of the overburden, the sediments are subjected to a corresponding pressure which reduces their volume by inducing closer packing, crushing, and deformation of the grains or, at extreme pressures, actual recrystallization of the particles. Evidently the amount of compaction will depend upon the composition of the particles as well as the initial porosity of assemblage.

The analyses, with respect to grain-size distribution, porosity, and permeability, of some typical natural sandstones and a few synthetic ceramic materials are given in Table 2.[2]

1.8. Compaction of Sand and Gravel.—Pebbles and sand grains are deposited in a more nearly stable state than are smaller particles. This is quite evident from porosity measurements made on freshly deposited sands as compared with clays or silts, or from laboratory measurements on artificially deposited particles having marked disparity in grain size (*cf.* Sec. 1.6). In fact, normal deposition of medium-sized sand grains results in an assemblage so nearly equal to the minimum porosity obtainable by any mode of natural packing that further agitation or compression will produce almost inappreciable decreases in porosity unless the applied pressure is sufficient to crush the grains and cause a more or less complete elimination of bridging

---

[1] FRASER, H. J., *Jour. Geol.*, **43**, 910, 1935.

[2] This table has been taken from the paper by G. H. Fancher, J. A. Lewis, and K. B. Barnes, *Bull.* 12, *Mineral Industries Exp. Sta.*, Penn State College, 1933.

TABLE 2.—SCREEN ANALYSES OF SANDS

Sand number

Mesh	Opening, inches	1	2	3	4	5	6	7	8	9	10	11	12
10	0.0787									3.41			
20	0.0331									17.35			
30	0.0232			0.24						24.34			
40	0.0165			0.62		0.59			0.17	30.14			
50	0.0117	3.81	0.15	1.39		12.84		0.29	2.44	14.59	0.07	0.15	
70	0.0083	57.30	6.57	11.74		46.67		18.43	23.23	4.92	6.90	9.41	0.29
100	0.0059	17.42	56.04	71.54		14.61		42.09	49.28	2.03	48.21	43.06	31.76
140	0.0041	12.59	17.99	6.90		8.77		31.93	22.45	1.75	25.83	30.12	61.84
200	0.0029	8.88	10.13	6.95	1.14	2.94	1.14	4.79	1.77	0.69	4.73	3.89	2.19
270	0.0021		9.12	0.62	20.32	2.60	20.32	1.31	0.44	0.78	3.31	2.57	2.31
Over 270	0.0021				78.54	10.98	78.54	1.16	0.22		10.95	10.80	1.61
Porosity, %		12.5	12.3	16.9	37.0	20.3	37.8	19.7	15.9	11.94	19.5	18.4	22.3
Av. grain diam., in.		0.00222	0.00220	0.00385	0.00113	0.00217	0.00113	0.00398	0.00550	0.00986	0.00214	0.00216	0.00355
Permeability, darcys		0.0026	0.0028	0.0444	0.0051	0.0295	0.0088	0.182	0.350	1.130	0.0638	0.0672	0.228

Sand number

Mesh	Opening, inches	13	14	15	16	17	18	19	20	21	22	23
10	0.0787											
20	0.0331											
30	0.0232											
40	0.0165	0.17			0.07			0.14			1.77	
50	0.0117	2.44	0.04	0.05	5.86		0.09	2.11	2.25		36.59	0.21
70	0.0083	23.23	7.33	0.92	49.28		1.68	10.01	30.78	0.98	53.92	8.25
100	0.0059	49.28	58.26	22.70	18.61		24.03	20.64	54.15	14.90	5.43	58.23
140	0.0041	22.45	20.05	64.70	9.00		60.83	42.17	12.15	68.99	1.25	27.80
200	0.0029	1.77	9.26	6.44	3.68	1.14	9.61	13.28	0.50	10.74	0.39	4.17
270	0.0021	0.44	5.06	3.10	2.46	20.32	2.27	6.74	0.11	3.00	0.35	1.09
Over 270	0.0021	0.22		2.09	11.04	78.54	1.49	4.91	0.06	1.39	0.30	0.25
Porosity, %		16.3	19.2	21.4	20.6	33.2	21.9	23.8	26.9	27.7	22.1	28.8
Av. grain diam., in.		0.00550	0.00256	0.0033	0.00216	0.00113	0.00357	0.00265	0.00658	0.00355	0.00638	0.00504
Permeability, darcys		0.555	0.139	0.241	0.126	0.0371	0.417	0.111	2.500	0.859	3.390	2.350

All analyses expressed as per cent weight retained on the respective sieve. The sources of the samples will be found in Table 8, the numbering of the sands in the two tables being the same.

and the collapse of arching across the voids. Thus the porosity of slightly cemented sands obtained from cores at depths of 4,000 ft. have been found to have even higher porosities than sands of similar character obtained at shallow depths.[1] Experimental evidence[2] also substantiates the minor importance of compaction of sands, subsequent to deposition, upon the porosity of such sediments. In most cases, postdepositional reductions in porosity of such materials are due to cementation by mineral matter.

1.9. Compaction in Clay.—The abnormally high porosity (*cf.* Sec. 1.6) obtainable in newly deposited clays and silts would indicate a high susceptibility of such materials to compaction. It is therefore not surprising to find that actual measurements indicate the porosities of shales at the same depths in any given locality to be approximately equal[3] and that the porosity may be taken as a measure of the compaction which the shale has undergone. Athy[4] has found the variations in porosity of a shale with thickness of the overburden to be given by the following formula:

$$f = f_0 e^{-bz},$$

where f is porosity, f_0 is the average porosity of surface clays, b is a constant and z is the depth of burial.

The same author finds that, while the average porosity of surface clays (not fresh deposits) is 45 to 50 per cent, the porosity of shale at 6,000 ft. is approximately 5 per cent. Thus it is seen that compaction equivalent to more than 20 per cent of the original deposit has occurred by the time the clay has been buried by 1,000 ft. of overburden, 35 per cent at 2,000 ft., and 40 per cent at 3,000 ft. It is difficult to determine to what extent this reduction is due to a physical rearrangement of the grains or to an actual recrystallization. The wide diversity in size of particles in clays and silts, ranging as they do down to particles of colloidal dimensions, is conducive to abnormally low porosity if sufficiently close packing can be obtained. It is not difficult to see that such materials when subjected to high

[1] ATHY, L. F., *Amer. Assoc. Petrol. Geol. Bull.*, **14**, 8, 1930.

[2] BOTSET, H. G., and D. W. REED, *Amer. Assoc. Petrol. Geol. Bull.*, **19**, 1053, 1935.

[3] HEDBERG, A. D., *Amer. Assoc. Petrol. Geol. Bull.*, **10**, 1035, 1926.

[4] *Loc. cit.*

pressures will tend to pack very closely because of their heterogeneous size and shape distribution. On the other hand, recrystallization of some of the material is also a distinct possibility. Thus, in contrast to sandstones and similar rocks, the compaction of clays is considerable and increases with pressure up to a limit as yet undefined. The conversion of shale into slate and phyllite is obviously a recrystallization process whereby porosity is practically eliminated.

This high susceptibility of clays and derived shales to compaction forces and the fact that at sufficiently high pressures the porosity may practically disappear is of primary geological significance as will be briefly discussed later.

1.10. Effect of Compaction and Alteration of Sediments on Permeability.—While we have stressed the fact that in general no relation exists or is to be expected between porosity and permeability, it is evident that any alteration of a *given* material which produces a decrease in porosity must necessarily result in a decreased permeability. For the reduction in porosity of a given material implies a decrease in size of the pores and hence an even greater percentage change in permeability. Therefore any postdepositional alteration in a clastic sediment, whether by compaction or cementation, will result in a marked decrease in permeability. Since the compaction of sands is almost negligible, major changes in porosity of such sediments are due to cementation, and likewise changes in permeability are to be accounted for by the same process. On the other hand, clays and shales which, even in their early stages, are considered practically impermeable, become after compaction almost completely so. This characteristic of clays and shales, as well as other impermeable materials, has important significance geologically in determining the distribution and accumulation of subsurface fluids.

1.11. Kinds of Rock and Their Fluid-bearing Properties.—Thus far the discussion of porous media has been confined to important clastic types wherein the porosity is determined by the geometry of the assemblage of particles with a brief mention of the alterations which may occur after deposition. It is necessary to review briefly also the origin and mode of occurrence of these materials, as well as others which may be encountered in subsurface studies.

Rocks constituting the earth's crust may be broadly classed with respect to origin as (1) *igneous rocks* which are produced by the cooling and solidification of molten materials; (2) *sedimentary rocks* which are produced by the deposition of materials weathered from older rocks of any type, derived from the remains of plant or animal life, or precipitated out of solution in water; (3) *metamorphic rocks* which are produced by the profound alteration of other rocks chiefly through the agencies of heat and extreme pressure.

The materials penetrated by a deep well in areas of normal geological sequence are unconsolidated sediments composed alternately of clays, sand, and gravel; consolidated sediments as shales, sandstones, limestones or conglomerates; and metamorphic and igneous rocks.

Igneous rocks differ widely in type and texture depending upon their composition and mode of origin. Molten material that remains deeply buried in large masses solidifies very slowly, forming the crystalline texture typical of the intrusive granites. Molten material extruded at the surface or in small intrusions cools rapidly, forming a glassy texture, and in the case of lava flows may be vesicular, owing to the escape of gases, and traversed by many large joints or fractures produced by shrinkage during rapid cooling. In general, except for the presence of fractures or severe weathering, intrusive igneous rocks are so relatively impermeable and lacking in porosity that they are of little importance in our present study. However, the extrusive igneous rocks (lava flows) may be highly pervious, as is the case of the basalts which cover large parts of the Northwestern United States and nearly all of the Hawaiian Islands, and are important water bearers.

It is of interest, however, to note the composition of igneous rocks since they are the basic source of all sedimentary rocks not organic in origin. The principal minerals found in fresh igneous rocks are quartz (crystalline silica); feldspar (silicates of aluminum with potassium, sodium, calcium, or barium); mica (silicates of aluminum and alkalies or iron); hornblende (silicate of calcium, magnesium, aluminum, etc., with varying compositions); pyroxene (silicate of calcium and magnesium chiefly); and olivine (silicate of magnesium and iron). Because of its hardness and chemical stability, quartz is not easily broken and worn, nor decomposed when the rest of the rock weathers. Furthermore, it is not readily dissolved by percolating waters. Thus when all other minerals of a granite rock are broken down to form clay or other fine materials, the grains of quartz will persist unbroken, and it is largely these particles that form the deposits of sand and sandstone. Mica, when thoroughly decomposed, produces a residuum of clay, but it breaks down very slowly and consequently is found in undecomposed remnants in many deposits of clay and sand. True clay is aluminum silicate formed by the decomposition of the more complex silicates found in igneous and metamorphic rocks and consists of such exceedingly small particles that it may be considered a colloidal material. And while most deposits of clay are mixtures with coarser materials, with porosities which may be as high as 50 to 60 per cent, their permeabilities are generally extremely low.

Metamorphic rocks are represented by such common varieties as quartzite, slate, marble, schist, and gneiss. Because of their common occurrence and their presence within sedimentary sections, we shall briefly outline the sources and characteristics of those types most frequently encountered.

Quartzite is usually the result of a complete consolidation of quartzose sandstone by precipitation of silica within the pores. It is the hardest and most durable of common rocks. It generally contains joints or fractures but otherwise is impervious.

Slate is derived from shale or related clayey materials by extreme pressure and thermal processes. It is harder than shale and weathers less readily but, because of its marked cleavage and susceptibility to fracture, may be pervious to fluids.

Marble is produced from limestone by induration and crystallization. It is a dense rock but, like limestone, can be slowly dissolved by water percolating through fractures or other openings.

Schist is the result of profound alterations of shale, slate, or other rock by intense pressure and deformation. It has an irregular foliated structure which is due primarily to the development of flakes of mica parallel to the planes of shearing.

Gneiss is a banded granular rock, much of which is derived from, and grades into, granite and other crystalline igneous rocks, but may be derived from sedimentary rocks.

Unless found at the surface or within the weathered layer where decomposition and pronounced fracturing can occur, it is evident that metamorphic rocks are in general too impervious to be of interest as fluid-bearing materials.

Sedimentary rocks may be grouped according to origin into three classes: (1) clastic; (2) organic; and (3) chemical deposits. As already discussed, the clastic deposits are composed of fragments derived from the erosion of older rocks, and include gravels, sand, silt, clay, and the consolidation products of these materials. The organic deposits include chiefly the calcareous and siliceous remains of animals such as shells, corals, etc., and the carbonaceous remains of plants. The calcareous materials form limestone, chalk, marl, and related rocks, while the carbonaceous materials form peat, coal, and kindred materials. The chemical deposits consist of minerals precipitated out of solution in water, and include deposits of silica such as flint, chert, vein quartz, ferruginous deposits as some of the iron ores, calcareous deposits such as caliche, gypsum, salt (halite), and other soluble alkali salts. In addition chemical deposits are of importance as comprising the source of cementing material whereby the clastic sediments are consolidated.

The clastic types of sedimentary materials have already been considered in some detail (*cf.* Secs. 1.6 to 1.10) as regards their porosity and permeability. Since the sediments of organic origin are also frequently important sources of underground fluids but yet differ from the other types in susceptibility to postdepositional alterations, we shall now briefly discuss their properties as carriers of fluids.

No rocks differ among themselves more radically with respect to their fluid-bearing properties than the limestones and related materials. Some limestones are among the best fluid sources, while others are wholly unproductive because of the lack of porosity or permeability. These differences are due only in part to the original texture of the rock. They are chiefly the result of differences in the extent to which it has been subjected to the solvent action of percolating waters. Thus, while in general postdepositional alterations of clastic sediments result in decreased porosity and permeability because of cementation and consolidation or metamorphosis, the exceptionally high porosity and permeability of some limestones are the results of solution by exposure to circulating water at some period in their history. This development of increased porosity subsequent to deposition and consolidation is aptly termed "secondary porosity."

Newly formed limestone may be quite porous and permeable, but because of the ease with which the calcareous materials are compacted, or are dissolved and again precipitated, the original pore space tends to be closed up or filled with secondary deposits. For this reason, and in addition, because of metamorphic alterations induced by deep burial, the older limestone rocks are generally of low porosity and quite impervious. However, such rocks may be uplifted and, by erosion of overlying material, brought near the surface. The presence of bedding planes or fractures formed by the intense folding during uplift provides ready access of circulating ground waters. Since these waters contain carbon dioxide, relatively rapid solution of the calcareous material will result along these channels, and in this manner are formed the "cavernous" openings which characterize those limestones which are particularly prolific sources of fluids. While in some cases these so-called "cavernous" openings may be true caverns of great dimensions, the term is also applied to limestone containing relatively small pores of the order of fractions of a millimeter to several millimeters in section. Such openings forming connected channels, augmented perhaps by fractures or fissures, constitute the major portion of the unusually prolific limestone reservoirs.

Since carbon dioxide in the circulating waters is the dissolving agent, and it is generally abundant only in surface waters, limestone formations which have always lain at considerable depths are likely to have low porosity and permeability. For at such depths the water will usually have a low carbon dioxide content as well as relatively sluggish circulation. Evidently, then, the most favorable circumstances leading to the development of very porous limestone formations involve the rise of the formation from its normal position to one above or within the zone of groundwater percolation where solvent action is rapid, followed later by subsidence and burial beneath other sediments. If the secondary porosity so formed is not subjected to subsequent deposition of precipitates partially or even completely filling the openings, the formation will retain its favorable porosity even at great depths of burial, for the consolidated rock is able to withstand the compressive forces arising from heavy overburdens.

Thus from the susceptibility of limestone to leaching action, which in turn is dependent upon the geological history of the formation, arise the radical variations in porosity and permeability already mentioned as being characteristic of limestone and related rocks. It is also evident that these same factors are responsible for very local variations within the same strata as well as between different strata which originally may have had identical composition and texture. Thus it is usual to find that the most porous area in a given limestone lies at the crests of domes or anticlinal folds which at some period lay within or above the ground-water zone.

This brief discussion will suffice to indicate the nature of the secondary porosity characterizing some limestones. Bearing in mind also the susceptibility of limestone to fracture when subjected to pronounced folding, it is clear that such formations may appear to deviate markedly from the ideal porous media assumed in this study and, while it is true that the pertinent characteristics of the material cannot be determined by an examination of small samples from the formation, nevertheless on a gross scale, such as involved in practical problems, the departure from the ideal porous media will not be of serious significance. Even in the case of a connected network of numerous fissures the analogy will be substantially retained and only where a relatively impermeable rock is traversed by a few extended fissures will serious departures be encountered.[1]

While particular emphasis has been placed upon secondary porosity in view of the fact that it accounts for the highly pervious formations frequently encountered, many limestones may have considerable porosity and permeability in their unaltered state. Thus limestones comprising cemented fragmentary calcareous particles are quite comparable to sandstones as regards the structure of their pore space. Others may be porous because of the presence of the skeletal remains of minute organisms or even larger structures such as characterize coral formations.

1.12. Structure of Rocks in Relation to Underground Fluids.— While the volume of fluid contained in rocks is determined by their texture or porosity, and the facility with which fluids may traverse them depends upon their permeability, the distribution of the fluids within the rock formations, as well as their migratory behavior within the strata, is dependent upon the areal distribution and configuration of the formations.

[1] Special problems involving such conditions are treated in Secs. 7.4 to 7.8.

The earth's crust comprises layers or strata of rocks of various kinds and massive or foliated bodies of rock that underlie or intersect these stratified series. Sedimentary rocks, because of their mode of deposition, are usually stratified. Igneous and metamorphic rocks may be stratified but usually form massive intrusions. Below this mantel of stratified sediments and intrusives lie the igneous rocks which were the original source of the sedimentaries—the so-called "igneous basement" or "basement complex" where our interest ceases.

Formations of stratified rocks range from a few feet to hundreds of feet in thickness, with an areal distribution of a few miles to thousands of square miles, and may be present either at the surface or deeply buried beneath other formations. Thus most sedimentary formations consist of strata that are very thin in comparison with their areal dimensions, as for example the St. Peter sandstone averaging 100 ft. in thickness, which is known to underlie an area in North-Central United States of some 300,-000 square miles in extent. Such persistence of a single formation with the uniformity in texture characterizing this sandstone is of course unusual, for it calls for a uniformity in depositional conditions not ordinarily encountered. For, while many sedimentary deposits laid down contemporaneously may be considered as part of the same formation, their character may differ widely within relatively short distances, depending upon the particular conditions existent in any local area at the time of deposition. Moreover, any given formation such as a sandstone may not be a comparatively solid and unbroken bed of sand of considerable thickness but when penetrated by a drill will show alternate layers of sand, shale, or calcareous sand and shale members of varying relative thicknesses. These individual strata may represent shale lenses within an otherwise homogeneous sand body, or the sands may be lenses within a more or less uniform body of shale strata. On the other hand, an essentially continuous sand may be traversed by relatively thin shale "breaks" extending over large areas.

These detailed features will be appreciated and their cause understood by observing the types of sediments and their mode of deposition in areas of recent or currently active sedimentation. Thus a stream flowing from high mountains out upon a plain generally deposits boulders and pebbles near the mountains and carries silt and clay far out upon the plain. Likewise, at some places along a shore there is a beach consisting of clean gravel, the movement of the waves and currents being adequate to agitate and ultimately carry away any finer materials that are washed into the water either by streams or by wave erosion of shore-line material. At other places the waves and currents may be less strong, and hence sand or mud is deposited. Outward from shore there is generally a gradual change from thick accumulations of coarse debris to progressively thinner deposits of finer and finer sediments, and where the water is deep and clear, limestone may be forming. It is evident in view of this type of sorting that a gradual encroachment of the sea, caused by subsidence of the land mass, will result in the gradual extension of each type of deposit in the direction of the encroachment. Thus the coarse debris on the shore will be gradually covered by sands, then silt and clays, and finally even limestone if the subsidence is sufficiently great. At the same time, each type of formation is extended more or less uniformly in the direction of the encroachment. As a result a sequence of limestone, shale, and sandstone strata will be produced having great lateral extent and uniformity. However, local variations in texture and composition appearing as lenses of shale in sand, or vice versa, are to be expected.

While it is impossible here to go further into the physiographic and geological factors involved in determining the type and distribution of the various sediments, the underlying principles should be evident. All stratification is the result of changes in the physical conditions under which deposition occurred. The outstanding broad-scale differences between successive formations are due to regional changes; the minor irregularities within a single stratum are due to local or temporary changes. As exemplifying the latter, the clayey partings between similar layers of limestone may be due to storms of exceptional violence that produced considerable turbidity in waters which were usually clear, but the innumerable partings in some shales may be the result of ordinary or cyclic changes in the weather.

The mechanism of deposition of the sediments necessarily implies that in their original state the strata were laid down horizontally. If the area is one of geologic inactivity, the formations may retain their original posture and exist at present as horizontal or slightly dipping beds. On the other hand, such inactivity over long geological periods is not the usual condition encountered. For accompanying the regional uplifts or subsidence effective over large areas, which is a necessary condition for the accumulation of thick sedimentary sections, local variations in the amount of activity or postdepositional diastrophy are almost inevitable. The result is local distortion of the otherwise horizontal strata, forming folds, domes, or monoclinal slopes which may have extremely steep dips. If the distortional forces are applied extremely slowly, the plasticity of most materials permits a gradual bending or distortion of the strata into relatively smooth anticlinal and synclinal features without serious fracturing of the beds. However, a very rapid action, geologically speaking, or very sharp folding will usually result in fracturing or major faulting of the formations. Obviously gentle folding or doming without major fracturing will not seriously disturb the continuity of the strata as far as fluid flow is concerned. On the other hand, major faults may have sufficient "throw" to cause a vertical displacement of a given member by an amount greater than its thickness. In such cases the continuity of formation may be interrupted at the fault for its fractured faces may be in juxtaposition with impervious strata, or the fault zone itself may be composed of impermeable debris resulting from pronounced local pressure and distortion. Similarly in some areas impermeable intrusive dykes even of small size may abruptly terminate the continuity of a pervious stratum.

When at some time during its history an area comprised of a strongly folded sedimentary section is exposed to erosion, anticlines may be partially or completely leveled. As a result of such scalping the tilted strata will terminate at surface outcrops. Subsequent sedimentation over the area, resulting from regional subsidence, will cover the original land surface with new formations of different geological age lying unconformably above the older rocks. Obviously such discontinuities in a geologic section, termed unconformities (cf. Fig. 3), may form the terminal boundary of some buried strata and, like faults, limit its continuity to the areas between its now buried outcrops. Thus unconformities, which are very common and far-reaching structural features, are nothing more than fossil land surfaces with pronounced irregularities such as characterize errosional surfaces. And evidently they have very important effects on the occurrence and regional migration of underground fluids.

This very elementary and cursory discussion of formations and structural features of the rock strata comprising the subsurface zone with which we will be concerned, is intended only to call attention to major features. Further and more detailed consideration of the subject is quite beyond the scope of this work.

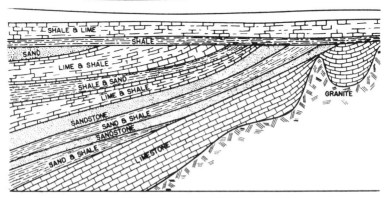

FIG. 3.—A typical geologic section traversed by unconformities.

The reader is advised to consult texts on structural geology to obtain a rather thorough understanding of the geological features essential to the proper application and discriminatory qualification of the analytical discussions presented in succeeding chapters.

1.13. Subsurface Fluids—Their Occurrence and Migratory Behavior.—Below the superficial mantel of surface soils the materials comprising the earth's crust are the unconsolidated gravels, sands, and clays gradually merging with depth into the sandstones and shales which result from the consolidation of these materials by cementation and compaction. These strata will also contain interspersed limestone beds if, during their geological history, relatively deep water teeming with organic life has overlain the area. The zone of unconsolidated or semiconsolidated material will normally range from a few feet to many hundreds of feet, and the zone of consolidated sedimentary rocks may be many thousands of feet thick followed by metamorphic rocks and finally the igneous basement.

From the previous discussion it will be recalled that these materials all have a certain amount of pore space or porosity which may vary from some 85 per cent of the total volume of surface mud and silt deposits, and 40 per cent for recent deposits of well-sorted sands, to the practically negligible porosity of the very deep metamorphosed sediments and igneous rocks. Even at depths as great as 10,000 ft. effective porosities of 5 to 10 per cent, or even greater, may be encountered in some rocks, and at intermediate depths, sandstones of 20 to 25 per cent porosity are not particularly uncommon.

Now the greater portion of these sediments have been deposited by the action of water or are of direct marine origin. And even though originally laid down in arid regions by wind or other agencies, or in the case of formations of any type which have been uplifted and exposed to surface erosion, during the final process of burial they have been exposed to saturation by ground waters. Thus it is to be expected that all subsurface rocks having appreciable pore space must be saturated with fluid, as is in fact found to be true. Usually in the case of deeply buried marine deposits they contain residual or connate water or, perhaps connate water forced into them from contiguous sources or transmitted over long distances through permeable connecting channels. However, in the case of oil and gas deposits, these fluids, formed subsequently to the deposition of the sediments, may form a part of the fluid content of the rocks.

Since the pore space or fractures of subsurface rocks are filled with fluids, it is pertinent to consider the pressures existing within these fluids, for there are two sources whereby pressure on the formation fluids might be sustained. On the one hand, as we have seen, all rocks are subject in varying degrees to compaction subsequent to deposition. Because compaction is caused by the weight of the overburden together with any additional diastrophic forces which may be imposed, it might be supposed that compressive forces of similar magnitude are imposed upon the fluids contained in deeply buried rocks. These pressures would be of the magnitude of the overburden weight, or approximately 1,000 lb./sq. in. per 1,000 ft. of depth. On the other hand, if the fluids are free to migrate over long distances, either along the strata to surface outcrops or through overlying semipermeable strata or minute fissures traversing impermeable material whereby

vertical access to the surface may be attained, the fluid pressure
ultimately existent within a given stratum cannot exceed the
hydrostatic head corresponding to the depth of burial or to the
outcrop of the permeable channel. It is therefore of considerable
interest to note that the virgin-fluid pressures encountered in
drilling are usually very near to the hydrostatic head corresponding to the depth, or approximately 433 lb./sq. in. per 1,000 ft.
depth.[1] Occasionally higher pressures are encountered, especially as in the case of artesian aquifers, when the formation has
an "active" outcrop where considerable quantities of water may
be supplied currently to the formation.

In view of these facts it is of interest to consider the volume
of fluids which lie below the surface in what may be called reservoir
rocks. Let us assume, for example, the case of a sandstone
stratum having a porosity of 20 per cent and a thickness of 1 ft.
Each square mile of such formation will contain 5,578,680 cu. ft.
of pore space and hence an equal volume of liquid. Moreover,
if the formation contains gas which, as mentioned above, will be at
a pressure approximately equal to the hydrostatic head corresponding to the depth of burial, many billion cubic feet of gas
may be stored within a relatively small area. Since the thickness
of a single porous stratum of this type may be many feet, it is
evident that vast quantities of water as well as other fluids are
contained even within the sedimentary rocks lying far below the
surface.

Considering fluids other than water as special cases, we may
classify subsurface waters into two broad classes: (1) ground water
(meteoric water) encompassing waters which have their source at
and usually remain relatively near the surface of the ground; and
(2) connate water, usually in the deeper lying beds, which represents fluid indigent to the sediments and trapped within them at
the time of deposition or migrated into them shortly thereafter.

1.14. Occurrence of Ground Water.—Of the water falling upon
the land one portion finds its way immediately via the surface
into drainage channels, a second portion is evaporated directly
from the surface, while the remainder enters the ground by

---

[1] In the Gulf Coast oil-producing area of Louisiana and Texas, remarkably
close agreement with this figure has been observed in "discovery" wells
since the recent advent of accurate reservoir-pressure measurements has
made such data available.

seepage. Eventually this seepage or ground water finds its way slowly via underground migration to topographically lower surface-drainage channels, while a portion of it is returned directly to the surface by capillarity or by the 'action of vegetation where it is slowly evaporated during dry periods. Ground water may thus be considered as occurring in two zones, one an unsaturated or so-called capillary zone[1] involving almost exclusively the very superficial layer of unconsolidated soils which lie above the surface of saturation and second, the zone lying below the surface of saturation or true ground-water level. A more quantitative differentiation of these two zones may be made in terms of pressure, the true capillary zone being defined as that in which the pressure of the water is less than atmospheric and the ground-water zone where the pressure is greater than atmospheric. Clearly the true ground-water level is then defined by the surface at which the fluid pressure is equal to the atmospheric pressure.

It should also be mentioned that the ground-water zone involves two regions characterized by differences in stratigraphic boundaries which influence the migratory behavior of the water. Thus, there is a *surface zone of flow* which in principle may be considered as extending from the level of the water table to the first effectively impervious stratum of large areal extent reached by the ground water in its downward percolation, and the second consisting of *deeper zones of flow* which lie below the first impervious stratum. The former, characterized by a free unconfined upper boundary (the water table) and an effectively impervious lower boundary, results in migratory behavior controlled to a considerable extent by the local topography or surface drainage

---

[1] The term is here used broadly to include the so-called vadose (aerated) zone as well as the true capillary zone since a distinct boundary between the two is impossible to define. Evidently water will rise above the normal water table by capillary action exactly as in a capillary tube, the height of rise being determined by the usual factors, but in addition, in an aerated zone such as exists in soils, drops, or even masses of water involving many pores, may remain suspended above the true capillary zone owing to capillary forces trapping the globule within the pores. A protracted period of evaporation would evidently eliminate such suspended water, and there would remain only the true capillary zone above the water table. Evaporation from the surface of this zone would then disturb the pressure equilibrium, and a vertical flow of water through this zone would result. Strictly speaking, only such flow may be termed "capillary flow."

system and governed by so-called "gravity-flow" characteristics.[1] The deep zones, involving strata having both upper and lower impervious boundaries, have no free or zero-fluid-pressure surface, so that the flow is governed by simpler laws. The distinction between these zones is frequently expressed in the terms surface seepage, and artesian flow, the former controlled primarily by local topography, and the latter by regional subsurface geologic features.

1.15. Water Content of Unconsolidated Surface Sediments.— The porosity of porous media has already been discussed in sufficient detail to indicate the magnitude of the pore space involved in such materials. Because of the possible wide variations in the type of material, grain size, and grain-size distribution, the porosity of even unconsolidated sediments may vary over a considerable range. Quoting from the work of King:[2]

A saturated sand carries from 20 to 22% of its dry weight of water while soils and clays range all the way from these values up to 40% or even 50% of their dry weights. Since a cubic foot of sand weighs from 102 to 110 pounds, while soils, clays, and gravels range from 79 to 110 pounds, it is apparent that vast quantities of water lie stored in underground reservoirs.

Actual measurements by King show that the storage capacity of soil is in round numbers equivalent to a sheet of water two feet deep for every five feet of soil lying below the saturation level.

Where the soil does not lie below the plane of saturation it usually contains 75% of the saturated value except during dry times in a surface layer 1 to 5' thick. Water thus stored in this zone ranges from 4% of the dry weight for coarse mixed sands up to as high as 32% in clays of fine texture.

1.16. Fluid Movements in the Capillary Zone.—The so-called capillary movements of ground water, by the usual definition, are limited to the superficial zone above the ground-water level. There can be no true capillary movement of water in sediments or rocks below the water table where the pore spaces are already

---

[1] This type of flow is discussed in Chap. VI where its unusually difficult analytical characteristics will become evident.

[2] KING, F. H., U. S. Geol. Surv., 19th Ann. Rept., Part II, 67, 1897–1898.

filled with water, but rather such capillary motion is limited to flow from regions of higher to those of lower saturation. Clearly such capillary movements may be upward, downward, or lateral, depending upon the saturation conditions existing at the moment. Thus rain falling upon a dry surface so saturates the top portion of the soil that capillary action tends to accentuate the downward gravity drainage or seepage into regions of low saturation. During dry periods the capillary flow is reversed, since evaporation from the surface continually releases water from the top of the capillary zone and in order to maintain equilibrium this water must be replaced from the lower saturated zone. Under equilibrium conditions, normal soil will be comparatively dry at the surface with a gradually increasing water content with depth until the zone of saturation or true ground-water level is reached. While obviously there is no abrupt change between the two zones in a homogeneous medium, the true ground-water level is nevertheless definitely indicated by the normal level of the water surface in an open trench or well which penetrates below the saturated zone. Clearly an open trench penetrating only the capillary zone will contain no free water, the only flow into the opening being that induced by evaporation from the walls of the trench. These circumstances follow directly from a consideration of the pressure conditions already mentioned as defining the two zones, and the definition of the ground-water level as that surface where the water pressure is equal to atmospheric pressure is obviously fulfilled by the free-water surface in an open trench or well. Regarding the source of energy responsible for capillary flow of the type referred to here, it is clear that it must eventually be traced to the source which causes evaporation of the water.

In addition to capillary flow of the type ultimately traceable to evaporation processes, it is clear that water may also be forced through this zone by the application of fluid pressure. Thus as already mentioned, water gathered on the surface during rains will, when the saturation condition is reached, be forced into the capillary zone and thence to the water table by the action of gravity. This apparently simple process is complicated, however, by the presence of air in the unsaturated zone. This air may be considered as being trapped to some extent, hence decreasing the cross-sectional pore area available to the flow of water, although such entrapped air will participate in the motion along with the water. As a result, the rate of seepage of water through this zone depends in a complex manner upon its degree of saturation.

Another cause of secondary or transient flow in the capillary zone is the normal variation in atmospheric pressure. Such variation necessarily results in changes of the fluid pressure within the soil, and the readjustment to equilibrium conditions causes flow of the ground water. While these barometric variations are of relatively small magnitude, yet a variation of 1 in. of mercury corresponds to a variation of some 3 or 4 per cent of the total atmospheric pressure. Since this pressure variation is imposed upon the gas masses trapped within the capillary zone they must respond by a change in volume of corresponding amount. Clearly this "breathing" of the absorbed air will induce a transient flow of small magnitude in the capillary zone. As a result the water table must show variations in level with changing barometric pressure.

As to the magnitude of the effects on ground-water flow traceable to changes in barometric pressure, King (*loc. cit.*, page 73) cites some interesting observations. Examples are given in which the rate of flow of a spring showed direct correlation with changes in barometric pressure, while almost exactly similar variations were observed in the level of an artesian well situated about ½ mile distant. In fact, he states that the influence of barometric changes is great enough to modify the rate of flow of the spring by as much as 8 per cent of the normal discharge, while the artesian well 979 ft. deep showed variations of at least 10 per cent. In the case of tile drains, necessarily very close to the surface, such observed changes amount to as much as 15 per cent. The primary cause of these surprisingly large variations induced by barometric-pressure changes may be traced to an interesting feature of the capillary zone. King has shown experimental data indicating that the addition of very small amounts of water to the capillary zone suffices to create relatively large changes in the height of the water table, a result less surprising if one considers the high degree of saturation existing in this zone immediately above the water table. Thus an increase in atmospheric pressure results in a downward flow of water in the capillary zone because of the reduction in volume of the

entrained air.   In view of the disproportionate increase in height
of the water table for a given volume of water added to the
capillary zone near the saturation level, a relatively large increase
in hydrostatic head is applied to the system, and a spring or
artesian well responds with an increased rate of flow.   This
performance is of interest primarily in illustrating the fact that
relatively small variations in the volume of air entrained in the
capillary zone may produce transient disturbances of considerably
greater magnitude in the normal equilibrium condition of this
zone.   In a somewhat similar manner temperature changes are
capable of contributing to the transient flow within this region.

While we shall not consider in further detail the motion of water
in the capillary zone, it should be mentioned that these processes
are of great economic importance.   Thus, during dry periods,
the maintenance of an adequate supply of moisture for plant
roots not reaching the water table is dependent upon capillarity
effects.   Furthermore, the rate of evaporation and hence the
lowering of the water table is profoundly affected by the texture
of the surface soil.   Moreover, the "breathing" of the soil, which
has been mentioned as being brought about by the downward
percolation of surface water together with barometric pressure
variations supplemented by temperature changes, serves to
provide a downward flow of air necessary to maintain organic
life below the surface of the ground.   Similarly, these processes
provide considerable aid to the normal diffusion processes in the
dissemination and eventual exhalation of the gaseous products
of organic reactions.   Clearly, without such aids the slow process
of normal diffusion alone would result in equilibrium conditions of
the gaseous content of the soils quite different from those
actually existent, with a resultant marked change in the abundance and type of organic life within the soil.

These problems which involve the flow of fluids within the
superficial capillary zone are obviously of great importance to
agriculture and form a major item of investigation in that field.
Furthermore, this zone is of importance to the hydrologist to the
extent that surface water reaching the saturation zone must
necessarily pass through the capillary region, since only water in
excess of equilibrium saturation can reach the true water table.
However, beyond the mere mention already made of the problems
involved in capillary flow, we will not concern ourselves further

with them, our real interest being in the regions of saturation starting at the true ground-water level or water table.

1.17. Fluid Movements below the Water Table. The Surface Zone.—As has been indicated, the flow of water within the capillary zone is governed by saturation characteristics and other factors which place it beyond the scope of our present work. At the water table or ground-water level and below it, complete saturation may be assumed, and this zone, being one of positive pressures where the flow is governed by pressure gradients alone, becomes reasonably amenable to analytical study. Henceforth our discussion will be confined to such regions.

It has already been mentioned (Sec. 1.14) that the ground-water zone involves (1) a surface zone of flow where gravitational influences govern the migration, and (2) the deep zones of flow where the porous medium may be considered as having definite upper and lower boundaries and the systems involve flow through well-defined porous conduits. The latter type constitutes by far the major discussions of this treatise since it covers not only deep ground-water zones but is equally applicable to the deeper zones involving connate waters.

While it will be impossible to discuss in detail the innumerable types of problems involved in the surface zone of flow, a brief consideration of the general characteristics will be of value. No only will it serve as an introduction to the practical aspects of the later analytical discussions pertaining to these special problems, but the performance of these surface waters is to a considerable extent directly observable. And such direct observations of their performance are not only interesting but provide an invaluable method of obtaining an intuitive insight into all practical problems of flow through porous media.

The unit of the surface zone of flow of ground waters is the river valley.[1] In the surface zone the rate and direction of motion of the underground water conform primarily to the slopes and grades of the land surface. The principal feature of flow in the surface zone is that the flow follows the trend and direction of the surface drainage. The direction taken by the surface waters in their course into the streams and drainage channels is in general the same as that taken by the seepage

[1] The following paragraphs quote liberally from "The Motions of Underground Waters," C. S. Slichter, *U. S. Geol. Surv. Pub.*, 1902.

waters of the upper zone of flow. Actual determinations show that the water table usually has a slope essentially similar to the slope of the surface of the ground, differing from the latter principally in being less steep. The surface divide or watershed usually coincides with the line of the underground water divide, and the motion of the underground seepage into the streams and rivers is similar to the lines followed by the surface drainage into the same streams.

Fig. 4.—Contour map showing position of water table (continuous lines), supposed lines of motion of ground water (arrowed lines), and the thalwegs or drainage lines (heavy lines). (*After C. S. Slichter, U.S. Geol. Survey Water Supply Papers.*)

The lowest line of drainage of the valley is known technically as the *thalweg*. Topographically it is a line upon a contour map which is a natural watercourse. Beneath the thalweg there is usually a similar drainage line for the underground current, in general coincident with the thalweg. For other parts of the valley the actual lines of motion of the underground water are represented by a set of curves which cut the contours of the water table at right angles, *i.e.*, the flow lines are in the direction of the pressure gradient and thus in general follow the slopes of the land surface, as indicated in Fig. 4.

The similarity between the contours of the land surface and subterranean thalwegs must not be taken too literally. The coincidence of the surface and subterranean thalwegs and of the watersheds is a common occurrence, but not a geologic necessity. For the surface topography is only one, and often not the most important, element in the control of the underground flow.

The horizontal distribution and motion of the ground water is influenced first of all by the form of the surface of the first effectively impervious layer encountered below the water table. It is also influenced in a marked degree by the varying altitude of the surface or receiving area, by the character and uniformity of the pervious layer, by the altitude and distance to the nearest thalweg or drainage channel, and finally by the amount of rainfall. These elements combined determine the depth to the water table at any given place and the direction and rate of motion of the underground current. They form a complicated system and it is not easy to describe the precise part which each plays in a given case. Fine material and large rainfall tend to make the water table stand high within the hills and elevated places and to give steep gradients to the water table. Conversely, coarse material and light rainfall tend to result in low water tables and small gradients. From these considerations it is evident that the form of the impervious stratum affects the water table much less in humid climates than in semiarid or arid climates.

As already indicated, the general trend of the moving underground water, under the influence of gravity, is into the neighboring streams and lakes. This motion must, however, be modified by many causes which frequently present most complex combinations. Thus while the return flow of ground water to the watercourses via diffused and almost imperceptibly slow seepage is the rule, yet geologic conditions may be such—for example, the outcropping of an impervious stratum—as to force the water table above the surface of the ground and converge and concentrate the flow into a strong current. In the latter case we have the phenomenon of the flowing spring. Such obvious evidence of ground-water motion tends to divert the layman's attention from the more important but less obvious diffused seepage.

The ground water, after starting on its journey toward the river valley, may not after all find its way immediately into the channel. Sometimes it takes a general course down the thalweg and toward the sea within the porous medium itself. This movement may be so great as to constitute a large underground stream, many feet in depth and miles in breadth. Such a moving sheet of water beneath the bed and banks of a stream is termed the underflow.

It is evident that a considerable underflow is impossible with material as fine as is sometimes found filling the valleys of rivers. Such material may be an important storage reservoir, but it cannot play the primary part in the regional drainage of the area. An entirely different condition may exist where the sands and gravel beneath the stream are sufficiently coarse, as is likely to be the case near the headwaters of a stream. Here the materials deposited are the coarser sands, and gravels, and boulders brought down by mountain torrents, the finer material being carried on until the stream bed reaches a gentle slope and the current has lost its high velocity. Thus, particularly near the headwaters, below the stream bed there is a vast body of water slowly percolating through the coarse material. Occasionally the stream may pass through a narrow valley or gorge where the rock walls converge, or the bedrock beneath the debris is closer to the surface. The result in either case is a marked increase in the amount of water in the river because the waters of the underflow are forced to the surface.

At the mouth of the canyon the valley broadens out and the bedrock sinks more deeply below the surface while the stream deposits more and more of its suspended matter. The underflow may also broaden and deepen to fill the greatly enlarged channel, the finer material and lesser slope decreasing the velocity with which it flows down the valley. The valley may be now so broad that if the rainfall is sufficient, seepage from the surface of the higher land will continually augment the volume of water in the underflow, giving rise to a regular flow into the perennial stream itself, causing its constant growth during its course to the sea. On the other hand, if the stream issues from its mountain canyon in an arid region, the surface stream may gradually disappear as its water seeps into the underlying valley fill, the course of the stream being marked only by a dry wash through the valley which is only occasionally swept by floods from the mountains.

Thus the broadening of the valley, as a stream flows out of its mountain canyon will have entirely different effects, depending upon whether the climate of the valley is humid or arid. In the one case the broadening of the valley will result in the constant growth of the river, and in the other in a constant diminution in the volume of the river as it loses water to the underflow where it spreads out in a broad belt and is there exposed to increased evaporation losses.

The relation of the underflow to the waters of the river channel presents many interesting phenomena and variations. In some cases silt may have rendered the river channel so impervious that for considerable distances little or no interchange can take place between the river waters and the underflow waters. As a result the two may be quite independent and of entirely different chemical composition. Evidently in such cases the relations of the underflow and river channel discussed above will no longer apply.

While the storage capacity of the material comprising a valley fill may be very great, the total volume of water transported daily by the underflow may be relatively small. Through very coarse material and where the slope of the water table is great the velocity of the percolating waters may be of the order of 10 ft. or more per day, whereas in less pervious material and under small gradients the velocity of the seepage water may be extremely slow. Because of the wide variations possible each case presents an individual problem and it will be unprofitable to discuss this feature in further detail.

1.18. Fluid Movement below the Water Table. The Deep Zones.—Ground water may find its way into deeper zones of flow either by percolation into exposed ends of the pervious strata or indirectly by seepage from streams and rivers which have cut their valleys through the outcrops. The water must leave the deep zone at a lower level than that at which it enters, and hence the outlets must be sought in the lower outcrops of the rocks where perhaps a river has eroded a channel through it or where the upper impervious stratum has been eroded away. Moreover, in addition to such obvious outlets, the strata may be traversed by faults and joints through which the water may escape to overlying beds, or even by slow seepage through the covering strata an influx of water to a given formation may be dissipated via overlying formations or vertically to the surface. For it must be remembered that imperviousness as applied to rocks is a relative term. The covering layer of shale or rock has an enormously greater area than the vertical transmitting section of a given stratum, so what the covering rock lacks in permeability can be counterbalanced by the vastly greater area through which the seepage may proceed. Thus upward seepage may cause motion of the water in a porous stratum, even though no actual outcrop of the transmitting rock exists.

Local upward seepage may therefore explain some apparently anomalous variations in pressure of underground water in artesian basins. And while the existence of an outcrop and hence a definite point of infiltration of ground water to a pervious formation would lead to the expectation of a uniform head over the areal extent of the stratum in the absence of a definite outlet, loss by vertical seepage through the overlying cover, augmented by losses through faults and joints, results in a general regional migration within the conducting strata and a corresponding loss in fluid head. Similar effects must be present in the very deep lying strata involved in the zone of connate waters, where the fluid pressures in totally unexposed formations are found to correspond closely to the hydrostatic head equivalent to the depth of burial, as mentioned in the introductory remarks of Sec. 1.13.

It will be evident, therefore, that while the strata involved in the deep zones of flow have been classified for convenience as closed porous conduits, this conception of them is a purely relative matter. Rather, such strata must be looked upon as pseudo-closed channels whose performance approaches the ideal in a manner depending upon the ratio of the permeability along the stratum to that through the overlying cover. For as long as the pressure within a given stratum exceeds the hydrostatic head corresponding to its depth of burial, seepage vertically upward is not only a possibility, but it is practically inevitable, at least within certain portions of the overlying cover.

These considerations should be sufficient to disclose the wide variations in performance which may be expected of artesian aquifers. Thus where a very permeable formation has a copious supply of water at its elevated outcrop or other intake area, and the permeability of the overlying cover is relatively low, and if, moreover, the stratum has no effective outlet at its lower boundary, the artesian head will be substantially uniform over the areal extent of the stratum. On the other hand, a low rate of influx at the elevated outcrop, a low permeability in the strata themselves, a relatively high rate of vertical seepage, or an effectively large outlet area will singly or collectively result in steep pressure gradients within the strata and a rapid decline in artesian head in the direction of migration.

Figure 5, showing a cross section through a hypothetical artesian basin, illustrates in principle some of the major features which have been mentioned both as regards the surface and deep zones of ground-water flow. Further study of details available in

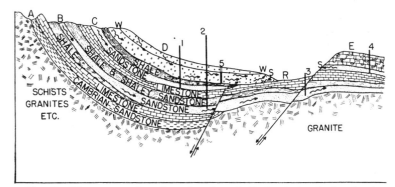

Fig. 5.—The cross section through a hypothetical artesian basin, showing surface and deep zones of ground-water flow. (Vertical scale exaggerated.) *A*. Outcrop of impervious sandstone and limestone—unproductive aquifer. *B*. Outcrop of pervious sandstone supplied by river at *B*—deep continuous artesian aquifer. *C*. Pervious sandstone supplied by outcrop with seepage in valley—shallow artesian aquifer losing water at fault zone to overlying till with resultant loss of head. *D*. Recent deposition of unconsolidated material with water table *WW*, being an underground source for river at *R*, and a source of shallow wells (5) and springs *S*. *E*. Cavernous limestone—source of prolific wells and seepage into underlying sandstone, the latter providing springs at escapement and artesian wells on eastern slope of plateau. 1, 2, 3, 4. Deep wells, artesian head indicated by height of solid line. 5. Shallow well below water table in zone of surface flow. In all cases the loss of head is due to migration, as indicated by arrows. Abrupt loss of head at well 3 is due to relatively low permeability across fault zones.

numerous hydrological papers and texts is recommended in order to develop familiarity with these visible evidences of sub-surface flow, and, coupled with actual field observations, the subject will be found most fascinating.

1.19. The Occurrence of Connate Waters.—It is evident that any marine deposit, which subsequently to deposition has not risen above the water table or has not been flushed by meteoric waters through exposure to ground-water percolation, must be saturated with water perhaps identical with, or at least derived directly from, the water of the ancient seas. The tendency for this retention over geologic time will be the more evident upon recalling that immediately after deposition the porosity of the sediments is a maximum and the trend is toward expulsion of fluid as the porosity is decreased by compaction and cementation. Thus, excepting the obvious cases where ground water has direct access to the deeply buried strata, with the resultant possibility of contamination or complete displacement by ground water, or in the case of surface exposure, the original or connate water (or more strictly speaking, a portion of it) is effectively trapped after burial.

However, it must be emphasized that the connate water present in a given stratum is not necessarily indigenous to that particular local area or stratum, for, in view of the complex migration possible in the subsurface rocks, the present water content may represent connate water forced into it from contiguous sources or even by migration along permeable strata from relatively great distances. Moreover, it is not to be expected that the water now recovered from wells penetrating such formations represents in chemical composition the ancient sea water. Changes in composition must have occurred during the geologic time elapsed between deposition and the present day. For during that period chemical activity in the form of solution of rock minerals, interchange of salts, and precipitation of minerals has taken place, as evidenced by cementation of sands and other porous materials. Furthermore, profound changes in chemical composition of some rocks have radically altered their character—such changes as could be effected only through the agency of migratory waters.

It is of interest to examine the composition of a few of the salt waters or brines recovered from deep wells and even from relatively shallow strata uncontaminated by meteoric waters. Typical examples are given in Table 3.

The markedly greater salinity of these brines as compared with sea water is the immediately striking feature. While it is unlikely that the salinity of our ocean water is lower than that of the ocean water of millions of years ago, many of the sediments now explored by the drill were no doubt laid down in inland seas where, by gradual evaporation, very high concentrations of dissolved minerals were attained. Vast deposits of almost pure sodium chloride forming a portion of the stratigraphic column in many areas provide definite evidence of the arid conditions prevailing in such areas at the time of deposition. On the other hand, the increased salinity may be due entirely to physical and chemical processes occurring after deposition and deep burial.

TABLE 3.—ANALYSES OF CONNATE WATERS

Parts per Million

Source	Depth	Chloride	Sulphate	Carbonate	Bicarbonate	Sodium	Potassium	Calcium	Magnesium	Silica	Iron, aluminum
1	Ocean water	19,410	2,700	70		10,710	390	420	1,300		
2	1,400'	77,340	730	...	.....	31,950	650	13,260	1,940		
3	4,242'	108,990	455	...	43	54,363	.....	10,560	2,390		
4	......	152,100	319	...	.....	73,620	.....	17,700	2,541	30	300
5	1,359'	90,540	140	...	.....	40,100	140	12,210	2,140	...	70
6	1,170'	82,350	.....	...	.....	36,300	260	11,400	1,950	10	150
7	6,300'	196,000	60	...	.....	78,200	6,250	30,500	3,000	...	190
8	1,500'	3,273	11	48	3,684	3,350	47	28	24	127	4
9	3,100'	23,553	18	626		13,253		1,089	685	65	11
10	2,450'	16,640	1,310	...	587	9,318		1,350	377	...	42
11	......	2,670	trace	696	5,120	4,125	.....	.....	16		
12	......	153	.....	...	2,866	1,148	.....	10	10		

1. Mean of 77 analyses of ocean water, Challenger Expedition.
2. Bradford sand, Bradford, Pa.
3. Wilcox sand, St. Louis Oil Field, Okla.
4. Layton sand, Garber Oil Field, Okla.
5. Butler sand, Oakland Twp., Pa.
6. Butler sand, Center Twp., Pa.
7. Oriskany sand, Washington, Pa.
8. Noncorrosive water from the Sunset-Midway Field, Calif.
9. A corrosive water from the Sunset-Midway Field.
10. Eldorado Field, Kan.
11. First Wall Creek sand, east side of field (basinward) Salt Creek Oil Field, Wyo.
12. First Wall Creek sand, west side of field (mountainward) Salt Creek, Wyo.

Variations in other chemical constituents are clearly the result of alterations incurred during prolonged contact with the reservoir rocks and mixing of waters during migration. Thus a peculiarity of the waters shown in Table 3 is the markedly low sulphate content which has been found to be characteristic of oil field waters[1] and is believed to be due to the reducing action of the oil.[2] The marked increase in calcium is equally striking. Other brines contain unusual quantities of bromine, iodine, and other elements and form profitable commercial sources of various chemicals.

[1] The analyses given in Table 3 are typical of those taken in and adjacent to the oil fields.

[2] ROGERS, C. S., U. S. Geol. Surv. Prof. Paper 117, Part 2, 1–103, 1919; E. S. BASTIN, Amer. Assoc. Petrol. Geol. Bull. 10, 1290, 1936.

On the other hand, such waters are subject to dilution by migrating ground water when stratigraphic configurations and outcrop exposures provide access to the deeper lying portions of the strata. For example, in Table 3 the waters 11 and 12 are from the same sand, but, as this sand outcrops about 15 miles west (mountainward) of the oil field, meteoric waters have partially flushed the sand and diluted the connate water. (It is of interest to note that the greatest dilution has occurred in the "upstream" side, showing the shielding influence of the dome wherein oil and gas have accumulated and thus forced the migrating water to circulate around the oil field.) Other examples could be cited as in a sand in California where practically fresh water was encountered at a depth of 6,000 ft. Outcrops of this stratum in the mountains provided ready access for ground water which, coupled with the steep slope of the stratum and evidently a ready outlet, induced a relatively rapid migration and an effective flushing of the connate water. Similarly, the Woodbine sand in North-Central Texas outcrops along an extended belt passing through the vicinity of Dallas, where it obtains a copious supply of ground water which has flushed out the connate water up to the Mexia fault zone. In this sand west of the fault zone, fresh water gradually becomes brackish as the faulted region is reached, while east of the faults and throughout the basin typical connate water occupies the sand.

This general discussion will serve to illustrate the distinction between connate and meteoric waters. Moreover, it will be clear that though the term connate implies that the water content of the rock is indigent to the rock itself, such a literal interpretation of the term is not contemplated. For, inasmuch as the rock structure in the connate-water zones is similar to that in the deep ground-water zones, complex and long-distance migration of fluids within them is to be expected. Indeed, the few examples cited wherein dilution by ground water has occurred provide ample evidence of such migration. This broad interpretation of the term connate water is emphasized in order to dispel any implication that such water has lain essentially stagnant within its associated strata since the time of deposition.

1.20. The Migration of Connate Waters.—In the case of deep ground-water zones it is easy to visualize the process of migration since the point of entrance of water at topographically high outcrops and its exit into lower surface drainage channels is directly evident. The existence of large springs emerging from strata which are buried deeply beneath adjacent highlands is obvious and conclusive evidence of the mechanism. Likewise, regional migration induced by the entrance of ground water at elevated outcrops is certainly to be expected in the case of strata outcropping on land and dipping gradually to outcrops which may lie far below the surface of the ocean (Fig. 6). In view of such plain facts the migration of connate water induced by exposure to the direct drive of encroaching ground water need not be considered further. On the other hand, in the case of closed basins or where major unconformities apparently preclude any effective outcrops

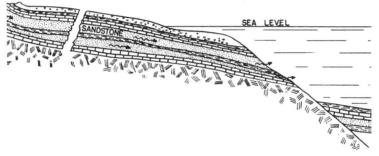

Fig. 6.— A typical section of dipping strata which outcrop below an ocean surface.

(*cf.* Fig. 3), certain strata may have no evident zone of intake or egress whereby circulation may be induced. Under such conditions the mechanism responsible for migration is not so apparent, though evidence of its existence over protracted geological periods is proved by diverse facts. To mention only one here, the precipitation of siliceous or calcareous cementing material within the pores of sand strata, amounting in some cases to 30 per cent or more of the total volume of the strata, is obviously not the effect of quiescent waters. Sufficient quantities of solids could not be present in solution in the trapped water in the immediate vicinity to effect such deposition. Evidently a more obscure mechanism than already discussed must be offered to explain the facts.

1.21. The Influence of Compaction of Sediments on Migration of Fluids.—The compaction of sediments has been discussed briefly in Secs. 1.7 to 1.9, and it is now pertinent to consider the importance of this feature as it affects geological processes through its relation to the migration of connate waters. Thus we have seen that through compaction, owing to the weight of overlying sediments, clays having porosities of 45 to 50 per cent, and containing therefore an equal amount of residue water, lose porosity progressively with depth of burial until at 5,000 to 6,000 ft. the residual porosity and water content is of the order of 5 per cent. Evidently the process of forming shales from the original clays and silts requires the expulsion of vast quantities of water originally contained within the sediments. The compaction of sands is less important since, as already disclosed, the total compaction even under very heavy overburdens amounts to a few per cent only. Nevertheless, the cementation of the sands which has reduced the porosities from original values as high as 40 per cent to values ranging from 30 per cent to as low as 5 per cent evidently requires the expulsion of an equivalent amount of water from these sediments. However, in the case of sands this process must be considered as a secondary effect rather than a primary cause, since precipitation of the cementing material would be impossible were it not for the circulation of vast quantities of water through the action of other agencies.

These facts clearly reveal not only the necessity for the circulation of enormous volumes of connate water through its associated permeable sediments, but the mechanism whereby migration has proceeded during the geological periods involved in the deposition and deep burial. Admittedly, however, the exact process whereby the water is expelled involves some speculation.

Not only does the magnitude of compaction required to form shales from clays and silt deposits indicate these sediments as the primary source of circulating connate waters, but the preponderating volume of shales in any extensive sedimentary section is further evidence of their dominant role in the process. Therefore, in the brief discussion which follows only the shales will be considered, bearing in mind the similar but less important contribution of other sediments.

In the initial formative process the shale-source sediments are gradually buried under successively thicker layers of additional sediments. Because of the enormously greater susceptibility to compaction during this formative stage and the relatively high vertical permeability of the overlying material, unquestionably the greater portion of the water squeezed out of the clay and silt is expelled vertically through the overburden. This slow process may be accompanied by the gradual precipitation of cementing material within the sands and similar sediments, a large portion of the precipitated material being derived from solution of the minerals comprising the clays and silts. As the depth of burial becomes greater, however, and the permeability of the shales becomes lower and lower, an increasing portion of the ejected water will be expelled into adjacent sand strata where, because of the relatively high permeability of such media, lateral rather than vertical migration of the water will result. For accompanying the transformation, the shale beds become more and more impervious and act as boundary walls tending to confine the expelled water within the contiguous sands where, even though migration over long distances may be involved, the water finds an easier exit to regions of lower pressure and finally escapes from the sedimentary reservoir. It should be kept in mind, however, that as explained in Sec. 1.18, vertical seepage upward through the rocks and particularly via joints and fault planes in the overlying cover must be involved in the process. This is in fact quite conclusively proved by the observation that in general the fluid pressure in such formations is closely equivalent to the hydrostatic head corresponding to the depth of burial, though in isolated instances, especially at great depths, excessive pressures may be encountered. This process of vertical seepage accompanying lateral migration through permeable strata is not an unreasonable picture when one contemplates the enormous span of time involved and the very great pressures that could be imposed if expulsion of the water were strongly resisted. Moreover, as previously mentioned, the vanishingly low permeability across the strata is offset, to some extent, by the exceedingly great area involved. In view of these factors even the vast quantity of water which must be expelled involves only an infinitesimally low average rate of flow vertically.

Considering, therefore, a typical sedimentary basin where at its center 10,000 to 20,000 ft. or more of sediment may be involved, the gradual compaction of the shales comprising a large proportion of the deposits results in the expulsion of water into adjacent permeable beds. These strata in turn serve as conduits whereby along the path of least resistance the water is finally forced through the overlying rocks. Fault zones or minute joints and fissures at points of folding are obvious zones of exit supplemented over vast areas by exceedingly slow seepage through the rocks themselves. In such a system it is clear that the general migration will be from the center or major source of water toward the edges of the basin, this general course being modified profoundly by geological features which may disrupt the continuity of the system. Moreover, the process is in any case not to be considered as a purely lateral migration such that the entire flow must eventually find exit at the edges of the basin, but rather, as has been emphasized, the general flow radially outward is accompanied by vertical seepage whereby gradual areal elimination is accomplished.

The above generalizations are far from sufficient to provide a complete and satisfactory picture of the process. A detailed explanation of some features would of necessity be highly speculative, for it is quite impossible to develop an unassailable detailed explanation when periods of time involving millions of years are involved. For our knowledge of physical and chemical processes involving a time factor of this magnitude is entirely inadequate to answer many of the questions which arise upon close scrutiny. Nevertheless, in the light of existing evidence and known facts regarding the underlying physical behavior of such systems, it is felt that the picture presented here must represent the true background. The additional details which must be added for completeness may clarify, but cannot profoundly modify, the mechanism portrayed.

1.22. The Occurrence of Gas and Oil.—Aside from their obvious importance in geological processes and in rather isolated cases where unusual chemical constituents warrant commercial exploitation, the connate waters themselves would hardly justify consideration in the present study. However, as is well known, economically important deposits of gas and oil are found and exploited by means of wells penetrating deep below the surface. Such deposits are not only intimately associated with connate waters, but their accumulation in local areas in such concentrations as to make their exploitation economically possible may be attributed directly to the migratory behavior of these waters. Moreover, in the process of withdrawing oil and gas from the reservoir rocks the water contributes in varying degrees to the reservoir performance. This latter role will be considered in the detailed analytical treatments presented in later chapters.

For the present attention will be confined to a rather cursory examination of the source and mode of occurrence of gas and oil.

1.23. The Source of Petroleum—Gas and Oil.—There is probably no mineral deposit of commercial importance, the source or mode of origin of which is more debatable than in the case of petroleum—gas and oil. Gas associated with coal or related deposits is patently derived from the vegetable matter during the process of transformation. Soil gases, principally nitrogen, carbon dioxide, and methane, are clearly products of decomposition of organic remains, though in their formation bacterial action is an important phase. Such gases are ordinarily so thoroughly dispersed that only careful analysis indicates their presence. However, in many areas, as for example, in some portions of the Mississippi delta, the thickness of the marshy deposits may be so great as to result in accumulations of considerable quantities of "marsh gas." Careful analysis of such gas indicates the combustible or hydrocarbon constituent to be methane (CH_4) with not more than 0.001 per cent of ethane (C_2H_6) and no higher hydrocarbons. On the other hand, petroleum gas, though varying over a wide range in composition, comprises a mixture of hydrocarbons ranging from methane to hexane. Methane may comprise 95 per cent of the total, but in some cases propane and butane may be the major constituents. This is a marked contrast to the ordinary decomposition products found in relatively shallow deposits where, as mentioned above, only methane with minute traces of ethane are found. Consequently, until recent years it has been contended by some investigators that petroleum gases were essentially inorganic in origin. While this conception has now been quite thoroughly refuted and the true source now conceded to be organic, the process whereby these gases are formed remains essentially speculative.

Even less satisfactory is our knowledge of the processes involved in the formation of the very complex mixtures of liquid hydrocarbons which comprise petroleum. But again diverse evidence indicates conclusively that, although the process of formation is uncertain, the original source material is organic.

Also pertinent to the question of origin is the fact that gas and oil are always intimately associated in the reservoir rocks. While it is true that large deposits of gas are found with but little associated oil and conversely some oil deposits contain only very small quantities of gas, such conditions may be attributed to circumstances distinct from the problem of origin. For it is possible that those gas deposits which are almost devoid of oil content exist as such because of fortuitous conditions favorable to the formation of gas alone or to its accumulation. On the other hand, no deposits of oil are found which are devoid of gas except where conditions are obviously conducive to the dissipation of gas because of leakage from the reservoir. It may be assumed, therefore, that the oil and gas have a common origin and are derived from organic reactions, the details of which are unknown.

In view of this accepted organic origin it is natural to assume a close generic relation between coal and oil. On the contrary, the great coal-producing strata generally are not oil-producing strata though in some areas such strata are associated with highly lignitic beds. Moreover, the most prolific oil-producing beds are found so distinctly separated from any lignitic deposits as to prove conclusively that while such deposits may have contributed to a small part of the known oil deposits, no necessary relation between coal and petroleum exists.

Briefly then it has been observed that to explain the existence of oil and gas the original source material must be found within the normal sedimentary deposits of limestone, sand, and shales. And further, that while petroliferous strata may exist within sediments known to have been deposited under arid conditions, contiguous thereto or within reach of migratory fluids, there must be sediments of marine origin. Thus it is assumed that while any permeable rock or sedimentary deposit may be a suitable reservoir, the important source materials are confined to the marine sediments and particularly to those deposited in the relatively shallow waters where organic life was most abundant. While undeniable proof of this general statement is lacking, it is the accepted hypothesis at the present time.

In summarizing a report covering a world-wide study of the organic content of present-day sediments Trask[1] states:

1. The organic content of sediments is influenced strongly by the configuration of the sea bottom. Deposits in depressions and closed basins contain more organic matter than do those on adjoining ridges and on slopes inclining more steeply than adjacent areas.

[1] TRASK, P. D., Origin and Environment of Source Sediments of Petroleum, 1932, (A. P. I. Research Program, Project 4).

2. The organic content of sediments increases as the texture becomes finer. Clays contain approximately twice as much organic matter as silt and silts about double that of fine sands.

3. The organic content of fine sediments may vary considerably within a few miles without appreciable macroscopic change in the texture of the deposits.

4. The organic content of typically marine sediments varies roughly with the supply of plankton in the surface water, but in the deposits of shallow bays and estuaries it may depend more on attached plants and inwashed organic matter than on plankton.

5. Near-shore sediments contain much more organic matter than do oceanic deposits.

6. The organic content of sediments in regions of upwelling of deep water to the surface is large. Upwelling is believed to be caused by off-shore movement of the surface water in coastal regions of considerable submarine relief.

On the basis of this work we may, therefore, assume that the clays and silts are the primary sources of organic material not only because of their relative richness, but because of the abundance of shale in any sedimentary section of considerable thickness. Also, ignoring the generic processes involved, we will tacitly assume that within these shales, and some considerable time after burial,[1] petroleum gas and oil was formed.

1.24. The Migration and Accumulation of Gas and Oil.—The effect of compaction upon the migration of the water contained within the shales has already been discussed (Sec. 1.21) in sufficient detail to indicate immediately that the gas and oil content of the shales will ultimately be expelled as was the case with water. Upon entrance into adjacent strata such as a sandstone having a sufficiently high permeability to permit lateral migration, the oil and gas will tend to take part in the migration along with the connate water. And in general this migration will be, as already described, in a direction away from the area of maximum compaction, *i.e.*, thickest sedimentary section, toward the regions of easiest exit. Here again we are faced with a difficulty involving the exact manner whereby free gas and oil which are immiscible in water, may traverse the porous media. Simple experiments indicate that small globules of oil or bubbles of gas comparable in size to the pore spaces will, to a considerable extent, be locked in

[1] Trask (*loc. cit.*) has shown that recent sediments contain no petroliferous material.

the pores.[1] If such were the case in nature, both oil and gas would be so broadly disseminated that no exploitable deposits would be found. If, however, the enormously great period of time involved in the process permits the accumulation of masses of gas and oil which are large in comparison with the pore spaces, then the surface tension effects responsible for locking of the small globules would be eliminated as far as the large body of fluid is concerned. Thus under the influence of the imposed pressure gradients one might picture the oil and gas migrating with the water as thin sheets of relatively large dimensions and effectively floating on the water in the uppermost portion of the pervious stratum because of their low density with respect to the water.

Now it is evident that irregularities in the upper boundary of the stratum, such as would present a pocket or dome-shaped re-entrant surface, must tend to trap the low-density fluids because of their buoyancy. In fact, in the absence of solution effects which might eliminate such accumulations, the lighter fluids will remain trapped within these boundary irregularities unless the pressure gradient in the zone of water flow is sufficient to overcome the buoyancy forces on the oil or gas and flush these accumulations out of the trap. Figure 7(a) shows schematically an idealized section through an anticlinal fold or dome where the trapping of low-density fluids carried along with the water is an obvious result of gravitational separation. Similarly, the possible accumulation against a fault, as shown in Fig. 7(b), or in a "pinchout" of the permeable stratum, as in Fig. 7(c), are typical of deposits actually occurring.

Of course, it should be observed that all oil and gas reservoirs are not located at the crests or in such obviously good traps as have been shown in Fig. 7. Occasionally the "closure" of the reservoir is due to a marked decrease in porosity of the stratum or a relatively large deposit or lens of sand may exist within highly impermeable beds of shale. In such traps accumulation of oil and gas may occur. However, whatever be the nature of the reservoir, field evidence indicates positively that in numerous cases the oil and gas present cannot be indigenous fluids but must have migrated into the trap from other sources. And, granting that migration over distances of a few miles is an obvious necessity, there is no reason whatever to place any limit upon the

[1] WYCKOFF, R. D., and H. G. BOTSET, *Physics*, **7**, 325, 1936.

distance over which such migration can occur except as lack of continuity of the permeable strata or intermediate zones of exit may limit the lateral circulation of the connate waters.

It is of interest to mention a consequence of long-distance migration. Since the buoyancy of gas in water is considerably greater than that of oil, gas will preferentially flush oil out of a

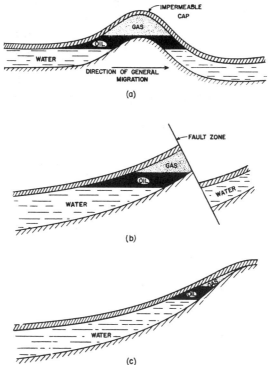

Fig. 7.—Types of traps that may form oil or gas reservoirs.

reservoir if a sufficient amount of it is available completely to fill the trap. One would, therefore, suspect that if a series of traps are present along the direction of migration, the surplus of gas will preferentially occupy the reservoirs nearest to the organic-source beds. Data are not available to substantiate this conclusion.

As to the relative abundance of the oil and gas which may be found in such reservoirs, no consistent relation has been found or should be expected. Practically all oil accumulations contain at least sufficient gas to saturate the oil at the pressure existent within the reservoir. Moreover, there is usually a considerable surplus of free gas. Other reservoirs contain only gas or very small quantities of oil. From these facts it is safe to conclude immediately that the major portion of the organic-source material has been converted into gas, and the few cases found wherein the oil is undersaturated are easily explained by leakage during, or subsequent to, the accumulation.

It will now be recalled that the mechanism of migration within a major sedimentary basin, as here proposed, calls for a continual loss of fluids vertically through overlying beds. The question naturally arises regarding the ultimate loss of oil and gas from the original reservoir. In some cases involving fault zones, such losses are evident. In other cases observed undersaturation of the oil or unusually small volumes of free gas are, inferentially, adequate proof that gas has been dissipated, although the overlying cover appears to have the normal degree of imperviousness. Assuming, therefore, that as long as abnormal pressures exist the gas accumulations are slowly expelled through the overburden, such leakage must stop when the reservoir pressure is in equilibrium with that in its surroundings. Complete equilibrium from the surface downward requires, therefore, that the fluid pressure at any depth be equal to the hydrostatic head of water at that depth—a condition which is actually observed in the majority of cases. Obviously, variations from this condition will exist if over a considerable area the overlying cover is truly impermeable. Likewise, if rapid subsidence or uplift is in progress and the pressure adjustments are insufficiently rapid to keep pace, abnormally high or low pressures will prevail.

Again we see a cause for the wide diversity in ratio of gas and oil accumulations. For quite apart from possible diversity in transformation ratio from the organic sediments themselves, fortuitous variations in geological and stratigraphic conditions will necessarily affect the accumulations within the reservoirs.

Numerous other evidences of the seepage of gas from apparently "tight" reservoirs are available. A most interesting problem is presented by the fact that escaping gas will carry with it a considerable amount of water vapor. If the volume of gas lost in this manner is sufficiently great, the concentration of solids in solution in the connate waters would necessarily increase and thereby an embarrassing question would be adequately explained. As a matter of fact, evidence of the paucity of such an hypothesis has been found in a study of the salt content of sands in the Bradford field and in the analysis of the connate waters adjacent thereto.[1] While no estimates are available regarding the probable volume of gas lost by seepage and hence the contribution of this process to the loss of water is purely speculative, it is not improbable that the actual volume far exceeds any cursory guess. Trask[2] has found that for recent "source" sediments, the average quantity of gas obtainable during distillation amounts to 500 cu. ft. per ton. And while only a portion of this might be released during natural processes, it is not unreasonable to assume that the actual volume of gas evolved from such sediments since the time of deposition is indeed enormous.

Further discussion of numerous interesting details regarding gas and oil accumulations is not warranted. Such details as have been presented are intended only to provide a reasonable supporting foundation for the general hypothesis that within the connate-water zone, migration of the waters may be induced by forces other than the encroachment of meteoric waters; that, in fact, compaction of shales may form a major driving source and through this process, oil and gas, derived from those sediments rich in organic matter, are ejected from their source beds to take part in the general migration process. In this manner a picture of the mode of accumulation of these fluids may be visualized, which provides an adequate background for the analytical investigations of the performance of these fluids when subjected to withdrawal from wells penetrating the reservoir rocks.

[1] NEWBY, J. B., P. D. TORREY, C. R. FETTKE, L. S. PANYITY, Structure of Typical American Oil Fields, *Amer. Assoc. Petrol. Geol.*, **2**, 432, 1929.
[2] *Loc. cit.*

CHAPTER II

DARCY'S LAW AND THE MEASUREMENT OF THE PERMEABILITY OF POROUS MEDIA

2.1. Darcy's Law.—In order to develop a quantitative representation of the behavior of fluids flowing through porous media it is necessary first to establish the physical principles determining this behavior. As will be explained in more detail in Chap. III, these principles must fundamentally be the same as those governing the motion of viscous fluids in ordinary free vessels and expressed in the Stokes-Navier equations of the classical hydrodynamics [*cf.* Eq. 3.2(1)], which impose on the velocity distribution in every flow system the requirement of the dynamical equilibrium between the inertial and viscous forces and those due to external body forces and the internal distribution of fluid pressures. Unfortunately, however, in spite of the justifiable simplification of neglecting the inertial forces—due to the small velocities generally characteristic of the flow through porous media—the mathematical difficulties of applying these equations to porous media are for practical purposes entirely unsurmountable.

It might, therefore, have been expected that when Darcy, in 1856,[1] was interested in the flow characteristics of sand filters, he had to resort to an experimental study of the problem, and was thereby led to the real foundation of the quantitative theory of the flow of homogeneous fluids through porous media. These classic experiments gave the very simple result—at present generally referred to as *Darcy's law*—that the rate of flow Q of water through the filter bed was directly proportional to the area A of the sand and to the difference Δh between the fluid heads at the inlet and outlet faces of the bed, and inversely proportional to the thickness L of the bed, or, expressed analytically, that

[1] DARCY, H., "Les fontaines publiques de la ville de Dijon," 1856.

$$Q = \frac{cA\,\Delta h}{L},\qquad (1)^1$$

where c is a constant characteristic of the sand.

In view of the fundamental character of this result, it is only natural that the study of its validity should have been the subject of numerous investigations. These have been of essentially two types, namely, (1) those with the object of either verifying Eq. (1) or of establishing the appropriate modification of it, and (2) those concerned with the nature of the constant c as determined by the properties of the sand or porous medium. In the next section we shall critically review the conclusions that may be drawn from the first type of investigation.

2.2. The Range of Validity of Darcy's Law.—In order to understand more clearly the general nature of the "law of flow" with which to represent the data of flow experiments such as those of Darcy, it is well to consider first the implications of the theory of dimensions.[2] Applying the well-known rules of this theory,[3] it is readily found that the pressure drop Δp over a column of sand of lengths Δs, carrying a fluid of density γ and viscosity μ with an average velocity v, should be related to these variables by a relation of the form

$$\Delta p = \text{const.}\ \frac{\mu^2}{\gamma d^2}F\!\left(\frac{dv\gamma}{\mu}\right)\phi\!\left(\frac{\Delta s}{d}\right),\qquad (1)$$

where the unknown functions F and ϕ are to be determined by the experiments, and d is a length characterizing either the size of the pore openings or the size of the sand grains. Now it is clear physically that the function ϕ should be simply the first power of its argument, and indeed this appears to be the only element of this analysis upon which there has been universal agreement

---

[1] While this result might have been intuitively anticipated from the classical hydrodynamics by analogy with Poiseuille's law, the nearest approach even at present to an analytical justification of Eq. (1) by the Stokes-Navier equations applies only to the extremely idealized case of the slow motion (neglect of inertia terms) of a viscous fluid between and parallel to the generators of a network of parallel circular tubes (*cf.* O. Emersleben, *Phys. Zeits.*, **26**, 601, 1925).

[2] MUSKAT, M., and H. G. BOTSET, *Physics*, **1**, 27, 1931.

[3] *Cf.*, for example, P. W. Bridgman, "Dimensional Analysis," Chaps. VI and VII, 1931.

among the numerous investigators of Darcy's law. This observation reduces Eq. (1) to the form

$$\frac{\Delta p}{\Delta s} = \text{const.}\ \frac{\mu^2}{\gamma d^3} F\left(\frac{dv\gamma}{\mu}\right),\qquad(2)$$

where now the left-hand side represents the pressure gradient in the linear system.[1]

The argument $dv\gamma/\mu = R$, of the function F, is a well-known term in the ordinary hydrodynamics and in engineering hydraulics in the discussion of the flow of fluids through sand-free pipes.[2] With d representing the diameter of the pipe, the term $dv\gamma/\mu$ is known as the "Reynolds number," and, as indicated by Eq. (2)— which must also apply to the flow through sand-free vessels as it is nothing more than a dimensional equation—determines the character of the flow. Thus, in particular, for low velocities, fluid densities, or pipe diameters, the function F is found to simply equal its argument, so that

$$\frac{\Delta p}{\Delta s} = \text{const.}\ \frac{\mu v}{d^2}.\qquad(3)$$

As this result is the same as that given by the classical hydrodynamics for viscous fluids, where it is known as Poiseuille's law, the constant coefficient having the value 32, and v representing the average velocity over the section of the tube,[3] Eq. (3) is

[1] Due to the constancy of γv over the length of a linear system in a condition of steady-state flow, it is clear that the pressure gradient $\frac{\Delta p}{\Delta s}$ in the case of liquid flow will be uniform along such a system regardless of the nature of the function F, and hence character of the flow, and similarly for $\gamma\frac{\Delta p}{\Delta s}$ in case the fluid is a gas.

[2] It has also more recently been applied with a generalized interpretation to problems in aerodynamics.

[3] One may note here again the fundamental difference between the viscous flow through a sand-free tube and what will be denoted here as the "viscous" flow—governed by Darcy's law—through a tube filled with a porous medium. For, while in the former the velocity traverse is essentially parabolic across the section (exactly so in the special case of a circular tube), falling from a maximum at the center to zero at the walls, the macroscopic velocity in a linear porous medium is uniform over the cross section. Thus, whereas for Poiseuille flow the total flux is proportional to the square of the cross-sectional area, that in a linear porous medium is proportional only to the first power of the area. The reason for this difference evidently lies in the

usually denoted as representing the pressure gradients in linear "viscous flow."

For larger values of d, v, γ, or $1/\mu$, or more specifically as the Reynolds number $dv\gamma/\mu$ increases so as to exceed a critical value—of the order of 2,000—the nature of the flow in sand-free pipes suddenly changes from a smooth streamline character to one permeated by an irregular and fluctuating distribution of eddies. The motion is then termed as "turbulent," the transition in the direction of either increasing or decreasing velocity (the most convenient parameter for continuous variation) being quite sharp, although the cycle of increasing and decreasing velocities across the transition region usually shows some hysteresis. This type of flow is characterized dynamically by the fact that the function F is now proportional to the square[1] of its argument—in particular the velocity—so that Eq. (2) takes the form

$$\frac{\Delta p}{\Delta s} = \text{const.} \frac{\gamma v^2}{d}, \tag{4}$$

the gradient here being independent of the viscosity of the fluid, while that for viscous flow is directly proportional to the viscosity.

By analogy with these well-known results of hydraulic engineering, it is natural to attempt similar representations of the data for fluid flow through sand columns, although owing to the tortuous and irregular channels of a porous medium it would be expected that the transition between the viscous and turbulent types of flow would not be so sharp as for the flow through sand-free vessels. This would suggest a representation of the gradient

---

tremendous surface exposed in a porous medium and its uniform dissemination throughout the medium, thus giving in a rough sense the equivalent of an enormous number of parallel capillaries, the fluid in each having the same *average* velocity, although within each there are undoubtedly microscopic velocity traverses similar to those in sand-free capillaries.

[1] This is strictly true only for rough pipes. For smooth pipes the power of the Reynolds number is usually closer to 1.75, although it is customary to separate out the factor of Eq. (4) and to denote the remainder correction factor as the "friction factor," which is further studied by plotting against the Reynolds number. For a complete review of the experimental data on turbulent flow in pipes, compare E. Kemler, *A.S.M.E. Trans.*, **55**, 7, 1933, and R. J. S. Pigott, *Mech. Eng.*, **55**, 497, 1933. An exhaustive theoretical discussion of the problem of turbulence will be found in Part III, by H. Bateman, of "Hydrodynamics," National Research Council, 1932.

$\frac{\Delta p}{\Delta s}$ as the sum of terms of several powers of v (the macroscopic velocity measured by the flux per unit area of the medium). And indeed this has been attempted, the "law of flow" being expressed by some investigators as

$$\frac{\Delta p}{\Delta s} = av + bv^n. \qquad (5)[1]$$

Furthermore, even the addition of a third-power term in v has been suggested.[2] Others have proposed a relation of the form

$$\frac{\Delta p}{\Delta s} = av^n, \qquad (6)[3]$$

where n is intermediate between 1 and 2.

Although we shall not enter into any attempt to explain the possible reasons for the discrepancies between the results of the many investigators, we shall simply state what appears to be the correct representation of the results of the most carefully conducted studies of the problem. These are: (1) At low velocities—low Reynolds numbers—the pressure gradient varies strictly linearly with the velocity v, as

$$\frac{\Delta p}{\Delta s} = \text{const. } v, \qquad (7)$$

as given originally by Darcy's law, and in contrast, in particular to Eq. (6). By analogy with the description common in the

---

[1] FORCHHEIMER, PH., *Zeits. V. deutsch. Ing.*, **45**, 1782, 1901. Here n is taken as 2.

[2] FORCHHEIMER, PH., "Hydraulik," 3d ed., p. 54, 1930.

[3] It should be noted that Eqs. (5) and (6) are, in principle, consistent with the dimensional requirements of Eq. (2), provided the constants a and b are adjusted to absorb the factor $\mu^2/\gamma d^3$ and the powers of $d\gamma/\mu$ left over on separating out v from the Reynolds number. In fact, as far as dimensional theory is concerned, F can be represented by any functional form including even such as have infinite power-series expansions. On the other hand, from a physical point of view it must be remembered that a choice of F carries with it an implication as to the variation of $\frac{\Delta p}{\Delta s}$ with the other variables d, γ, and μ. Thus, in particular, in the case of Eq. (6), values of n exceeding 2 would, by Eq. (2), imply that $\frac{\Delta p}{\Delta s}$ would *decrease* with *increasing* viscosity, which clearly is physically unreasonable.

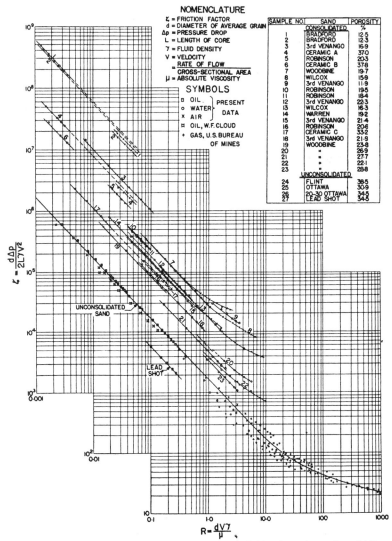

FIG. 8.—Friction-factor chart for the flow of fluids through sands. (*After Fancher, Lewis, and Barnes, Bull. Pa. State Expt. Sta.*)

ordinary hydrodynamics, this type of flow—governed by Eq. (7)—will, therefore, be termed "viscous." (2) As the Reynolds number increases, the gradient $\dfrac{\Delta p}{\Delta s}$ begins to increase faster than v and assumes a variation that is best described by Eq. (5). Representative graphical evidence for these conclusions as derived from flow experiments with liquids is given in Figs. 8 and 9.

Figure 8, published by Fancher, Lewis, and Barnes,[1] gives a representation of the flow data for a number of consolidated and unconsolidated sands in the form of a "friction-factor chart." Here the quantity $\zeta = d\Delta p/2L\gamma v^2$, which is dimensionally equivalent to the "friction factor" (*cf.* footnote on page 58) used extensively in the study of the fluid flow in pipes, is taken for the ordinates. The abscissas are the Reynolds numbers $dv\gamma/\mu$, the d here and in ζ being the "average grain diameter"[2] defined by the equation

$$d = \sqrt[3]{\frac{\Sigma n_s d_s{}^3}{\Sigma n_s}}, \tag{8}$$

where d_s is the arithmetic mean of the openings in any two consecutive sieves of the Tyler or U. S. Standard series, and n_s is the number of grains of the diameter d_s, as found by a sieve analysis. Regardless of the physical significance of this definition of d, the important implication of Fig. 8 is that up to a Reynolds number (as defined above) of the order of 1, the data strictly obey the relation

$$\log \zeta = a - \log R, \tag{9}$$

from which it follows at once that

$$\frac{\Delta p}{\Delta s} = \text{const. } v \tag{10}$$

in accordance with Darcy's law.

[1] FANCHER, G. H., J. A. LEWIS, and K. B. BARNES, *Min. Ind. Exp. Sta., Penn State College Bull.* 12, 1933. The units and notation used by these authors have been changed so as to correspond to those used here.

[2] Physically, of course, the term d should represent the average *pore* rather than grain diameter. However, as the former can be directly measured only by microscopic examination of a cross section of the porous medium itself, all attempts to define or use a value of d to enter into the Reynolds number have referred to the averages of the actual grain diameters.

While this law admittedly loses its strict validity and the flow becomes partially or completely turbulent[1] as the Reynolds number or velocity is increased, there appears to be no unique modification which is assumed by the relation between $\dfrac{\Delta p}{\Delta s}$ and v for large values of the latter. About all that appears to be definitely established is that the empirical data can be expressed in all cases by an equation of the form of Eq. (5), with n having a value in the neighborhood of 2. Perhaps the most striking example of this type of representation, with n exactly 2, is that recently published by Lindquist,[2] whose results on the flow of water through columns of shot of uniform size are summarized in the curve of Fig. 9. Here the ordinates $R\zeta$ are the product of the Reynolds number and the friction factor, and are, there-

[1] It may be noted that Lindquist and Nemenyi (cf. infra, and also Forchheimer, loc. cit., p. 66) attribute the deviation from the condition of viscous flow, as the velocities are increased, to the rise in importance of the inertial forces as compared to the viscous forces, rather than to a real turbulence. This view, however, seems open to question, since there is no net kinetic energy gained in a linear channel, so that the pressure drop is consumed only in supplying the friction losses. Furthermore, as pointed out by Fancher, Lewis, and Barnes (loc. cit.), the turbulent character of the flow can be readily demonstrated by injecting a stream of fluorescein solution into the main current of liquid flowing through a column of shot. At low velocities the dye passes around the shot in streamers with but little diffusion, whereas as a certain velocity is exceeded the streamers begin to bounce from one shot to another in a chaotic manner and become completely diffused. Finally, as Nemenyi has already observed, the fact that the Reynolds numbers at which the deviations appear are so much lower than those at which turbulence develops in sand-free vessels is in itself not surprising when account is taken of the fact that the actual velocities in the pores of a porous medium are higher by a factor of 4–8 than the macroscopic velocities v, and that the flow actually takes place in channels of sharply varying cross section, a condition which, as is well known, is conducive to the early establishment of turbulence. It, therefore, is not unlikely that the deviations from Darcy's law appearing at the low Reynolds numbers of the order of 1–10 (cf. Figs. 8, and 9) are manifestations of a real turbulence disseminated through the flow system.

[2] LINDQUIST, E., Premier Congres des Grands Barrages, Communication Stockholm, 1933. This figure is reproduced by P. Nemenyi in Wasserkraft und Wasserwirtschaft, 29, 157, 1934. Similar results obtained from flow experiments with graded gravel are reported by F. Schaffernak and R. Dachler, Die Wasserwirtschaft, 1, 145, 1934, who also find the transition to occur at Reynolds numbers of 3–6.

fore, inversely proportional to the velocity, the data for $R > 4$ being well represented by the linear relation

$$R\zeta = a + bR, \qquad (11)$$

which is clearly equivalent to the relation

$$\frac{\Delta p}{\Delta s} = a'v + b'v^2. \qquad (12)$$

The Reynolds number at which the break is shown in Fig. 9 is evidently hardly more than an indication of the order of magni-

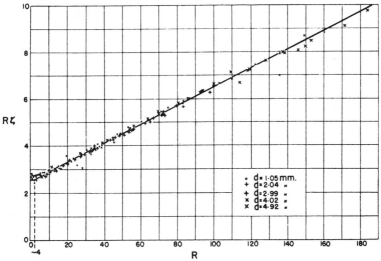

FIG. 9.—Plot of Reynolds number times friction factor $(R\zeta)$ *vs.* the Reynolds number for the flow of water through shot of uniform size. (*After Lindquist, Wasserkraft u. Wasserwirtschaft.*) d = diameter of shot.

tude of the region of transition between the viscous flow characterized by Darcy's law $(R\zeta = \text{const.})$ and that described by Eq. (12). For owing to the irregular and capillary nature of the flow channels in a porous medium, the deviation from the viscous type of flow—following Eq. (10)—will develop gradually and not appear suddenly as in the case of sand-free vessels.[1] In fact, the absence

[1] It is, indeed, not unlikely that the transition between strictly viscous and complete turbulent flow in a porous medium consists essentially in the gradual dissemination of the turbulence throughout all the pores of the medium. Thus the first deviations from Darcy's law would correspond

of such a sharp transition has undoubtedly been the reason why
Eq. (5) has been proposed by some investigators as governing
the flow through porous media even for very small velocities.
Indeed, it is clear that Eq. (5) must inherently approach equivalence to a linear relation as v is decreased (in the case of Lindquist's curve, $b/a \sim \frac{1}{60}$), and unless the data are taken with
considerable accuracy and for sufficiently low velocities, the
existence of a real viscous-flow region governed by Eq. (10) may
be easily missed.

It is important to observe that insofar as one interprets the
Reynolds number as being the essential factor in determining
the nature of the flow through a porous medium and uses the
basic representation of Eq. (2), the "law of flow" should be the
same for both liquids and gases within the same range of Reynolds
number. The variability of the density along the flow column
in the case of gas.flow may be conveniently taken into account by
taking $\gamma \frac{\Delta p}{\Delta s}$ or its equivalent $\frac{\Delta p^2}{\Delta s}$, when the gas flow is isothermal
($\gamma = \gamma_0 p$), for the dependent variable and the *mass* velocity γv
for the independent variable, or the combinations of the friction
factor S as the dependent variable plotted against the Reynolds
number, or finally the pressure gradient $\frac{\Delta p}{\Delta s}$ plotted as a function
of the volume outflow reduced to the mean pressure in the flow
column. Figure 8, in which the second representation is used,
also includes results for the flow of gases, as indicated in the
legend, and shows at once that within the range of Reynolds
numbers [with the d defined by Eq. (8)] for which Eq. (9), and

---

to the beginning of appreciable eddy losses in the larger pores, the local
regions of turbulence spreading to the smaller pores as the velocity is
incı :ased, until finally the turbulence is disseminated throughout the whole
of the medium. The persistence of the linear term $a'v$, as in Eq. (12), under
such conditions of "complete" turbulence, in contrast to the case of sand-
free pipes where the turbulent state is characterized by the single quadratic
term, $b'v^2$, is probably to be attributed to the tremendous wall surface
exposed to the fluid in a porous medium and which gives rise to the viscous
drag on the fluid, as compared to the free pore volume in which the eddy
motion can give rise to friction losses. Thus, whereas in a free capillary
of 1 cm. radius the ratio of the wall surface to the free volume is 2 cm.$^{-1}$,
the ratio of wall surface to pore volume in a 20 per cent porosity sand
composed of uniform spherical grains of 0.1 mm. radius would be 1,200 cm.$^{-1}$.

hence Darcy's law, holds for liquids, this same equation is accurately obeyed by the gas-flow data. In fact for this range—

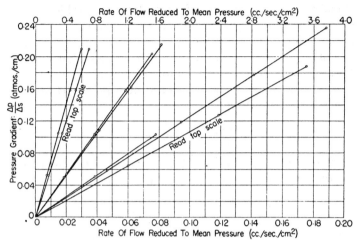

FIG. 10.—Flow curves for air through consolidated sands. (*After Fancher, Lewis, and Barnes, Pa. State Exp. Sta. Bull.*)

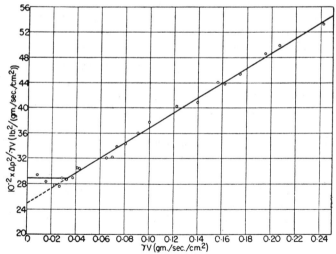

FIG. 11.—Flow data for air flowing through glass beads of average diameter 0.063 cm. Δp^2 = differential of squares of pressures (over a length of 92 cm); γv = mass velocity. Break in curve corresponds to $R \sim 12$.

viscous flow—the density formally disappears from the equation, and a form identical with Eq. (10) is obtained. A still more striking verification of Darcy's law for gas flow is that shown in

Fig. 10, replotted from another figure by the authors of Fig. 8, in which the last of the above representations is used.

Finally, as to the deviations from Darcy's law for higher velocities, the situation is exactly similar to that discussed above for the flow of liquids. For example, a parallel to the results of Lindquist is shown in Fig. 11, in which are replotted some data of Muskat and Botset,[1] obtained for the flow of air through glass beads. The linearity of the curve for $\gamma v > 0.03$ evidently implies that in this range

$$\frac{\Delta p^2}{\Delta s} = a(\gamma v) + b(\gamma v)^2. \tag{13}$$

On the other hand, Chalmers, Taliaferro, and Rawlins[2] find that their data can be expressed best by an equation of the form

$$\frac{\Delta p^2}{\Delta s} = a(\gamma v) + b(\gamma v)^n, \tag{14}$$

where n lies, for their various porous media, between 1.753 and 2.018. As these authors plotted their data on log-log paper, the curves are not very sensitive to the true behavior at very low velocities.[3] It is, therefore, not unlikely that a closer study of their data by plotting on rectangular coordinate paper would have confirmed Darcy's law for the low velocities. In fact, an examination of their numerical expressions of the form of Eq. (14) shows that in the case of none of the 16 samples of porous media studied by these authors will the second term in Eq. (14) exceed 1 per cent of the contributions of the linear term until the velocity is as high as 0.09 cm./sec.

While the above considerations have led to the conclusion that Darcy's law asserting the proportionality between the macroscopic velocity in a porous medium and the pressure gradient does give an accurate representation of the "law of flow" for small velocities, it is rather difficult to give a precise formulation of its range of validity. The difficulty lies essentially in the inherent ambiguity in the definition of the quantity d entering

[1] MUSKAT, M., and H. G. BOTSET, *Physics*, **1**, 27, 1931.

[2] CHALMERS, J., D. B. TALIAFERRO, and E. L. RAWLINS, *Pet. Dev. Tech. A.I.M.E.*, 375, 1932.

[3] Owing to the very low viscosities of gases appreciable deviations from Darcy's law naturally occur for much lower mass velocities than for the case of liquids.

in the Reynolds number, which appears to be the natural variable which should determine the nature of the flow.   Thus the use for d of the diameter of the uniform shot in Lindquist's experiments leads, as indicated in Fig. 9, to a limiting range for the validity of Darcy's law given by $R = 4$, while the curve of Fig. 11 obtained with uniform glass beads leads to a limiting R of approximately 12.[1]   On the other hand, the definition of d of Eq. (8) for heterogeneous and consolidated porous media results in a Reynolds number of the order of 1 at which appreciable deviations from Darcy's law begin to appear (*cf.* Fig. 8).   Although this lower value is undoubtedly due in part to the larger variations in pore size in consolidated media or heterogeneous sands, it is doubtful if any formal definition of d, as by Eq. (8), which leaves out of account the angularity of the grains and degree of cementation, in the case of consolidated sands, can possess .any precise physical significance.   Nevertheless, for the present purposes, where we are interested essentially in establishing the range of applicability of Darcy's law rather than in the detailed specification of the region of deviation, *it will suffice to accept as a safe lower limit where deviations from Darcy's law will become appreciable as given by a Reynolds number of* 1, *with d chosen as any reasonable average diameter of the sand grains.*

There remains now the question as to the extent to which the range of Reynolds number up to the value 1 is inclusive of actual flow systems of practical interest.   While very high rates of flow in exceptional cases of practical flow systems will undoubtedly correspond to Reynolds numbers appreciably exceeding 1, it is unlikely that the macroscopic velocities in actual flow systems carrying liquids will frequently exceed that at the sand face of a well of 6-in. diameter producing 1,000 bbl./day ($\sim$30 gal./min.) from a 10-ft. sand, and which is readily found to be 0.126 cm./sec.[2] Taking for the density and viscosity the values 1 gm./cm.[3] and

---

[1] Mention may also be made of the experiments of R. Ehrenberger (*Zeits. Osterr. Ing. Arch. Ver.*, p. 71, 1928) on the flow of water through heterogeneous sands, with grain diameters as high as 3.0 mm., in which Darcy's law was found to hold up to a fluid velocity of 0.3 cm./sec., corresponding to a Reynolds number of the order of 5.

[2] The general flow of water in artesian aquifers is still much slower, the average velocity ranging from 1 to 3 miles/year, or from 0.005 to 0.015 cm./sec. (*cf.* C. S. Slichter, p. 26, The Motion of Underground Waters, *U. S. Geol. Surv., Water-supply Paper* 140, 1902).

0.01 poises, corresponding to water (those for oil would make the resulting Reynolds number still smaller), and for d the value 0.05 cm., which should be an upper limit to those for consolidated sands,[1] the Reynolds number for the above case would have the value 0.63. Similarly for a gas well producing at 20°C., from the same type of sand and size of well, 100,000 cu. ft./day[2] of a gas of 0.7 the density of air, the mass velocity at the sand face would be $18.9 \cdot 10^{-4}$ gm./cm.[2]/sec. If the gas has the viscosity of methane, the corresponding Reynolds number would be 0.79, again less than the upper limit 1 chosen above, and hence within the range of validity of Darcy's law.

When it is noted that owing to the geometrical convergence of the radial flow about a well (*cf.* Sec. 4.2) the above velocities and Reynolds numbers will be reduced to one-tenth of their values at the sand face as one recedes to only $2\frac{1}{2}$ ft. from the center of the well bore (of radius $\frac{1}{4}$ ft.), it is clear that it is safe to conclude that in the great majority of flow systems of physical interest the flow will be strictly governed by Darcy's law, except possibly in very localized parts of the porous medium of very limited dimensions.

Finally, it may be observed that while we have attempted here to give a strict justification of the use of Darcy's law in the quantitative discussion of flow problems on the basis of fundamental experiments designed to test this law, one may offer in addition the pragmatic argument of the fertility of Darcy's law as compared to the various other modifications that have been proposed either for the complete range of fluid velocities or for the region of larger Reynolds numbers. For, as will be seen in the following chapters, the simplicity of the analytical formulation of Darcy's law and its immediate consequences, as for example, the Laplace equation for the steady-state pressure distributions in the interior of a porous medium, makes it possible to treat quantitatively a very great variety of specific flow problems of practical importance. On the other hand, even the simplest modifications, such as those of Eqs. (5) and (6), immedi-

---

[1] *Cf.* Table 2.

[2] While gas wells with capacities of millions of cubic feet per day are not at all uncommon, the actual daily outputs of most gas wells are usually limited to values of the order of 10^5 ft.³/day. Furthermore, those with abnormally high capacities in general drain sands appreciably thicker than 10 ft.

ately lead to differential equations for the pressure distribution within the porous medium which are effectively intractable for all but the simplest cases, so that it is impossible to generalize and extend the empirical results of the basic experiments with linear channels to flow systems of different geometry. It would, therefore, be necessary to return to the experiments and study in detail empirically each specific flow problem separately for complete ranges of geometrical dimensions and imposed pressure differentials. Although one cannot, of course, condone the use of incorrect basic laws simply on the basis of analytical simplicity, it is the feeling of the author that the pragmatic argument should serve to justify the treatment of flow problems by the methods given in the following chapters even for cases where the Reynolds number may somewhat exceed the limiting value 1, and in which the strict validity of Darcy's law may be open to question. A discussion of a problem which is not quantitatively exact because the fundamental assumptions are only approximately correct should certainly be preferable to the strict adherence to basic laws which are so complex when expressed accurately as to forbid the deduction of further implications or generalizations.

2.3. The Constant in Darcy's Law. The Permeability of a Porous Medium.—Although the general analytical formulation of Darcy's law will be deferred to Chap. III, we shall examine here the nature of the constant of proportionality between the pressure gradients and velocity in a linear flow channel, and review the second type of investigation mentioned at the close of Sec. 2.1. Denoting the pressure gradient in the linear channel by $\frac{dp}{dx}$, and the macroscopic velocity by v, this constant may be most conveniently defined by the relation

$$v = \text{const.}\frac{dp}{dx}. \tag{1}$$

Returning to the dimensional considerations of the last section, it is seen by inverting Equation 2.2(4) that the constant in Eq. (1) can be resolved as indicated in the equation

$$v = \text{const.}\frac{d^2}{\mu}\frac{dp}{dx}, \tag{2}$$

where d represents the "effective" diameter of the sand grains or pore openings,[1] μ is the viscosity of the fluid, and the remaining constant must be a dimensionless quantity. Thus the physical dimensional elements of the composite system of porous medium and fluid appear only in the square of d for the former and the viscosity for the latter, the fluid density being absent, so that the coefficient of $\dfrac{dp}{dx}$ will be different for different liquids or between gases and liquids only because of their differences in viscosity.[2] The remaining dimensionless constant must, therefore, contain the dimensionless characteristics of the flow system, which clearly can involve only the geometrical properties of the porous medium such as its porosity, shape of the grains of the medium, and the degree of cementation, the size distribution of the grains being supposed for the moment to be included in the definition of d.

In principle, therefore, a knowledge of the detailed nature of the porous medium and the viscosity of the fluid should permit an a priori prediction of the numerical value of the constant in Eq. (1). And indeed the determination of a relation by means of which this prediction can be effected has been the subject of numerous investigations. Unfortunately, however, the only feature in common to all the results is that the constant must vary as the square of some average of the diameters of the grains of the porous medium, as is explicitly required by Eq. (2).[3] These investigations have ranged from the classic analysis of Slichter,[4] in which

[1] *Cf.* footnote on p. 61.

[2] *Cf.* Sec. 3.3 for a discussion of the reason for this absence of the density.

[3] In fact, the viscosity μ has not been expressly represented in most of the analytical expressions for the constant of Eq. (1), although this omission has been probably due to the use by most of the investigators of a single fluid in their experiments, so that μ was simply absorbed in their numerical constants. The first explicit separation of the term μ as giving the sole effect of the fluid (gas or liquid) on the constant of Eq. (1), and the attribution of the whole remainder of the constant as giving a complete dynamical characterization of the porous medium, appears to be that of Wyckoff, Botset, Muskat, and Reed, *Rev. Sci. Instr.*, **4**, 394, 1933; *Bull. Amer. Assoc. Petrol. Geol.* **18**, 161, 1934.

[4] SLICHTER, C. S., U. S. Geol. Surv., 19th Ann. Rept., 1897–1898, pt. II p. 295. A critique of this work, as well as an extended qualitative discussion of the nature of the pores in assemblages of spheres and their effect on the permeability, will be found in the papers by L. H. Graton and H. J. Fraser, *J. Geol.*, **43**, 785, 1935, and H. J. Fraser, *J. Geol.*, **43**, 910, 1935.

the flow in the individual pores of an assembly of uniform spheres
was computed by means of Poiseuille's law, modified so as to take
into account the detailed shape and length of the pores, to the
extensive experiments of Fancher, Lewis, and Barnes,[1] in which
an attempt was made to classify various sands by the position
of the line giving the variation of the friction factor with the
Reynolds number among similar lines on a single-friction-factor
chart (cf. Fig. 8).

In view, however, of the complete failure to attain in these
investigations any result that even has a moderate range of
validity and generality, it seems that the most reasonable point of
view of the problem is the following: One should lump together
the constant and the term d^2 in Eq. (2) as a single constant characterizing the porous medium completely and uniquely with
respect to the flow of fluids through it.   This resultant constant,
which completely defines dynamically the porous medium as the
carrier of a homogeneous fluid in viscous motion, will be denoted
by the symbol k, and termed the "permeability"—frequently
denoted with or without the factor $1/\mu$ as the "coefficient of
permeability", the "transmission constant"—or in German as the
"Durchlässigkeit" or "Hydrogeologische Konstante"— so that
Eq. (2) finally assumes the form

$$v = -\frac{k}{\mu}\frac{dp}{dx}.   \qquad (3)[2]$$

The formal definition of the permeability of a porous medium may,
therefore, be stated as *the volume of a fluid of unit viscosity passing
through a unit cross section of the medium in unit time under the
action of a unit pressure gradient*.   It is thus a constant *determined
only by the structure of the medium in question and is entirely*

---

[1] *Loc. cit.* In this paper will also be found a great number of references to
other such studies as well as a general review of the literature on permeability
from the historical point of view.   Still other references will be found in the
books of P. Nemenyi, "Wasserbauliche Strömungslehre," pt. V, 1933, and
Ph. Forchheimer, "Hydraulik," Chap. III, 3d ed., 1930, and a recent bibli-
ography prepared by C. V. Fishel, *U. S. Geol. Surv.*, February, 1935.

[2] The algebraic sign of v is arbitrary.   In the analytical developments of
the following chapters it will be chosen so that a minus sign will precede
the right side of Eq. (3), with the convention that v should be positive if the
fluid is flowing toward increasing values of the coordinates, and conversely
for negative values of v (*cf.* footnote on p. 128).

independent of the nature of the fluid. Its dimensions, as readily seen by comparing Eqs. (2) and (3), are those of an area, or

$$k = [L^2]. \tag{4}$$

It should be noted that the dimensions $[L^2]$ for k are the direct consequence of the separation of μ from k and the expression of the driving agent in terms of a pressure differential. Although this has been necessary in order to make the analysis sufficiently general to include all types of homogeneous fluids and all modes of driving agent, as gravity heads, pressure drives, or a combination of both, it may be expedient from a practical point of view to specialize the definition of k when dealing exclusively with a restricted class of fluids and driving agents. Thus in the branch of civil engineering dealing with the seepage of water through dams, or of agricultural engineering concerned with the seepage of water out of canals or ditches, or in any phase of homogeneous fluid flow through porous media in which gravity is the predominant driving agency, the pressure gradient in Darcy's law may be justifiably re-expressed in terms of the hydraulic gradient. Moreover, since water is exclusively the fluid of interest, its viscosity, except for temperature variations, may be also absorbed in the permeability. Darcy's law may then be rewritten as

$$v = \frac{k}{\mu}\frac{dp}{dx} = \frac{k\gamma g}{\mu}\frac{d\bar{h}}{dx} \equiv \bar{k}\frac{d\bar{h}}{dx}, \tag{5}$$

where $\bar{h}$ is the fluid head and $\dfrac{d\bar{h}}{dx}$ is the hydraulic gradient. The coefficient $\bar{k}$ thus corresponds to the constant c in Darcy's original formula Eq. 2.1(1) and may be termed the "effective" permeability for the composite system of porous medium and liquid. Indeed it will be found that $\bar{k} = k\gamma g/\mu$ separates itself naturally in all the strictly gravity-flow problems, such as those treated in Chap. VI, where it is explicitly denoted by $\bar{k}$. Since the hydraulic gradient is dimensionless, the dimension of $\bar{k}$ will clearly be that of v, *i.e.*, centimeter per second, and in magnitude a unit value of $\bar{k}$ will be equivalent to 1,040 darcys, as indicated in Table 5.

Returning now to the question of the direct computation of the permeability from the structural properties of the porous medium, such as its porosity, grain-size distribution, shapes of the grains, and degree of cementation, it appears that from a practical point of view any particular formula will at best be even approximately valid only within a very limited range of types of sand. Thus, for example, it is a priori clear that not only will the porosity not be a determining factor in establishing the value of k, but that moreover it will in general enter in a very complex manner, when it is observed that a porous medium may have a high percentage of pore space and yet be quite impermeable because of the lack of interconnection between the pores, while, on the other hand, the permeability of a sand of spherical grains of uniform size in a given mode of packing will vary with the square of the grain size even though the porosity be kept constant.[1]

That the structure of the medium cannot be simply taken into account by an averaging of the grain-size distribution, such as is implied by Eq. 2.2 (9), for example, so as to give an "effective" d, is clearly shown by the fact that the samples 2 and 16 of Fancher, Lewis, and Barnes[2] (*cf.* Table 2), whose values of d differ by only 2 per cent, have values of k differing by as much as 45 fold. Other similar instances can be readily found among the rest of their data. Finally, the degree of cementation, which would naturally be a difficult factor to take into account, also materially affects the permeability of a consolidated sand. This is clearly shown in Table 4, taken from Howe and Hudson[3] who

TABLE 4

Bond content, per cent	Porosity	Permeability, darcys
5.6	46.2	375
10.5	40.8	317
15.0	35.2	243
19.0	32.7	200
22.7	27.4	134

studied the effect of artificially varying the clay-bond content of some porous plates composed of aluminous abrasive grain. The

[1] A striking example of the failure of theoretical and even semiempirical formulas to give even approximately correct values of k is shown by the data of Kozeny (*Wasserkraft und Wasserwirtschaft*, **22**, 86, 1927) giving a comparison of measured and computed values of k. Thus the formula of Hazen shows discrepancies ranging from -36 to $+180$ per cent, that of Kruger from -50 to $+84$ per cent, and that of Kozeny from -68.7 to $+85.7$ per cent.

[2] *Loc. cit.*

[3] HOWE, W. L., and C. J. HUDSON, *J. Amer. Ceram. Soc.*, **10**, 443, 1927.

permeabilities of these authors, expressed as (cu. ft. air/min.)/sq. ft. in. for a pressure· differential of 2 in. of water, have been recomputed into terms of darcys (*cf.* the next section). The average grain size in each of the plates was 0.787 mm.

Although it is not inconceivable that there may ultimately be found a relation between the permeability of a porous medium and the porosity, grain sizes, their distribution, and a shape factor for the grains, it is at present certainly far simpler and more accurate to make a direct permeability measurement by means of a flow experiment than to carry through a laborious sieve analysis and a porosity determination and then make a permeability calculation by means of a relation that has at best only a limited range of approximate validity. In fact, as will be seen below, this direct measurement of k can be made so simply, once the technique has been mastered, that from purely practical grounds it seems difficult to justify the lengthy and indirect procedure involving a sieve and grain analysis.

2.4. Permeability Units and Nomenclature.—As pointed out in the last section, the permeability of a porous medium has the same dimensions as an area, namely, L^2. Its formal definition, however, implies an expression of k as $\dfrac{\text{vol./area}}{\text{time (pressure gradient)}}$.[1]

While the choice of the units with which to specify the elements of this expression is in principle perfectly arbitrary, experience has shown that perhaps the most convenient set is:

> Time:second; length:cm.; volume:cc.
> Pressure:atm. (76.0 cm. Hg); viscosity:centipoise.

Although a consistent absolute c.g.s. system would require that the unit of pressure be the dyne per square centimeter, this would lead to inconveniently small numerical values for the permeability of naturally occurring porous media. This may be seen from Table 5 which gives the conversion factors between the various other permeability units which have been suggested, including some that have been frequently used in the hydraulic engineering and hydrology literature. Most consolidated sands

---

[1] The formally equivalent expression of k as velocity/(pressure gradient) is not so satisfactory, as the term velocity might be confused with the velocity of advance of the fluid in the medium, which would be given by $\dfrac{\text{vol./area}}{\text{time} \cdot \text{porosity}}$.

TABLE 5.—CONVERSION TABLE FOR PERMEABILITY UNITS

	Darcy 1 cc. Sec. cm.² (atm./cm.)	1 cc. Sec. cm.² (dyne/cm.²)/cm.	1 cu. ft. Sec. ft.² (atm./ft.)	1 cu. ft. Sec. ft.² (lb./in.²)/ft.	1 cc. H₂O (20°C.) Sec. cm.² (1 cm. H₂O/cm.)	1 gal. H₂O (20°C.) Min. ft.² (1 ft. H₂O/ft.)
$\dfrac{\text{Darcy } 1 \text{ cc.}}{\text{Sec. cm.}^2\,(\text{atm./cm.})} =$	1	$9.8697 \cdot 10^{-7}$	$1.0764 \cdot 10^{-3}$	$7.3243 \cdot 10^{-6}$	$9.6130 \cdot 10^{-4}$	$1.4156 \cdot 10^{-2}$
$\dfrac{1 \text{ cc.}}{\text{Sec. cm.}^2\,(\text{dyne/cm.}^2)/\text{cm.}} =$	$1.0132 \cdot 10^{6}$	1	$1.0906 \cdot 10^{3}$	74.210	$9.7399 \cdot 10^{2}$	$1.4342 \cdot 10^{4}$
$\dfrac{1 \text{ ft.}^3}{\text{Sec. ft.}^2\,(\text{atm./ft.})} =$	$9.2904 \cdot 10^{2}$	$9.1693 \cdot 10^{-4}$	1	$6.8046 \cdot 10^{-2}$	0.89302	13.151
$\dfrac{1 \text{ ft.}^3}{\text{Sec. ft.}^2\,(\text{lb./in.}^2)/\text{ft.}} =$	$1.3653 \cdot 10^{4}$	$1.3475 \cdot 10^{-2}$	14.696	1	13.124	$1.9327 \cdot 10^{2}$
$\dfrac{1 \text{ cc. H}_2\text{O (20°C.)}}{\text{Sec. cm.}^2\,(1 \text{ cm. H}_2\text{O/cm.})} =$	$1.0403 \cdot 10^{3}$	$1.0267 \cdot 10^{-3}$	1.1198	$7.6195 \cdot 10^{-2}$	1	14.725
$\dfrac{1 \text{ gal. H}_2\text{O (20°C.)}}{\text{Min. ft.}^2\,(1 \text{ ft. H}_2\text{O/ft.})} =$	70.644	$6.9724 \cdot 10^{-6}$	$7.6041 \cdot 10^{-2}$	$5.1742 \cdot 10^{-3}$	$6.7910 \cdot 10^{-2}$	1

First four units refer to a fluid of 1 centipoise viscosity; viscosity of water at 20°C. is taken as 1.005 centipoise.

occurring in underground reservoirs have permeabilities of 5.0 — 0.0005 in the above units.

In view of the fundamental role played by Darcy's law in the quantitative discussion of all problems of the flow of homogeneous fluids through porous media, it seems to be very appropriate to name the unit of permeability the "darcy" after this pioneer investigator.[1] With the above units we have, therefore, that

$$k = 1 \text{ darcy} \sim 1(\text{cc./sec.})/\text{cm.}^2/(\text{atm./cm}).$$

All numerical values of the permeability will hereafter be expressed in darcys, or in self-explanatory subdivisions of this unit.

2.5. The Principle of the Method of Measurement and of the Calculation of the Permeability of Porous Media.—As will be clear from Sec. 2.3, the principle of a laboratory measurement of the permeability of a porous medium consists simply in the measurement of the volume flux per unit area of a fluid of known viscosity through a linear sample[2] of the medium, together with the pressure gradient inducing that flux, and then calculating k by the equation

$$k = \frac{\mu v}{dp/dx}. \tag{1}$$

---

[1] This name was originally suggested in a paper by Wyckoff, Botset, Muskat, and Reed (*Rev. Sci. Instr.*, **4**, 394, 1933). It has since been adopted, together with the above set of units, by the American Petroleum Institute. However, the above units, except that the pressure was expressed in 10^6 dynes, or 0.99 atm., seem to have been first suggested by P. G. Nutting, *Amer. Assoc. Petrol. Geol.*, **14**, 1337, 1930.

[2] F. Tolke [*Ing. Archiv.*, **2**, 428 (1931)] has given the complete theory for the measurement of the permeability of consolidated porous media (concrete) with radial disc samples, in which the liquid is injected over part of one face, and leaves through the other face and cylindrical surface, and also, in one case, through the remainder of the injection face. Such measurements are proposed with the argument that it is difficult to properly seal off the peripheral surface of a linear sample. The technique for linear flow experiments proposed below has, however, been found to be not only accurate but simple and rapid as well. Furthermore, the method of Tolke is open to the objection that the capillary back pressure at the periphery and faces of the radial sample will disturb the flow unless it is immersed in a liquid. Finally, it may be noted that the resulting formulas of Tolke for the permeability constant involve complex functions of the dimensions of the sample which, while subject to calculation, do not afford as simple and as easily interpretable formulas as those for linear-flow experiments.

Now in an actual experiment, the direct measurements are those of the total flux Q through the sample, of length L and area A, and the terminal pressures P_1 and P_2. These, however, may be simply translated into terms equivalent to those in Eq. (1) as follows: Thus, first it is obvious that v is related to Q as

$$v = \frac{Q}{A}. \qquad (2)$$

Furthermore, if the fluid is a liquid, the velocity v must evidently be uniform along the length of the sample (neglecting the compressibility effects which are entirely unimportant in such experiments). Hence by Eq. (1), the pressure gradient $\frac{dp}{dx}$ must also have the same value along the length of the flow column and must equal its average value, $i.e.$,

$$\frac{dp}{dx} = \text{const.} = \frac{P_1 - P_2}{L}, \qquad (3)[1]$$

provided that k is uniform over the length of the sample. Inserting these values of v and $\frac{dp}{dx}$ into Eq. (1), it follows that k may be calculated by means of the formula]

$$k = \frac{\mu Q L}{A(P_1 - P_2)}. \qquad (4)$$

In the case of gases, however, the velocity v will no longer be constant along the length of the flow column, but will rather increase as the outflow end is reached, owing to the expansion of the gas. $\frac{dp}{dx}$ will, therefore, also increase as the outflow end is approached, and indeed it must increase in just such a way that the ratio $\frac{v}{dp/dx}$ remains constant. The product γv, on the other hand, will be uniform along the flow column. Assuming the gas

[1] An immediate consequence of this result is that the pressure varies linearly along the flow column—a conclusion valid for any type of liquid flow in a uniform linear porous medium (uniformity of v), whether it be viscous or turbulent. Similarly p^2 will vary linearly in any isothermal linear system carrying a gas which may be considered as "ideal." The linear variations of p or p^2 are, therefore, no criteria for the viscous character of the flow in a porous medium.

to expand isothermally as an ideal gas, as it will except for very high velocities of flow, it follows that the product $p\frac{dp}{dx}$ is also uniform. Hence,

$$2p\frac{dp}{dx} = \frac{dp^2}{dx} = \text{const.} = \frac{P_1{}^2 - P_2{}^2}{L} = 2\bar{P}\frac{(P_1 - P_2)}{L}, \quad (5)$$

where $\bar{P}$ is the algebraic mean pressure, $(P_1 + P_2)/2$. Applying this to Eq. (1), k takes the form

$$k = \frac{2\mu v p L}{P_1{}^2 - P_2{}^2} = \frac{\mu v p L}{(P_1 - P_2)\bar{P}}, \quad (6)$$

where v refers to the velocity at the pressure p. Observing now that vp is proportional to the mass velocity in the system, and that $vp/\bar{P}$ is the velocity at the *mean* pressure $\bar{P}$, and denoting the total *mass* outflow rate by Q_m and the *volume* outflow rate as measured at the mean pressure by $\bar{Q}$, it follows that k can be finally expressed either as

$$k = \frac{2\mu Q_m L}{\gamma_0 A (P_1{}^2 - P_2{}^2)} \quad (7)[1]$$

or

$$k = \frac{\mu \bar{Q} L}{A(P_1 - P_2)}, \quad (8)$$

where γ_0 is the density of the gas at atmospheric pressure, and is given for an ideal gas by w/RT, w being its molecular weight, R the gas constant per mole, and T the absolute temperature. Thus the permeability can be computed from gas flow experiments by exactly the same formula—Eq. (8)—as from liquid-flow experiments, provided only that the volume flux in the former is reduced to the algebraic mean pressure in the flow channel.

While the above formulas make it possible to calculate k from a single measurement of the rate of volume outflow and the pressure differential, it must be remembered that they are based

---

[1] It may be noted from a comparison of Eqs. (7) and (8) that if a "gas permeability" k_g be computed simply by the value of $\frac{\mu Q_m L}{A(P_1{}^2 - P_2{}^2)}$, it will be related to the true permeability by the relation $k = 2k_g/\gamma_0$.

upon the assumption that the flow experiment is being carried out in the viscous region. It is, therefore, advisable to verify this in each determination by measuring the rates of flow for several pressure differentials and plotting the one against the other or computing k by the above equations from each measurement. A constancy in the results of the latter computation, or a straight line on Cartesian paper passing through the origin[1] resulting from a plot of the flow rates Q against $\Delta P(P_1 - P_2)$ in the case of liquids, and $\bar{Q}$ against ΔP or Q_m against $\Delta P^2(P_1^2 - P_2^2)$ in the case of gases, is a necessary and sufficient condition for the viscous character of the flow. The slope of these lines will give the quantity $kA/\mu L$ for the first two types of plotting, and $\gamma_0 kA/2\mu L$ in the last. If the flow under the conditions of the experiments is even partially turbulent the curves will bend toward the ΔP or ΔP^2 axes.

Finally, it may be explicitly observed that in order to obtain the numerical value of k by the above formulas—Eqs. (4), (7), and (8)—directly in terms of darcys, it is necessary to express μ in centipoises, Q, $\bar{Q}$ in cubic centimeters per second, Q_m in grams per second, L in centimeters, A in centimeters squared, P_1, P_2 in atmospheres, and in Eq. (7) γ_0 in grams per cubic centimeter or, if its ideal gas equivalent is used, w in grams, T in degrees Kelvin, and R must be given the value 82.07.

2.6. The Measurement of the Permeability of Unconsolidated Sands.—As has been observed by everyone who has attempted to measure the permeability of an unconsolidated porous medium or sand, the value obtained will in general vary over a wide range as the packing of the sand in the sample holder is varied. For, even though the effective grain diameter of a given sand, regardless of its exact definition, remains unchanged by variations in the packing of the sand, the effects of the packing on the porosity[2] may be reflected in greatly magnified changes in the permeability of the sample. This effect of the porosity on the permeability of an unconsolidated sand may be demonstrated by reference

---

[1] This is the condition that the fluid is not plastic and it does not refer to the viscous character of the flow.

[2] For an analysis of the relation of packing to porosity in the case of homogeneous spheres, cf. W. O. Smith, P. D. Foote, and P. F. Busang, *Phys. Rev.*, **34**, 1271, 1929, and the very exhaustive discussion by L. C. Graton and H. J. Fraser, *J. Geol.*, **43**, 785, 1935; *cf.* also sec. 1.6.

to the theoretical formulas for the permeability given by Slichter[1] and Kozeny[2] which, while undoubtedly entirely unreliable for absolute values of the permeability, should nevertheless give the order of magnitude of the effect of varying the porosity in a given sand. Thus Slichter's formula, derived for an assembly of uniform spheres for various types of packing, gives a ratio of the permeability at the highest porosity possible for spheres of uniform size, 47.64 per cent (cubic packing), to that at the lowest possible porosity, 25.95 per cent (close hexagonal packing), equal to 7.5. And Kozeny's formula, which explicitly gives a variation of k with the porosity f defined by the function $\dfrac{f^3}{(1-f)^2}$, will give a variation in the above range—26 to 47 per cent—of 11.5-fold.

A measurement of the permeability of an unconsolidated sand has, therefore, but little significance *per se*, unless one can at the same time specify a further characterization of the sand whereby the permeability value becomes unique and reproducible. While it appears probable that the porosity of a given heterogeneous sand will be the primary property determining the final value of the permeability[3] of the given assembly of sand grains, it is nevertheless not unlikely that the porosity will still not give a *completely* sufficient specification,[4] or that at least small changes in the porosity will result in quite appreciable variations in the permeability.

For cases where the porosity does turn out to be sufficiently definitive with respect to the permeability, the latter may be measured by the following procedure: The sand, well mixed, is poured into a glass tube—of dimensions appropriate to the volume of sand available—fitted with screens to hold it in place,

---

[1] Slichter, C. S., U. S. Geol. Surv., 19th Ann. Rept., 1897–1898, pt. II. p. 295.

[2] Kozeny, J., *Wasserkraft und Wasserwirtschaft*, **22**, 67, 86, 1927.

[3] This interrelation between the porosity, as affected by the packing, and the permeability of a given assembly of sand grains is not to be confused with that which has frequently been claimed as implying a unique correlation between porosity and permeability among *different* sands (*cf.* Sec. 2.3).

[4] In the case of assemblies of uniform spheres, media with different permeabilities but the same porosities can be obtained by "twinning" or "tripling" some of the fundamental systems of cubic or hexagonal packing *cf.* Graton and Fraser, *loc. cit.*).

and then tamped to the desired porosity.[1] Manometers attached
to the inflow and outflow ends of the sand column will serve to
give the pressure differential acting on the fluid. In all cases
it is best to set the flow tube vertically in order to avoid a settling
of the sand and the formation of a by-pass channel at the upper
part of the tube if placed horizontally. Further details of the
procedure and precautions to be taken to insure accuracy should
be essentially the same as those to be described in Secs. 2.7 to
2.9 for measurements with consolidated media. A diagrammatic
representation of a convenient arrangement of the apparatus

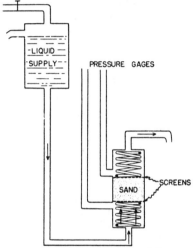

Fig. 12.—Diagrammatic arrangement of permeability apparatus for unconsolidated sands.

for determinations with liquids is that shown in Fig. 12. The
modifications required for gas-flow experiments will be obvious.

In general, however, it is necessary to know the value of the
permeability of an unconsolidated porous medium as it occurs
in its original undisturbed state. This is especially true in
hydrological problems involving the flow of water in surface
soils. In such cases, it is clear that even if the porosity were a
unique index by which the "state" of the soil could be identified,

[1] This can be most readily computed from the length of the sand column
occupied by a given weight of the sand of previously determined average
grain density, or from the volume of water required to saturate the sand as
it rises vertically into it.

one would still be faced with the problem of determining at least this porosity without disturbing the soil in order to verify for the permeability measurement that the soil in the experiment is in the same "state" as it was in the ground. It is, therefore, necessary either to measure the permeability of such consolidated media in place, or to remove a sample of the soil by a method that is certain to leave it undisturbed. The former alternative, as applied to permeability determinations by means of flow experiments with artesian wells or wells drilled for the production of oil or gas, will be discussed in Sec. 2.11.[1]

Typical of the techniques that have been developed for obtaining soil samples without disturbing the soil—the latter of the above alternatives—is that described by Terzaghi.[2] The essential features of the method are as follows: One first digs away an annulus about the particular soil sample to be tested, leaving it as a column of somewhat larger size than that to be finally used. A metal cylinder composed of two hinged halves is then pressed down upon the central column until the top of

---

[1] An extended discussion of the measurement of the permeability of soils by observations on the seepage of water into the soil has been given by J. Kozeny, *Die Wasserwirtschaft*, pp. 555, 589, 1931. Aside from the approximate character of the analysis leading to Kozeny's final formulas for the permeability calculation, such methods appear to be inherently unsatisfactory for getting results of any accuracy. For not only do they explicitly involve the effect of capillary- forces, but the nature of the flow is such that no provision can be made to avoid the trapping of air bubbles in the pores of the soil. Although the method of taking soil samples and measuring their permeability by the same technique as that suitable for consolidated sands may not be entirely free of objections, it does seem to be definitely preferable to those depending on measurements of the direct seepage into the surface soil.

[2] Terzaghi, K., *Die Wasserwirtschaft*, Nos. 18/19, 1930. A somewhat similar sampling procedure has been outlined by H. J. Fraser, *Amer. J. Sci.*, **22**, 9, 1931, for determining the porosities of soils. However, his method of impregnating the dry soils with paraffin would make it impossible to obtain satisfactory permeabilities from such samples. Still other samplers are described by E. B. Powell, *Soil Sci.*, **21**, 53, 1921, and F. J. Veihmeyer, *Soil Sci.*, **27**, 147, 1929, that the latter being designed for securing samples appreciably below the top surface soil. In neither case, however, has an adaptation been made of these sampling devices so as to be used with an apparatus for the determination of the permeability of the soil sample; *cf.* also section *B*, vol. I, of *Proc. Intern. Conf. Soil Mechanics and Foundation Eng.*, 1936.

the soil appears at the top of the cylinder. The soil about the cylinder is next removed and the sample cut away from the underlying soil, a plate being inserted at the same time to give a base to the sample. The sample and cylinder are then set on a table, the cylinder opened and removed, and a larger metal sheath placed over the sample. Molten paraffin is then poured into the space between the sample and sheath and also over the ends of the latter. In this condition the sample is trans-

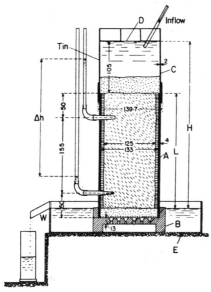

FIG. 13.—Apparatus for measuring the permeability of soils. (*After Terzaghi, Die Wasserwirtschaft.*)

ported to the laboratory. Here the sheath and paraffin coatings are carefully removed. The sample is then once more coated with a thin layer of hot paraffin, and the cylindrical holder of the permeability apparatus is set over it, the free space again being filled with molten paraffin. The cylinder is next inverted, about 2 cm. of the soil is removed and replaced with a sand filter, and the perforated foot piece *B* is screwed on (*cf.* Fig. 13). *A* is then righted, the top face of the soil is cleaned, the container *C* is screwed on, and another sand filter added. The whole apparatus is now set into the basin *E* fitted with an overflow *W* held above the bottom of the sample.

The pressure readings are taken with two manometer tubes inserted into the sample through the walls of the container A. The flow rates are measured, with downward percolation,[1] at the overflow W.

For the case of samples of low permeability, Terzaghi has used a method in which the rate of fall of the upper water level[2] extending into a tube of small section is observed, rather than the steady flux rate with a constant head. In such experiments, however, a decrease in the rate of fall due to decreased head may be masked by a decrease in the permeability of the sample due to the percolating water. That this latter effect does not enter can only be tested by plotting the log of the ratio of the total head h_t to its initial value h_0, as a function of the time. A linear relation will verify that the flow is not disturbed by changes in the medium. The equation between the fluid head and the time t, until the water level leaves the upper tube, is readily shown to be given by

$$\log_e \frac{h_t}{h_0} = -\frac{k\gamma g A_s t}{\mu L A_t}, \qquad (1)^3$$

where A_s is the area of the sample, L is its length, and A_t the area of the tube in which h_t is observed. With the aid of these constants, as well as the density γ and viscosity μ of the liquid, the permeability can be readily computed from the slope of the line given by Eq. (1).

If the samples are originally free of moisture, the use of air, with the sample prepared and taken as outlined above, should be preferable to that of water, and will be particularly suited to such cases where the permeability is quite low. The detailed technique for the use of air will be similar to that described below (Sec. 2.9) in the discussion of permeability measurements with consolidated porous media. However, if the water does disturb the sample by loosening or causing to swell some of its constituents, it may be advisable also to use water in the measure-

---

[1] As in the case of consolidated-sand measurements, it should probably be safer to flow the water upward—with obvious changes in the apparatus—in order to avoid trapping air as the water first penetrates the sample.

[2] Here again an upward percolation of the water should in general be preferable.

[3] The method of deriving this equation is similar to that outlined in the footnote of p. 642, for finding the permeability of a producing sand from the rate of fluid rise in the well bore.

ments, provided sufficient time is given for these disturbances
to assume a constant magnitude before the observations are
made.

In general, it should be noted that the only essential difference
between the problems of determining the permeability of a
consolidated sample—the detailed technique for which will be
described in the following sections—and that of an unconsolidated
porous medium occurring naturally, is that of obtaining a sample
of the latter without disturbing its structure. In this connection
it may be observed that while the method of sampling proposed
by Terzaghi should insure a freedom of the sample from disturbance, a more satisfactory and simpler procedure would
seem to be one in which the sampler itself is made an integral
part of the permeability apparatus. This would avoid changing
the holder so frequently and would be particularly advantageous
if the sample has such poor cohesive properties that it would fall
apart as the original sampling cylinder is opened. However,
once the sample is obtained and introduced into the permeability
apparatus, the further technique and precautions should be similar to that described below for consolidated media determinations.

**2.7. The General Technique for Permeability Determinations
of Consolidated Porous Media.**[1]—Whether the final measurements of the permeability of a sample of a consolidated porous
medium are to be made with liquids or gases there are certain
features of the procedure, such as those concerning the preparation and mounting of the sample, that must be followed in all
cases. Only by strictly adhering to certain definite precautions
can reproducible and significant values of the permeability be
obtained.

In order to minimize erratic effects due to very local inhomogeneities as concretions and small shale streaks, it is advisable to
use samples of appreciable size—of the order of 3 cm. in diameter

---

[1] The techniques described in this and the succeeding sections follow
closely those developed by and used at the Gulf Research & Development
Co. and reported in papers by R. D. Wyckoff, H. G. Botset, M. Muskat,
and D. W. Reed in the *Rev. Sci. Instr.*, **4**, 394, 1933, and the *Bull. Amer.
Assoc. Petrol. Geol.*, **18**, 161, 1934. Essentially the same procedures are
given by Fancher, Lewis, and Barnes (*loc. cit.*) and W. L. Horner (*Petrol.
Eng.*, **5**, May, 1931). While the details of the technique given here have
been developed from experience with cores from oil sands, they should be
applicable to all consolidated porous media.

and 1 cm. in length.   When possible, samples should be cut with axes both perpendicular and parallel to the bedding planes of the sand in order to get values along both these directions.   This cutting may be effected by means of a slotted brass tube mounted in a slow-speed drill press fed with coarse carborundum and a small amount of water.   The core is then mounted in a special machine such as that of Fig. 14, which is fitted with a Norton

FIG. 14.—Core-cutting machine.   (*From Rev. Sci. Instr.*, **4**, 397, 1933.)

"chrystolon" bakelite-bonded carborundum cutting disk of 7-in. diameter and 0.0625 in. in thickness.[1]   A convenient speed is 1,700 r.p.m., a small amount of water being fed continuously to clean the cut.   It is necessary to soak the cores in water, or in some other liquid which does not affect the cementing material, before cutting, in order to avoid the fine cuttings from being carried into the face of the core by capillary action, thus mudding off the faces and plugging the pores.   That this precaution is actually a sufficient safeguard against the plugging action of the cutting disk is shown by the examples in Table 6 giving the

---

[1] The method of fracturing has been preferred by some investigators, but as shown below, the same results can be obtained by the cutting method if the proper precautions are taken.

vealus of the permeability of two adjacent portions of a sandstone block which were determined in one case after cutting by the cutting disk and in the other after fracturing in a specially constructed press where the faces were necessarily free of contamination.

TABLE 6

Sample cut by	Cutting wheel	Fracture
k (sample 1), darcys............	0.65	0.67
k (sample 2), darcys............	4.83	4.83

If the sample of porous medium is a core from an oil sand, the oil it contains must be thoroughly removed before the permeability measurement. For this purpose one may use a large Soxhlet extractor with carbon tetrachloride, benzol, or chloroform as a solvent. The extraction can be accelerated by pressing the sample into the bottom end of a brass tube some 15 cm. long, the pressed fit being sufficiently good so that when this tube is inserted in the extractor with the sample end down, a head of fluid which may be as high as 15 cm. is maintained to force the solvent through the sample. The difficulty of removing all traces of oil makes it advisable to use thin samples of cores from oil sands.

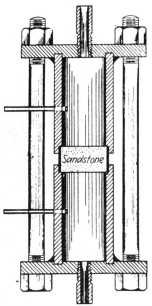

While the details of mounting the sample may vary with the personal taste of the experimenter, a safe and simple procedure is the following: The periphery of the dried and slightly heated sample is first coated with hot pitch or tar so as to cover this sur-

FIG. 15.—Sample holder for permeability measurements of consolidated sands. (*From Rev. Sci. Instr.* **4**, 398, 1933.)

face with an impermeable layer of the sealing material.[1] It is then inserted in the ends of the tube (*cf.* Fig. 15), the through

[1] Any nonpermeable cement will do, or the specimen may be sealed into the flow tube by means of a soft-rubber plug which is drawn up to a tight seal on the walls of the sample by suitable compression glands.

bolts tightened slightly, and just sufficient heat is applied to
form a good bond.   A gap is left between the tubes to permit
detection of an imperfect seal which would allow by-passing
around the sample.   Care must be taken to prevent any sealing
material from covering the faces of the sample.   The pitch or tar

FIG. 16.—Complete apparatus for measurements of consolidated sands.   (*From
Rev. Sci. Instr.*, **4**, 398, 1933.)

is quite satisfactory for measurements with air or water as it is
somewhat less brittle than other sealing materials.   If oil or
other fluids are used which dissolve the pitch even very slightly,
they must be replaced by an insoluble material, such as ordinary
sealing wax or litharge and glycerin for tests with oil or carbon
tetrachloride.

A convenient form for arranging the apparatus is shown in
Fig. 16.   *A* is the sample holder; *B* is a calcium chloride drying

tube; C is a wet-test gas meter for measurements with gases; D is a differential manometer giving the pressure drop across the sample; E is a manometer indicating the pressure at the outflow end of the sample; F are reservoirs to be used with liquid measurements; and G are gauges for determining roughly the pressures existing in the reservoirs. Manometers are necessary to obtain satisfactory results, as gauges are not sufficiently accurate. For low pressures water manometers may be used, while for higher pressures mercury manometers should be used. For gas measurements one must also take care to determine the absolute pressures in the system as well as the total pressure differential.

When measuring permeabilities with liquids it is advisable, as in the case of unconsolidated sands, to mount the sample holder vertically with the input end at the bottom, thus insuring that the top surface is always covered with liquid and eliminating the possibility of an air film forming over the outlet face. If imperfect wetting exists at this face, the results may be seriously affected by the capillary forces.

The only precaution to be particularly observed in the measurements of the rates of flow is that the pressure differentials remain steady during the measurement. In the case of small rates of gas flow the gas may be collected over water, the final measured volume being that at the known atmospheric pressure. Higher flow rates with gases can be measured with an accurately calibrated meter or gasometer. The temperature of the fluid should be noted in all cases, in order to have the correct viscosity. This is especially important in the case of measurements with liquids, as a variation in the temperature of 1°C. will cause changes in the viscosity of many liquids at room temperature of the order of 2 per cent.

2.8. Further Details for Permeability Measurements with Liquids.—Perhaps the most serious difficulty associated with the early measurements of the permeability of consolidated porous media with liquids was that of the plugging of the sand as the flow was continued, with a consequent continuous decrease in the permeability with time. In fact, in much of the early literature on the subject the values of the permeability refer to measurements taken after a certain time following the beginning of the flow experiments, with explicit acknowledgment of the variation of the measured value with the time. Such results, however, are of no value from the point of view of giving a dynamical characterization of the porous medium rather than a numerical determination peculiar to the particular experimental details by which it is obtained.

One very common cause of the plugging of a porous medium upon continuous flow of a liquid, as water, through it, is the evolution of air or gas dissolved in the liquid. This gas becomes trapped in the pores of the medium in the form of small bubbles and may decrease the resultant permeability for the liquid to a small fraction of its original value.[1] An effective means for eliminating this difficulty is that of using distilled water and, in addition, prefiltering it before its entry into the sample to be tested.

A less obvious reason for plugging of a consolidated porous medium, even when distilled water is used, is that due to the solution of silica from the glass in which the water is handled[2] and the deposit of the solute in the pores of the sample. Here, again, prefiltering through sandstone or alundum filters immediately before using will prevent the occurrence of this difficulty, or the water container may be coated with paraffin so as to avoid entirely the solution of the silica. On the other hand, when oil is used as the liquid, plugging may occur owing to the gummy substances formed by the oxidation of the unsaturated hydrocarbons. Here reproducible and reliable results can be nevertheless obtained if the unsaturates are removed by well-known chemical means and the oil is kept in a nonoxidizing atmosphere of nitrogen or natural gas.[3]

[1] While the presence of both the gas and liquid phases in the porous medium gives rise to a "heterogeneous-fluid" system, the discussion of which is beyond the scope of the present work, it may be mentioned that the development of such systems by the evolution of gas from the liquid phase consists in an initial stage of the trapping and accumulation of the evolved gas in the pores of the medium with a continuous decrease in the permeability to the liquid until a condition of dynamical equilibrium is reached in which the additional gas leaving the liquid is flushed out, the gas and liquid phases occupying, on the average, definite fractions of the pore space determined by the initial permeability of the medium. The permeability to the liquid no longer declines after this condition is attained, unless the amount of gas dissolved in the liquid is increased; cf. R. D. Wyckoff and H. G. Botset, *Physics*, **7**, 325, 1936.

[2] Botset, H. G., *Rev. Sci. Instr.*, **2**, 84, 1931.

[3] Botset, *loc. cit.*

A further precaution to be observed is that air is not trapped
in the sample as the testing liquid enters it. A safe procedure
is to exhaust the system by means of a vacuum pump attached
to the outflow end and then to permit the liquid to enter after
the pressure has been considerably reduced. The liquid used
must, of course, be such that it does not loosen the cementing
material of the sample. Carbon tetrachloride will in many
cases be found suitable when water will attack the cementing
material.

**2.9. Further Details for Permeability Measurements with
Gases.**—Analogous to the requirement that a sample, through
which a liquid is flowing for a permeability measurement, must
be free of trapped air or gas, is the precaution that the sample be
free of moisture when the testing fluid is a gas. The liquids that
may be originally in the pores can be removed by long careful
drying in a suitable oven—although the temperature should not
be raised beyond 215°F.—or by the application of a vacuum to
aid in volatilizing the liquid. And the danger of contamination
by moisture or other foreign substances from the flowing gas can
be eliminated by prefiltering and drying the air (by passing
through a calcium chloride tube) just before it enters the sample
to be tested.

As already mentioned in Sec. 2.7, the apparatus for a gas
measurement must be provided with a manometer giving the
absolute pressure in the system in addition to a differential
manometer indicating the pressure drop across the sample.
This requirement follows from the fact that the formula for
computing the permeability from a gas-flow measurement
involves either the squares of the absolute pressures [Eq. 2.5(7)]
or the volume rate of flow reduced to the mean pressure in the
system [Eq. 2.5(8)].

2.10. Gas versus Liquids for Permeability Measurements.—
Although, as has been seen in the previous sections, the permeability of a porous medium is a constant determined only by
the structure of the medium and is independent of the nature of
the homogeneous fluid passing through it, it has been only
recently that this has been recognized in the literature. In fact,
a perusal of the literature leaves one with the impression that if
one is concerned in a practical problem with the flow of a particular fluid through a porous medium, the permeability must be

determined with that fluid, the permeability being usually expressed by the terminology of "permeability to or for the particular fluid of interest." Thus, in practically all the literature describing permeability measurements made in connection with problems arising in the ceramic industry, the use of air is explicitly assumed in the procedure, without any suggestion as to the applicability of the result (if corrected for the fluid viscosities) to the flow through the media of any other fluid. Similarly, in the hydrological literature permeability determinations are almost universally reported in terms of the permeability to water, the fluid used almost without exception.

However, in view of the fact that the measurement of the permeability of a porous medium by means of a gas-flow experiment gives the dynamical characterization of the medium applicable to the flow through it of either gases or liquids, it is of practical importance to note that there actually are definite advantages to measurements with gases.[1] These may be summarized as follows: (1) Elimination of the difficulties resulting from plugging the sample by materials carried by the liquids or swelling of the cementing material in a consolidated porous medium; (2) freedom from the error due to air trapped within the sample and the necessity for evacuation and filling with liquid under a vacuum; (3) freedom from the danger of disintegrating a consolidated sample by the loosening of the cementing material; and (4) the ease of attaining measurable flows without the use of excessive pressures even for very "tight" samples. From the points of view, therefore, of simplicity, speed, and accuracy, it appears that measurements with air or any conveniently available gas should be preferable to those with liquids.

That the fundamental assumption that the permeability of a porous medium is a constant of the medium and is independent of the fluid used in its determination is actually valid, is clearly shown by the data of Table 7. These, giving the results[2] of measurements with air and a liquid on the same samples, and as calculated by the formulas of Sec. 2.5, show that within the

---

[1] If, however, it is impossible completely to remove the oil or other liquids originally present in the sample, it will be safer to use for the test fluid a liquid that is miscible with that in the sample.

[2] Wyckoff, Botset, Muskat, and Reed, *loc. cit.*

TABLE 7

Sample	Permeability, darcys	
	k (air measurements)	k (liquid measurements)
40- to 45-mesh sand...................	139.13	139.40 (H_2O)
80- to 100-mesh sand..................	24.90	22.00 (H_2O)
No. 1 sandstone (Woodbine) E. Texas....	1.18	1.20 (CCl_4)
No. 2 sandstone (Woodbine)...........	1.56	1.57 (H_2O)
No. 3 sandstone (Woodbine)...........	1.63	1.63 (H_2O)
No. 4 sandstone (Berea)...............	1.54	1.50 (H_2O)

limits of the experimental errors of such measurements the final value for the permeability is actually the same regardless of the fluid used. The discrepancy in the case of the 80 to 100-mesh sand was undoubtedly caused by a slight change in packing when water was flowing, together with incomplete removal of trapped air. It may be mentioned here, too, that the similar discrepancies appearing in Table 8 below are due to the hydration and swelling in the clay in the sand.[1] For the sands containing no clay or easily hydrating material, as the Wilcox sand, the agreements between the values found by gas and water measurements are about the same as those indicated in Table 7.

2.11. Permeability Measurements in the Field.—Although a laboratory measurement of the permeability of a porous medium with suitable apparatus and with the proper technique will undoubtedly give the most accurate results in a determination of this fundamental constant, it is frequently necessary to be able to determine its value in the field from the general flow characteristics of the flow system. One typical situation in which this necessity arises is that where the fluid-bearing sand is unconsolidated and at the same time occurs at such a depth that the method of getting samples from surface soils described in Sec. 2.6

[1] When the practical problem in which the value of the permeability is to be used involves the flow of water, and the sand does show a hydration and swelling when exposed to water, one must, of course, nevertheless use the low permeability value given by the measurements with water. In fact, since the tendency of natural rocks to suffer hydration or swelling when wetted with water is not uncommon, it is advisable to test for this effect when making measurements with rock samples from previously unexplored strata.

cannot be easily applied. This is often the case in the study of artesian aquifers drained by wells or other artificially constructed outflow surfaces. Another such typical situation very frequently encountered in the production of fluids from underground reservoirs is that where, although the porous medium is consolidated, there are no cores or samples available for the determination of its permeability in the laboratory. For not only has the saving of cores from wells been considered as a costly luxury until within the last few years, but even at present, when deliberate attempts are being made to obtain core samples as new wells are drilled, it is often found that the process of drilling disintegrates the core. If the consolidation of the sand is not very firm, or, if it retains its structure, it may be damaged by the penetration of the drilling fluids, which may leave in the sand appreciable solid matter not present in the medium in its virgin state.

It is, therefore, of considerable practical importance to have available at least the principles of a procedure by which the permeability of the porous medium as it is in place can be determined from the flow characteristics of the composite system of sand and well without further disturbing the medium. As such methods will find their widest application in flow systems in which a large region of a porous medium is drained by a well of small diameter, the formulas to be presented here will apply only to such systems, the nature of the flow being taken as radial about the well of interest. The particular formulas will be anticipated from derivations to be given in later chapters in order not to confuse the principles of the calculations with analytical details.

It must be pointed out, however, that the application of any formulas to the computation of the permeability of the sand from flow experiments with a well draining the sand requires a preliminary knowledge of the dimensional constants characterizing the system. These are, in particular, the well radius r_w, the sand thickness h, and the penetration of the well into the sand. In addition, the viscosity of the fluid produced from the well must be known or measured. Finally, all the formulas involve a radius r_e which is to be considered as defining the "effective external boundary" for the sand-well system. In the derivation of the formulas this radius enters as the distance from the center of the well at which the fluid pressure is known to be P_e, the reservoir pressure. This latter, on the other hand, corresponds to the pressure at the well face after the well has been closed in for a long enough period for equilibrium to be attained. While the radius r_e at which the fluid pressure, when the well is flowing, is approximately equal to this reservoir pressure will not be known a priori with any certainty, it will be noted that it enters in the formulas in a logarithmic manner. Hence an appreciable error in the estimation of r_e will still, in general, lead to a rather small error in the resulting value of k. For the purpose of actual computation it should suffice to use for r_e half of the average distance of the well under consideration from its immediate neighbors since, except when the production rates from the other wells are widely different from the one in question, the pressure at the mid-points between the wells should be approximately equal to the reservoir pressure as defined above. When no specific information is available as to the positions or production rates of the other wells, one may choose for r_e the value 500 ft., the error in the resultant computed value for k being only 10 per cent if the correct magnitude of r_e is either half or twice 500 ft.

Assuming now that the constants discussed above have been determined, the value of the permeability for a liquid-bearing sand may be finally computed from the relation

$$k = \frac{\mu Q \log r_e/r_w}{2\pi h(P_e - P_w)}, \tag{1}[1]$$

where Q is the volume rate of flow in unit time when the bottom-hole pressure at the well surface is P_w.

In case the well produces gas, the appropriate formula is

$$k = \frac{\mu \bar{Q} \log r_e/r_w}{2\pi h(P_e - P_w)} = \frac{\mu Q_m \log r_e/r_w}{\pi h\gamma_0(P_e{}^2 - P_w{}^2)}, \tag{2}[2]$$

where $\bar{Q}$ is the volume rate of outflow as reduced to the algebraic mean pressure in the sand, $(P_e + P_w)/2$, Q_m is the mass rate of

[1] This equation is derived in Sec. 4.2. It may be noted here that the value of the logarithm in all the formulas in this work refer to the base $e = 2.71828$, the relation between that to the base e and that to the base 10 being given by: $\log_{10} x = 0.43429 \log_e x$.

[2] These equations are derived in Sec. 11.3

outflow, and γ_0 is the gas density at atmospheric pressure (and temperature of the measurements), with the value w/RT for an ideal gas.

In both these equations the resultant value for k will be in terms of darcys if μ is expressed in centipoises, Q and $\bar{Q}$ in cubic centimeters per second, h in centimeters, P_e and P_w in atmospheres. Q_m in grams per second, γ_0 in grams per centimeter cubed, and in case this is computed by the ideal gas value, w must be expressed in grams per mole, T in degrees Kelvin, and R must be given the value 82.07. As r_e and r_w enter only in ratio form, the unit used for them is immaterial, provided both are expressed in the same units.

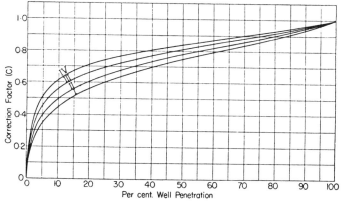

Fig. 17.—Correction factors for the computation of the permeability from flow data of partially penetrating wells with sand thicknesses of: I, 50 ft.; II, 75 ft.; III, 125 ft.; IV, 200 ft. (*From Rev. Sci. Instr.*, **4**, 405, 1933.)

These formulas are valid only if the wells completely penetrate the sand, and furthermore—in the case of Eq. (1)—when the fluid head equivalent to the flowing bottom-hold pressure, P_w, is sufficient to exceed the sand thickness, so that the flow system is one of horizontal radial flow with no gravity component. When the well does not completely penetrate the sand, and h in Eqs. (1) and (2) is taken simply as the thickness of the sand exposed to the well bore, a correction must be applied, or else the use of the formula will give a value for k that is too high. The correction factors by which the values computed from Eqs. (1) and (2) are to be multiplied in order to obtain the true values of the permeability are given graphically in Fig. 17 as functions of the well penetration for various sand thicknesses, for a well radius of $\frac{1}{4}$ ft., as obtained from the detailed analysis of the flow problem of a partially penetrating well (*cf.* Secs. 5.3 and 5.4).[1] This correction factor applies both for wells producing liquids and those producing gas, but, of course, it cannot be used unless the actual well penetration, as well as the total sand thickness, is known.

When the sand is anisotropic—the permeability is different in different directions, as parallel and perpendicular to the bedding planes—a further correction should be applied to the apparent value of k given by Eqs. (1) and (2). This correction can be readily computed from the results developed in Sec. 5.5, especially Fig. 87. However, as the magnitude of the anisotropy will, in general, not be known a priori, we shall not give any explicit representation of the correction factor.

On the other hand, one must be careful to use the proper modification of Eq. (1) if the fluid height in the well during the test does not exceed the sand thickness. This situation will frequently arise in the case of artesian wells, where it may not be possible to obtain appreciable flow rates without pumping down the fluid level until it falls below the top of the producing sand. In addition to the actual fluid height in the well, it is then also necessary to know if the flow into the well is strictly one of gravity flow with a fluid height at distant points in the sand also below the top of the sand, or if it is a composite system of artesian drive plus gravity flow with fluid heads at distant points exceeding that equivalent to the sand thickness.

In the former case—strict radial gravity flow—it has been found empirically (*cf.* Sec. 6.18) that the permeability will be given by the equation

$$k = \frac{\mu Q \log r_e/r_w}{\pi \gamma g (h_e{}^2 - h_w{}^2)}, \tag{3}[2]$$

[1] Perhaps the simplest way of computing these correction factors, C, is by the formula $C = $ (fractional penetration)$/(Q/Q_0)$, Q/Q_0 being the ratio of the production capacity of the partially penetrating well to that of the completely penetrating well, and which is plotted in Figs. 264 and 265.

[2] Mention may also be made in this connection of the so-called "Thiem method" for computing k. Here, using the original Dupuit interpretation of the h's in Eq. (3) as actual fluid *heights* (*cf.* Sec. 6.17 for a more detailed discussion of this point), and taking the points r_e, r_w (r_1, r_2) as representing any two points in the sand with corresponding fluid *heights*, h_1, h_2, the term $h_1{}^2 - h_2{}^2$ is factored as $(h_1 + h_2)(h_1 - h_2)$, or the product of twice the

where γ is the density of the liquid, g the acceleration of gravity, h_w the fluid head in the well above the impermeable base of the producing zone, and h_e is the head at the distant radius r_e.

If, however, h_e exceeds the sand thickness h, so that there is a "drive" of magnitude $\gamma(h_e - h)$ superposed on the gravity-flow component, the resultant flow in the system will be such that k will be given by the equation (*cf.* Sec. 6.19)

$$k = \frac{\mu Q \log r_e/r_w}{\pi \gamma g(2hh_e - h^2 - h_w{}^2)}. \tag{4}$$

Again it is assumed, in the case of both Eqs. (3) and (4), that the well completely penetrates the producing sand, as will practically always be the case when the fluid pressures are so low that the gravity component of the flow is an important part of the whole production from the sand. If, however, the true sand thickness h exceeds the well penetration h_e, the formula of Eq. (3) should be corrected by the factors of Fig. 17 exactly as described for Eqs. (1) and (2) to take account of the partial penetration of normal pressure-drive wells. Likewise in the case of Eq. (4), if the true sand thickness exceeds the well penetration or apparent thickness h, and if h_e, h_w represent fluid heads above the bottom of the well, the application of the correction factors of Fig. 17 will again take care of the lack of complete penetration of the well.[1]

The procedure to be followed in the actual determinations of k by the above formulas should consist essentially of (1) a measurement of the reservoir bottom-hole pressure P_e, or fluid head h_e, with the well shut in long enough for the attainment of equilibrium between the fluid in the well and that in the surrounding

---

average thickness of saturation and the difference in "drawdowns"—depression below the undisturbed fluid level—between r_1 and r_2. Combining an independent determination of the former with a measurement of the latter, for a given rate of pumping Q, in two probe wells drilled at distances r_1 and r_2 from the pumping well, a computation is then made of k. However, in view of the questionable validity of the Dupuit interpretation of the h's in Eq. (3) as fluid *heights*, and the much greater cost of making such determinations, it seems that the procedure suggested here, in spite of its limitations, should in general be both more reliable and practical (*cf.* also Sec. 6.18 for a brief review of the application of this method in actual field tests by L. K. Wenzel).

[1] The analytical justification for the use of these correction factors will be given in Sec. 6. 20.

sand; and (2) a measurement of the rate of flow Q, $\bar{Q}$, or Q_m, corresponding to a value of the bottom-hole pressure P_w or fluid head h_w, determined while the well is flowing at the rate Q. These latter measurements, too, should be made only after a steady-state condition of flow has been established and the transient following the opening of the well after its shut-in period,[1] or change from its previous rate of flow, has died out. This, of course, can be ascertained by noting when the bottom-hole pressure or fluid head assumes a steady value while Q is kept fixed, or conversely.

In addition to these steps in the actual flow experiment one must also estimate or measure the sand thickness h. While h does not enter explicitly in Eq. (3), it must nevertheless be known even in that case in order to be certain that the flow is really by gravity alone and that the equation is applicable. The most accurate determination of h—or the actual penetration —will in general be that given by the drilling log of the well. If this is not available, or is otherwise unsatisfactory, data from neighboring wells may be used.[2] In any case, however, it is clear that as far as the particular well under consideration is concerned, a single measurement of Q and Δp or $h_e{}^2 - h_w{}^2$ will in general give sufficient data for the computation of both the maximum production capacity of the well and its capacities at any preassigned value of P_w or h_w, which frequently will be the real purpose of the measurement. For as long as the flow is viscous, the production rates will be directly proportional to $P_e - P_w$ or $h_e{}^2 - h_w{}^2$ (for simple gravity flow). A single measurement of Q and $P_e - P_w$ or $h_e{}^2 - h_w{}^2$ will, therefore, give the proportionality constant, including the geometrical and physical

---

[1] A different type of permeability field measurement in which use is made of the rate of rise of the fluid in a well bore after the fluid head has been depressed below its equilibrium height is outlined in the footnote of page 642.

[2] If the measurements could be made sufficiently accurate and for a continuous range of the bottom-hole pressure, the top of the producing sand for nongravity-flow systems could, at least in principle, be found by plotting $Q/(P_e - P_w)$ against P_w and noting the value at which this ratio begins to increase (owing to the addition of the gravity-flow component as the fluid height in the well drops below the top of the sand). This value will give the fluid head equivalent of the sand thickness (for a completely penetrating well).

constants of the system, which will enable the prediction of the production capacities for other values of $P_e - P_w$ or $h_e{}^2 - h_w{}^2$.

As to the value of the reservoir pressure P_e or fluid head h_e, it may be noted that a combination of two observations of Q for two values of P_w or h_w will permit the elimination of P_e or h_e. For denoting the difference in the two values of Q by ΔQ and that in P_w or $h_w{}^2$ by ΔP_w and $\Delta h_w{}^2$, it is readily verified from Eqs. (1), (2), (3), and (4) that k will be given in terms of such pairs of observations by

$$k = \frac{\mu \Delta Q \, \log r_e/r_w}{2\pi h \Delta P_w} \tag{5}$$

for liquids in radial flow; by

$$k = \frac{\mu \Delta \bar{Q} \, \log r_e/r_w}{2\pi h \Delta P_w}, \tag{6}$$

for gases in radial flow; and by

$$k = \frac{\mu \Delta Q \, \log r_e/r_w}{\pi \gamma g \Delta h_w{}^2} \tag{7}$$

for both simple and composite gravity-flow systems. Of course, if P_e or h_e are known, the values of k computed from each experiment by Eqs. (1) to (4) should agree, a discrepancy being indicative either of errors of measurement or the nonviscous character of the flow. For this same reason it is advisable, in general, to measure the values of Q for at least two values of P_w or h_w in the determination of the permeability.

A further point to be noted in connection with the use of the above formulas for determining k by field experiments, is that although they have been derived for cases of strictly radial flow they may still be safely used even when the flow is no longer exactly radial. Thus, in the case of Eqs. (1) and (2), it may be rigorously proved (*cf.* Sec. 4.5) that even if the pressures over the circular boundary at the radius r_e are far from uniform, the equations remain correct, provided only that one uses for the reservoir pressure P_e the average of that actually to be found over the boundary. Likewise, if the well is not at the center of the circular boundary, the radial-flow expressions will still be valid unless the well location is within a very short distance from the boundary (*cf.* Sec. 4.6). In fact, one may retain the above equations for k even when the boundary itself is no longer circular, provided one replaces r_e by any reasonable average distance of the well from the actual boundary (*cf.* Sec. 4.16). While these results can be proved rigorously only for nongravity-flow systems, it is clear that since they are essentially consequences of the geometry of the flow system they will undoubtedly be at least approximately valid also for gravity-flow systems. In fact, on the basis of the approximate theory of the gravity flow into a well to be developed in Chap. VI (*cf.* Sec. 6.20), the above generalizations of the validity of Eqs. (1) and (2) can be shown to hold under the same conditions also in the case of the gravity-flow formulas given in Eqs. (3) and (4).

Finally, it must be remembered that underlying all the above discussion has been the implicit assumption[1] that the sand is homogeneous. Only when this assumption happens to be strictly valid should one expect the value of k determined by the field experiment in a given well to agree with that obtained from a laboratory measurement with a core that may have been taken from the same well. If the sand is stratified or has spots or streaks of varying permeability, no such agreement is to be expected. The field measurement will in such cases give an "effective" permeability which, while characteristic of no single element of the medium, will nevertheless be of greater practical significance than the accurate value for a small sample obtained by a laboratory determination. Indeed, it is because of this difference that field measurements will give a more accurate means of predicting the actual production capacity of a well under various back pressures than the inherently more precise laboratory determination of the permeability of a single sand sample, even when the latter is available.

2.12. Typical Permeability Values.—Although permeability measurements have been made by scores of experimenters in the last 40 years, the most systematic published list of results of measurements of the permeability of oil sands carried out with

---

[1] Of course, the additional assumption is being made throughout the whole of this work that the fluids are homogeneous, so that one must be careful not to apply the formulas given here to such cases, for example, where appreciable amounts of gas are being liberated in the sand as' the liquid (oil) flows to the well. The "effective" permeabilities obtained by applying the above formulas to such cases will not be constants of the flow systems but will in general vary with the rate of flow and the amount of gas liberated in the sand.

all the precautions necessary to insure accuracy is that of Fancher, Lewis, and Barnes.[1]   This list is reproduced in part in Table 8. While the earlier determinations of Barb and Branson were all made with water, many of those of the authors of the paper, which were made after the equivalence of permeability measurements with gases and liquids became recognized, were carried out by air-flow experiments.   The great preponderance of the data for sands from Pennsylvania is due not only to the proximity of these investigators to the Pennsylvania fields, but also to the fact that it has become current practice to core and determine the permeabilities of the producing sands for all new wells drilled for the water-flooding operations in the northwestern Pennsylvania oil fields.   However, enough data for other fields are given in this table to give at least a general idea as to the order of magnitude to be found elsewhere.   The successive data listed for samples in the same field show furthermore the variability to be found in the same producing sand.

Some particularly interesting examples of the variability of the permeability of sand samples from the same well bore are given in Table 9, also taken from the paper of Fancher, Lewis, and Barnes. The existence of impermeable shale breaks is strikingly shown by the data in this table obtained by the analysis of continuous cores.

Thus it will be noticed in the case of the first core that, although the porosities are as a whole quite uniform, there are two distinct parts to the sand with respect to its permeability.   Similarly, while the second sample as a whole shows a fair degree of uniformity even in permeability, there are two definite "tight" zones in this core.   The third sample is an example of remarkable uniformity in the porosity with only a moderate variation in the permeability.   Core No. 4 was supposed to be quite representative of the Bradford sand and shows the erratic character of the permeability and the marked stratification in the sand.[2]   That the same is true of the Speechley sand is shown by the data of core No. 5.   The rather high permeability of a

---

[1] Fancher, Lewis, and Barnes, *loc. cit.*

[2] It is of interest to note that the marked stratification of the Bradford sand follows also as a necessary consequence of the discrepancy between the time required for the water to travel from the input to the output wells in typical flooding networks as computed theoretically and observed in field practice (*cf.* Sec. 9.33, and M. Muskat and R. D. Wyckoff, *A.I.M.E., Tech. Pub.* 507, 1933).

TABLE 8.—THE PERMEABILITY AND POROSITY OF OIL SANDS

Sample number	Sand[1]	State	Field or locality	Per cent porosity	Permeability, millidarcys (0.001 darcys)	
					2	3
1	Bradford	Pennsylvania	Bradford	12.5	2.60	3.13
2	Bradford	Pennsylvania	Bradford	12.3	2.78	3.48
3	3rd Venango	Pennsylvania	Oil City	16.9	44.4	65.9
4	Ceramic A	20% bond	....	37.0	5.13	5.93
5	Robinson	Illinois	S. E. Illinois	20.3	29.5	43.5
6	Ceramic B	10% bond	....	37.8	8.82	11.4
7	Woodbine	Texas	East Texas	19.7	182	192
8*	Wilcox	Oklahoma	Seminole	15.9	350	344
9	3rd Venango	Pennsylvania	Venango	11.9	1130	
10	Robinson	Illinois	S. E. Illinois	19.5	63.8	88.7
11	Robinson	Illinois	S. E. Illinois	18.4	67.2	99.3
12	3rd Venango	Pennsylvania	Oil City	22.3	228	356
13	Wilcox	Oklahoma	Seminole	16.3	555	556
14	Warren	Pennsylvania	Warren	19.2	139	139
15	3rd Venango	Pennsylvania	Oil City	21.4	241	315
16	Robinson	Illinois	S. E. Illinois	20.6	126	201
17	Ceramic C	5% bond	....	33.2	37.1	33.7
18	3rd Venango	Pennsylvania	Oil City	21.9	417	
19	Woodbine	Texas	East Texas	23.8	111	
19*	Woodbine	Texas	East Texas	23.6	19.8	
20	Woodbine	Texas	East Texas	26.9	2500	
21	Woodbine	Texas	East Texas	27.7	859	1020
22	Woodbine	Texas	East Texas	22.1	3390	3000
23	Woodbine	Texas	East Texas	28.8	2350	
23*	Woodbine	Texas	East Texas	29.0	341	
24	Bradford	Pennsylvania	Bradford	12.1	2.55	
25	Bradford	Pennsylvania	Bradford	12.0	2.48	
26	Bradford	Pennsylvania	Kane	8.4	†	
27	Bradford	Pennsylvania	Kane	13.5	2.93	
28	Bradford	Pennsylvania	Kane	14.3	5.02	
29	Bradford	Pennsylvania	Kane	14.1	2.53	
30	Bradford	Pennsylvania	Kane	9.8	†	
31	Bradford	Pennsylvania	Kane	11.3	0.073	
32	Bradford	Pennsylvania	Kane	12.1	0.094	
33	Bradford	Pennsylvania	Kane	13.3	....	0.699
34	Bradford	Pennsylvania	Bradford	14.3	....	4.53
35	Bradford	Pennsylvania	Bradford	13.4	....	3.28
35*	Bradford	Pennsylvania	Bradford	12.8	....	2.98
36	Bradford	Pennsylvania	Bradford	9.6	....	0.515
37	Bradford	Pennsylvania	Bradford	13.2	....	3.35
38	Bradford	Pennsylvania	Bradford	13.2	....	1.56
38*	Bradford	Pennsylvania	Bradford	13.2	....	0.826

* Sample cut perpendicular (across) to the planes of bedding.
† Sample impermeable under the conditions of test, namely, no flow obtained in 2 hr. under a pressure gradient of 80 lb./sq. in./in.
[1] Name of geologic formation. Throughout this table the term sand is used in the sense customary to petroleum engineering and geology, i.e., an oil-bearing formation, usually a sandstone, but frequently another rock or relatively unconsolidated sand.
[2] Data obtained from flow of water through sample.
[3] Data obtained from flow of air through sample.

TABLE 8.—THE PERMEABILITY AND POROSITY OF OIL SANDS.—(Continued)

Sample number	Sand[1]	State	Field or locality	Per cent porosity	Permeability, millidarcys (0.001 darcys)	
					[2]	[3]
81	Speechley	Pennsylvania	Oil City	11.0	51.0	
82	Speechley	Pennsylvania	Oil City	15.7	24.8	
83	1st Venango	Pennsylvania	Pleasantville	20.7	....	14.3
84	1st Venango	Pennsylvania	Pleasantville	12.4	....	0.531
85	2nd Venango	Pennsylvania	Pleasantville	17.2	53.3	
86	3rd Venango	Pennsylvania	Oil City	3.4	....	0.016
87	3rd Venango	Pennsylvania	Oil City	8.5	....	58.9
88	3rd Venango	Pennsylvania	Oil City	14.6	....	13.4
89	3rd Venango	Pennsylvania	Oil City	9.9	....	60.4
90	3rd Venango	Pennsylvania	Oil City	19.6	....	171.4
91	3rd Venango	Pennsylvania	Oil City	12.4	....	171.0
92	3rd Venango	Pennsylvania	Oil City	11.3	....	69.5
93	3rd Venango	Pennsylvania	Oil City	17.6	....	60.5
94	3rd Venango	Pennsylvania	Oil City	11.8	....	56.2
95	3rd Venango	Pennsylvania	Oil City	11.0	....	541.0
96	3rd Venango	Pennsylvania	Oil City	7.7	....	40.8
97	3rd Venango	Pennsylvania	Oil City	13.3	....	101.0
98	3rd Venango	Pennsylvania	Oil City	8.7	....	0.904
99	3rd Venango	Pennsylvania	Grand Valley	22.5	70.7	
100	3rd Venango	Pennsylvania	Grand Valley	23.6	100	
101	3rd Venango	Pennsylvania	Grand Valley	23.3	157	
102	3rd Venango	Pennsylvania	Grand Valley	22.1	235	
103	3rd Venango	Pennsylvania	Grand Valley	24.9	148	
104	3rd Venango	Pennsylvania	Grand Valley	12.0	....	0.991
105	3rd Venango	Pennsylvania	Grand Valley	15.4	2.48	
106	Salt	Pennsylvania	Pleasantville	15.0	....	14.7
107	Salt	Pennsylvania	Pleasantville	13.7	....	14.4
108	Salt	Pennsylvania	Pleasantville	13.8	....	10.6
109	Windfall	Pennsylvania	Eldred	8.3	0.250	
110	Windfall	Pennsylvania	Eldred	14.3	6.04	
111	Windfall	Pennsylvania	Eldred	16.0	6.65	
112	Warren	Pennsylvania	Warren	18.6	0.134	
113	Berea grit	Pennsylvania	Bessemer	14.2	....	0.614
114	Berea grit	Pennsylvania	Bessemer	22.6	....	18.4
115	Clarendon	Pennsylvania	Warren	2.8	*	
116	Clarendon	Pennsylvania	Warren	25.6	32.3	
117	Clarendon	Pennsylvania	Warren	....	27.2	
118	Clarendon	Pennsylvania	Warren	10.1	*	
119	Clarendon	Pennsylvania	Warren	10.0	0.595	
120	Clarendon	Pennsylvania	Warren	10.1	*	
121	Clarendon	Pennsylvania	Warren	5.3	*	
122	Clarendon	Pennsylvania	Warren	7.1	*	

* Sample impermeable under the conditions of test, namely, no flow obtained in 2 hr. under a pressure gradient of 80 lb./sq. in./in.

[1] Name of geologic formation. Throughout this table the term sand is used in the sense customary to petroleum engineering and geology, i.e., an oil-bearing formation, usually a sandstone, but frequently another rock or relatively unconsolidated sand.

[2] Data obtained from flow of water through sample.

[3] Data obtained from flow of air through sample.

TABLE 8.—THE PERMEABILITY AND POROSITY OF OIL SANDS.—(*Continued*)

Sample number	Sand[1]	State	Field or locality	Per cent porosity	Permeability, millidarcys (0.001 darcys)	
					[2]	[3]
123	Woodbine	Texas	East Texas	23.4	387	
124	Woodbine	Texas	East Texas	24.1	365	
125	Woodbine	Texas	East Texas	8.1	†	
126	Onondaga[4]	Canada	Bothwell	19.6	....	1280
127	Onondaga[4]	Canada	Petrolia	26.5	....	873
			Data of Barb and Branson			
128	Bradford	Pennsylvania	Kane	16.4	2.81	
128*	Bradford	Pennsylvania	Kane	14.4	1.26	
129	Bradford	Pennsylvania	Bradford	12.8	3.01	
129*	Bradford	Pennsylvania	Bradford	12.9	2.28	
130	Bradford	Pennsylvania	Bradford	12.7	0.807	
131	Bradford	Pennsylvania	Bradford	8.8	1.18	
132	Bradford	Pennsylvania	Bradford	12.6	2.04	
133	Bradford	Pennsylvania	Bradford	22.7	131	
134	Bradford	Pennsylvania	Bradford	11.0	0.945	
135	Bradford	Pennsylvania	Bradford	13.5	1.16	
136	Bradford	Pennsylvania	Bradford	12.9	3.65	
136*	Bradford	Pennsylvania	Bradford	13.4	1.93	
137	Bradford	Pennsylvania	Bradford	14.8	20.5	
137*	Bradford	Pennsylvania	Bradford	14.6	3.18	
138	Bradford	Pennsylvania	Kane	11.8	3.81	
139	Bradford	Pennsylvania	Kane	10.7	1.87	
140	Bradford	Pennsylvania	Kane	9.6	0.453	
141	Bradford	Pennsylvania	Kane	4.7	0.063	
142	Bradford	Pennsylvania	Kane	11.0	12.4	
143	Bradford	Pennsylvania	Kane	8.8	2.21	
144	Bradford	Pennsylvania	Kane	9.7	1.17	
145	Bradford	Pennsylvania	Kane	11.5	4.33	
146	Bradford	Pennsylvania	Kane	11.7	5.27	
147	Bradford	Pennsylvania	Kane	12.0	3.88	
148	Bradford	Pennsylvania	Kane	12.3	3.20	
149	Bradford	Pennsylvania	Kane	3.0	0.453	
150	Bradford	Pennsylvania	Kane	11.9	1.56	
151	Bradford	Pennsylvania	Kane	11.3	1.13	
152	Bradford	Pennsylvania	Kane	11.6	11.1	
153	Bradford	Pennsylvania	Kane	11.6	9.28	
154	Bradford	Pennsylvania	Kane	11.0	15.4	
154*	Bradford	Pennsylvania	Bradford	10.3	0.365	
155	Bradford	Pennsylvania	Crystal	10.9	5.73	
156	Bradford	Pennsylvania	Crystal	10.9	2.61	

* Sample cut perpendicular (across) to the planes of bedding.

† Sample impermeable under the conditions of test, namely, no flow obtained in 2 hr. under a pressure gradient of 80 lb./sq. in./in.

[1] Name of geologic formation. Throughout this table the term sand is used in the sense customary to petroleum engineering and geology, *i.e.*, an oil-bearing formation, usually a sandstone, but frequently another rock or relatively unconsolidated sand.

[2] Data obtained from flow of water through sample.

[3] Data obtained from flow of air through sample.

[4] Granulated dolomitic limestone.

TABLE 8.—THE PERMEABILITY AND POROSITY OF OIL SANDS.—(Continued)

Sample number	Sand[1]	State	Field or locality	Per cent porosity	Permeability, millidarcys (0.001 darcys) [2]
232	2nd Venango	Pennsylvania	Oil City	20.1	30.3
232*	2nd Venango	Pennsylvania	Oil City	19.7	45.7
233	2nd Venango	Pennsylvania	Oil City	15.5	18.3
233*	2nd Venango	Pennsylvania	Oil City	15.6	17.9
234	2nd Venango	Pennsylvania	Oil City	13.6	23.3
235	2nd Venango	Pennsylvania	Oil City	10.6	0.843
236	2nd Venango	Pennsylvania	Oil City	8.2	12.7
236*	2nd Venango	Pennsylvania	Oil City	8.5	11.7
237	2nd Venango	Pennsylvania	Oil City	5.8	3.34
238*	2nd Venango	Pennsylvania	Oil City	18.4	66.4
244	2nd Venango	Pennsylvania	Bullion	23.6	136
257	3rd Venango	Pennsylvania	Vrooman	15.3	36.8
258	Clarendon	Pennsylvania	Saybrook	13.4	0.076
259	Clarendon	Pennsylvania	Saybrook	11.5	0.214
260	Clarendon	Pennsylvania	Warren	14.9	0.467
261	Clarendon	Pennsylvania	Warren	8.8	†
262	Clarendon	Pennsylvania	Warren	18.5	3.38
263	Clarendon	Pennsylvania	Warren	13.5	0.582
264	Clarendon	Pennsylvania	Warren	16.1	3.17
265	Gray	Pennsylvania	Cranberry	6.0	0.504
266	Glade	Pennsylvania	Warren	10.9	613
267	Kane	Pennsylvania	Kane	15.1	23.2
268	Kane	Pennsylvania	Kane	18.7	28.3
269	Kane	Pennsylvania	Kane	14.2	1.37
270	Kane	Pennsylvania	Kane	18.1	87.8
271	Kane	Pennsylvania	Kane	22.1	216
272*	Kane	Pennsylvania	Kane	21.9	60.9
273*	Kane	Pennsylvania	Kane	22.4	33.3
274	Gordon	Pennsylvania	Greene Co.	12.9	
275*	Richburg	New York	Bolivar	6.7	0.718
276	Olean	New York	Rock City	15.8	259
277	Olean	New York	Rock City	14.4	30.5
278	Berea	West Virginia	Hancock Co.	22.2	261
279	Berea	West Virginia	Hancock Co.	16.7	287
280	Berea	Ohio	Monroe Co.	11.7	1.39
281	Wilcox	Oklahoma	Oklahoma City	16.9	677
282	Wilcox	Oklahoma	Oklahoma City	8.0	78.8
283	Wilcox	Oklahoma	Seminole	15.6	356
283*	Wilcox	Oklahoma	Seminole	15.6	75.6
284	Wilcox	Oklahoma	Bowlegs	12.1	84.9
284*	Wilcox	Oklahoma	Bowlegs	12.8	88.0
285*	Below Wilcox	Oklahoma	Oklahoma City	9.3	0.328

* Sample cut perpendicular (across) to the planes of bedding.
† Sample impermeable under the conditions of test, namely, no flow obtained in 2 hr. under a pressure gradient of 80 lb./sq. in./in.
[1] Name of geologic formation. Throughout this table the term sand is used in the sense customary to petroleum engineering and geology, i.e., an oil-bearing formation, usually a sandstone, but frequently another rock or relatively unconsolidated sand.
[2] Data obtained from flow of water through sample.

Table 8.—The Permeability and Porosity of Oil Sands.—(Continued)

Sample number	Sand[1]	State	Field or locality	Per cent porosity	Permeability, millidarcys (0.001 darcys) [2]
298	Johnson	Oklahoma	Oklahoma City	11.7	24.7
298*	Johnson	Oklahoma	Oklahoma City	11.7	0.793
299	Johnson	Oklahoma	Oklahoma City	21.3	692
299*	Johnson	Oklahoma	Oklahoma City	21.3	278
300	Johnson	Oklahoma	Oklahoma City	15.3	88.8
300*	Johnson	Oklahoma	Oklahoma City	15.3	65.8
301	Johnson	Oklahoma	Oklahoma City	15.5	464
301*	Johnson	Oklahoma	Oklahoma City	15.5	147
302	Johnson	Oklahoma	Oklahoma City	15.7	258
303	Johnson	Oklahoma	Oklahoma City	5.3	0.818
304	School Land	Oklahoma	Oklahoma City	11.4	36.1
304*	School Land	Oklahoma	Oklahoma City	11.4	128
305	School Land	Oklahoma	Oklahoma City	14.8	518
305*	School Land	Oklahoma	Oklahoma City	14.8	223
306	Layton	Oklahoma	Oklahoma City	17.6	170
306*	Layton	Oklahoma	Oklahoma City	17.6	118
307	Prue	Oklahoma	Oklahoma City	11.4	0.466
307*	Prue	Oklahoma	Oklahoma City	11.4	3.40
308	Hammer-Haindel	Oklahoma	Oklahoma City	14.1	464
308*	Hammer-Haindel	Oklahoma	Oklahoma City	14.1	144
309	Gilcrease	Oklahoma	Holdenville	16.8	62.1
309*	Gilcrease	Oklahoma	Holdenville	16.8	60.8
310	Gilcrease	Oklahoma	Francis	27.3	559
310*	Gilcrease	Oklahoma	Francis	27.5	800
311	Gilcrease	Oklahoma	Sasakwa	18.3	202
311*	Gilcrease	Oklahoma	Sasakwa	18.1	149
312	Cromwell	Oklahoma	Konawa	16.6	172
312*	Cromwell	Oklahoma	Konawa	16.7	405
313	Cromwell	Oklahoma	Konawa	22.6	68.1
313*	Cromwell	Oklahoma	Konawa	21.6	108
314	Cromwell	Oklahoma	Konawa	17.9	772
314*	Cromwell	Oklahoma	Konawa	18.8	140
315	Cromwell	Oklahoma	Little River	16.9	53.8
315*	Cromwell	Oklahoma	Little River	16.2	
316	Cromwell	Oklahoma	Little River	19.9	96.8
316*	Cromwell	Oklahoma	Little River	19.4	144
317	Cromwell	Oklahoma	Little River	21.3	359
317*	Cromwell	Oklahoma	Little River	21.5	134
318	Cromwell	Oklahoma	Little River	20.9	159
318*	Cromwell	Oklahoma	Little River	21.2	68.0
319	Cromwell	Oklahoma	Little River	23.2	314

* Sample cut perpendicular (across) to the planes of bedding.
[1] Name of geologic formation. Throughout this table the term sand is used in the sense customary to petroleum engineering and geology, i.e., an oil-bearing formation, usually a sandstone, but frequently another rock or relatively unconsolidated sand.
[2] Data obtained from flow of water through sample.

TABLE 8.—THE PERMEABILITY AND POROSITY OF OIL SANDS.—*(Continued)*

Sample number	Sand[1]	State	Field or locality	Per cent porosity	Permeability, millidarcys (0.001 darcys) [2]
319*	Cromwell	Oklahoma	Little River	22.8	74.3
320	Cromwell	Oklahoma	Little River	21.6	289
320*	Cromwell	Oklahoma	Little River	21.5	164
321	Cromwell	Oklahoma	Little River	17.8	246
321*	Cromwell	Oklahoma	Little River	18.2	220
322	Cromwell	Oklahoma	Little River	19.4	154
322*	Cromwell	Oklahoma	Little River	19.4	459
323	Cromwell	Oklahoma	Little River	19.1	772
323*	Cromwell	Oklahoma	Little River	19.1	378
324	Cromwell	Oklahoma	Little River	18.8	360
324*	Cromwell	Oklahoma	Little River	18.8	670
325	Cromwell	Oklahoma	Little River	20.8	261
325*	Cromwell	Oklahoma	Little River	20.8	38.7
326	Wanette	Oklahoma	Wanette	22.7	482
326*	Wanette	Oklahoma	Wanette	22.2	37.4
327	Wanette	Oklahoma	Wanette	10.6	
328	Wanette	Oklahoma	Wanette	13.0	
329	Wanette	Oklahoma	Wanette	17.4	199
329*	Wanette	Oklahoma	Wanette	18.5	164
330	Wanette	Oklahoma	Wanette	14.2	18.3
330*	Wanette	Oklahoma	Wanette	14.5	7.85
331	Wanette	Oklahoma	Wanette	14.8	11.4
332	Wanette	Oklahoma	Wanette	12.6	13.4
332*	Wanette	Oklahoma	Wanette	13.0	2.51
333	Wanette	Oklahoma	Wanette	15.9	21.2
333*	Wanette	Oklahoma	Wanette	16.4	7.15
334	Wanette	Oklahoma	Wanette	18.1	52.7
334*	Wanette	Oklahoma	Wanette	18.6	33.2
335*	Wanette	Oklahoma	Wanette	14.4	8.03
336	Tokio	Louisiana	Pine Island	3.2	†
337	Tokio	Louisiana	Pine Island	18.4	†
337*	Tokio	Louisiana	Pine Island	18.4	†
338	Glen Rose	Louisiana	Caddo	8.9	
339	Glen Rose	Louisiana	Caddo	24.8	1460
340	Glen Rose	Louisiana	Caddo	24.3	341
340*	Glen Rose	Louisiana	Caddo	24.3	125
341	Glen Rose	Louisiana	Caddo	7.2	†
342	Glen Rose	Louisiana	Caddo	16.2	3.89
343	Pechelbroon	France	Pechelbroon	17.8	3.28

* Sample cut perpendicular (across) to the planes of bedding.

† Sample impermeable under the conditions of test, namely, no flow obtained in 2 hr. under a pressure gradient of 80 lb./sq. in./in.

[1] Name of geologic formation. Throughout this table the term sand is used in the sense customary to petroleum engineering and geology, *i.e.*, an oil-bearing formation, usually a sandstone, but frequently another rock or relatively unconsolidated sand.

[2] Data obtained from flow of water through sample.

TABLE 9.—PERMEABILITY AND POROSITY OF CONTINUOUS CORES

Depth, ft. in sand	Porosity, per cent	Permeability, millidarcys	Depth, ft. in sand	Porosity, per cent	Permeability, millidarcys
1	Bradford sand, Kane		**2**	Bradford sand, Bradford	
1.6	8.4	*	4.5	14.3	4.53
4.8	13.5	2.93	13.5	13.4	3.28
8.5	14.3	5.02	23.5	9.6	0.515
13.0	14.1	2.53	28.5	13.2	3.35
17.5	9.8	*	45.5	13.2	1.56
29.7	11.3	0.073	49.5	12.9	1.15
37.7	12.1	0.094	65.5	9.7	0.430
41.8	13.3	0.699	68.5	16.1	4.35
			91.5	12.8	3.23
3	Bradford sand, Kane		**4**	Bradford sand, Bradford	
1.2	2.0	*	2.5[1]	7.1	1.36
3.0	12.1	1.27	2.5	17.8	61.6
4.0	12.9	2.51	5.0	16.5	38.6
8.4	13.4	4.02	8.1	16.7	55.8
13.7	13.0	3.77	13.6	14.8	14.7
25.8	13.0	1.54	18.2	9.9	0.427
36.5	12.1	1.60	28.9	11.7	3.67
41.4	12.5	0.342	46.5	14.7	10.0
			60.7	12.8	2.65
			62.3	12.5	3.51
			63.5	14.2	6.58
			89.3	5.6	0.106
			91.3	11.6	0.660
			93.1	13.0	0.464
			100.1	13.7	6.901
			[1] Shale break.		
5	Speechley sand, Oil City		**6**	3rd Venango, Oil City	
3.3	10.3	2.45	1.0	3.4	0.0161
4.0	13.6	36.6	3.6	8.5	58.9
5.0	10.7	0.887	5.6	14.6	13.4
5.7	7.6	0.156	15.2	9.9	60.4
6.7	14.8	16.3	18.1	19.6	171.4
10.5	3.7	0.021	21.1	12.4	171.0
11.3	4.7	*	24.1	11.3	69.5
12.8	14.5	33.4	27.2	17.6	60.5
29.8	7.1	0.563	30.2	11.8	56.2
33.1	11.0	51.0	33.2	11.0	541.0
34.5	15.7	24.8	35.5	7.7	40.8
			38.2	13.3	101.0
			40.1	8.7	0.904

Depth, ft. in sand	Porosity, per cent	Permeability, millidarcys
7	Clarendon sand, Warren	
0.9	2.8	*
5.5	25.6	32.3
8.9	....	27.2
15.8	10.1	*
18.8	10.0	0.595
23.2	10.1	*
28.5	5.3	*
36.6	7.1	*

* Sample impermeable under the conditions of test, namely, no flow obtained in 2 hr. under a pressure gradient of 80 lb./sq. in./in.

typical Venango sand is shown by the analysis of the sample No. 6, and the tremendous variation that may be observed in some continuous cores is illustrated by the case of the Claredon sand sample. Here only three samples in a thickness of some 37 ft. had a measurable permeability under conditions of the tests, and these showed a variation of 54-fold.

While these results show clearly the interesting information that may be obtained by systematic permeability measurements, it may be suggested that at the same time they prove the assumption of homogeneity in the porous media which constitute underground fluid-bearing reservoirs to be entirely fallacious. Although one could nevertheless find considerable justification for the assumption of homogeneity in the argument that by far the majority of the problems of practical importance would not be amenable to analytical treatment without this assumption and hence could be given no quantitative discussion, there are, however, sound physical grounds for believing that the inhomogeneities illustrated in Table 9 do not, in most cases, seriously affect the validity of the results derived on the basis of this assumption. For certainly in such cases where the wells completely penetrate the sand, the flow is two-dimensional, and its fundamental characteristics, including the pressure and streamline distributions, are independent of the numerical value of the permeability. Indeed, these will be the same in two parallel and adjacent layers of different permeability, provided only that the boundary pressures are the same. The only physical difference is that in the numerical values of the fluid velocities which, for the corresponding points in the two layers, will always be in the ratio of the permeabilities. All the theoretical deductions developed on the assumption that the layered sand is equivalent to a single homogeneous porous medium will be strictly correct, provided only that the permeability entering in the expressions for the velocities or total fluxes is taken as the average of those in the various layers, weighted according to their thickness. With this adjustment the total flux will come out correctly, and the velocity at any point will be the weighted average of those in the various layers. Hence, as long as the horizontal variations in permeability are not of appreciable magnitude, those in the vertical direction may be entirely neglected in a theoretical treatment of such problems which are essentially two-dimensional and whose planes of flow are horizontal.

With regard to the horizontal variations of permeability, it should be noted that, except when the variations occur in the immediate vicinity of a well bore or a convergent outflow surface, only those are of practical significance which are of appreciable areal extent. The effects of scattered localized spots of high or low permeability will be averaged out in a flow system of large dimensions and need not be explicitly taken into account in the analytical treatments. In the case of problems such as those of partially penetrating wells or seepage under dams, in which an appreciable part or all of the flow takes place in the direction of large permeability variations, the latter may, of course, seriously influence the fundamental features of the flow. In such cases accurate results can only be attained by taking into account the variations in permeability. However, in view of the tremendous analytical difficulties of doing this except in certain simple cases, one is here forced to return to the assumption of homogeneity as the only means of even getting specific semiquantitative results. In most problems the qualitative nature of the effect of the inhomogeneity will be rather easy to deduce from physical arguments, and the results derived on the assumption of a uniformity in the permeability will represent limiting cases with respect to those in which the permeability is variable.

One may consider from the same point of view the significance of the anisotropy of the sand, as indicated by the relative values of the permeability measured parallel and perpendicular to the bedding planes. Of the 65 pairs of data originally included in Table 8 more than two-thirds (46) gave permeabilities parallel to the bedding planes greater than those normal to it, the ratio between the two values ranging from 1 to as high as 42. On the other hand, among the cases in which the permeability normal to the bedding plane exceeded that parallel to it, the maximum ratio of the two values is 7.3. Here again it is to be observed that the assumption of isotropy, underlying the treatment of practically all the problems discussed in the following chapters, will still suffice to give, in most cases of practical interest, a true representation of the significant features of the flow. This is indeed strictly correct when the flow systems are two-dimensional with the planes of flow parallel to the bedding planes. For under such conditions there simply is no component of the velocity normal to the bedding planes, so that the permeability in that direction does not enter the problem.[1] On the other hand, when the problem involves components of flow along more than one direction with different permeabilities, the anisotropy can be strictly taken into account by applying a transformation of the coordinates as outlined in Sec. 4.15 and which is illustrated by the treatment in Sec. 5.5 of the problem of wells partially penetrating an anisotropic producing sand. As will be seen there, however, the analytical problem in the transformed coordinate system is equivalent to one corresponding to the flow in an isotropic medium with appropriately modified boundaries. From the analytical point of view, therefore, one returns in the treatment of such anisotropic systems to the solution of problems for isotropic systems with somewhat modified geometry, so that a complete discussion of the latter will implicitly include at the same time the solution for similar problems in which one may actually wish to take the anisotropy into account. Hence, in most cases, it will suffice to consider the problem from the beginning as one in an isotropic medium, and only at the very end introduce the appropriate transformation of coordinates if the effects of the anisotropy are to be studied.

Typical permeability data for unconsolidated porous media are listed in Table 10. The porosities are included in this table in order to give a closer specification of the nature of the porous medium, in addition to that indicated by the grain size. For, as pointed out in Sec. 2.6, the permeability of an unconsolidated porous medium of a given assembly of sand grains may vary manyfold, depending upon the porosity to which it is packed. As unconsolidated media do not involve the extra variable of cementation, the variations of which may cause large differences in the permeability of different sands of approximately the same average grain size and porosity, only a very few of the many measurements available are listed here. These should nevertheless suffice to give at least the order of magnitude of the numerical results that may be anticipated from permeability measurements of unconsolidated porous media.

[1] *Cf.* Sec. 5.5, especially Fig. 87, showing that for a completely penetrating well the production capacities are independent of the permeability normal to the bedding planes.

Table 10.—Permeabilities and Porosities of Unconsolidated Sands

Sand, mesh	Porosity, per cent	Permeability, darcys
30 to 40	40.0	344.81
40 to 50	40.0	65.89
50 to 60	40.0	43.46
60 to 70	40.0	30.96
70 to 80	40.0	26.36
80 to 100	40.0	10.49
100 to 120	40.0	9.93
120 to 140	40.0	9.26
Fine heterogeneous..............	30 to 35	1 to 10
Silts.........................	36 to 41	5 to 180*
Fine powders..................	37 to 70	0.01 to 0.1†

* Vaidhianathan, V. I., and H. R. Luthra, Punjab Irrigation Research Institute. *Research Pub.*, **2**, no. 2, 1934.
† Traxler, R. N., and L. A. H. Baum, *Physics*, **7**, 9, 1936.

2.13. Porosity Measurements.—Although, as already stressed in previous sections, the homogeneous-fluid-carrying capacity of a porous medium is completely described by its permeability and is only indirectly affected by the porosity, which is essentially a static property, the latter is nevertheless of tremendous importance in the evaluation of the economic value of porous media bearing oil or gas, as it determines the total fluid content of such reservoirs. We shall, therefore, very briefly outline some typical methods of measuring the porosity of porous media, pointing out only the essential features so as to indicate the principles involved.[1]

Not only in interpretation, but also in the method of measurement, must one distinguish between the "total" porosity of a porous medium and its "effective" porosity. The former, which gives the total fractional void space of a porous medium as expressed in per cent of the bulk volume, will include, in the case of consolidated media, the volume of all pores, whether or not they are open to the flow of fluids. The latter, on the other hand, refers only to that part of the pore space which is available to fluid flow. Although it is clear that only such open

[1] A more complete review of these methods will be found in the article by Fancher, Lewis, and Barnes (*loc. cit.*); *cf.* also O. E. Meinzer, *U. S. Geol. Surv. Water-supply Paper* 489, 1923.

pores can affect the permeability of the medium, as indirect and involved as the relation may be, we shall, for completeness, outline methods both for measuring the total and effective porosity of a porous medium.

What has been generally considered as the most accurate method for determining the total porosity of a sample of a porous medium is that of Melcher and Nutting.[1] This method is completely gravimetric and is based on the relation

$$f = 100\left(1 - \frac{V_g}{V_b}\right) = 100\left(1 - \frac{\gamma_b}{\gamma_g}\right), \tag{1}$$

where f is the porosity, V_g, γ_g the volume and density of the sand grains of the sample, and V_b, γ_b the bulk volume and apparent (bulk) density of the sample. After thoroughly extracting any fluid contained in the specimen and drying, as described in the discussion of permeability measurements, the density γ_b is determined by the following steps: (1) Weighing the sample as it is; (2) weighing again after coating with a thin layer of paraffin or collodion of known density, the gain in weight divided by the density giving the volume of the coating[2]; and (3) weighing again after immersion in a water-filled pycnometer previously weighed both when empty and filled with water. The loss in weight found in step (3) divided by the density of the water gives the volume of the coated sample, and when corrected by subtracting the volume of the coating and then divided into the weight of (1), γ_b will be the quotient. The sand grain density is then determined by crushing the specimen in a mortar, and measuring the volume of a weighed quantity of grains[3] by displacement of a

---

[1] MELCHER, A. F., *A.I.M.E.*, **65**, 469, 1921; *Amer. Assoc. Petrol. Geol.*, **8**, 716, 1924; *ibid.*, **9**, 442, 1925; P. G. NUTTING, *Amer. Assoc. Petrol. Geol.*, **14**, 1337, 1930.

[2] Except in the case of samples with large pores, this step may be omitted if mercury is used for the immersion liquid. In the latter case a modified Jolly balance devised by Athy (*Amer. Assoc. Petrol. Geol.*, **14**, 1, 1930) will be found convenient.

[3] It should be noted, however, that when experimenting with oil- or water-bearing sands, one must not dry the grains after crushing the sample to get the grain density, as the γ_g in Eq. (1) must be considered to include the mass content of the fluids in the sealed-off pores of the sample. If the absolute density of the dried grains is used for γ_g, Eq. (1) will give an error in the total porosity of magnitude $\Delta f = \gamma_1 \delta f / \gamma_g$, where γ_1 is the average density of

liquid (tetrachlorethane or tetralin) in the pycnometer in a manner similar to that used in finding γ_b. Equation (1) then gives f. A considerably simpler procedure is that involved in the Russell method,[1] which is based on the first part of Eq. (1) and requires only volume measurements. The bulk volume is determined by means of a glass volumeter, indicated in Fig. 18, which consists essentially of two bulbs of approximately equal volume, held together by two parallel graduated tubes. One bulb is joined to a small sample container—the "stopper"—by means of a ground joint. The apparatus is inverted and the upper bulb is filled with tetrachlorethane to a calibrated zero mark. The specimen, well saturated with tetrachlorethane, is then placed in the stopper, the two bulbs joined again, and the apparatus returned to its normal position, the displacement of the previously introduced liquid by the sample being noted on the calibrated connecting tubes, thus giving the bulk volume directly. The same procedure, with obvious modifications, is repeated with the crushed sand grains of the sample, and their volume is determined. The application of Eq. (1) then gives the porosity. Although this method is, in principle, quite direct and straightforward it demands skillful technique to avoid appreciable errors both in the transference of the saturated sample to the volumeter and in reducing the sample to its individual grains without any loss. Furthermore, a failure completely to saturate the sample, which may well occur in the case of tight sands, will lead to additional errors owing to the

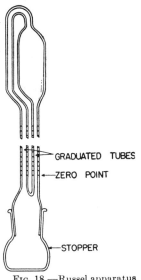

FIG. 18.—Russel apparatus for porosity determinations. *(After Russell, Amer. Assoc. Petrol. Geol.)*

GRADUATED TUBES
ZERO POINT
STOPPER

---

the fluids in the sealed-off pores, and δf of the difference between the total and effective porosities. (M. Muskat, *Rev. Sci. Instr.*, **7**, 503, 1936). Thus if $\gamma_1 = 1$, $\delta f = 1$ per cent, $\gamma_g = 2.6$, $\Delta f = 0.38$ per cent, which would entirely vitiate the high accuracy inherent in the technique of determining γ_b and γ_g.

[1] RUSSELL, W. L., *Amer. Assoc. Petrol. Geol.*, **10**, 931, 1926; *cf.* also H. R. Brankstone, W. B. Gealy and W O. Smith, *Amer. Assoc. Petrol. Geol.*, **16**, 915, 1932.

subsequent penetration of the volumeter liquid into the sample after its introduction into the stopper and immersion.

One of the first methods developed for measuring the *effective* porosity, and the main principle of which is the basis of most forms of "gas porosimeter," is that of Washburn and Bunting.[1] This basic principle is that of measuring directly the volume of air or gas contained in the connected pore space of the sample at

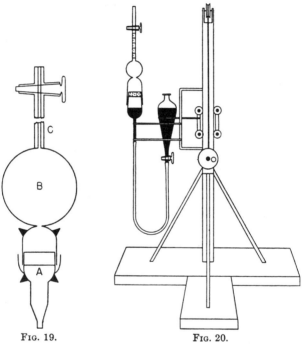

FIG. 19.   FIG. 20.

FIG. 19.—Washburn-Bunting type of porosimeter. (*After Fancher, Lewis, and Barnes, Bull. Pa. State Exp. Sta.*)

FIG. 20.—Diagrammatic representation of an elevating device for the leveling bottle of the Washburn-Bunting-type porosimeter.

atmospheric pressure, after subjecting the gas to expansion. The determination is carried out with a porosimeter of the type indicated in Figs. 19 and 20, in which *A* is the sample chamber, *B* the expansion chamber, and *C* a graduated capillary fused to the top of *B* and fitted with a stopcock at its extremity. The air in the specimen is first trapped at atmospheric pressure by

[1] WASHBURN, E. W., and E. N. BUNTING, *J. Amer. Ceram. Soc.*, **5**, 48, 112, 1922.

raising the leveling bulb until the mercury level in the porosimeter rises above the stopcock when the latter is closed. The leveling bulb is then lowered to diminish the pressure in the porosimeter and permit the air trapped in the sample to expand and escape into the graduated tube. The volume at atmospheric pressure of this air which has escaped is measured by raising again the leveling bulb until the mercury in it and the graduated tube are at the same height. The air remaining in the sample at the reduced pressure corresponding to that at which the first measured volume escaped is determined by a repetition of the original procedure, after expelling the air gathered in the graduated tube by opening the stopcock and once more filling the tube with mercury to the level of the stopcock. The process is thus repeated until the remanent air in the specimen is of too small volume to be of consequence. The sum of the volumes obtained in the various repetitions then gives the total effective pore volume of the sample. The bulk volume is then measured with either the displacement or gravimetric method, and the ratio will give the porosity.[1]

A different method is that of Barnes.[2] The principle of this method, which also leads to a value of the effective porosity, is that of determining the open-pore volume by weighing the sample when dry and when saturated with tetrachlorethane, the difference divided by the density of the liquid giving this volume. The saturation of the sample is obtained by first evacuating the dried core of the air filling its open pores while immersed in a flask of tetrachlorethane. When the flask is opened to the atmosphere, the liquid penetrates and saturates the evacuated pores. The bulk volume is then determined by the Russell or an equivalent method. Here again the questions of the completeness of the saturation of the core, especially of tight sands, within a reason-

---

[1] The procedure described here really represents the modification of the original Washburn-Bunting method devised by Fancher, Lewis, and Barnes (*loc. cit.*). Further variations of the original principle involve the expansion of the gas in the sample and sample bulb, filled at a high pressure, into another bulb of fixed volume (C. J. Coberly and A. B. Stevens, *A.I.M.E. Pet. Dev. Techn.* **103**, 261, 1933), or into a burette against atmospheric pressure (K. B. Barnes, p. 50, *Production Bull.* 217, *Amer. Petrol. Inst.*, 1936). The grain volumes of the samples are then obtained by a direct application of Boyle's law.

[2] BARNES, K. B., *Bull. Mineral Ind. Exp. Sta., Penn State College*, **10**, 1931.

TABLE 11.—POROSITY OF CONSOLIDATED SANDS AND ROCKS

Sample number	Sand[1]	Locality[2]	Total porosity, per cent		Effective porosity, per cent	
			Melcher-Nutting	Russell	Washburn-Bunting	Barnes
1	Stockton	Stockton, N. J.	....	1.3	....	1.2
2	Stockton	Stockton, N. J.	....	2.9	....	2.7
3	Stockton	Stockton, N. J.	....	7.9	....	7.6
4	Stockton	Lumberville	....	3.7	....	3.5
5	Stockton	Lumberville	....	4.3	....	4.3
6*	Chickies	Narvon Station	....	3.8	....	3.6
7*	Chickies	Narvon Station	....	4.3	....	4.1
8*	Chickies	Valley Forge	....	4.9	....	4.8
9*	Chickies	Valley Forge	....	6.0	....	5.9
10*	Chickies	Valley Forge	....	7.8	....	7.2
11	3d Venango	Oil City	....	7.5	....	4.8
12	3d Venango	Oil City	....	8.5	8.3	
13	3d Venango	Oil City	11.1	9.8		
14	3d Venango	Oil City	11.4	11.3		
15	3d Venango	Oil City	....	12.7	12.6	12.4
16	3d Venango	Oil City	....	....	13.3	12.9
17	3d Venango	Oil City	13.2	13.0		
18	3d Venango	Oil City	....	13.3		13.3
19	3d Venango	Oil City	....	....	14.6	14.3
20	3d Venango	Oil City	14.3	14.5		
21	3d Venango	Oil City	....	15.3	14.9	14.5
22	3d Venango	Oil City	....	15.5	14.9	14.4
23	3d Venango	Oil City	....	16.0	15.7	
24	3d Venango	Oil City	16.4	16.3		
25	3d Venango	Pleasantville	....	17.1	14.7	
26	3d Venango	Oil City	....	17.7	....	17.4
27	3d Venango	Oil City	....	17.8	15.3	
28	3d Venango	Oil City	17.7	17.8		
29	3d Venango	Oil City	....	....	20.9	20.8
30	3d Venango	Oil City	....	....	21.6	21.3
31	Clarendon	Warren	....	10.0	9.9	9.3
32	Clarendon	Warren	....	11.8	11.8	11.6
33	Clarendon	Warren	....	12.1	11.8	11.4
34	Clarendon	Warren	....	12.2	11.8	11.5
35	Clarendon	Warren	....	12.2	11.8	11.7
36	Clarendon	Warren	....	12.6	12.2	11.8
37	Bradford	Kane	....	13.0	....	10.2
38	Bradford	Kane	....	13.0	....	12.3
39	Bradford	Kane	....	14.1	....	13.9
40	Bradford	Kane	....	16.4	....	14.4
41	Bradford	Kane	....	17.4	....	13.6
42	3d Bradford	Bradford	....	13.1	12.2	12.1
43	3d Bradford	Bradford	14.1	13.4	12.8	12.5
44	3d Bradford	Bradford	....	13.4	13.3	13.0
45	3d Bradford	Bradford	13.6	13.5	13.3	12.8
46	3d Bradford	Bradford	....	13.6	13.4	13.1
47	3d Bradford	Bradford	....	13.7	13.2	13.0
48	3d Bradford	Bradford	16.1	15.7	15.4	14.9
49	3d Bradford	Bradford	....	16.1	15.6	15.5
50	3d Bradford	Bradford	....	16.2	15.2	15.0
51	3d Bradford	Bradford	....	16.9	15.3	15.0

* Sample is quartzite, all others are sandstone.
[1] Name of geologic formation. Throughout this paper the term sand is used in the sense customary to petroleum engineering and geology, i.e., an oil-bearing formation, usually a sandstone, but frequently another rock or relatively unconsolidated sand.
[2] In the State of Pennsylvania unless otherwise designated.

TABLE 12.—COMPARISON OF METHODS FOR DETERMINATION OF POROSITY

Consideration	Melcher-Nutting	Russell	Washburn-Bunting	Barnes
Type of porosity	Total	Total	Effective	Effective
Apparatus	Expensive*	Low cost	Low cost	Expensive*
Accuracy	To 0.005%†	To 0.25%‡	To 0.075%	To 0.05%†‡
Time required	4 hr.	1 hr.	0.75 hr.	1 hr.
Size and shape	1 piece 1 in. diam. by 0.5 in. thick	1 or several small pieces equal to 10 to 14 cc. Shaped for insertion in apparatus	1 piece preferable, 10 to 14 cc., regular contours, shaped for insertion in apparatus	1 piece 10 to 14 cc., r e g u l a r contours, shaped for insertion in apparatus
Only suited to	Consolidated s a n d , any texture	Consolidated s a n d , any texture	Any reservoir rock, small pore openings	Any reservoir rock, small pore openings
Conditions after analysis	Reduced to grain size	Reduced to grain size	Some pores filled with mercury, l a t t e r clinging to surface	Easily restored to orig- inal condition

* The major item of expense is an analytical balance. If this is purchased solely for this use the apparatus is expensive. If a balance is available, the cost is comparable to the other methods.
† With precision apparatus.
‡ With skillful operation.

able time. and of the proper wiping of the saturated surfaces are those leading to the greatest uncertainty as to the absolute accuracy of the method.

An interesting tabular comparison given by Fancher, Lewis, and Barnes,[1] of the four methods outlined above, and a table of typical measurements made by these authors with the various methods are reproduced in Tables 11 and 12. It will be seen that in many of the samples the total porosity exceeds the effective porosity by 0.5 to 1.0 per cent, and in one case by as much as 3.8 per cent.[2] Such differences may clearly represent large quantities of available reservoir fluids in the estimation of the reserves in underground reservoirs.

Other typical porosity values of consolidated cores than those of Table 11 are given together with their permeabilities in Table 8.[3] In the case of unconsolidated sands, it may be recalled from Sec. 1.6 that the limits of porosity possible for homogeneous spheres are 26.0 and 47.6 per cent, respectively. While the upper limit will generally remain valid for nonspherical grains and heterogeneous size distributions, except when there is extensive bridging of the grains throughout the porous medium,[4] the lower limit of 26 per cent may be appreciably reduced by mixing sands of various grain sizes, and the porosity may even be made to practically vanish by the addition of fines and silts.

[1] *Loc. cit., Bull.* 12, 1933.

[2] In the case of such sands as that in the Kettlemen Hills field in California differences have been reported which are so high as to make the total porosity almost three times as great as the effective porosity (16.2 − 5.6), *cf.* C. J. Coberly and A. B. Stevens, *loc. cit.*

[3] Many other results with relevant data concerning the wells from which the samples were obtained will be found in a paper by A. F. Melcher, *Amer. Assoc. Petrol. Geol.*, **8**, 716, 1924.

[4] These exceptions will be frequently found in the case of powders where the bridging may be so extensive as to raise the porosities to as high values as 70 per cent (*cf.* R. N. Traxler and L. A. H. Baum, *Physics*, **7**, 9, 1936).

CHAPTER III

GENERAL HYDRODYNAMICAL EQUATIONS FOR THE FLOW OF FLUIDS THROUGH POROUS MEDIA

3.1. Fundamental Hydrodynamic Relations.—Before entering upon the analysis and quantitative discussion of special problems of fluid flow through porous media, it will be well first to review and summarize several well-known principles of hydrodynamics such as would apply to any flow system. We shall then formulate the particular relations characterizing the flow of fluids through porous media, and from these develop the solutions corresponding to specific problems of practical interest.

In analyzing the fundamental principles of hydrodynamics it is readily found that they are essentially nothing more than reformulations of the corresponding principles of mechanics in such a way as to be appropriate to the flow of fluids. Thus, first of all, it must be noted that although fluids are nonrigid systems, they are subject to the law of conservation of matter. This law states that the fluid mass in any closed system can be neither created nor destroyed.

For analytical purposes, however, it is convenient to restate this law, when referred to a fluid in motion, in the form: The net excess of mass flux, per unit of time, into or out of any infinitesimal volume element in the fluid system is exactly equal to the change per unit of time of the fluid density in that element multiplied by the free volume of the element. When this statement is further analyzed, it becomes equivalent to the equation

$$\text{div } (\gamma \bar{v}) = \frac{\partial}{\partial x}(\gamma v_x) + \frac{\partial}{\partial y}(\gamma v_y) + \frac{\partial}{\partial z}(\gamma v_z) = -\frac{f \partial \gamma}{\partial t}, \qquad (1)[1]$$

[1] The symbol div (or $\nabla \cdot$) is taken from the symbolic notation of vector analysis and is defined by the intermediate expression in Eq. (1), when $\gamma \bar{v}$ is considered as a general vector function (*cf.* A. P. Wills, "Vector Analysis," 2d ed., 1923).

where $\bar{v}$ is the vector velocity of the fluid at the point (x, y, z), and has the components v_x, v_y, v_z; γ is the fluid density at (x, y, z), and f is the porosity of the medium.

Equation (1) may be derived as follows: Choosing a rectangular parallele-piped of edges, dx, dy, dz, with center at (x, y, z), the mass flow into the side $dydz$ perpendicular to the x axis at a distance $x - dx/2$ from the yz plane will evidently be (*cf.* Fig. 21)

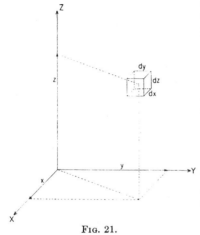

$$\left[\gamma v_x - \frac{\partial}{\partial x}(\gamma v_x)\frac{dx}{2} \right]dydz,$$

where γv_x refers to the point (x, y, z). As the mass flux out of the parallel side will be

$$\left[\gamma v_x + \frac{\partial}{\partial x}(\gamma v_x)\frac{dx}{2} \right]dydz,$$

the net flux leaving the parallelepiped through these two sides will be the difference of the flux at the two sides, or

$$\frac{\partial}{\partial x}(\gamma v_x)dxdydz.$$

Fig. 21.

Treating the other sides of the parallelepiped similarly, and adding them together to give the resultant flux per unit time out of the volume element $dxdydz$, it is found to have the value

$$\left[\frac{\partial}{\partial x}(\gamma v_x) + \frac{\partial}{\partial y}(\gamma v_y) + \frac{\partial}{\partial z}(\gamma v_z) \right]dxdydz.$$

Now the mass of fluid in the volume element is evidently $f\gamma dxdydz$, where γ is the instantaneous value of the density at (x, y, z). Hence the loss of mass in the element per unit time can also be expressed as

$$-\frac{f\partial\gamma}{\partial t}dxdydz.$$

Equating this to the other expression for the loss of mass from the element according to the requirement of the principle of the conservation of mass, and canceling the volume differential $dxdydz$, Eq. (1) is obtained at once.

Equation (1)—the so-called "equation of continuity"—or an equivalent of it appears in almost all branches of physics. It expresses, here, the law of the conservation of matter. In the theory of electricity, slightly different forms express the law of conservation of charge. In the theory of heat conduction, the conservation of heat energy is expressed by a similar equation of continuity. In fact, every conservation law of physics can be expressed in a form equivalent to Eq. (1).

When the state of flow is independent of the time, *i.e.*, when it is "steady," the right side of Eq. (1) vanishes. The equation of continuity then reduces to

$$\text{div } (\gamma \bar{v}) = \frac{\partial}{\partial x}(\gamma v_x) + \frac{\partial}{\partial y}(\gamma v_y) + \frac{\partial}{\partial z}(\gamma v_z) = 0. \tag{2}$$

It will be presently found that this equation may be still further simplified in certain cases.

Thus far it has been specifically noted that we are dealing, in our development of hydrodynamics, with a *material* fluid. To proceed further, it is clear that one must also specify the *nature* of the fluid involved and the thermodynamic character of its flow. In other words, one must know the "equation of state" defining the fluid. In general terms this will be a quantitative relation between its density γ, the pressure p, and the absolute temperature T, which may be expressed by an equation of the form

$$\phi(p, \gamma, T) = 0, \tag{3}$$

where p, γ, and T refer to the same element of the fluid.

As an illustration of Eq. (3) it may be observed that when it is stated physically that the fluid is a strictly incompressible liquid, one implies that

$$\gamma = \text{const.} \tag{4}$$

is the equation of state. Or, if one is interested in an ideal gas, Eq. (3) would take the form

$$p - \frac{\gamma RT}{w} = 0, \tag{5}$$

where w is the molecular weight of the gas, and R is the gas constant per mole. In any case, Eq. (3) is the analytical definition of the nature of the fluid to be treated, and hence must be given as the starting point of any hydrodynamical discussion.

The definition of the thermodynamic character of the flow will, in general, require another equation of the form of Eq. (3) and will, therefore, permit an elimination of one of the variables p, γ, T so as to give a combined description of the thermodynamic nature of the fluid and of the flow in terms of two variables alone, such as $\gamma = \gamma(p)$. Thus, if the fluid is an ideal gas— with Eq. (5) for the equation of state—in isothermal flow, the analytical characterization of this type of flow would simply be

$$T = \text{const.}, \tag{6}$$

which would simplify Eq. (5) to the form

$$\gamma = cp; \qquad c = \frac{w}{RT}. \tag{7}$$

On the other hand, if the flow be adiabatic, we should have in place of Eq. (6)

$$T = \text{const. } \gamma^{m-1}, \tag{8}$$

where m is the ratio of the specific heat at constant pressure to that at constant volume. In this case Eq. (5) will reduce to

$$p = \left(\frac{\gamma}{\gamma_0}\right)^m, \tag{9}$$

where γ_0 is the density for $p = 1$.

The relevance of the equation of state to hydrodynamical problems may be illustrated by a direct application of Eq. (4). Thus, when one is interested in incompressible liquids, Eq. (4) must be inserted into the equation of continuity, Eq. (2), reducing it to

$$\frac{\partial v_x}{\partial x} + \frac{\partial v_y}{\partial y} + \frac{\partial v_z}{\partial z} = 0, \tag{10}$$

which now involves only the velocity components.

Equation (10) may now be considered as a condition upon the velocity distribution in any incompressible fluid system. Analytically, however, it is clear that it does not suffice to determine the individual velocity components, and physically it does not discriminate between one incompressible fluid and another. Neither does it distinguish between systems exposed to external body forces, such as gravity, and those flowing only under the action of pressure differentials, nor finally between fluids flowing through porous media and fluids flowing in unobstructed vessels.

It appears, then, that one has to characterize the fluid of interest *dynamically* as well as thermodynamically, and state explicitly how it reacts to pressure gradients and external forces. Specifically, we have to formulate the hydrodynamical equivalent to Newton's law that the force acting on any body is equal to the product of the mass of the body and its acceleration. The details of this formulation will depend on the nature of the fluid and the conditions under which it is flowing. Although we shall ultimately be interested only in fluid flow through porous media, it will be useful to consider first the dynamical characterization of a fluid as given by the classical hydrodynamics of viscous flow.

3.2. The Classical Hydrodynamics.—It has been seen that in addition to the equation of continuity and an equation of state a dynamical definition of the nature of the flow is required before we can have a complete system of hydrodynamics. Considering a unit volume element of the fluid, it is readily found that in general it will be subject to three types of forces. These are (1) the pressure gradients of components $\dfrac{\partial p}{\partial x}, \dfrac{\partial p}{\partial y}, \dfrac{\partial p}{\partial z}$; (2) external "body forces," such as gravity, of components F_x, F_y, F_z acting on each volume element of the fluid; and (3) the forces opposing the motion of the fluid and due to the internal resistance or friction experienced by the fluid. In texts on hydrodynamics[1] it is shown that these latter forces, for the case of "viscous flow," are given by the Cartesian components

$$\mu\nabla^2 v_x + \frac{1}{3}\mu\frac{\partial\theta}{\partial x}, \qquad \mu\nabla^2 v_y + \frac{1}{3}\mu\frac{\partial\theta}{\partial y}, \qquad \mu\nabla^2 v_z + \frac{1}{3}\mu\frac{\partial\theta}{\partial z},$$

where μ is the viscosity of the fluid—defined as the shearing stress in the fluid set up by a unit velocity gradient perpendicular to the plane of shear—∇^2 is the operator

$$\nabla^2 \equiv \frac{\partial^2}{\partial x^2} + \frac{\partial^2}{\partial y^2} + \frac{\partial^2}{\partial z^2}$$

and θ is a function appearing in the equation of continuity and given by

$$\theta = \operatorname{div} \bar{v} = \frac{\partial v_x}{\partial x} + \frac{\partial v_y}{\partial y} + \frac{\partial v_z}{\partial z},$$

with the physical significance that it represents the rate of volume dilatation of the fluid.

---

[1] See, for example, H. Lamb "Hydrodynamics," 6th ed., p. 577, 1932.

As has already been suggested, the dynamical equation of motion is to be obtained by equating the sum of these forces to the product of the mass and acceleration of the volume element to which the forces refer. Since the velocity in the fluid will, in general, vary from point to point, it must be noted, in computing its acceleration, that not only will the velocity of the fluid element change during any time interval at the position it had originally, but it will experience an additional change owing to the fact that in the time interval it has moved to another region of the fluid. The acceleration must, therefore, be expressed by the *total* time derivative of the velocity, *i.e.*, by the operator

$$\frac{D}{Dt} \equiv \frac{\partial}{\partial t} + \frac{dx}{dt}\frac{\partial}{\partial x} + \frac{dy}{dt}\frac{\partial}{\partial y} + \frac{dz}{dt}\frac{\partial}{\partial z} = \frac{\partial}{\partial t} + v_x\frac{\partial}{\partial x} + v_y\frac{\partial}{\partial y} + v_z\frac{\partial}{\partial z}.$$

Combining these results, one finally obtains the following dynamical equations of motion—originally due to Navier and Stokes:[1]

$$\left.\begin{array}{l}
\gamma\dfrac{Dv_x}{Dt} = -\dfrac{\partial p}{\partial x} + F_x + \mu\nabla^2 v_x + \dfrac{1}{3}\mu\dfrac{\partial\theta}{\partial x} \\[2mm]
\gamma\dfrac{Dv_y}{Dt} = -\dfrac{\partial p}{\partial y} + F_y + \mu\nabla^2 v_y + \dfrac{1}{3}\mu\dfrac{\partial\theta}{\partial y} \\[2mm]
\gamma\dfrac{Dv_z}{Dt} = -\dfrac{\partial p}{\partial z} + F_z + \mu\nabla^2 v_z + \dfrac{1}{3}\mu\dfrac{\partial\theta}{\partial z}.
\end{array}\right\} \qquad (1)$$

From physical considerations it has been seen that a complete system of hydrodynamics requires the formulation of the equation of continuity, the equation of state of the fluid, and the dynamical equations of motion of the fluid. Without entering into the analytical theory of the sufficiency of these equations, it may be noted that at least from an elementary point of view the above formulation may be considered complete, since we now have five independent equations [Eqs. 3.1(1), 3.1(3), and (1)] for the five essential unknowns, γ, p, v_x, v_y, v_z.[2] And indeed it is true that these five equations suffice in principle to predict all the details of the motion of a viscous fluid flowing in or through a vessel of any shape whatever.

[1] Navier, C. L. M. H., *Ann. chim. phys.*, **19**, 234, 1821; G. G. Stokes, *Trans. Cambridge Phil. Soc.*, **8**, 287, 1845.

[2] The variable T is supposed to have been eliminated by means of the equation for the thermodynamic character of the flow, such as Eq. 3.1(6) or 3.1(8).

Strictly speaking, the flow of a viscous fluid through a porous medium is but a special case of the general problem of the viscous flow of fluids between impermeable boundaries. Insofar as the pores of the medium are fixed and their bounding surfaces are geometrically describable, the flow through these pores is, in principle, subject to detailed description by means of the classical equations of hydrodynamics—Eqs. 3.1(1), 3.1(3), and (1). However, even a cursory inspection of treatises on hydrodynamics will disclose that except for certain cases of relatively simple geometry the mathematical difficulties in the solution of the classical equations are quite unsurmountable. The treatment by these equations of problems in which a fluid flows through channels as irregular and tortuous as those in a sand are, therefore, entirely out of the question, and one must resort to another method of attack.

3.3. Generalized Form of Darcy's Law.—From the considerations of Sec. 3.1 it will be clear that insofar as a hydrodynamics of flow through porous media may be formulated differently from the classical theory of viscous flow, the difference must lie essentially in the expression of the dynamical equations, which the classical theory presents in the form of Eqs. 3.2(1). Certainly the law of conservation of matter and the thermodynamic definitions of a fluid must be retained in any hydrodynamic system. However, it is not unreasonable that the dynamical reactions of a fluid passing through the fine channels of a porous medium may, from a macroscopic point of view, appear in quite different form than when analyzed microscopically and represented by Eqs. 3.2(1).

It is this difference that has been empirically established by the early experiments of Darcy for the case of liquids and the more recent experiments with gases cited in Sec. 2.2, and has been formulated as Darcy's law. Recalling this law, it states that *macroscopically*, the velocity of a fluid flowing through a porous medium is directly proportional to the pressure gradient acting on the fluid. By the qualification "macroscopically" we mean that the volume elements to which the velocity and pressure refer are supposed to contain a large number of pores and the dynamical variables are really averages over a large number of pores, although in detail they may show large variations within the individual cells, as indeed it is certain they would according to the detailed description implied by Eqs. 3.2(1), if one could solve them explicitly. In other words, Darcy's law is of the nature of a statistical result giving the empirical equivalent to the Stokes Navier Eqs. 3.2(1) as averaged over very large numbers of individual pores.

Since, as pointed out in Chap. II, the direct experiments relating to Darcy's law have been restricted to columns or beds of porous materials in which the macroscopic flow is necessarily of a one-dimensional character, it is necessary to generalize these empirical results before a complete theory applicable to any type of flow system can be developed.[1]

Thus, first, we shall suppose that in a general three-dimensional flow system the resultant velocity at any point is directly proportional to, in magnitude, and in the same direction as the resultant pressure gradient at that point. This is, of course, equivalent to the supposition that the resultant velocity may be resolved into three component velocities parallel to the coordinate axes, each reacting to the pressure gradients independently of the others. Darcy's law then takes the form

$$
\left.
\begin{aligned}
v_x &= -\frac{k_x}{\mu}\frac{\partial p}{\partial x} \\[4pt]
v_y &= -\frac{k_y}{\mu}\frac{\partial p}{\partial y} \\[4pt]
v_z &= -\frac{k_z}{\mu}\frac{\partial p}{\partial z}
\end{aligned}
\right\}
\qquad (1)^2
$$

where μ is the viscosity of the fluid. In these equations the permeability k may vary from point to point and may be different for the three components v_x, v_y, v_z as is indicated by the subscripts. However, it will generally suffice to consider the medium to be isotropic,[3] and k shall hereafter be taken as independent of direction unless otherwise specified.

But Eq. (1) is still not sufficiently general to cover all the cases in which we shall be interested. For, when the fluid velocity

---

[1] Some specific empirical evidence justifying these generalizations will be given in Secs. 4.7 and 4.11 (*cf.* Figs. 37 and 55).

[2] The minus sign is chosen in Eq. (1) so that the velocity components will be positive when the fluid flows toward increasing values of the coordinates (*cf.* footnote on p. 71).

[3] *Cf.* Sec. 2.12.

has a component along the vertical, it is clear that gravity must be taken into account either explicitly or implicitly.   Here again we shall generalize and assume that, if any "body" force, of components F_x, F_y, F_z per unit volume, is acting on the fluid, it will affect the velocity just as the pressure gradients; in such cases, then, Darcy's law will become

$$
\left.
\begin{aligned}
v_x &= -\frac{k}{\mu}\left(\frac{\partial p}{\partial x} - F_x\right) \\
v_y &= -\frac{k}{\mu}\left(\frac{\partial p}{\partial y} - F_y\right) \\
v_z &= -\frac{k}{\mu}\left(\frac{\partial p}{\partial z} - F_z\right)
\end{aligned}
\right\}
\tag{2}
$$

If the force $\bar{F}$ has a potential[1] V, one may introduce the function Φ defined by

$$
\Phi = \frac{k}{\mu}(p + V),
\tag{3}
$$

so that

$$
\left.
\begin{aligned}
v_x &= -\frac{\partial \Phi}{\partial x} \\
v_y &= -\frac{\partial \Phi}{\partial y} \\
v_z &= -\frac{\partial \Phi}{\partial z}
\end{aligned}
\right\}
\tag{4}
$$

From this it is seen that Φ is really a "velocity potential"—a function whose negative gradient gives the velocity vector. Equations (4) may be conveniently combined into the single vector equation:

$$
\bar{v} = -\nabla\Phi.
\tag{5}
$$

Equations (3) and (5) may be considered as the "generalized Darcy's law," and may be taken as the *dynamical* basis for all problems of viscous fluid flow through porous media and for all types of *homogeneous* fluids.   They are our substitute for the

---

[1] A vector $\bar{F}$ is said to have a "potential" Φ when it can be represented as the gradient (positive or negative, depending on convention) of Φ, *i.e.*, if $\bar{F}$ can be expressed as $\bar{F} = -\nabla\Phi$, where ∇ is the vector differential operator with components $\dfrac{\partial}{\partial x}, \dfrac{\partial}{\partial y}, \dfrac{\partial}{\partial z}$.

Stokes-Navier Eqs. 3.2(1) and may be considered as their macroscopic equivalent.

It is to be noted that the dependence of the potential function Φ on the fluid viscosity μ is expressed explicitly and is not to be included in the permeability k, even when both will be taken as constant. As already pointed out in Sec. 2.3, this separation frees k from any dependence on the nature of the fluid and leaves it determined only by the porous medium. In fact, this one constant k suffices completely to describe a uniform porous medium as the carrier of any homogeneous fluid.

Another point to be observed with respect to the generalized Darcy equations, Eq. (3) and (4), is that they do not involve the density γ. For they are not only different in *form* from the classical hydrodynamic equations, Eqs. 3.2(1), but they also differ from the classical equations in that they do not contain at all the dependent variable γ. Now in the Stokes-Navier equations γ enters in the term $\gamma\dfrac{D\bar{v}}{Dt}$ which represents the inertia or acceleration forces in the fluid. The absence of γ in the Darcy equations, Eqs. (3) and (4), must, therefore, imply that the inertia forces are being neglected. This fundamental difference is, however, to be expected when it is observed that owing to the very large surface exposed to a fluid in a porous medium the viscous resistance will greatly exceed any acceleration forces in the fluid unless turbulence sets in. Under such conditions it would be justified even a priori to neglect the inertia terms, as is indeed frequently done in treatments of the classical equations for cases in which the predominating forces are those due to the viscous resistances. The absence of the variable γ in the Darcy equations is, therefore, to be ascribed to the comparative unimportance of the acceleration forces as compared to the internal friction resistances in a porous, fluid-bearing medium, while the differences in *form* of the equations are to be attributed to the effect of the statistical averaging of the classical equations over the minute and detailed variations occurring in the individual pores so as to give a simplified representation in terms which are of macroscopic significance.[1]

[1] Although it would indeed be very satisfying to be able to derive Darcy's law as a strict consequence of the equations of the classical hydrodynamics, Eqs. 3.2(1), it is felt that such derivations as given by N. K. Bose (*Mem.*

3.4. The Equations of Motion.—Now that the dynamical laws characterizing the flow of fluids through porous media have been formulated, we must return and add to them the equation of continuity and the equations of state to make the system complete. Thus applying first the dynamical equations, Eqs. 3.3(5), to the equation of continuity, Eq. 3.1(1), we obtain

$$\text{div} \ (\gamma\nabla\Phi) = \nabla \cdot \left[\gamma\nabla\frac{k}{\mu}(p + V) \right] = f\frac{\partial\gamma}{\partial t}, \tag{1}$$

where, under all conditions of practical interest, μ may be taken as independent of the pressure and may, therefore, be taken out of the bracket. For homogeneous media k may also be taken out of the bracket, and as we shall confine ourselves to homogeneous media except when dealing with certain special problems —Chap. VII—k will hereafter be taken as constant[1] unless otherwise specified. Equation (1) then takes the form

$$\nabla \cdot [\gamma\nabla(p + V)] = \frac{f\mu}{k}\frac{\partial\gamma}{\partial t}. \tag{2}$$

There is now left simply the specification of the nature of the fluid to obtain a differential equation in the single variable γ or p. As a sufficiently general equation including all homogeneous fluids of practical interest and all types of viscous flow we may take[2]

$$\gamma = \gamma_0 p^m e^{\beta p}. \tag{3}$$

The particular fluids of physical significance may now be classified as follows:

Punjab Irrigation Res. Lab., **2**, no. 1, 1929, and *Paper* no. 140, *Punjab Eng. Cong.*, 1930), in which the Eqs. (4) are obtained as solutions of Eqs. 3.2(1) with the inertia terms dropped and with each side of Eqs. (4) varying exponentially with the time, are fallacious. For aside from the extraneous time variations, the solutions of Eqs. 3.2(1) in the form of Eqs. (4) must necessarily refer to the *microscopic* pore velocities and pressures, and cannot be identified with the macroscopic velocities and pressures of Eqs. (4) until detailed account is taken of the boundary conditions to be satisfied at the boundaries of all the pores of the medium.

[1] Cf. Sec. 2.12 for a justification of this assumption in spite of the large variations in k that have been found in cores from the same well bore.

[2] MUSKAT, M., *Physics*, **5**, 71, 1934.

Liquids: $m = 0$.

Incompressible liquids: $\beta = 0$,

Compressible liquids: $\beta \neq 0$;

Gases: $\beta = 0$;

Isothermal expansion: $m = 1$,

Adiabatic expansion: $m = \dfrac{\text{specific heat at const. vol.}}{\text{specific heat at const. pressure.}}$

Applying these to Eq. (2) with the assumption that gravity is the only body force acting on the fluid, so that $V = -\gamma gz$, it is found that for *incompressible liquids*

$$\nabla^2\Phi = \frac{\partial^2\Phi}{\partial x^2} + \frac{\partial^2\Phi}{\partial y^2} + \frac{\partial^2\Phi}{\partial z^2} = 0 = \nabla^2 p, \qquad (4)[1]$$

where Φ is defined by Eq. 3.3(3). For *compressible liquids*

$$\nabla \cdot \left[\left(\frac{1}{\beta} - \gamma gz \right)\nabla\gamma - \gamma^2 g\nabla z \right] = \frac{f\mu}{k}\frac{\partial\gamma}{\partial t} \qquad (5)$$

Observing that for normal liquids β is of the order of $10^{-4}/\text{atm.}$, whereas γg has a value of the order of 10^{-3} atm./cm., it is clear that the term γgz can be entirely neglected as compared to $1/\beta$. The terms $\gamma^2 g\nabla z$ and $\dfrac{1}{\beta}\nabla\gamma$, on the other hand, are in the ratio of the vertical body force due to gravity to that due to the fluid-pressure gradients. When this ratio is of appreciable magnitude, as it will be, for example, in the case of gravity-flow systems (Chap. VI), and the compressibility of the liquid is of physical significance, an accurate treatment would indeed require the solution of Eq. (5). However, such a treatment will in general be very difficult owing to the nonlinearity of the equation. For practical purposes, therefore, it is necessary to discuss the phases of the compressibility and the gravity component of the flow separately, *i.e.*, the former by Eq. (5) with the neglect of the terms involving g, and the latter by the solution of Eq. (4). The errors in this approximation to the rigorous treatment by

---

[1] The derivation of this equation as governing the flow of liquids through porous media was first given by C. S. Slichter, *U. S. Geol. Surv.*, 19th Ann. Rept., 1897–1898, p. 330, where a number of solutions for specific problems are also presented.

Eq. (5) will, however, be of no major consequence from a practical point of view, as in those cases where the pressure gradients will be of sufficiently large magnitude to lead to significant variations of the liquid density over small distances, these same gradients will also be very large as compared to the gravity-head gradient γg.* If the density variations are of appreciable magnitude essentially because of the extended dimensions (horizontal) of the flow system (*cf.* Sec. 10.1), the flow will be largely confined to horizontal planes and the gravity component will again be of negligible importance. Hence in such problems where gravity will play a significant role the effects of the compressibility will be of minor importance and the analysis can be safely based upon Eq. (4), whereas in those where the effects of the liquid compressibility are the predominant features, the force of gravity can be left out of consideration. Where the liquid flow will actually be considered as that of a compressible liquid, we shall, therefore, drop the terms with g in Eq. (5) and take the system to be governed by the equation

$$\nabla^2\gamma = \frac{\partial^2\gamma}{\partial x^2} + \frac{\partial^2\gamma}{\partial y^2} + \frac{\partial^2\gamma}{\partial z^2} = \frac{f\beta\mu}{k}\frac{\partial\gamma}{\partial t}. \tag{6}$$

Gases:

$$\nabla^2\gamma^{\frac{1+m}{m}} = \frac{\partial^2\gamma^{\frac{1+m}{m}}}{\partial x^2} + \frac{\partial^2\gamma^{\frac{1+m}{m}}}{\partial y^2} + \frac{\partial^2\gamma^{\frac{1+m}{m}}}{\partial z^2} = \frac{(1+m)f\mu\gamma_0^{\frac{1}{m}}}{k}\frac{\partial\gamma}{\partial t}. \tag{7}$$

These are the fundamental differential equations which we shall take as the basis of the treatment of the various problems of flow through porous media of practical significance. It will be seen at once that for incompressible liquids the time variations disappear so that there can be no time transients or nonsteady states within the system unless the conditions at the boundaries vary with the time. The pressure obeys the so-called "Laplace's equation," which also frequently occurs in other branches of physics.[1]

* Even if $\dfrac{1}{\gamma_0}\dfrac{\partial\gamma}{\partial x} = 10^{-5}$, the associated pressure gradients will be larger than the gravity gradient γg by a factor of 100.

[1] *Cf.* Sec. 3.6.

In the case of compressible liquids the fundamental equation, Eq. (6), does involve the time and is in fact identical in form with the "Fourier heat-conduction equation."[1] Its steady-state[2] form is, however, identical with that for incompressible liquids with the density playing the role of the pressure or velocity potential. It will be discussed in detail in Part III.

Equation (7) for gases also contains the time and therefore permits nonsteady states. However, it is nonlinear in that it involves the dependent variable γ to a power other than the first, and cannot be solved rigorously in closed form. An approximation theory for the flow of gases will, therefore, be developed in Part IV. On the other hand, it is to be noted that here, again, the steady state is governed by the Laplace equation in the dependent variable $\gamma^{\frac{1+m}{m}}$.

The Laplace equation (4) will be taken as the basis of all the analysis of Part II, in which will be considered problems of the steady-state flow of liquids. While this equation resulted from the assumption that the liquid is strictly incompressible, it will in general represent a very good approximation for actual liquids except when the liquid has an abnormally high compressibility or when the dimensions of the flow system are very large (*cf.* Sec. 10.1). On the other hand, the use of Eq. (4) may be considered as involving only a formal simplification of the problem of a really compressible liquid, for if desired, the distribution of the liquid density γ in a *compressible-liquid* steady-state system can be obtained from that for Φ or p derived for the incompressible-liquid system of the same geometry by simply reinterpreting the Φ or p as γ, provided only that the effects of gravity are neglected and the boundary conditions (*cf.* Sec. 3.5) are expressed in terms of the boundary density or *mass* flux. The role of the equipressure or equipotential surfaces or curves will be played by those of constant density, and the streamlines will be tangential to the vector gradient of γ, $\nabla\gamma$, or the mass

---

[1] CARSLAW, H. S., "Mathematical Theory of the Conduction of Heat in Solids," 2d ed., 1921.

[2] The term "steady state" is, in this work, taken as defining the condition of flow in which the significant dynamical variables—the pressure or density and velocity—do not vary with the time, so that all terms in the fundamental differential equations involving $\dfrac{\partial}{\partial t}$ may be set equal to zero.

velocity vector[1] $-\frac{\gamma k}{\mu}\nabla p$. For practical purposes, however, it will always suffice, in view of the very low compressibility of normal liquids, to treat liquid-flow systems in the steady state explicitly as problems in the flow of incompressible liquids, and hence as governed by Eq. (4).

Furthermore, except when the compressibility inherently enters as a significant feature of the problem, the conditions for which are discussed in Sec. 10.1, these steady-state solutions may be considered as subject to synthesis in continuous sequences to correspond to variations with time in the boundary conditions, the time entering in all the expressions as a parameter rather than as an independent variable. Each instantaneous pressure distribution and associated flux will correspond to the boundary conditions at the same instant, as if these conditions had been maintained previously for an indefinitely long time. Although this treatment of time variations will be rigorous only if the liquid is strictly incompressible, it will also suffice, from a practical point of view, in the discussion of all such problems where the compressibility could be neglected if the system were actually in a steady state.

It is also to be noted that the fundamental differential equations (4), (6), and (7) have been based upon the implicit assumption that the flow systems are of fixed geometry. If, however, as in the case of certain gravity-flow systems, the liquid leaving the original body of porous medium is not replaced, the geometrical boundaries of the region of interest will vary in such a way as to give it a continuously decreasing volume. Although this type of problem is explicitly involved in the study of fluctuations in ground-water levels, which is of great practical interest when applied to the questions of water logging, irrigation, etc., it is unfortunately beset with such analytical difficulties that no satisfactory solutions for even the simplest cases are yet available. The extension of the Dupuit-Forchheimer theory which has been frequently used to treat such problems, and which is outlined in Sec. 6.17, involves so many questionable assumptions that reproductions of the analyses based on this theory hardly seem to be warranted, even though nothing better has yet appeared.

[1] *Cf.* footnote on p. 129 and Eqs. 3.3(4) and 3.3(5) for the interpretation of the symbol ∇.

We shall, therefore, omit in this work any discussions of ground-water-level fluctuations, with the hope that this very omission will stimulate future research on this important problem.

Finally, it may be observed that although the analytical bases —Eqs. (4), (6), and (7)—of the treatments to be given in the following chapters of specific flow problems have been derived on the explicit assumption of the validity of the generalized Darcy's law—Eq. 3.3(5)—or that the flow is "viscous," the range of validity of this law will cover almost all problems of practical interest, as pointed out in Sec. 2.2. In fact, we may define the scope of this work as consisting of those flow problems which are governed by Darcy's law.

3.5. Boundary and Initial Conditions.—As is well known from the theory of partial differential equations, Eqs. 3.4(4), 3.4(6), and 3.4(7) have infinite numbers of solutions. Furthermore, the solutions of the linear Eqs. 3.4(4) and 3.4(6) may be combined linearly with arbitrary constant coefficients to give still other solutions. The question naturally arises as to how one may choose, among these infinities of solutions, those applying to any particular problem. Without any detailed analysis it is clear that the choice of the solution must be related to that which is "particular" to the problem of interest. Since all the problems we shall consider are of the same dynamical character and are all governed by the Eqs. 3.4(4) to (7), their differences must necessarily refer to the differences in the boundaries defining the fluid system and to the detailed physical conditions that are to be imposed at these boundaries and at the initial instant when the "boundary conditions" are introduced.

These boundaries, it should be noted, are not necessarily impermeable walls confining the fluid to a given region in space. Rather they are, in general, geometrical surfaces at all points of which either the fluid velocity or the velocity potential or a given function of both may be considered as known. Only in the special cases where over parts of these geometrical surfaces the normal[1] velocity vanishes do these parts correspond to physically impermeable boundaries. Further, these surfaces may lie entirely in the finite regions of space, in which cases the physical problem is the determination of the velocity and potential dis-

---

[1] By the "normal" velocity at the surface is meant the component perpendicular to the surface.

tribution in the interior for given "boundary conditions" on the boundary surface; or the region of interest may extend to infinity, being bounded by closed surfaces in finite regions of space. Thus we may be interested in one case in the velocity and potential distribution in the interior of a spherical region for a given potential or velocity distribution on the surface of the sphere, or one may focus his attention on the space exterior to the sphere extending to another surface enclosing the sphere, or even to infinity. In the last case the conditions to be imposed on the solution at infinity might consist, for example, in the requirement that the potential or velocity vanish in a preassigned manner.

When the problem is one of steady states, as will necessarily be the case when strictly incompressible liquids are treated, in view of the absence of the independent variable t from Eq. 3.4(4), a specification of the "boundary conditions"—the preassigned values of the pressure or velocity potential,[1] or of the normal velocities, or of a linear combination of them, at all points of the boundaries of the system—suffices to determine uniquely the pressure or potential distribution in the interior of the region defined by the boundaries. However, when the system is one of unsteady states, with the density distribution in the system changing with the time, one must also specify the "initial conditions"—the initial density distribution with which the system begins its history. For clearly the densities, and hence pressures, at any finite time in two systems with the same boundary conditions will be quite different if, for example, in one case the system has a uniform density at an arbitrary initial instant while in the other case the density, at that same initial instant, has some arbitrary nonuniform distribution.

On the other hand, it is to be observed that every nonsteady-state system with boundary conditions tending to fixed values will ultimately approach, as the time progresses, a steady-state distribution, independent of the initial conditions and determined

---

[1] The terms pressure and velocity potential are used here rather loosely and interchangeably as both satisfy Laplace's equation, Eq. 3.4(4), and either may be taken as the fundamental physical dependent variable characterizing the incompressible-liquid system. The final choice in the actual analysis is to be made on the basis of convenience, which in turn will be largely determined by the role played by gravity in affecting the fluid motion.

only by the limiting values of the boundary conditions.[1] That is, after an infinitely long time all flow systems with the same geometry and with the same limiting fixed boundary conditions, and not containing in its interior any fluid sources or sinks (regions of infinitesimal volume where fluid is introduced or removed from the system), will have the same steady-state density and pressure distribution. Now the rate at which the effect of the initial conditions disappears and the steady-state distribution is established is essentially determined in a given porous medium by the effective compressibility of the fluid, the rate increasing as the compressibility decreases. It is really because of this fact that "initial conditions" are not necessary for the treatment of incompressible liquids. For an incompressible liquid may be considered as a liquid with vanishing compressibility. Hence, if there were an initial nonsteady-state pressure distribution in the system, it would be immediately removed and the steady-state distribution established. The consideration of nonsteady-state problems of incompressible liquids with fixed-boundary conditions would thus automatically transform itself immediately into the steady-state problem for the given boundary conditions.

For the same reason the pressure distribution in an incompressible-liquid system with boundary values of the pressure or velocities varying with the time will go through a continuous succession of steady-state distributions, each corresponding to the instantaneous conditions at the boundaries. In the case of compressible liquids or in the case of gases, however, there can be no steady states unless the boundary conditions are permanently fixed, since the changes in the boundary values will not be immediately transferred to the remainder of the system. Owing to the compressibility of the fluids, it will in fact require an infinite time for the rigorously *complete* redistribution of the densities and the reestablishment of steady-state conditions.

It is, therefore, seen that ultimately a physical problem in the flow of fluids through a porous medium is in general defined by (1) a geometrical statement of the boundaries of the region in space for which a solution is desired; (2) a specification of the "boundary conditions" to be satisfied on the boundary; and (3)

[1] It is assumed here that the total *algebraic* flux through the system tends ultimately to a vanishing value.

an assignment of the density and hence pressure distribution at the initial instant. Denoting the boundaries by S and the dependent variable p, Φ, or γ by Φ, the boundary conditions of physical interest may take one of the following forms:

$$
\left.
\begin{aligned}
&(a) \quad \Phi \text{ is given on } S \\
&(b) \quad \frac{\partial \Phi}{\partial n} \text{ is given on } S \ (n \text{ is normal to } S) \\
&(c) \quad \frac{\partial \Phi}{\partial n} + h\Phi \text{ is given on } S \ [h \text{ may be a function of } (xyz)],
\end{aligned}
\right\} \quad (1)
$$

where these boundary values may or may not depend on the time.

The physical problem then reduces to the analytical one of finding a function $\Phi(x, y, z)$ which will satisfy Eq. 3.4(4), 3.4(6), or 3.4(7), depending on the nature of the fluid, and at the same time the particular boundary condition (1) that has been chosen, and the initial condition, if the problem is one in the nonsteady-state flow of a compressible liquid or of a gas. Fortunately, it may be proved that once such a function has been found, there are no others that will satisfy all the conditions. One can, therefore, be certain that whatever the method used to find the solution, every other method must necessarily lead to the same result.

3.6. Analogies with Other Physical Problems.—It has been seen that the general problem of the steady-state flow of fluids through porous media may be reduced to the solution of Laplace's equation, in the dependent variables p, γ, or $\gamma^{\frac{1+m}{m}}$, with appropriate boundary conditions, such as are given by Eq. 3.5(1). Since this equation is so well known in other branches of physics, the general method of solution and in some cases even the explicit solutions themselves for a number of flow problems of practical interest may be taken over from those that have already been derived for other purposes in other branches of physics by simply translating them into their proper hydrodynamic equivalents.

We, therefore, present in Table 13 the formal correspondence between the hydrodynamics of the steady-state fluid flow through porous media and the problems of steady-state heat flow, electrostatics, and current flow in continuous conductors. These analogies may be of assistance in visualizing the flow problems to those already familiar with similar problems in heat flow, electrostatics, or current conduction. There are still other problems such as certain cases in the theory of the torsion of elastic rods or in the flow of viscous liquids according to the classical hydrodynamics which are also governed by Laplace's equation, but the

TABLE 13.—CORRESPONDENCES BETWEEN THE FLOW OF AN INCOMPRESSIBLE LIQUID THROUGH A POROUS MEDIUM AND HEAT CONDUCTION, ELECTROSTATICS, AND CURRENT CONDUCTION

Hydrodynamics of steady-state flow through porous media (incompressible liquids)	Heat conduction	Electrostatics	Current conduction
Pressure: p	Temperature: u	Electrostatic potential: Φ	Voltage (potential): V
Negative pressure gradient: $-\nabla p$	Negative temperature gradient: $-\nabla u$	Field-strength vector: $\bar{E} = -\nabla\Phi$	Negative potential gradient: $-\Delta V$
$\dfrac{\text{Permeability}}{\text{Viscosity}}: \dfrac{k}{\mu}$	Thermal conductivity: k	Dielectric constant $\dfrac{\epsilon}{4\pi}$: $\dfrac{\epsilon}{4\pi}$	Specific conductivity: σ
Velocity vector: $\bar{v} = -\dfrac{k}{\mu}\nabla p$ (Darcy's law)	Rate of heat transfer: $\bar{q} = -k\nabla u$ (Fourier's law)	Dielectric displacement: $\dfrac{\epsilon}{4\pi}\bar{E} = -\dfrac{\epsilon}{4\pi}\nabla\Phi$ (Maxwell's law of dielectric displacement)	Current vector: $\bar{I} = -\sigma\nabla V$ (Ohm's law)
Equipressure surface: $p = $ const.	Isothermal surface: $u = $ const.	Equipotential surface: $\Phi = $ const.	Equipotential surface: $V = $ const.
Impermeable boundary or streamline: $\dfrac{\partial p}{\partial n} = 0$	Insulated surface or line of heat flow: $\dfrac{\partial u}{\partial n} = 0$	A tube or line of force: $\dfrac{\partial \Phi}{\partial n} = 0$	Free or insulated surface or tube or line of flow: $\dfrac{\partial V}{\partial n} = 0$

more familiar examples of the table will suffice to illustrate the general nature of the analogies. For convenience the specific case of incompressible-liquid flow has been chosen in which the dependent variable is the pressure p, gravity effects being neglected.

There is one feature, however, of the hydrodynamics of flow through porous media which has no immediate analogue in the other physical problems outlined in the table. It is the potential V in Eq. 3.3(3) which, when it enters at all in our problems of flow, will represent a gravity potential, and hence have the form

$$V = -\gamma g z, \tag{1}$$

where z represents the vertical axis directed downward.   Although it is formally taken account of by the introduction of the velocity potential Φ, it must be considered in formulating the boundary conditions which, physically, present themselves frequently more naturally in terms of the pressure.   Thus, if a liquid is flowing into a cavity that is kept drained of the liquid, the pressure will be uniform over the surface of the cavity, whereas Φ will vary linearly with z.   On the other hand, if the cavity is kept filled with the liquid in hydrostatic equilibrium, the pressure over the surface of the cavity will increase linearly with z, while the potential function Φ will be constant over the surface.   If gravity be neglected, however, the pressure will be uniform over the surface of the cavity in both cases.   In general, either the pressure p or the potential Φ may be used, and in the following chapters we shall freely change from one to the other, depending on the particular problems under consideration.

3.7. Other Coordinate Systems.—Since the fluid systems to be studied in the following chapters will frequently possess special types of symmetry, it will be convenient to express the basic Laplace's equation in coordinate systems in which the symmetry of the system can be appropriately expressed.   We shall, therefore, rewrite the equation

$$\frac{\partial^2 \Phi}{\partial x^2} + \frac{\partial^2 \Phi}{\partial y^2} + \frac{\partial^2 \Phi}{\partial z^2} = 0$$

in cylindrical and spherical coordinates, give their relations to the rectangular Cartesian coordinates (x, y, z), and, on the assumption that Φ represents a velocity potential, we shall give the velocity components in the new coordinate systems.

Cylindrical coordinates (r, θ, z), (Fig. 22).

$$\left. \begin{array}{lll} r = \sqrt{x^2 + y^2}; & \theta = \tan^{-1} \dfrac{y}{x}; & z = z, \\[2mm] x = r \cos \theta; & y = r \sin \theta; & z = z. \end{array} \right\} \tag{1}$$

$$v_r = -\frac{\partial \Phi}{\partial r}; \qquad v_\theta = \frac{-1}{r} \frac{\partial \Phi}{\partial \theta}; \qquad v_z = -\frac{\partial \Phi}{\partial z}. \tag{2}$$

$$\nabla^2 \Phi = \frac{1}{r} \frac{\partial}{\partial r} \left(r \frac{\partial \Phi}{\partial r} \right) + \frac{1}{r^2} \frac{\partial^2 \Phi}{\partial \theta^2} + \frac{\partial^2 \Phi}{\partial z^2} = 0. \tag{3}$$

Spherical coordinates (r, θ, χ), (Fig. 23).

$$r = \sqrt{x^2 + y^2 + z^2}; \quad \theta = \tan^{-1}\frac{\sqrt{x^2 + y^2}}{z}; \quad \chi = \tan^{-1}\frac{y}{x}, \bigg\}(4)$$
$$x = r \sin \theta \cos \chi; \quad y = r \sin \theta \sin \chi; \quad z = r \cos \theta. \bigg\}$$

$$v_r = -\frac{\partial\Phi}{\partial r}; \quad v_\theta = -\frac{1}{r}\frac{\partial\Phi}{\partial\theta}; \quad v_\chi = -\frac{1}{r \sin \theta}\frac{\partial\Phi}{\partial\chi}. \quad (5)$$

$$\nabla^2\Phi = \frac{1}{r^2}\frac{\partial}{\partial r}\left(r^2\frac{\partial\Phi}{\partial r}\right) + \frac{1}{r^2 \sin \theta}\frac{\partial}{\partial\theta}\left(\sin \theta\frac{\partial\Phi}{\partial\theta}\right) + \frac{1}{r^2 \sin^2 \theta}\frac{\partial^2\Phi}{\partial\chi^2} = 0.$$
$$(6)$$

These equations show that if the fluid system is symmetrical about an axis, the symmetry is readily expressed in the system

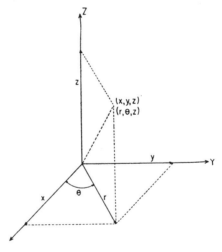

Fig. 22.

of cylindrical coordinates by taking the z axis along the axis of symmetry and setting $\frac{\partial\Phi}{\partial\theta} = 0$. On the other hand, if it possesses spherical symmetry, this is naturally expressed by taking the origin of a system of spherical coordinates at the center of symmetry and dropping the terms in $\frac{\partial\Phi}{\partial\theta}$ and $\frac{\partial\Phi}{\partial\chi}$ in Eq. (6).[1]

---

[1] The transformation to general curvilinear coordinates is given in Appendix II.

3.8. Summary.—Before one can solve any specific problems in the flow of fluids through porous media, one must develop a general formulation of the hydrodynamics of this type of fluid flow. Any such development may be considered as the reformulation of the well-known basic definitions and laws of mechanics in hydrodynamic terms so as to be applicable to the flow of fluids. These imply, in the first place, an explicit requirement of the flow system that it obey the law of the conservation of matter. It must, therefore, satisfy the equation of continuity [*cf.* Eq. 3.1(1)], which is the analytical statement of the law of the conservation of matter.

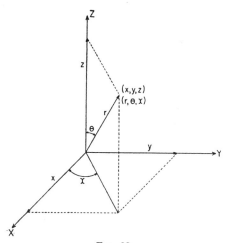

F**IG**. 23.

Secondly, one must define thermodynamically the nature of the fluid one is interested in, and the character of the flow. The nature of the fluid may in general be defined by a relation between the pressure, density, and temperature of the fluid [*cf.* Eq. 3.1(3)], which may be termed its "equation of state." Thus the statement that the density is constant is the "equation of state" defining an incompressible fluid, or Boyle's law may be taken as the equation of state when one is interested in the flow of an ideal gas. The thermodynamic character of the flow may be similarly defined by a relation between the pressure, density, and temperature, such as the statements: The temperature is constant for isothermal flow; or, it varies as a certain power of the density, for adiabatic flow.

Finally, one must specify the dynamical reactions of the fluid to pressure gradients or external forces. This involves essentially a hydrodynamical restatement of Newton's first law of motion. This characterization is the most specific of all the requirements. Thus, whereas all fluids must satisfy the equation of continuity, and large classes may be governed by a single equation of state, one and the same fluid may be given different dynamical characteristics, depending on the conditions under which it is flowing and the medium in which it is flowing. The classical hydrodynamics, for example, defines the dynamical reaction of a fluid by means of the Stokes-Navier equations [*cf.* Eq. 3.2(1)]. These involve the resolution of the forces to which the fluid is subject into those of the pressure gradients, the external "body forces," as gravity, and forces due to the internal viscous friction of the fluid, this latter being determined by its viscosity.

This analysis, in principle, applies to all kinds of fluid motion, and even when the fluid is passing through a porous medium. However, the resulting equations can only be solved in relatively simple cases, and present unsurmountable mathematical difficulties when applied to anything as involved as the flow through the labyrinths of passages in a porous medium. Fortunately these analytical difficulties are not as serious as they seem, since we are after all not interested in the details of the flow as applied to the individual pores. Of practical interest, rather, are only the macroscopic features of the flow as averaged over a large number of pores.

These resultant average effects of the flow through the individual channels have been studied empirically; these studies have led to the exceedingly simple result—Darcy's law—that the velocity of the fluid at any point in a porous medium is directly proportional to the pressure gradient at that point, all quantities representing averages over a large number of pores [*cf.* Eq. 3.3(1)].[1] This, then, is the empirical macroscopic equiva-

---

[1] This result, of course, is valid only within certain limits of the fluid velocity, depending on the average grain or pore size of the medium and the density and viscosity of the fluid, and becomes increasingly less exact as these limits are exceeded. However, in almost all problems of practical interest the fluid velocities actually turn out to be definitely smaller or at the most of the same order as the limiting velocities, and hence are in the "viscous" or macroscopically "streamline" range (*cf.* Sec. 2.2).

lent of the dynamical definition of the nature of the flow of viscous fluids underlying the classical hydrodynamics, and forms the dynamical basis for the hydrodynamics of the viscous flow of homogeneous fluids through porous media.

Generalizing the original results of Darcy so as to have a three-dimensional form and to include the effect of gravity on the flow [*cf.* Eq. 3.3(2)], and then applying the equation of continuity and the equations of state for the various types of fluids, it is found that the pressure and densities must satisfy the following equations:

Incompressible liquids:

$$\frac{\partial^2 \Phi}{\partial x^2} + \frac{\partial^2 \Phi}{\partial y^2} + \frac{\partial^2 \Phi}{\partial z^2} = \frac{\partial^2 p}{\partial x^2} + \frac{\partial^2 p}{\partial y^2} + \frac{\partial^2 p}{\partial z^2} = 0;$$

$$\Phi = \frac{k}{\mu}(p + V), \qquad [cf. \text{ Eq. } 3.4(4)]$$

Compressible liquids:

$$\frac{\partial^2 \gamma}{\partial x^2} + \frac{\partial^2 \gamma}{\partial y^2} + \frac{\partial^2 \gamma}{\partial z^2} = \frac{f\beta\mu}{k}\frac{\partial \gamma}{\partial t}, \qquad [cf. \text{ Eq. } 3.4(6)]$$

Gases:

$$\frac{\partial^2 \gamma^{(1+m)/m}}{\partial x^2} + \frac{\partial^2 \gamma^{(1+m)/m}}{\partial y^2} + \frac{\partial^2 \gamma^{(1+m)/m}}{\partial z^2} = \frac{(1+m)f\mu\gamma_0^{1/m}}{k}\frac{\partial \gamma}{\partial t},$$

$$[cf. \text{ Eq. } 3.4(7)]$$

where Φ and p are the "velocity potential" and the fluid pressure at (x, y, z), V the potential of the body forces, as gravity, acting on the fluid, γ the fluid density, t the time variable, k the permeability of the porous medium, f its porosity, β the compressibility of the liquid, μ the fluid viscosity, γ_0 the gas density at unit pressure, and m is a constant defining the thermodynamic nature of the flow, having the value 1 for isothermal flow and 0.71 for air in adiabatic flow. In all the cases governed by the above equations the porous medium is supposed to be isotropic and homogeneous.

All problems of the flow of homogeneous fluids through porous media under viscous conditions are governed by one or another of these equations. Specific problems are characterized by the geometry of the region in which the flow is taking place, the "boundary conditions" that are preassigned at the boundaries of these regions, and the initial distributions of pressure or density with which the system begins its history, if the problem is one of nonsteady states. The "boundary conditions" consist in specifications of the value of the dependent variable over the boundaries, or the normal derivatives of the dependent variable which represent the normal fluid velocities at the boundaries. When these conditions are stated, and they may be preassigned quite arbitrarily, the problem is analytically determinate, and there exists an unique solution which will satisfy the differential equation and both the boundary and initial conditions (Sec. 3.5).

Now that the hydrodynamics of fluid flow through porous media has been generally formulated in the form of the partial differential equations for the pressure or density, methods have to be developed for their solution. It is, therefore, of interest to observe that the Laplace equation, which governs all cases of steady-state flow, is already well known in other branches of physics such as the theories of steady-state heat conduction, electrostatics, and steady-state electrical conduction. Since many problems have already been solved in studies of these subjects, these solutions can be carried over to the subject of fluid flow through porous media if we know how to reinterpret the quantities of interest from one subject to another. Formal analogies are, therefore, presented, relating such quantities as temperature, voltage, current, dielectric constant, etc., with the corresponding quantities in our hydrodynamic system (Sec. 3.6). Finally, anticipating that some of the problems of interest will possess special types of symmetry, Laplace's equation has been rewritten in other coordinate systems in which certain kinds of symmetry find a more natural expression than in the Cartesian coordinate system (Sec. 3.7).

PART II
STEADY-STATE FLOW OF LIQUIDS

CHAPTER IV

TWO-DIMENSIONAL FLOW PROBLEMS AND POTENTIAL-THEORY METHODS

4.1. Introduction.—Next to the most elementary type of flow problem—the case of linear flow, used to establish Darcy's law and discussed in Chap. II[1]—the simplest is that in which the flow is of a two-dimensional character. By this is meant that the velocity distribution vector $\bar{v}$ in the fluid system varies only with two of the rectangular coordinates and is independent of the third. Physically, of course, all fluid systems necessarily extend in three dimensions, but the significant characteristic of two-dimensional systems is that all the features of the motion may be observed in a single plane, the motion being identically the same in parallel planes.

The two-dimensional problems of practical interest are of two general types. The first is concerned with horizontal two-dimensional motions in which $\bar{v}$ is independent of the vertical coordinate z. Such problems arise in the discussion of sands of uniform thickness which are completely filled with fluid and pierced by wells penetrating the whole sand thickness. The flow must then necessarily be two-dimensional. Furthermore, it follows that even though gravity be acting on each fluid element, the fluid either moves bodily in a vertical direction or it has no vertical velocity anywhere. It is clear, therefore, that in either case gravity is of no significance in this type of motion and one may use, quite rigorously, the pressure p as the equivalent[2] of the velocity potential. The corresponding problems in which the fluid levels at the outflow surfaces fall below the top of the sand do explicitly involve the effect of gravity and will be treated in Chap. VI.

[1] While no formal treatment of the case of linear flow as such was given, all the significant features are actually contained in the discussion of Sec. 2.5.

[2] The factor k/μ must be, of course, multiplied into p in order to get the correct *numerical* values of the internal velocities in any particular flow system.

From the above considerations it follows that the fundamental equations in rectangular coordinates governing horizontal two-dimensional fluid motions in the steady state will be:

$$\frac{\partial^2 p}{\partial x^2} + \frac{\partial^2 p}{\partial y^2} = 0, \tag{1}$$

$$v_x = -\frac{k}{\mu}\frac{\partial p}{\partial x}; \qquad v_y = -\frac{k}{\mu}\frac{\partial p}{\partial y}. \tag{2}$$

The second type of two-dimensional flow problem to be treated in this chapter is characterised by a great extension of the flow system in one horizontal direction without any changes in its dynamical behavior along that direction, so that except near the ends of the region the nature of the flow is the same in all vertical planes cutting the direction of great extension at right angles. Examples of this type arise in the discussion of the seepage and uplift pressure under dams of lengths that are great compared with their thickness. Here the appropriate dynamical variable will be the potential function Φ, as owing to the fact that the flow will take place in vertical planes, the streamlines will be orthogonal to the equipotentials rather than to the equipressure lines. However, either p or Φ can be used in the development of the solutions as both will satisfy the Laplace equation

$$\frac{\partial^2 \Phi}{\partial x^2} + \frac{\partial^2 \Phi}{\partial y^2} = \frac{\partial^2 p}{\partial x^2} + \frac{\partial^2 p}{\partial y^2} = 0; \qquad \Phi = \frac{k}{\mu}(p \pm \gamma g y), \tag{3}$$

with

$$v_x = -\frac{\partial \Phi}{\partial x} = -\frac{k}{\mu}\frac{\partial p}{\partial x}; \qquad v_y = -\frac{\partial \Phi}{\partial y} = -\frac{k}{\mu}\left(\frac{\partial p}{\partial y} \pm \gamma g\right), \tag{4}$$

where y will here be taken as the vertical coordinate, the $\pm$ corresponding to the upward or downward direction of $+y$.

4.2. The Radial Flow into a Well.—By "radial flow" is meant a two-dimensional fluid motion symmetrical about an axis, its details varying only with the distance from the axis of symmetry. This is clearly the simplest case of a two-dimensional motion. However, to take advantage of the symmetry of the problem one must use a coordinate system appropriate to it. This is evidently the cylindrical-coordinate system reduced to two dimensions.

From Eqs. 3.7(2) and 3.7(3) of the previous chapter the fundamental equations in cylindrical coordinates applying to two dimensions, and the equivalents of Eqs. 4.1(1) and 4.1(2), are:

$$\frac{1}{r}\frac{\partial}{\partial r}\left(r\frac{\partial p}{\partial r}\right) + \frac{1}{r^2}\frac{\partial^2 p}{\partial \theta^2} = 0; \tag{1}$$

$$v_r = -\frac{k}{\mu}\frac{\partial p}{\partial r}; \qquad v_\theta = -\frac{k}{\mu r}\frac{\partial p}{\partial \theta}. \tag{2}$$

For the more particular case of radial flow, it follows from the above definition that

$$\frac{1}{r}\frac{\partial}{\partial r}\left(r\frac{\partial p}{\partial r}\right) = 0; \tag{3}$$

$$v_r = -\frac{k}{\mu}\frac{\partial p}{\partial r}; \qquad v_\theta = -\frac{k}{\mu r}\frac{\partial p}{\partial \theta} = 0. \tag{4}$$

Integrating Eq. (3), there results first

$$r\frac{\partial p}{\partial r} = \text{const.} = c_1, \tag{5}$$

and then

$$p = c_1 \log r + c_2, \tag{6}$$

as the general expression for the pressure distribution in a system of radial flow.

It was seen, however, in the last chapter that to the differential equations for the pressure or velocity potential one must add a description of the boundaries of the flow system and a statement of the boundary conditions. As was suggested there, these boundaries and boundary conditions are in general quite arbitrary, provided no previous assumptions are made as to the nature of the flow. But, since in the present case it has already been assumed that the flow is radial, we are no longer free as to the corresponding boundaries and boundary conditions. In fact, it is not difficult to show that the only system giving radial flow is that bounded by two concentric circles over each of which the pressure is uniform.

Thus the radial-flow system may be characterized by the conditions (*cf.* Fig. 24)

$$\left.\begin{array}{ll} r = r_w; & p = p_w \\ r = r_e; & p = p_e, \end{array}\right\} \tag{7}$$

where, in practical problems, r_w may correspond to the radius of a well, into which the liquid—oil or water—is flowing. p_w, the pressure at the sand face in the well bore, is to

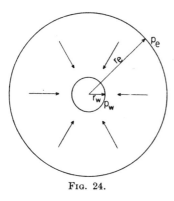

some extent arbitrary and may be controlled within limits by the fluid head maintained in the well bore.[1] The pressure p_e, on the other hand, cannot be manipulated at will. It depends, in general, upon the depth of the liquid-bearing sand below its outcrop, and the history of the production from the sand. The distance r_e is quite arbitrary and is to be chosen as the distance to such points where it is felt a reasonable estimate of p_e can be made.

Fig. 24.

Introducing the conditions (7) into Eq. (6), there result two equations for c_1, c_2:

$$p_w = c_1 \log r_w + c_2,$$
$$p_e = c_1 \log r_e + c_2,$$

whence

$$c_1 = \frac{(p_e - p_w)}{\log r_e/r_w},$$
$$c_2 = \frac{p_w \log r_e - p_e \log r_w}{\log r_e/r_w},$$

so that

$$p = \frac{p_e - p_w}{\log r_e/r_w} \log r + \frac{p_w \log r_e - p_e \log r_w}{\log r_e/r_w}$$
$$= \frac{p_e - p_w}{\log r_e/r_w} \log \frac{r}{r_w} + p_w. \tag{8}$$

Applying now Eq. (4), v_r takes the form

$$v_r = -\frac{k}{\mu} \frac{\partial p}{\partial r} = -\frac{k}{\mu r} \frac{(p_e - p_w)}{\log r_e/r_w}. \tag{9}$$

Finally, the total outflow per unit time from the sand into the well is evidently given by

[1] It must, however, always be kept above the top of the producing sand, or else the effect of gravity must be taken into account (*cf.* Chap. VI).

$$Q = -h \int_0^{2\pi} r v_r d\theta = \frac{2\pi k h(p_e - p_w)}{\mu \log r_e/r_w}, \tag{10}$$

where h is the sand thickness. In terms of Q, p and v_r may, therefore, be written as

$$p = \frac{Q\mu}{2\pi k h} \log \frac{r}{r_w} + p_w, \tag{11}$$

$$v_r = -\frac{Q}{2\pi r h}. \tag{12}$$

As the above equations show, the pressure in a radial-flow system varies with the logarithm of the distance from the center of the well, the velocity varies directly as the net pressure drop, $p_e - p_w$, over the system, and inversely as the radius r, and the production rate or flux through the system is also proportional to the net pressure drop, $p_e - p_w$. All these functions vary inversely with the logarithm of the ratio of the boundary radii. The equipressure curves are circles concentric with the well boundary, and the streamlines are the radii vectors from the center of the internal boundary. These results may be illustrated by the following numerical example, in which it is assumed that

$$p_w = 0; \qquad p_e = 10 \text{ atm.} \cong 150 \text{ lb./sq. in.},$$
$$r_w = \tfrac{1}{4} \text{ ft.}; \qquad r_e = 500 \text{ ft.}; \qquad k = 1 \text{ darcy},$$
$$\mu = 1 \text{ centipoise.}$$

It follows then from Eqs. (8) and (9) that

$$p = \frac{10 \log r/r_w}{\log 2000} \doteq 1.32 \log 4r \text{ atm.,} \tag{13}$$

$$v_r = -\frac{10}{r \log 2000} = -\frac{1.32}{r} \text{ cm./sec.,} \tag{14}[1]$$

$$Q = \frac{20\pi}{\log 2000} = 8.27 \text{ cc./sec./cm. of sand}$$
$$= 136.93 \text{ bbl./day/ft. of sand.} \tag{15}$$

The graphs of Eqs. (13) and (14) are given in Fig. 25.

Although the above development was based upon the direct analytical treatment of the problem of radial flow, it is not with-

---

[1] The negative value of v_r here, as well as in Eqs. (9) and (12), when $p_e > p_w$, simply means that the liquid is flowing towards decreasing values of r.

out interest to outline a somewhat more physical method which is perhaps more illuminating than that of manipulating the differential equation (3). Here one begins with the integrated equation of continuity, which states that the total flow through any circle concentric with the boundaries is a constant, so that

$$2\pi r v_r h = \text{constant} = -Q.$$

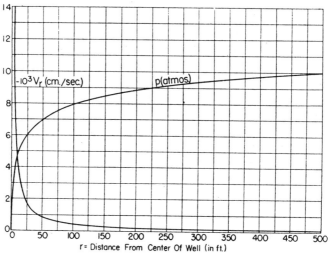

Fig. 25.—Radial velocity (v_r) and pressure (p) distributions about a well completely penetrating a homogeneous sand of uniform thickness, with $k/\mu = 1$, and the pressures at $r = \frac{1}{4}$ ft., 500 ft., being 0 and 10 atm., respectively.

Hence there follows at once that

$$v_r = -\frac{Q}{2\pi r h}. \tag{12}$$

Applying now Eq. (4),

$$v_r = -\frac{k}{\mu}\frac{\partial p}{\partial r} = -\frac{Q}{\cdot 2\pi r h},$$

so that again

$$p = \frac{Q\mu}{2\pi k h}\log\frac{r}{r_w} + p_w. \tag{11}$$

These results are, of course, identical with those obtained before. This derivation, however, shows clearly that the first integration of the differential equation for p, as given by Eq. (5), simply takes care of the equation of continuity, whereas the second integration corresponds to the application of Darcy's law to the velocity distribution and the derivation therefrom of the pressure distribution.

Returning now to the pressure and velocity distributions, one may note a few of the salient features indicated in Fig. 25. Thus it is seen that the pressure rises very rapidly, whereas the velocity falls steeply with increasing distance from the well for small values of the latter. On the other hand, at great distances from the well the variations in pressure and velocity are very slow, the pressure slowly approaching its maximum value at the external boundary,[1] while the velocity falls still more gradually to its minimum at the external boundary.

With respect to the absolute values of the velocity, it is to be noted that for the numerical case just considered, which corresponds to a production rate of 1,370 bbl./day for a 10-ft. sand, the velocity even at the well bore is not greater than 0.17 cm./sec. Hence for a well producing water at the above rate from a sand of effective grain diameter of 0.05 cm. the Reynolds number even at the sand face will be only 0.85, and thus will still be less than the critical value at which Darcy's law begins to fail (*cf.* Sec. 2.2).

With regard to the production rate Q, it is seen from Eq. (10) that it is directly proportional to the permeability of the sand. Its logarithmic variation with the well and external-boundary radii implies that large variations in these are required in order

---

[1] The fact that the p of Eq. (11) continues to increase indefinitely as r is taken increasingly larger than r_e must not be interpreted as an indication of an error in the fundamental theory leading to Eq. (11). For all that can be demanded of any analytical theory is that it give physically correct results *within* the boundaries of the regions for which the solutions are derived. The extrapolation of the solution beyond these boundaries need have no physical significance. Hence, if it is desired that at $10r_e$, for example, the pressure should be p_e, this requirement must be explicitly taken as a boundary condition for the problem, and the value of p at r_e will then be uniquely fixed by the solution. Similarly, the fact that the p of Eq. (11) becomes infinitely large ($-\infty$) as r is made to vanish is of no significance, as Eq. (11) can inherently claim no validity for $r < r_w$ and must not be used for $r < r_w$. In fact, the freedom of the well bore from sand necessarily invalidates the solution derived for flow in a sand from being applicable within the well bore.

appreciably to affect the total production rate. Thus, to double the production rate by increasing the well radius, the latter must be increased to

$$r_w' = \sqrt{r_e r_w},$$ (16)

and to get the same effect by decreasing the external boundary radius one must shrink it to

$$r_e' = \sqrt{r_e r_w}.$$ (17)

For the above example, this means that to double the production rate it is necessary either to enlarge the well from $\frac{1}{4}$ to 11.18 ft. radius or to shrink the external boundary from 500 to 11.18 ft.

The practical significance of this slow variation in Q with r_w and r_e is that relatively small uncertainties in the values of r_w and r_e for any actual case will introduce inappreciable errors in the final estimates for Q. Hence any reasonable assumption for r_w and r_e, such as $\frac{1}{4}$ and 500 ft., can be made without fear of introducing large uncertainties in the estimates for Q (*cf.* Sec. 2.11). There is also the further interest in the above result in that it shows that to increase appreciably the production capacity of an artesian or oil well by increasing the size of the well is rather impractical since, as in the above example, it may require an enlargement as high as 40-fold just to double the production capacity.

It may be mentioned finally that according to the analogies of the last chapter, Eqs. (8) and (9) correspond to the potential and current-density distributions between two coaxial cylinders separated by a medium of conductivity k/μ. Equation (10) gives the total current flowing through the system.

4.3. Fourier Series.—In order to generalize the discussion of the last section to cases where the flow into the well is not exactly radial, it will be necessary to digress and outline the essential facts of the theory of Fourier series. Since we are interested only in its application to the solution of Laplace's equation 4.2(1), all proofs will be omitted, and the discussion will be confined to statements of the theoretical results and a few examples which illustrate them.

The fundamental existence theorem defining a Fourier series may be stated as:

Any function $f(x)$ with only a finite number of finite discontinuities and of maxima and minima in the interval $-c \leqslant x \leqslant c$ may be expressed in that interval by the Fourier series

$$f(x) = \frac{1}{2}a_0 + \sum_1^\infty \left(a_n \cos \frac{n\pi x}{c} + b_n \sin \frac{n\pi x}{c} \right), \qquad (1)$$

where

$$a_n = \frac{1}{c} \int_{-c}^{+c} f(x) \cos \frac{n\pi x}{c} dx; \qquad b_n = \frac{1}{c} \int_{-c}^{+c} f(x) \sin \frac{n\pi x}{c} dx. \qquad (2)^1$$

For values of x where $f(x)$ is continuous, the series on the right side of Eq. (1) equals $f(x)$. For values of x where $f(x)$ is discontinuous, the series in Eq. (1) has a sum equal to the algebraic average of $f(x)$ at the two sides of the discontinuity.

If the function with the above restrictions is defined only in the range of $0 \leqslant x \leqslant c$, it may be expressed in the interval $0 \leqslant x \leqslant c$ by either of the series

$$\left. \begin{aligned} f(x) &= \frac{1}{2}a_0 + \sum_1^\infty a_n \cos \frac{n\pi x}{c}, \\ a_n &= \frac{2}{c} \int_0^c f(x) \cos \frac{n\pi x}{c} dx, \end{aligned} \right\} \qquad (3)$$

or

$$\left. \begin{aligned} f(x) &= \sum_1^\infty b_n \sin \frac{n\pi x}{c}, \\ b_n &= \frac{2}{c} \int_0^c f(x) \sin \frac{n\pi x}{c} dx. \end{aligned} \right\} \qquad (4)$$

[1] Assuming Eq. (1), the Eqs. (2) can be readily derived by multiplying both sides of Eq. (1) by a term as $\cos \frac{m\pi x}{c} dx$ and integrating from $-c$ to $+c$. This will give the coefficient a_m and a similar procedure with $\sin \frac{m\pi x}{c} dx$ will give b_m. For a rigorous presentation of the theory of Fourier series the reader may consult "Fourier's Series and Integrals," 3d ed., 1930, by H. S. Carslaw, or Chap. IX of "Modern Analysis," 4th ed., 1927, by E. T. Whittaker and G. N. Watson, while the practical applications of the theory will be found well illustrated in "Fourier's Series and Spherical Harmonics," (1893) by W. E. Byerly. The above phraseology, which is admittedly loose, has been used here in order to avoid introducing the extended terminology required for a rigorous discussion.

The process of deriving these expressions in actual cases will be illustrated by a few examples.

1. *General Fourier Series.*—Suppose

$$f(x) = 0: \quad -\pi \leqslant x \leqslant 0;$$
$$= x: \quad 0 \leqslant x \leqslant \pi,$$

as is indicated in Fig. 26.

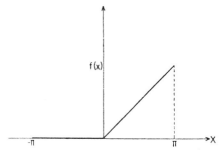

FIG. 26.—Plot of $f(x) = 0 : -\pi \leqslant x \leqslant 0; f(x) = x : 0 \leqslant x \leqslant \pi$.

Referring to Eqs. (1) and (2), it is clear that for this example $c = \pi$. Hence the coefficients are

$$a_n = \frac{1}{\pi}\int_{-\pi}^{+\pi} f(x) \cos nx dx = \frac{1}{\pi}\int_0^{\pi} x \cos nx dx,$$

$$= -\frac{2}{n^2\pi}: \quad n \text{ is odd};$$

$$= 0: \quad n \text{ is even}.$$

a_0, however, must be computed separately as

$$a_0 = \frac{1}{\pi}\int_0^{\pi} x dx = \frac{\pi}{2}.$$

Similarly, b_n has the value:

$$b_n = \frac{1}{\pi}\int_{-\pi}^{+\pi} f(x) \sin nx dx = \frac{1}{\pi}\int_0^{\pi} x \sin nx dx = \frac{(-1)^{n+1}}{n}.$$

Putting these into Eq. (1), $f(x)$ takes the form:

$$f(x) = \frac{\pi}{4} - \frac{2}{\pi}\left(\cos x + \frac{\cos 3x}{3^2} + \frac{\cos 5x}{5^2} + \cdots\right)$$
$$+ \left(\sin x - \frac{\sin 2x}{2} + \frac{\sin 3x}{3} - \frac{\sin 4x}{4} + \cdots\right). \quad (5)$$

It is well to point out here the following precaution: It is true that for every value of x between $-\pi$ and 0 the series in Eq. (5) will actually sum to the value 0, and for all values of x between $x = 0$ and $x = +\pi$ it will sum to the value of x. However, for values of $x < -\pi$, it will not continue to equal 0, nor will it continue to equal x for values of $x > \pi$. Rather it will reproduce in the region outside of the interval $-\pi \leqslant x \leqslant \pi$ its behavior in that interval. In particular, it has the property that:

and in fact:

$$f(x + 2\pi) = f(x),$$
$$f(x + 2n\pi) = f(x),$$

(6)

as is shown in Fig. 27.

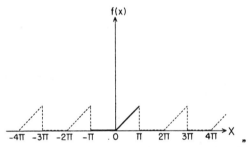

FIG. 27.—Plot of the general Fourier-series representation of Fig. 26.

Thus it is seen that whereas a Fourier series is numerically equivalent at every point to an arbitrary function within the interval in which the function is defined, it will reproduce that function *periodically* in equal intervals outside of the range of definition. Hence, if one wishes a Fourier series to represent a function about a point x_0, the function must be defined over an interval that at least includes x_0. The next example will be that of a "cosine series," that is, a Fourier series containing only cosine terms, as Eq. (3).

2. *Cosine Series.*—Suppose

$$f(x) = b; \qquad 0 \leqslant x \leqslant \pi$$
$$= 0; \qquad \pi \leqslant x \leqslant 2\pi,$$

as is indicated in Fig. 28.

For this case, $c = 2\pi$, and by Eq. (3),

$$
\begin{aligned}
a_n &= \frac{1}{\pi}\int_0^{2\pi} f(x)\cos\frac{nx}{2}dx \\
&= \frac{b}{\pi}\int_0^{\pi} \cos\frac{nx}{2}dx \\
&= 0, \qquad n \text{ even;} \\
&= \frac{2b(-1)^{\frac{n-1}{2}}}{\pi n}, \qquad n \text{ odd.}
\end{aligned}
$$

Further, $a_0 = b$.

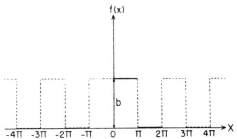

f(x)

b

-4Π -3Π -2Π -Π 0 Π 2Π 3Π 4Π   X

Fig. 28.—Plot of the cosine-series representation of the function: $f(x) = b : 0 \leqslant x \leqslant \pi; f(x) = 0 : \pi \leqslant x \leqslant 2\pi$.

Hence,

$$
f(x) = \frac{b}{2} + \frac{2b}{\pi}\left(\cos\frac{x}{2} - \frac{1}{3}\cos\frac{3x}{2} + \frac{1}{5}\cos\frac{5x}{2} \\
- \frac{1}{7}\cos\frac{7x}{2} + \cdots\right). \tag{7}
$$

There are two points of interest to be noted about this expression. First, it is to be observed that for $x = \pi$ all the trigonometric terms in Eq. (7) vanish, and

$$
f(\pi) = \frac{b}{2}.
$$

Since this is the average value of the function, as was defined above, at the discontinuity, it is thus seen that at a point of discontinuity a Fourier series actually assumes a value equal to the algebraic average of the defining function on the two sides of the discontinuity.

The other point to be noted is that since cos x is an even function of x—its value is unchanged on changing x to $-x$—the Fourier series of Eq. (7) shown in Fig. 28 not only reproduces itself periodically for $x > 2\pi$, but for $x < 0$ it gives an exact reflection of its values in the region $x > 0$. Although the behavior of a Fourier series is of significance only in the region of its definition, it is of interest to observe that the expansion of a function in a cosine series for $0 \leqslant x \leqslant c$ is equivalent to assuming that it is an even function in the double range $-c \leqslant x \leqslant c$. It will presently be seen that in the case of the sine series the situation is quite different.

3. *Sine Series.*—The same function as just analyzed will now be expanded in a sine series. That is, supposing again that

$$f(x) = b; \quad 0 \leqslant x \leqslant \pi,$$
$$= 0; \quad \pi \leqslant x \leqslant 2\pi,$$

Eq. (4) gives

$$b_n = \frac{1}{\pi} \int_0^{2\pi} f(x) \sin \frac{nx}{2} dx = \frac{b}{\pi} \int_0^{\pi} \sin \frac{nx}{2} dx$$
$$= \frac{2b}{n\pi}: \quad n \text{ odd,}$$
$$= 0: \quad n = 4m,$$
$$= \frac{4b}{n\pi}: \quad n = 2m.$$

Hence,

$$f(x) = \frac{2b}{\pi} \left(\sin \frac{x}{2} + \sin x + \frac{1}{3} \sin \frac{3x}{2} + \frac{1}{5} \sin \frac{5x}{2} \right.$$
$$\left. + \frac{1}{3} \sin 3x + \cdots \right) \Bigg\} \quad (8)$$

Here again it follows that for $x = \pi$:

$$f(\pi) = \frac{2b}{\pi} \left(1 - \frac{1}{3} + \frac{1}{5} - \frac{1}{7} + \cdots \right)$$
$$= \frac{b}{2},$$

as in the case of the cosine series.

However, since sin x is an odd function of x—it changes in sign on changing x to $-x$—the series in Eq. (8) gives the negative reflection of $f(x)$ when x is changed to $-x$. The result is a representation as given in Fig. 29. It is thus seen that an expansion in a sine series for $0 \leqslant x \leqslant c$ is equivalent to assuming that $f(x)$ is an odd function in the range $-c \leqslant x \leqslant c$.

It also follows from the general sine series of Eq. (4) that a sine series sums to the value 0 at the ends of the interval of definition, even though the function is defined to have non-vanishing values at the end points. Thus Eq. (8) gives $f(0) = 0$, although the definition of $f(x)$ required that $f(0) = b$. The

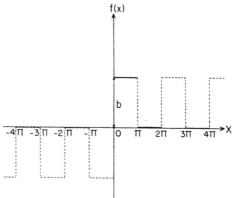

FIG. 29.—Plot of the sine-series representation of the function:
$$f(x) = b : 0 \leqslant x \leqslant \pi; f(x) = 0 : \pi \leqslant x \leqslant 2\pi.$$

cosine series, however, do reproduce at the ends of the intervals the values defined by the original function. Thus it is readily verified that the cosine series of Eq. (7) gives $f(0) = b$, since the

series $\displaystyle\sum_{0}^{\infty} \frac{(-1)^n}{2n + 1} = \frac{\pi}{4}.$

Finally, the following important property of uniqueness should be noted: If two Fourier series with the same interval of definition are equal for all points of that interval, the two series are identically equal and the coefficients of the corresponding trigonometric terms must all be equal. This is equivalent to the statement that if a Fourier series vanishes at all points of the interval of its definition, it vanishes identically.

4.4. Unsymmetrical Flow into a Well.—From a practical point of view the case of strictly radial flow, which implies an exactly uniform pressure imposed on a circular boundary concentric with the well surface, is evidently too idealized to correspond to situations which are likely to occur in practice. Rather it is to be anticipated that even such flow systems which contain but a single well will, in general, have nonuniform pressure distributions over their external boundaries; that the wells will not, in general, lie at the centers of their external boundaries; and that finally the boundaries themselves, over which the pressure distributions are preassigned and known, will be other than circular in shape. In all these cases the flow into the wells will be unsymmetrical and the pressure distributions will depend upon both the azimuthal and radial coordinates. In the following three sections three such typical situations are treated. In the first, we shall still retain for the external boundary a circle concentric with the well, but permit the pressure on this boundary, as well as that over the well surface, to vary in an arbitrary manner, the treatment being based on the theory of Fourier series. The next case will be that of circular but non-concentric boundaries, corresponding to a well displaced from the center of its external boundary; the method of Green's functions will be applied here. Finally, a problem will be treated in which the external boundary is no longer circular, but rather a straight line as in a line drive or edge-water encroachment.

4.5. Arbitrary Pressure Distributions over the Boundaries.— When the flow into a well is not perfectly symmetrical, the problem can no longer be simplified as was done in Sec. 4.2. Rather one must use the more general Eqs. 4.2(1) and 4.2(2) and their solutions, which in general will depend on the angle θ as well as on the radius r. These solutions, it is readily verified, may have one of the following forms:

$$\text{constant}; \quad \log r; \quad r^\alpha \cos \alpha\theta; \quad r^\alpha \sin \alpha\theta; \quad r^{-\alpha} \cos \alpha\theta;$$
$$r^{-\alpha} \sin \alpha\theta,$$

α being a real constant for all problems of practical significance.

Now, since Eq. 4.2(1) is linear—it contains no terms in p or its derivatives to a power higher than the first—every linear combination of individual solutions will also be a solution. The general solution of Eq. 4.2(1) can, therefore, be written in the form

$$p = c_0 \log r + \Sigma r^\alpha (a_\alpha \sin \alpha\theta + b_\alpha \cos \alpha\theta)$$
$$+ \Sigma r^{-\alpha}(c_\alpha \sin \alpha\theta + d_\alpha \cos \alpha\theta). \quad (1)$$

In this equation c_0, a_α, b_α, c_α, d_α, and the α's themselves are all

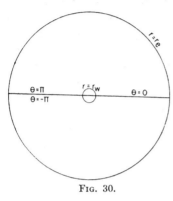

FIG. 30.

constants, independent of r and θ, and are to be chosen so that p or $\dfrac{\partial p}{\partial r}$ assumes preassigned values along two circles $r = r_w, r = r_e$, which bound the region of interest. This can be done by applying the theory of the Fourier series as outlined in Sec. 4.3.

For this purpose it is convenient to suppose that the circle of total angle 2π is divided at the radius $\theta = \pi$, as shown in Fig. 30, so that the interval is $-\pi \leqslant \theta \leqslant \pi$. Then if the values of p at the boundaries are expressed by their Fourier series, the boundary conditions may be written in the form

$$r = r_w: \quad p = p_w = \Sigma (w_n \sin n\theta + x_n \cos n\theta):$$
$$-\pi \leqslant \theta \leqslant \pi,$$
$$r = r_e: \quad p = p_e = \Sigma (e_n \sin n\theta + f_n \cos n\theta):$$
$$-\pi \leqslant \theta \leqslant \pi. \quad \Big\} \quad (2)$$

Before applying these, however, it must be observed first that, from the physical nature of the problem, p itself must be periodic in θ with the period 2π. Hence one can immediately conclude that the α's in Eq. (1) must necessarily take on integral values only, which may be denoted by n, where n is positive. Knowing the values of the α's, the conditions of Eq. (2) can now be imposed on Eq. (1). These give

$$\sum_0^\infty (w_n \sin n\theta + x_n \cos n\theta) = c_0 \log r_w + \sum_0^\infty r_w{}^n(a_n \sin n\theta + b_n \cos n\theta)$$
$$+ \sum_1^\infty r_w{}^{-n}(c_n \sin n\theta + d_n \cos n\theta),$$

$$\sum_0^\infty (e_n \sin n\theta + f_n \cos n\theta) = c_0 \log r_e + \sum_0^\infty r_e^n (a_n \sin n\theta + b_n \cos n\theta)$$
$$+ \sum_1^\infty r_e^{-n} (c_n \sin n\theta + d_n \cos n\theta).$$

Since these equations must be satisfied for all values of θ in the range $-\pi \leqslant \theta \leqslant \pi$, it follows from the properties of the Fourier series outlined in Sec. 4.3 that the coefficients of $\sin n\theta$ and $\cos n\theta$ on the two sides of the equations must be equal. This gives the equations:

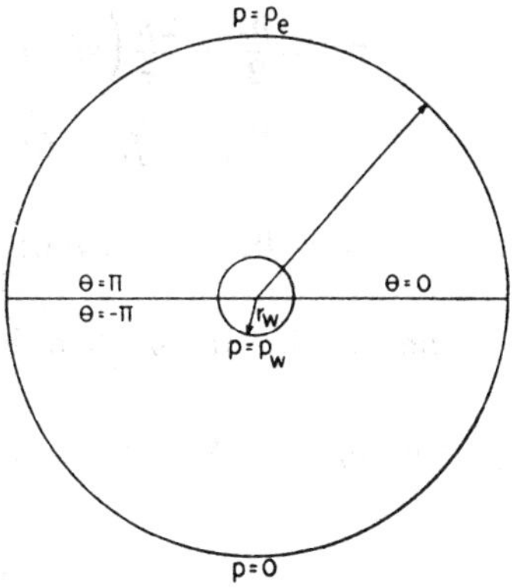

FIG. 31.

$$\begin{aligned} w_n &= a_n r_w^n + c_n r_w^{-n} & n > 0, \\ x_n &= b_n r_w^n + d_n r_w^{-n} & n > 0, \\ e_n &= a_n r_e^n + c_n r_e^{-n} & n > 0, \\ f_n &= b_n r_e^n + d_n r_e^{-n} & n > 0, \\ x_0 &= c_0 \log r_w + b_0, & \\ f_0 &= c_0 \log r_e + b_0, & \\ a_0 &= 0. & \end{aligned}$$

Or, solving for a_n, c_n, b_n, d_n:

$$\left.\begin{aligned} a_n &= \frac{w_n r_e^{-n} - e_n r_w^{-n}}{D_n}, \\ b_n &= \frac{x_n r_e^{-n} - f_n r_w^{-n}}{D_n}, \\ c_n &= \frac{e_n r_w^n - w_n r_e^n}{D_n}, \\ d_n &= \frac{f_n r_w^n - x_n r_e^n}{D_n}, \\ D_n &= r_w^n r_e^{-n} - r_w^{-n} r_e^n, \\ c_0 = \frac{(f_0 - x_0)}{\log r_e/r_w}; &\qquad b_0 = \frac{x_0 \log r_e - f_0 \log r_w}{\log r_e/r_w} \end{aligned}\right\} \tag{3}$$

Hence if the w_n, x_n, e_n, and f_n are known, Eq. (3) gives the constants a_n, b_n, c_n, and d_n, which determine the pressure distribution at all points between the boundaries r_w and r_e. As an example, the following problem will be chosen (*cf.* Fig. 31):

$$\left.\begin{aligned} r = r_w: &\quad p = p_w = \text{const.}: -\pi \leqslant \theta \leqslant \pi, \\ r = r_e: &\quad p = p_e = \begin{cases} p_e: & 0 \leqslant \theta \leqslant \pi \\ 0: & -\pi \leqslant \theta \leqslant 0. \end{cases} \end{aligned}\right\} \tag{4}$$

Although these conditions are a sufficient definition of the problem, they must first be expressed in the form of Fourier series.

Thus from Eqs. (2) it is evident that the expansion of p_w is simply

$$p_w = x_0; \qquad x_n = w_n = 0; \qquad n > 0. \tag{5}$$

p_e, on the other hand, has the expansion

$$
\left.
\begin{aligned}
p_e &= \frac{p_e}{2} + \frac{2p_e}{\pi}\left(\sin\theta + \frac{\sin 3\theta}{3} + \frac{\sin 5\theta}{5} + \cdots\right) \\[2mm]
f_0 &= \frac{p_e}{2}; \qquad f_n = 0: \qquad n > 0 \\[2mm]
e_n &= 0; \qquad n:\text{even}; \qquad e_n = \frac{2p_e}{n\pi}; \qquad n:\text{odd.}
\end{aligned}
\right\} \tag{6}
$$

The solution for p, then, has the form

$$p = c_0 \log r + \sum_0^\infty r^n(a_n \sin n\theta + b_n \cos n\theta)$$

$$+ \sum_1^\infty r^{-n}(c_n \sin n\theta + d_n \cos n\theta), \tag{7}$$

where, by Eqs. (3), (5), and (6),

$$c_0 = \frac{\dfrac{p_e}{2} - p_w}{\log r_e/r_w}; \qquad b_0 = \frac{p_w \log r_e - \dfrac{p_e}{2}\log r_w}{\log r_e/r_u},$$

$$a_0 = 0.$$

$$
\begin{aligned}
a_n &= 0 & n:\text{even,} \\
&= -\frac{2p_e r_w^{-n}}{n\pi D_n} & n:\text{odd,} \\
b_n &= 0 & n > 0, \\
c_n &= 0 & n:\text{even,} \\
&= \frac{2p_e r_w^{\,n}}{n\pi D_n} & n:\text{odd,} \\
d_n &= 0; & D_n = r_w^{\,n} r_e^{-n} - r_w^{-n} r_e^{\,n}.
\end{aligned}
$$

Hence p can be finally expressed as:

$$p = \frac{\dfrac{p_e}{2}\log\dfrac{r}{r_w} + p_w \log\dfrac{r_e}{r}}{\log r_e/r_w} + \frac{2p_e}{\pi}\sum_{n:\text{odd}}\frac{\sin n\theta}{n D_n}\left[\left(\frac{r_w}{r}\right)^n - \left(\frac{r}{r_w}\right)^n\right]. \tag{8}$$

It is readily verified that when $r = r_w$, Eq. (8) gives at once $p = p_w$, and when $r = r_e$, it reduces to Eq. (6), as it should. The velocity distribution corresponding to the pressure distribution of Eq. (8) is readily found on applying Eqs. 4.2 (2) to Eq. (8). Thus

$$
\left.
\begin{aligned}
v_r &= -\frac{k}{\mu}\frac{\partial p}{\partial r} = -\frac{k\left(\dfrac{p_e}{2} - p_w\right)}{\mu r \log r_e/r_w} + \frac{2p_e k}{\pi\mu r}\sum_{n:\,odd}\frac{\sin n\theta}{D_n} \\
&\qquad\qquad\qquad\qquad\qquad\qquad\left[\left(\frac{r_w}{r}\right)^n + \left(\frac{r}{r_w}\right)^n\right] \\
v_\theta &= -\frac{k}{\mu r}\frac{\partial p}{\partial \theta} = -\frac{2p_e k}{\pi\mu r}\sum_{n:\,odd}\frac{\cos n\theta}{D_n}\left[\left(\frac{r_w}{r}\right)^n - \left(\frac{r}{r_w}\right)^n\right]
\end{aligned}
\right\} \quad (9)
$$

The flux[1] into the well becomes

$$
Q = -\int_{-\pi}^{+\pi} r v_r d\theta = \frac{2\pi k\left(\dfrac{p_e}{2} - p_w\right)}{\mu \log r_e/r_w}. \quad (10)
$$

Comparing this result with Eq. 4.2(10), giving Q for strict radial flow, it is seen at once that the effect of applying the pressure p_e over only half of the external boundary is equivalent to applying half of p_e over the whole boundary. And similarly it is easily shown that if the pressure p_e is imposed only upon an arc of angle s of the circle at r_e the flux is

$$
Q_s = 2\pi k\frac{\left(\dfrac{p_e s}{2\pi} - p_w\right)}{\mu \log r_e/r_u}, \quad (11)
$$

which is equivalent to imposing the pressure $\dfrac{p_e s}{2\pi}$ over the whole of the external boundary. And still more generally, since for the general pressure distribution of Eq. (1) Q is given by

$$
\begin{aligned}
Q &= -\int_{-\pi}^{+\pi} r v_r d\theta = \frac{2\pi k c_0}{\mu} = \frac{2\pi k(f_0 - x_0)}{\mu \log r_e/r_w} \\
&= \frac{2\pi k(\bar{p}_e - \bar{p}_w)}{\mu \log r_e/r_w},
\end{aligned} \quad (12)
$$

[1] Here and in the following the term "flux" will be used to represent the production rate from a given outflow surface per unit thickness of the porous medium.

it is seen that *if $\bar{p}_e$, $\bar{p}_w$ are the averages of any arbitrary pressure distributions over the radii r_e and r_w, respectively, the total flow through the sand is the same as if the average pressures were applied uniformly over their respective boundaries.*

This interesting result, which is hardly self-evident physically, shows the power of the more advanced methods of analysis.[1] It means that if only the regional pressure about a well is known, one may disregard the irregularities in the pressure distribution and compute the production from the well as if it came in as a strictly radial flow. As in the latter case, uncertainties in the choice of the outer boundary, r_e, will involve rather small errors in the estimate of Q, since the radii enter in Q in a logarithmic manner.

The same analysis can be carried through when the velocity distribution, rather than the pressure distribution, is given over one or both of the radii r_e, r_w. Since, however, in such cases the total flux is already given from the outset and all that is left to find are the details of the pressure and velocity distribution in the interior of the sand body, the problem does not have as much practical interest. It is usually sufficient to know the average pressure conditions, and Eq. (12), when rewritten as

$$\bar{p}_e = \frac{\mu Q \log r_e/r_w}{2\pi k} + \bar{p}_w, \qquad (13)$$

suffices to give the average pressure at the distance r_e from the well when its average pressure $\bar{p}_w$ and production rate Q are known.

Returning to the original problem defined by Eq. (4), it should be pointed out that depending on whether p_w is positive or negative, the flow in the system will correspond to a drive from the upper half of $r = r_e$ $(0 \leqslant \theta \leqslant \pi)$ into the well and a leakage from the well to the lower half of $r = r_e$ $(-\pi \leqslant \theta \leqslant 0)$, or a

---

[1] It may be noted that the method of the Fourier series may also be applied in the treatment of Laplace's equation in Cartesian coordinates Eq. 4.1(1), as

a consequence of the fact that $\frac{\cosh}{\sinh} \alpha y \frac{\cos}{\sin} \alpha x$ and $\frac{\cosh}{\sinh} \alpha x \frac{\cos}{\sin} \alpha y$ are particular solutions of this equation. These solutions, however, are particularly useful only when the region is limited by parallel straight lines along either the x or y axes. Applications of Fourier-series expansions in the solution of problems treated in later chapters will be given in Secs. 5.3, 6.20, 7.5, 7.10 and 10.14.

drive from the whole of the external boundary into the well, the flux density being higher for $\theta > 0$ than for $\theta < 0$.

Finally, it will be clear that exactly the same analysis will hold if $\bar{p}_w > \bar{p}_e$, except that the velocities will be reversed and the system will correspond to a drive exerted at the well into the sand around it.

4.6. The Flow between Nonconcentric Circular Boundaries. Green's Function.—It has just been seen that the assumption of a uniform pressure over the external circular boundary surrounding a well may afford a very close approximation for the purpose of computing the flux into the well, provided only that the assumed uniform pressure represents the average of the actual pressure distribution over the boundary. The next question to be discussed is the approximation involved in the assumption that the well is at the center of the external boundary. One must, therefore, analyze the flow conditions in a system where the well and external boundary are not concentric, and see to what extent they differ from the ideal situation where the well center coincides with that of the external circular boundary.

For this purpose it is convenient to introduce the concept and method of the Green's function. For two-dimensional problems, Green's function may be defined as: A solution G of Laplace's equation, symmetrical in two points $(x, y; x', y')$, possessing a logarithmic singularity when $(x, y) = (x', y')$ and vanishing when (x, y) is a point on the boundary of the region in question. In potential theory a similar function is studied which has the property that its normal derivative vanishes over all points of the boundary, but for the present purposes only the function defined first will be needed here. The reason for introducing this function is that, once it is found, for a given region, the potential distribution in the interior of the region may be shown to be given by

$$p(x, y) = -\frac{1}{2\pi} \oint \bar{p}(x', y') \frac{\partial G}{\partial n'}(x, y; x', y') ds, \tag{1}$$

where $\bar{p}(x', y')$ is the value of p at the boundary of the region, the elements of which are denoted by ds, n' is the exterior normal to ds, and the integral is extended over the whole boundary.

Although, in principle, the Green's function is one of the most powerful analytical tools available in the solution of potential problems, it suffers from the disadvantage that to find the Green's function is often as difficult a task as to solve the potential problem by direct methods. However, for the problem of this section the Green's function can be found without difficulty. The procedure is as follows:

Suppose there are two unit point sources (line sources from a three-dimensional point of view) of opposite sign at distances δ

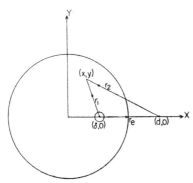

and d from the center of the circle of radius r_e, both lying along the same radial line and such that $\delta < r_e < d$. (cf. Fig. 32.) The resultant pressure at (x, y) may then be expressed as

$$p = \log \frac{r_2}{r_1} + \text{const.} = \log \frac{cr_2}{r_1},$$

(2)

where r_1 is the distance of (x, y) from the source at $(\delta, 0)$ and r_2 that from the sink at $(d, 0)$. This expression is a solution of Laplace's equation and has a logarithmic singularity in the interior of the circle $r = r_e$, at $r_1 = 0$. Further, it vanishes on the locus of the equation

Fig. 32.—A source at $(\delta, 0)$ and its image at $(d, 0)$ in the circle: $r = r_e$.

$$\frac{cr_2}{r_1} = 1,$$

(3)

or, over the curves

$$\left(x - \frac{\delta - c^2 d}{1 - c^2}\right)^2 + y^2 = \frac{c^2(d - \delta)^2}{(1 - c^2)^2}.$$

(4)

For this to represent the boundary $r = r_e$ with center at the origin it is necessary that

$$d = \frac{r_e^2}{\delta}, \qquad c = \frac{\delta}{r_e}.$$

(5)

Hence the function

$$G_0 = \log \frac{\delta}{r_e} \frac{r_2}{r_1}$$

satisfies the requirements of the Green's function for the region bounded by the circle $r = r_e$, except that its singularity is at the fixed point $(\delta, 0)$. If this point is replaced by the variable point $z' = (x', y')$, it is easily verified that the corresponding function, which will be the Green's function of interest, will have the form

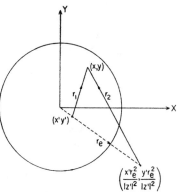

$$G(x, y; x', y') = \log \frac{|z'|r_2}{r_e r_1}, \quad (6)^1$$

where

FIG. 33.—The image of an arbitrary point (x', y') in the circle: $r = r_e$.

$$|z'| = \sqrt{x'^2 + y'^2},$$

$$\left.\begin{array}{l} r_2 = \sqrt{\left(x - \dfrac{x'r_e^2}{|z'|^2}\right)^2 + \left(y - \dfrac{y'r_e^2}{|z'|^2}\right)^2}, \\ r_1 = \sqrt{(x - x')^2 + (y - y')^2}, \end{array}\right\} \quad (7)$$

since in the general case the image in the circle $r = r_e$ of the point (x', y') will be the point $\left(\dfrac{x'r_e^2}{|z'|^2}, \dfrac{y'r_e^2}{|z'|^2}\right)$ as indicated in Fig. 33.

If now a well of small radius r_w is placed about (x', y') for which $|z'| = \delta$, it is clear that if

$$\frac{r_w}{r_e} \ll \frac{\delta}{r_e} < 1,$$

the function G will be, for all practical purposes, constant over the circle of radius r_w, and will in fact have the value

$$\log \frac{\delta(d - \delta)}{r_e r_w} = \log \frac{r_e^2 - \delta^2}{r_e r_w}.$$

---

[1] The lack of symmetry in this form for G is only apparent, as a slight manipulation will show $|z'|r_2$ to be actually symmetrical in (x, y) and (x', y'). A form for G that exhibits the symmetry more explicitly is $G = \log \left|\dfrac{r_e^2 - \bar{z}'z}{r_e(z - z')}\right|$, where z, z' are the complex vectors to (x, y) and (x', y') and the bar denotes the complex conjugate.

Hence a solution, p, for which

$$p = p_w: \quad \sqrt{(x - \delta)^2 + y^2} = r_w \Big\}$$
$$p = p_e: \quad r = r_e \tag{8}$$

is given by

$$p = p_e - (p_e - p_w) \frac{G(x, y; \delta, 0)}{\log \dfrac{r_e^2 - \delta^2}{r_e r_w}}. \tag{9}$$

This, then, gives the pressure distribution between a well and a circular boundary of radius r_e when the well is displaced by a distance δ from the center of the external boundary, and the boundary conditions are given by Eq. (8). To find the flux into the well, it is convenient to take its center as the origin of a polar coordinate system (r, θ). Q is then given by

$$Q = -\int_{-\pi}^{+\pi} r v_r d\theta = -\frac{k(p_e - p_w)}{\mu \log \dfrac{r_e^2 - \delta^2}{r_e r_w}} \int_{-\pi}^{+\pi} r_w \left(\frac{\partial G}{\partial r_1}\right)_{r_1 = r_w} d\theta$$

$$= \frac{2\pi k(p_e - p_w)}{\mu \log \dfrac{r_e^2 - \delta^2}{r_e r_w}}, \tag{10}$$

the term $\log r_2$ evidently giving no net contribution.

It is of interest to compare this value of Q with that for strict radial flow, Q_0, given by Equation 4.2(10). The ratio

$$\frac{Q}{Q_0} = 1 - \frac{\log \left(1 - \dfrac{\delta^2}{r_e^2}\right)}{\log \dfrac{r_e}{r_w} + \log \left(1 - \dfrac{\delta^2}{r_e^2}\right)}, \tag{11}$$

is plotted in Fig. 34 as a function of δ/r_e for $r_e/r_w = 2{,}000$. It is seen that the effect of the displacement of the well from the center of the region is so small that, even when it is halfway between the center of the outside boundary and the boundary itself the increase in flux is less than 5 per cent of that of Q_0.

From a practical point of view this result shows that the error introduced on assuming that a well is at the exact center of a circular boundary is entirely negligible unless its displacement from the center is of the order of the boundary radius itself. Further, this result justifies the use of the simple radial-flow

formula when the well bore is inclined to the plane of the sand
for any reasonable values of the inclination, since the resultant
displacements from the center of the formation will cause only
inappreciable changes in the value of Q.

To see more clearly the particular advantage of the method of
the Green's function once the latter has been found, the pressure
distribution will be derived for the case where the pressure over

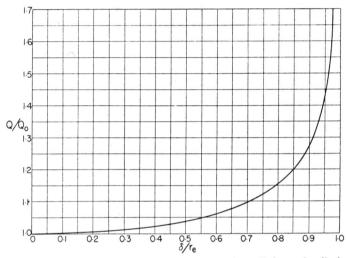

Fig. 34.—The effect on the production capacity of a well due to its displacement δ from the center of the external circular boundary. Q/Q_0 = (production
capacity of the displaced well)/(production capacity of central well); (well
radius)/(external boundary radius, r_e) = $\frac{1}{2000}$.

the external boundary is no longer uniform. Thus if the boundary conditions are restated as

$$p = p_w: \quad \sqrt{(x - \delta)^2 + y^2} = r_w \Big\}$$
$$p = p_e(\theta): \quad r = r_e, \qquad\qquad\qquad (12)$$

Eqs. (1) and (9) give at once the solution

$$p = -\frac{1}{2\pi} \int_{-\pi}^{+\pi} p_e(\theta') r_e \left(\frac{\partial G}{\partial n'}\right)_{r'=r_e} d\theta'$$
$$+ \frac{p_w - u_0}{\log \dfrac{r_e^2 - \delta^2}{r_e r_w}} G(x, y; \delta, 0), \quad (13)$$

where

$$u_0 = -\frac{1}{2\pi} \int_{-\pi}^{+\pi} p_e(\theta') r_e \left\{ \frac{\partial G(\delta, 0; x', y')}{\partial n'} \right\}_{r'=r_e} d\theta', \qquad (14)$$

and evidently is the value of the first term in Eq. (13) at the center of the well and hence, by Gauss' theorem of the mean (*cf.* Sec. 4.16), is the average pressure over the well surface due to the distribution over the external boundary. Thus the numerical determination of the pressures at any point between the well and the external boundaries is reduced to the evaluation of the single integral in Eq. (13).

To get the flux into the well under these conditions, the problem is equally simple. In the notation of Eq. (10) it is given by

$$Q = \frac{k}{\mu} \int_{-\pi}^{+\pi} r_w \left(\frac{\partial p}{\partial n} \right)_{r=r_w} d\theta.$$

Now it is readily verified that the integral in Eq. (13) does not contribute a net flux into the well, since it gives a term to p that is finite at all points within the external boundary. Hence Q becomes

$$Q = \frac{k r_w (p_w - u_0)}{\mu \log \dfrac{r_e^2 - \delta^2}{r_e r_w}} \int_{-\pi}^{+\pi} \left[\frac{\partial G}{\partial r_1} (x, y; \delta, 0) \right]_{r_1 = r_w} d\theta$$

$$= \frac{2k\pi (u_0 - p_w)}{\mu \log \dfrac{r_e^2 - \delta^2}{r_e r_w}}. \qquad (15)$$

The value of u_0, the average pressure over the well surface due to the distribution on the boundary—also the value at its center due to this surface pressure distribution—may be obtained by simply evaluating the integral of Eq. (14) on noting that

$$G(\delta, 0; x', y') = \frac{1}{2} \log \frac{r_e^4 + \delta^2 r'^2 - 2\delta r' r_e^2 \cos \theta'}{r_e^2 \{ r'^2 + \delta^2 - 2r'\delta \cos \theta' \}}, \qquad (16)$$

where $|z'|$ has been replaced by r', and θ' is the angle between z' and the x axis. The result is readily found to be the Poisson integral:

$$u_0 = \frac{1 - \delta^2 / r_e^2}{2\pi} \int_{-\pi}^{+\pi} \frac{p_e(\theta') d\theta'}{1 - 2\dfrac{\delta}{r_e} \cos \theta' + \dfrac{\delta^2}{r_e^2}}. \qquad (17)$$

As the method of using the Green's function requires first the determination of the Green's function appropriate to the geometry of the problem being studied, the Green's functions for several two-dimensional regions are listed in Appendix III. These may be useful for the solution of other two-dimensional problems.

4.7. The Flow from an Infinite Line Source into a Well. The Line Drive. The Method of Images.—Another two-dimensional problem of practical interest involving the flow into a single well is that in which a linear rather than a circular source represents the external boundary. Such a system corresponds to the

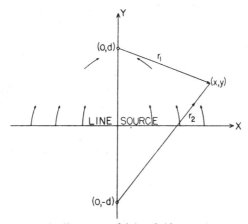

Fig. 35.—An infinite line source driving fluid toward a well at $(0, d)$.

simplest case of an edge-water encroachment in which the advancing water forms a "line drive," displacing and driving the oil into a well situated near the water-oil boundary. Or, it may correspond to the flow into an artesian well drilled into a permeable sand supplied with water by a neighboring river or canal. The latter will then be represented by the line source maintained at a uniform pressure above that at the well face (*cf.* Fig. 35).

Thus it will be supposed that the infinitely extended linear source of fluid is represented by the x axis, and that at a distance d from it there is a well of radius r_w. Furthermore, it will be assumed, for the moment, that the pressure along the line source is maintained at the value zero, and that over the well the pressure is p_w. What will be the fluid flow into or out of the well, and what will be the nature of the pressure distribution?

From the discussion of the previous section it will be clear that because their radii are small compared to the other dimensions of practical interest, one may, for analytical purposes, represent any well of uniform pressure p_w by a point source or sink (from a three-dimensional point of view, it will be a line source or sink with axis perpendicular to the x, y plane) at the center of the well. The potential due to such a well will then have the form

$$p = q \log \frac{r}{r_w} + p_w, \tag{1}$$

where r_w is the well radius and q is an arbitrary coefficient. Further, in a flow system containing a number of wells each will contribute just this term to the resultant pressure distribution, the r's being, of course, measured from the centers of the individual wells. It is true that Eq. (1) makes p become infinite at $r = 0$, the center of the well, but this is no difficulty since the region of interest is only that *between* the well and its external boundaries. Since there is supposed to be no sand or porous medium in the well itself, the fundamental Eq. 4.2(1) does not apply to the well interior. One need, therefore, feel no concern for what happens to solutions of Eq. 4.2(1) when they are extrapolated into regions which are not governed by that equation.[1]

In the same sense, one may add to terms as in Eq. (1), due to actual wells in the half plane $y > 0$—the region of interest— other terms due to hypothetical wells in the lower half plane, $y < 0$. Here again, these terms would become infinite at the centers of the corresponding wells, but since they would behave quite properly in the *region of interest*, $y > 0$, they can cause no difficulty when applied there.

Thus in particular, the actual well at $(0, d)$ may be reflected in the x axis, or one may suppose that there is an "image" in the lower half plane of the well in the upper half plane. Furthermore, it will be supposed that it is a *negative* image so that its potential term is just the negative of that in Eq. (1). The resultant effect of these two wells will then be

$$p = q \log \frac{r_1}{r_w} + p_w - q \log \frac{r_2}{r_w} - p_w$$

$$= q \log \frac{r_1}{r_2}, \tag{2}$$

---

[1] *Cf.* footnote on p. 154.

where r_1 and r_2 are the distances from the actual- and image-well centers, respectively.

We shall now see what this equation implies. To find its value over the well surface it is to be observed that over the surface of the actual well, $r_1 = r_w$, and r_2 varies from $2d - r_u$ to $2d + r_w$. But since r_w is much smaller than d and by Eq. (2) p varies logarithmically with r_2, it is certainly justified to set $r_2 = 2d$ over the whole of the well surface, so that Eq. (2) becomes

$$p = q \log \frac{r_w}{2d}; \qquad r_1 = r_w, \tag{3}$$

which, on the adjustment of q, may be made to assume any arbitrary preassigned value.

Of more significance, however, is the observation that by Eq. (2)

$$p = 0, \qquad \text{for:} \qquad r_1 = r_2,$$

and that the locus of $r_1 = r_2$ is the x axis. In other words, the solution

$$p = \frac{p_w}{\log r_w/2d} \log \frac{r_1}{r_2} \tag{4}$$

vanishes over the x axis and has the value p_w over the surface of the well. It is, therefore, the solution to the physical problem stated at the beginning of this section.

In a practical case it may be more convenient to specify the pressure over the line source as being p_e rather than 0. The solution, then, is readily seen to be

$$p = \frac{p_u - p_e}{\log r_w/2d} \log \frac{r_1}{r_2} + p_e = \frac{1}{2} \frac{p_e - p_w}{\log 2d/r_w} \log \frac{x^2 + (y - d)^2}{x^2 + (y + d)^2} + p_e \tag{5}$$

when expressed in Cartesian coordinates. The curves of constant pressure are given by the relations

$$\frac{r_1^2}{r_2^2} = \frac{x^2 + (y - d)^2}{x^2 + (y + d)^2} = \text{const.} = c^2, \tag{6}$$

or:

$$x^2 + \left[y - d\left(\frac{1 + c^2}{1 - c^2}\right) \right]^2 = \frac{4d^2c^2}{(1 - c^2)^2}, \tag{7}$$

which evidently are circles of radii $2dc/(1 - c^2)$ with their centers at $\left[0, d\left(\frac{1 + c^2}{1 - c^2}\right) \right]$.

Several equipressure curves and streamlines—the orthogonal trajectories of the equipressure curves giving the macroscopic paths of the fluid particles[1]—are drawn in Fig. 36. The theoretical streamlines of this figure—circles with centers at $(\bar{c}, 0)$ and radii $\sqrt{\bar{c}^2 + d^2}$—may be compared with the traces of those in the photograph of Fig. 37 obtained by Kozeny[2] with actual sand models. The complete agreement may well be considered

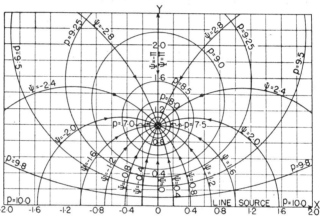

Fig. 36.—Equipressure curves and streamlines for the flow into a well from an infinite line source at a uniform pressure.

as a strong confirmation of the generalization of Darcy's law, originally derived from experiments with linear columns of sand, to general flow systems in which the fluid motion may have components along more than one of the Cartesian axes.

The total flux from the line source to the well is given by

$$Q = -\frac{k}{\mu} \int_{-\infty}^{+\infty} \left(\frac{\partial p}{\partial y}\right)_{y=0} dx = \frac{2kd(p_e - p_w)}{\mu \log 2d/r_w} \int_{-\infty}^{+\infty} \frac{dx}{x^2 + d^2}$$

$$= \frac{2\pi k(p_e - p_w)}{\mu \log 2d/r_w}. \tag{8}$$

It is easily verified that this result may be also obtained from the equation

[1] *Cf.* p. 183 for a more formal definition.

[2] Kozeny, J., *Wasserkraft u. Wasserwirtschaft*, **22**, 103, 1927.

The vertical character of the flow in Kozeny's experiments is of no consequence with respect to the streamlines; its only effect is that the equipressure curves of Fig. 36 are to be considered as equipotential curves over which $p - \gamma g z$ rather than p is constant.

$$Q = -\frac{kr_w}{\mu} \int_0^{2\pi} \left(\frac{\partial p}{\partial n}\right)_{r_1=r_w} d\theta, \qquad (9)$$

where n is the normal to the circle $r_1 = r_w$, defining the well.

It is of interest to note that according to Eq. (8) the flow into a well at a distance d from a line source at a pressure p_e is the same as that into the same well when surrounded by a circle of radius $2d$ at pressure p_e. Thus the flow is proportional to the pressure

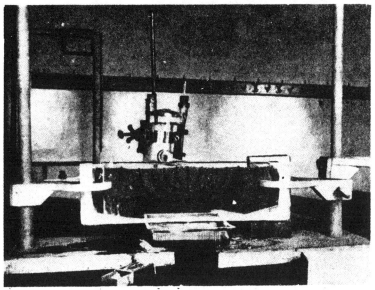

Fig. 37.—Photograph of a sand-model experiment verifying the streamline distribution of Fig. 36. (*After Kozeny, Wasserkraft u. Wasserwirtschaft*).

differential, $p_e - p_w$, and varies logarithmically with the distance of the well from the source.

Equation (8) may be applied to the problem of computing the capacity of an artesian well drilled near a river or canal (*cf.* Fig. 38).[1] In particular it shows that the capacity varies but slowly—logarithmically—with the distance of the well from the source of the water. Furthermore, the capacity is directly

[1] The fact that the sand is sloping in Fig. 38 will clearly have no effect upon the validity of the above solution. For as long as the well penetrates the sand completely, the flow in it will be strictly two-dimensional except for the negligible distortion due to the fact that the well axis is not perpendicular to the bedding planes of the sand.

proportional to the excess of the fluid head in the river or canal over that maintained in the well. Choosing for a numerical example a fluid-head differential of 10 ft., a sand permeability of 1 darcy, and a 6-in. well drilled 100 ft. from the bank (assumed vertical), cf. Fig. 38, *i.e.*,

$$p_e - p_w = 0.295 \text{ atm.,} \qquad k = 1, \qquad \mu = 1, \qquad r_w = \tfrac{1}{2} \text{ ft.,}$$
$$d = 100 \text{ ft.,}$$

Eq. (8) gives

$$Q = 0.309 \text{ cc./sec./cm. sand} = 0.149 \text{ gal./min./ft. sand.}$$

For a well that is drilled 200 ft. from the bank the capacity would be reduced to 0.134 gal./min./ft. sand.

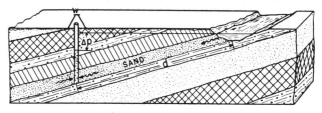

Fig. 38.—An artesian well draining a sand fed by a river or canal.

By a simple extension of the above analysis one may treat the interference effects due to the presence of other wells along the bank of the river or canal. The details of such a treatment will be given in Chap. IX where the general theory of multiple-well systems will be developed.

It may be mentioned that the method of images is essentially a method for developing the proper Green's function for the particular problem of interest. This will be clear on comparing Eq. (4) with the Green's function of Sec. 4.6, and Eq. (5) with the Green's function for the semiinfinite plane given in Appendix III. In fact the Green's functions listed in Appendix III may be most readily derived as the resultants of appropriate sets of finite or infinite numbers of images.

Finally it may be observed that Eq. (7) which shows that for the pressure function

$$p = q \log \frac{r_1}{r_2}$$

the circles of radii $\dfrac{2dc}{1 - c^2}$ and centers at $\left[0, \ d\dfrac{(1 + c^2)}{1 - c^2}\right]$ are equi-potentials, permit an exact solution[1] of the problem of Sec. 4.6 without the assumption that $r_u \ll r_e$. Thus it may be verified that with the notation $\alpha = r_e/r_w$, $\beta = \delta/r_e$, the constants c_e, c_w characterizing the external boundary and the well are given by

$$\beta c_w{}^2 + c_w\left(\alpha\beta^2 - \alpha + \frac{1}{\alpha}\right) + \beta = 0,$$

$$c_e = \beta + \frac{c_w}{\alpha}$$

and

$$p(x, y) = \frac{\Delta p}{\log c_e/c_w} \log \frac{r_1}{r_2} - \frac{p_e \log c_w - p_w \log c_e}{\log c_e/c_w},$$

where:

$$r_1{}^2 = x^2 + \left[y - \frac{r_e(1 - c_e{}^2)}{2c_e}\right]^2; \qquad r_2{}^2 = x^2 + \left[y + \frac{r_e(1 - c_e{}^2)}{2c_e}\right]^2$$

and the circular boundaries of the radii r_e, r_w have their centers at

$$\left[0, \frac{r_e}{2c_e}(1 + c_e{}^2)\right], \qquad \left[0, \frac{r_w}{2c_w}(1 + c_w{}^2)\right].$$

4.8. The Flow from a Finite Line Source into an Infinite Sand. The Method of Conjugate Functions.—In the last section a discussion was given of the problem of a single well draining a sand supplied with liquid by an infinite line source. Such a situation would arise if the line source, as a canal or river, runs parallel to and in fluid contact with the sand over a large distance compared to other dimensions of the system,

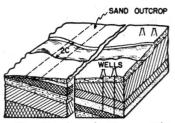

Fig. 39.—A sand outcrop cutting across a river or canal bed, forming a finite line source.

such as the distance of the well from the line source. If, however, the canal or river cuts across the sand outcrop in a mannér

[1] Still another method of treating this problem rigorously is by the use of "bipolar" coordinates (cf. H. Bateman, "Partial Differential Equations," p. 260, 1932.

indicated in Fig. 39, the effective source will simply be the width of the canal or river, which may even be considerably smaller than the distance of the well from the outcrop. The source must then be considered as a finite source and must be treated analytically as such.

In developing such a treatment it will be convenient to make use of the theory of conjugate functions, which will now be briefly outlined. The physical significance of the theory of conjugate functions consists essentially in the observation that both the real and imaginary parts of any analytic function[1] of the complex variable $z = x + iy$, defined as

$$\omega = \Phi + i\Psi = f(z) = f(x + iy),$$ (1)

satisfy the two-dimensional Laplace equation. Thus from Eq. (1) it follows at once that

$$\frac{\partial \Phi}{\partial x} + i\frac{\partial \Psi}{\partial x} = f'(z); \qquad \frac{\partial \Phi}{\partial y} + i\frac{\partial \Psi}{\partial y} = if'(z),$$ (2)

so that

$$\frac{\partial \Phi}{\partial x} = \frac{\partial \Psi}{\partial y}; \qquad \frac{\partial \Phi}{\partial y} = -\frac{\partial \Psi}{\partial x},$$ (3)

and

$$\frac{\partial^2 \Phi}{\partial x^2} + \frac{\partial^2 \Phi}{\partial y^2} = \frac{\partial^2 \Psi}{\partial x^2} + \frac{\partial^2 \Psi}{\partial y^2} = 0.$$ (4)

Hence both Φ and Ψ are potential functions. Furthermore, they form a mutually orthogonal network, since from Eq. (3) it follows at once that

$$\frac{\partial \Phi}{\partial x}\frac{\partial \Psi}{\partial x} + \frac{\partial \Phi}{\partial y}\frac{\partial \Psi}{\partial y} = 0.$$ (5)

If now it be supposed that Φ is a velocity potential, as defined by Eq. 3.3(5), so that

$$v_x = -\frac{\partial \Phi}{\partial x}; \qquad v_y = -\frac{\partial \Phi}{\partial y},$$ (6)

one obtains

$$\frac{v_y}{v_x} = \frac{\partial \Phi/\partial y}{\partial \Phi/\partial x} = -\frac{\partial \Psi/\partial x}{\partial \Psi/\partial y}.$$ (7)

[1] A function $f(z)$ is said to be "analytic" at a point $z = \alpha$ if it has a uniquely determined derivative at α and at every point in its immediate neighborhood.

Thus the direction of the fluid at any point coincides with the tangent, at that point, to the curves $\Psi(x, y) = $ const. These curves are, therefore, known as the "streamlines" for the fluid system for which the curves $\Phi(x, y) = $ const. are the equipotentials, the two functions together being called "conjugate functions." The functions $\Psi(x, y)$ are also known as "stream functions."

It will be seen, therefore, that if one sets up any equation of the form of Eq. (1) and takes Φ and Ψ as the real and imaginary parts, respectively, then Φ may be considered as the velocity potential of a physical problem with equipotentials $\Phi = $ const. and streamlines $\Psi = $ const. Recalling that for horizontal two-dimensional problems the velocity potential Φ is simply the pressure multiplied by k/μ, it follows that if in a particular case one of the curves $\Phi = $ const. coincides with a boundary of physical interest, the general values of Φ and Ψ correspond to the physical problem where that boundary is kept at a uniform pressure.

In illustrating this theory with a physical application it will be supposed, in this section, that the finite line source at a uniform pressure simply supplies the infinite sand. The case of a well draining this sand will be treated in the next section. Thus, as a particular case of Eq. (1) it will be assumed that

$$f(z) = p + i\Psi = \cosh^{-1}\frac{(x + iy)}{c}, \tag{8}$$

where c is a constant and the potential function Φ has been replaced by the fluid pressure p. Separating real and imaginary parts, it is readily seen that

$$\left.\begin{array}{l} x = c \cosh p \cos \Psi \\ y = c \sinh p \sin \Psi, \end{array}\right\} \tag{9}$$

so that:

$$\left.\begin{array}{l} \dfrac{x^2}{c^2 \cosh^2 p} + \dfrac{y^2}{c^2 \sinh^2 p} = 1 \\[2ex] \dfrac{x^2}{c^2 \cos^2 \Psi} - \dfrac{y^2}{c^2 \sin^2 \Psi} = 1. \end{array}\right\} \tag{10}$$

Equations (10) show that the equipressure curves $p = $ const. are the confocal ellipses with foci at $x = \pm c$ and semiaxes $c \cosh p$,

$c \sinh p$; the streamlines Ψ = const. are confocal hyperbolas with semiaxes $c \cos \Psi$ and $c \sin \Psi$, as shown in Fig. 40. Physically, these curves may be taken as giving the pressure and streamline

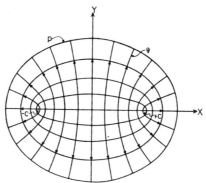

FIG. 40.—Equipressure curves and streamlines about a finite line source.

distributions in a system where the uniform pressure p_e is maintained over the elliptical boundary

$$\frac{x^2}{a_e^2} + \frac{y^2}{b_e^2} = 1,$$

and the pressure p_w over the interior confocal elliptical surface

$$\frac{x^2}{a_w^2} + \frac{y^2}{b_w^2} = 1.$$

When $b_w = 0$, the interior ellipse degenerates into the finite line source of length $2c$ given by

$$y = 0; \qquad |x| \leqslant a_w = c; \qquad b_w = 0. \tag{11}$$

The pressure and streamline distributions are then given explicitly by[1]

$$p = \cosh^{-1} H; \qquad \Psi = \cos^{-1} H, \tag{12}[2]$$

[1] These equations—(12)–(14)—have also been given by C. S. Slichter, *U. S. Geol. Surv.*, *19th Ann. Rept.*, 295, 1897-1898.

[2] A form of Eqs. (12) showing directly that p, Ψ = const. are confocal ellipses and hyperbolas is given by $p = \cosh^{-1} \frac{(r_1 + r_2)}{2c}$; $\Psi = \cos^{-1} \frac{(r_1 - r_2)}{2c}$, where r_1, r_2 are the distances of (x, y) from the foci ($\pm c$, 0), the extremities of the line source.

where

$$H = \left[\frac{x^2 + y^2 + c^2 \pm \sqrt{(x^2 + y^2 + c^2)^2 - 4x^2c^2}}{2c^2} \right]^{\frac{1}{2}}, \quad (13)$$

the plus sign referring to p and the minus sign to Ψ.

In particular, along the x axis, along which lies the line source, Eqs. (12) and (13) give

$$\left. \begin{array}{l}
p = 0: \quad |x| \leqslant c; \quad p = \cosh^{-1}\left|\frac{x}{c}\right|: \quad |x| \geqslant c; \\
\Psi = 0: \quad x > c; \quad \Psi = \pi: \quad x < -c; \\
\qquad\qquad\qquad \Psi = \cos^{-1}\frac{x}{c}: \quad |x| < c.
\end{array} \right\} \quad (14)$$

To find the total flux, Q, entering the sand from the line source, it is simply necessary to observe the following property of the stream function Ψ, namely,

$$\Delta\Psi = \int d\Psi = \int \frac{\partial\Phi}{\partial n}ds = \frac{k}{\mu}\int \frac{\partial p}{\partial n}ds, \quad (15)$$

where the integrals are extended over a curve, the element of which is ds. Hence the difference in Ψ between any two points is equal to the net flux passing across any curve joining these points. It therefore follows from Eq. (14) that the flux Q, leaving the source, is given by

$$Q = \frac{2\pi k}{\mu}. \quad (16)$$

This, however, corresponds to a pressure of 0 at the line source and a pressure: $p_e = \log \frac{(b_e + a_e)}{c}$ on the external confocal ellipse of semiaxes a_e, b_e. Hence for a general pressure differential Δp, maintained between the source and the external ellipse of semiminor axis b_e, the flux will be

$$Q = \frac{2\pi k \Delta p}{\mu \log \dfrac{a_e + b_e}{c}} = \frac{2\pi k \Delta p}{\mu \log \dfrac{b_e + \sqrt{c^2 + b_e^2}}{c}}, \quad (17)$$

$2c$ being the length of the line source.[1]

[1] It should be observed that Eqs. (16) and (17) give the flux leaving *both* sides of the source.

It may be finally observed that for distances from the line source which are large as compared to the length of the source, the pressure assumes a logarithmic radial form, as should be expected. For if $x^2 + y^2 = r^2 \gg c^2$, we have $H \sim r/c$, or x/r, as the upper or lower signs are taken in Eq. (13); hence,

$$
\left.\begin{array}{l}
p \cong \cosh^{-1} \dfrac{r}{c} \cong \log \dfrac{2r}{c}: \quad x^2 + y^2 \gg c^2, \\[2mm]
\Psi \cong \cos^{-1} \dfrac{x}{r} = \theta,
\end{array}\right\}
\tag{18}
$$

θ being the polar angle with origin at the center of the source.

4.9. The Flow from a Finite Line Source into a Well. Conjugate Function Transformations. Infinite Sets of Images.— In the last section it was seen that every equation of the form of Eq. 4.8(1) gives a pair of "conjugate functions," one of which may be interpreted physically as giving a system of equipotential curves and the other the corresponding streamlines. These will correspond to those in an actual physical system if the preassigned distribution of its boundary pressures or fluxes are proportional (to an additive constant) to those assumed by the conjugate function network on the curves bounding the physical system. Thus, in the case of the problem treated in Sec. 4.8, the solution really consisted in nothing more than the observation that one of the equipotentials of the conjugate function network defined by Eq. 4.8(8), namely, the degenerate ellipse of Eq. 4.8(14), coincided geometrically with the finite line source of the physical system under consideration. Similar problems may be treated in like manner.

For the solution of the present problem, however, use will be made of a different type of property of conjugate functions. That is, instead of deriving from them directly solutions of Laplace's equation, they will be used in transforming the *independent* variables (x, y). Here one simply observes that the conjugate-function *transformation*

$$
\zeta = \xi + i\eta = f(z) = f(x + iy)
\tag{1}
$$

leaves Laplace's equation invariant, so that, if

$$\frac{\partial^2 p}{\partial x^2} + \frac{\partial^2 p}{\partial y^2} = 0,$$

then also

$$\frac{\partial^2 p}{\partial \xi^2} + \frac{\partial^2 p}{\partial \eta^2} = 0. \tag{2}$$

The transformation of Eq. (1) may, therefore, be used to transform the *geometrical boundaries* of the physical system to others which are more suitable for analytical discussion. Thus, in the original (x, y) coordinate system, *i.e.*, in the z plane, the physical system corresponding to the flow from a finite line source into a well may be represented diagrammatically by Fig. 41 where, since the sand is supposed to outcrop along its intersection with the river bed, it may be assumed that there is no flux across the x axis and into the sand except along the line source, which is the outcrop of the sand in the river bed (*cf.* Fig. 39). This requirement, in itself, could be readily satisfied if one used the pressure distribution of the last section and added a logarithmic term due to the well at (x_0, y_0), and then one due to its positive image at $(x_0, -y_0)$. However, these two logarithmic terms would destroy the uniformity of the pressure over the line source,

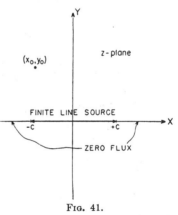

FIG. 41.

and the problem would arise of finding such an additional flux distribution over the line source as would correct for the distortion of the pressure distribution caused by the actual well and its image. Rather than attempting to solve the integral equation to which this problem would lead, the z-plane diagram will be transformed to the (ξ, η) or "ζ plane" as follows:

Using again the complex variable relation of Sec. 4.8,

$$\zeta = \xi + i\eta = \cosh^{-1}\frac{(x + iy)}{c}, \tag{3}$$

it follows as before that

$$x = c \cosh \xi \cos \eta \\ y = c \sinh \xi \sin \eta \Big\} \tag{4}$$

and

$$\left. \begin{array}{l} \dfrac{x^2}{c^2 \cosh^2 \xi} + \dfrac{y^2}{c^2 \sinh^2 \xi} = 1 \\[2mm] \dfrac{x^2}{c^2 \cos^2 \eta} - \dfrac{y^2}{c^2 \sin^2 \eta} = 1. \end{array} \right\} \tag{5}$$

In a manner similar to the interpretation of Eq. 4.8(14), it is to be noted that the x axis in the z plane goes over into

$$\left. \begin{array}{llll} y = 0: \\ \quad |x| < c: \rightarrow: & \xi = 0, & 0 \leqslant \eta \leqslant \pi \\ \quad x > c: \rightarrow: & \eta = 0, & 0 \leqslant \xi < \infty \\ \quad x < -c: \rightarrow: & \eta = \pi, & 0 \leqslant \xi < \infty \\ \quad (x_0, y_0) \rightarrow (\xi_0, \eta_0). \end{array} \right\} \tag{6}$$

Thus the upper half of the z plane is "mapped" onto the semi-infinite strip of the ζ plane, as shown in Fig. 42. ·

The equation for the pressure distribution in this ζ plane is again the Laplace's equation (2) and the boundary conditions are formally the same as before, namely,

$$\left. \begin{array}{llll} p = p_e: & \xi = 0, & \eta \leqslant \pi \text{ (line source)}, \\[2mm] \dfrac{\partial p}{\partial \eta} = 0: & \eta = 0, & \eta = \pi \text{ (remainder of } x \text{ axis)}. \end{array} \right\} \tag{7}$$

Although the differential equations and boundary conditions have remained formally the same, it is important to observe that in the ζ plane a *single* condition is imposed over each of the boundary segments, whereas in the z plane there were imposed over the single boundary, $y = 0$, the "mixed" conditions of a specification of p over part of it and of $\dfrac{\partial p}{\partial y}$ over the remainder.

It is really this reduction of the "mixed" boundary conditions that is the purpose of introducing here, as in other similar problems, the conjugate-function transformation.[1]

[1] An even simpler method for treating such "mixed" boundary-value problems is that of combining the method of conjugate-function trans-

To solve now the transformed problem of Eqs. (2) and (7), it is only necessary to extend and apply the method of images already outlined in Sec. 4.7. Thus considering again the well at (ξ_0, η_0) as a mathematical sink, a vanishing flux over the boundaries $\eta = 0, \pi$ can be attained by setting up plus images at $(\xi_0, -\eta_0)$ and $(\xi_0, 2\pi - \eta_0)$. However, the former will cause a nonvanishing flux over $\eta = \pi$ and the latter over $\eta = 0$, and hence must themselves be corrected for by their own images, etc. In addition, this infinite system must be reflected in the η axis in order to give a constant (vanishing) pressure on the η axis (the line source of the z plane).

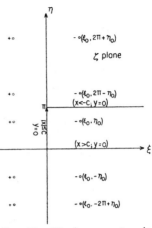

Fig. 42.—The image system in the ζ plane required to satisfy the boundary conditions indicated in Fig. 41.

The final complete set is then given by (*cf.* Fig. 42)

$$-q: (\xi_0, \pm 2n\pi \pm \eta_0); \qquad +q: (-\xi_0, \pm 2n\pi \pm \eta_0), \qquad (8)$$

where q represents the "strength" of the physical well at (ξ_0, η_0).

The pressure distribution which is the resultant of these image systems may, therefore, be written as

$$p = \frac{q}{2} \sum_{-\infty}^{+\infty} \log \frac{(\xi + \xi_0)^2 + (\eta - 2n\pi - \eta_0)^2}{(\xi - \xi_0)^2 + (\eta - 2n\pi - \eta_0)^2}$$

$$+ \frac{q}{2} \sum_{-\infty}^{+\infty} \log \frac{(\xi + \xi_0)^2 + (\eta - 2n\pi + \eta_0)^2}{(\xi - \xi_0)^2 + (\eta - 2n\pi + \eta_0)^2}. \qquad (9)$$

---

formations with that of "mixed" Green's functions, which are defined so that they vanish over the parts of the boundary where the pressure is specified and their normal derivatives vanish over those parts where the normal derivative is specified. (*cf.* M. Muskat, *Physics*, **6**, 27, 1935, in particular, Section B.) In fact, in such a solution the pressure distribution of Eq. (11) would be found simply as the Green's function itself, though the (ξ, η) coordinates would be defined by a transformation different from that of Eq. (3).

These sums may be simplified on noting that

$$\sum_{-\infty}^{+\infty} \log\left[(\xi - h)^2 + (na - \eta)^2\right] = \log \prod_{-\infty}^{+\infty}\left[(\xi - h)^2 + (na - \eta)^2\right]$$

$$= \log^* \prod_{-\infty}^{+\infty}{}' n^2 a^2 + \log \eta^2 + 2 \log \prod_{1}^{\infty}\left(1 - \frac{\eta}{na}\right)\left(1 + \frac{\eta}{na}\right)$$

$$+ \log \prod_{-\infty}^{+\infty}\left\{1 + \frac{(\xi - h)^2}{(na - \eta)^2}\right\}$$

$$= \log \sin^2 \frac{\eta\pi}{a} + \log \frac{\cosh 2\pi(\xi - h)/a - \cos 2\pi\eta/a}{1 - \cos 2\pi\eta/a}$$

$$= \log\left[\cosh 2\pi(\xi - h)/a - \cos 2\pi\eta/a\right] \qquad (10)^1$$

where the infinite constant $\log \prod_{-\infty}^{+\infty}{}' n^2 a^2$ and other finite constants have been dropped.

Equation (9) can, therefore, be reduced to

$$p = p_e + \frac{q}{2} \log \frac{\left[\cosh (\xi + \xi_0) - \cos (\eta - \eta_0)\right]}{\left[\cosh (\xi - \xi_0) - \cos (\eta - \eta_0)\right]}$$
$$\cdot\frac{\left[\cosh (\xi + \xi_0) - \cos (\eta + \eta_0)\right]}{\left[\cosh (\xi - \xi_0) - \cos (\eta + \eta_0)\right]}, \qquad (11)$$

where the constant p_e has been added to make $p(\xi = 0) = p_e$.

That Eq. (11) satisfies the second boundary condition of Eq. (7) and the differential equation (2) may be verified by direct substitution. If now (ξ, η) is expressed in terms of (x, y), by means of Eqs. (4) and (5), and put in Eq. (11), the pressure distribution for the original problem in the z plane will be obtained at once. To get the flux in the system, however, one may avoid this rather complex substitution. Thus, at the well radius one may set $\eta = \eta_0$, $\xi = \xi_0 = \epsilon \ll 1$, so that

$$p_w = p_e + \frac{q}{2} \log \frac{2 \sinh^2 \xi_0 (\cosh 2\xi_0 - \cos 2\eta_0)}{\epsilon^2 \sin^2 \eta_0}$$

$$= p_e + q \log \frac{2c \sinh \xi_0 (\sinh^2 \xi_0 + \sin^2 \eta_0)}{r_w \sin \eta_0}, \qquad (12)$$

---

* The prime indicates that the term $n = 0$ is to be omitted.
[1] *Cf.* Whittaker and Watson, "Modern Analysis," 4th ed., p. 137, 1927.

where r_w is the well radius, since

$$\overline{ds}^2(z \text{ plane}) = c^2 (\sinh^2 \xi + \sin^2 \eta)\overline{ds}^2(\zeta \text{ plane}). \qquad (13)$$

Now Eq. (12) implies that very near the well center the pressure distribution is given by

$$p = p_e + \text{const.} - q \log \rho, \qquad (14)$$

where ρ is the distance from the well center. The flux into the well is, therefore,

$$Q = -\frac{2\pi k q}{\mu}, \qquad (15)$$

so that, returning to Eq. (12), the flux will be, in general

$$Q = \frac{2\pi k(p_e - p_w)}{\mu \log \dfrac{2c \sinh \xi_0 (\sinh^2 \xi_0 + \sin^2 \eta_0)}{r_w \sin \eta_0}}$$

$$= \frac{2\pi k(p_e - p_w)}{\mu \log \dfrac{4y_0}{r_w}\left\{\dfrac{\sqrt{(c^2 - r_0^2)^2 + 4y_0^2 c^2}}{c^2 - r_0^2 + \sqrt{(c^2 - r_0^2)^2 + 4y_0^2 c^2}}\right\}} \qquad (16)$$

where

$$r_0^2 = x_0^2 + y_0^2.$$

Some limiting cases of this formula are of interest. Thus, if the well is on the perpendicular bisector of the line source, Q takes the form

$$Q_1 = \frac{2\pi k(p_e - p_w)}{\mu \log \dfrac{2y_0}{r_w}\left(1 + \dfrac{y_0^2}{c^2}\right)} = \frac{2\pi k(p_e - p_w)}{\mu \log \left\{\dfrac{2y_0}{r_w} \csc^2 \dfrac{\theta}{2}\right\}}, \qquad (17)$$

where θ is the angle subtended by the line source at the well. Equation (17) is of the same form as that found for an infinite line source [*cf.* Eq. 4.7(8)], except for the factor $1 + \dfrac{y_0^2}{c^2}$; hence it becomes identical with it if the well is close to the finite source ($y_0/c \ll 1$), as should be expected.[1] If, however, the well is very distant from the line source ($y_0/c \gg 1$), the production rate decreases as if the line source were equivalent to an external

[1] In fact, Eq. (16) reduces directly to Eq. 4.7(8) if c is made indefinitely large.

circular boundary of radius inversely proportional to the cube of the angle subtended by the line source at the well.

If the well is in line with the edges of the line source, *i.e.*, if $x_0^2 = c^2$,

$$Q_2 = \frac{2\pi k(p_e - p_w)}{\mu \log \dfrac{4y_0}{r_w} \dfrac{\sqrt{4c^2 + y_0^2}}{\sqrt{4c^2 + y_0^2} - y_0}}, \qquad |x_0|^2 = c^2, \qquad (18)$$

which is evidently less than that for $x_0 = 0$, since the argument of the logarithm here is larger than that in Eq. (17). Thus, if $y_0 = c$,

$$\frac{Q_1}{Q_2} = \frac{\log 7.24 y_0/r_w}{\log 4y_0/r_w}.$$

If the well is at a distance from the center of the line source equal to the half-width of the source, Eq. (16) reduces to

$$Q_3 = \frac{2\pi k(p_e - p_w)}{\mu \log 4y_0/r_w}, \qquad (19)$$

which is independent of x_0.

Finally, it is to be observed from Eq. (17) that since the correction factor $\csc^2 \dfrac{\theta}{2}$ enters in the logarithmic term, the assumption of an infinite line source $(\theta = \pi)$ will cause but little error in practical calculations of Q unless it is definitely known that the distance of the outcrop from the well is much larger than the width of the river or canal feeding the outcrop.

4.10. The Uplift Pressure on a Dam of Extended Base Width. No Sheet Piling.—A very interesting practical question arising in the consideration of the seepage of water underneath a dam of appreciable width is the nature of the uplift pressure acting on the base of the dam associated with the seepage. This is a problem of considerable importance in the design of dams, and may be solved by a direct application of the potential-theory methods outlined in the preceding sections. As the main interest here will be the nature of the pressure distribution at the base of the dam—assumed to set at the top of the porous basement without appreciable penetration—it must be considered as being of finite width, the length again being taken as infinite. Furthermore, the sand underlying the dam will be taken to be of infinite depth, both here and in the next section, in order to simplify the calculation of the uplift pressures and moments[1] (*cf.* Fig. 43). For the problem of the actual seepage flux, however, the rigorous solution for a bed of finite thickness will be given (*cf.* Secs. 4.12 to 4.14), as the infinitely thick sand will give an infinite seepage flux.

When there are no pilings at the base of the dam, the nature of the pressure distribution may be obtained by a simple reinterpretation of the results of Sec. 4.8. For, if we interchange the functions Ψ and p in Eqs. 4.8(10), and interpret the pressures p as the velocity potential Φ, it will be seen on recalling Eq. 4.8(14) that

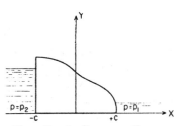

Fig. 43.—A diagrammatic cross section of an impervious dam setting on an infinitely thick porous bed.

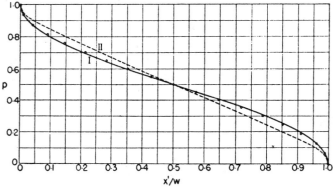

Fig. 44.—Pressure distribution at the base of a dam with no sheet piling. p = pressure at base for a unit total-pressure differential. x'/w = (distance from heel of dam)/(width of base). I: thickness, h, of underlying permeable stratum is infinite; II: $w/h = 5.0$; circles: $w/h = 1.0$. (*From Physics,* **7**, 119, 1936.)

the solutions given in Sec. 4.8 will correspond to a potential distribution in which on the x axis, $\Phi = \pi$ for $x < -c$ and $\Phi = 0$ for $x > c$, the region of the x axis in between corresponding to

[1] This approximation will not involve serious errors except when the dam width is large compared to the sand thickness, or when the sheet piling, if present, penetrates to appreciably more than half of the depth of the sand (*cf.* Figs. 44 and 68).

the streamline $\Psi = 0$. Furthermore, the potential and hence pressure distributions at the base will be given by

$$p = \frac{\Delta p}{\pi} \cos^{-1} \frac{x}{c} + p_1, \tag{1}$$

where Δp is the total pressure drop between the upstream and downstream sides of the dam, the latter being at the pressure p_1. The plot of Eq. (1) in Fig. 44 shows that while the usual assumption of a linear variation is not badly in error, it does lead to too high pressures near the upstream end and too low pressures near the downstream end of the dam. The difference is particularly marked when the velocities, proportional to the slopes of the curves of Fig. 44, are considered, Eq. (1) leading to infinitely high velocities[1] at the ends, and thus indicating greater dangers of erosion and sand flows than would be concluded from the linear pressure distribution.

The total upward force F per unit of dam length is given by

$$F = \int_{-c}^{+c} \left[\frac{\Delta p}{\pi} \cos^{-1} \frac{x}{c} + p_1 \right] dx = c\Delta p + 2p_1 c = (p_2 + p_1)\frac{w}{2}, \tag{2}$$

where w is the width of the dam and p_2 the upstream pressure. This is evidently the same value as would be given by the assumption of a linear pressure variation.

The total moment of the uplift forces with respect to the heel of the dam is:

$$M = \int_{-\frac{w}{2}}^{\frac{w}{2}} \left(x + \frac{w}{2} \right) \left[\frac{\Delta p}{\pi} \cos^{-1} \frac{2x}{w} + p_1 \right] dx = \frac{w^2}{16}(3p_2 + 5p_1), \tag{3}$$

whereas the linear pressure distribution gives $M = \dfrac{w^2(p_2 + 2p_1)}{6}$, which is always less than the correct value of Eq. (3), the error amounting to 11.1 per cent when $p_1 = 0$.

[1] Physically, of course, strictly infinite velocities will never occur owing to the breakdown at such points of Darcy's law (cf. Sec. 2.2). However, the tendency toward the development of such high velocities will be correctly described by the analytical theory based on Darcy's law, even though the fine details of the velocity and pressure distributions about the singular points of the solutions will not be accurately reproduced by the physical flow systems.

4.11. The Uplift Pressure on a Dam with Sheet Piling Present.[1] The Schwarz-Christoffel Theorem.—A more practical problem of uplift pressure on dams is that referring to the case when there is a sheet pile attached to the base of the dam. However, before deriving the explicit solutions for this case it will be useful to bring out more clearly the real analytical significance of the transformations of Sec. 4.8 which were applied in the last section.

Thus the original analytical problem simply consisted in the requirement of finding a potential distribution (satisfying Laplace's equation) which assumed a constant value on the x axis for $x < -c$, a different constant value for $x > c$, and such that for $|x| < c$ the x axis corresponded to a streamline, so that $\Psi = \text{const.}$ $\left(\dfrac{\partial \Phi}{\partial y} = 0 \right)$. This requirement, however, in which both the value of Φ and its normal derivative $\dfrac{\partial \Phi}{\partial y}$ are specified over parts of the same rectilinear boundary (the x axis), is difficult to satisfy by means of elementary solutions derived directly from Laplace's equation.[2] It is for this reason that the complex variable transformation equivalent to

$$\zeta = \xi + i\eta = \cosh^{-1} \frac{(x + iy)}{c} \tag{1}$$

was introduced.[3] For in the ζ plane the boundary of the original system (the x axis of Fig. 45) is mapped on that of the semi-infinite strip of Fig. 46. But for such a geometrical region with the indicated boundary conditions, one may write down at once for the potential distribution

$$\Phi = \frac{\Delta \Phi}{\pi} \eta + \Phi_1. \tag{2}$$

---

[1] The essential analytical results of this section up to Eq. (14), derived in a somewhat modified form, including Figs. 50 to 53 were given by W. Weaver, *J. Math. and Physics*, **11**, 114, 1932.

[2] Compare, however, the footnote on p. 188 for a method by which such types of problem can be solved even when the potential and flux distributions over the several segments are not uniform.

[3] The same procedure was also applied in Sec. 4.9, although the analysis subsequent to the transformation by Eq. (1) was quite different from that given here.

Observing now that along the x axis where $\Psi = 0$, $\eta = \cos^{-1} \dfrac{x}{c}$, Eq. 4.10(1) is obtained at once.

This same procedure can be applied to the more complex problem of the dam with sheet pilings attached, represented dia-

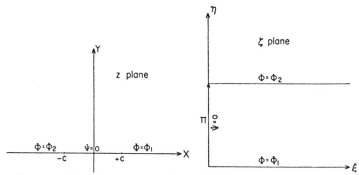

FIG. 45.—The z-plane map of a dam with no sheet piling and setting on an infinitely thick porous bed.

FIG. 46.—The ζ-plane map of Fig. 45.

grammatically in Fig. 47. For it is clear that the boundary $BCDEF$ is a streamline of the system. Hence if this boundary segment is mapped on a strip of the real axis of an auxiliary

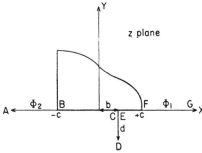

FIG. 47.—The z-plane representation of a dam with sheet piling of depth d setting on an infinitely thick porous bed.

plane and the segments AB and FG on the rest of the real axis, the problem will become formally equivalent to that of Fig. 45, which can then be solved by means of the further transformation of the type of Eq. (1). It is this mapping which is the essentially new analytical step required for the solution of the case of the dam with sheet piling.[1]

This mapping may be carried out by means of the Schwarz-Christoffel theorem, which provides a procedure for mapping any polygonal region onto the upper half of a complex plane, the boundary of the region being transformed into the real axis of the plane. In fact, the transformation of Eq. (1) used to map the z plane of Fig. 45 onto the ζ plane of Fig. 46 may itself be derived from this theorem by first mapping the infinite rectangle of Fig. 46 onto the z plane and then inverting. The theorem may be stated as follows: A polygon of vertices $z_k = x_k + iy_k$ in the $z = x + iy$ plane and internal angles $\alpha_1 \ldots \alpha_n$ will be mapped on the upper half of the $\zeta = \xi + i\eta$ plane, with the points $\zeta = \xi_k$ corresponding to the vertices $z = z_k$, provided z and ζ are related by the equation

$$ z = c_1 \int \frac{d\zeta}{\Pi(\zeta - \xi_k)^{1 - \alpha_k/\pi}} + c_2, \qquad (3) $$

where c_1 and c_2 are constants determined by the size and orientation of the polygon. If $\xi = \pm \infty$ is to correspond to a corner of the polygon, the corresponding factor in the product of the integrand is to be omitted. The correctness of this theorem may be readily confirmed by observing that the arguments of $\frac{dz}{d\zeta}$ corresponding to Eq. (3) actually change by $\pi - \alpha_k$ as the points $\zeta = \xi_k$ are crossed.

To apply this theorem to the problem at hand one need simply list the angles α_k of the "polygon" $ABCDEFGA$. These are evidently $\alpha = \angle FED = \pi/2; \alpha = \angle EDC = 2\pi; \alpha = \angle DCB = \pi/2; \alpha = \angle BAG = 0$. Choosing now $\zeta = 1$ to correspond to E, $\zeta = 0$ to D, $\zeta = -1$ to C, and $\zeta = -\infty$ to A, Eq. (3) takes the form

[1] It may be noted that the problem of finding the conjugate-function transformations with which to map one region upon another is closely related to that of finding the Green's function for the region. In particular, it may be shown that if $G(x, y; \xi, \eta)$ is the Green's function for a region and $G_1(x, y; \xi, \eta)$ is its conjugate, then the function $\zeta = f(z) = e^{-G - iG_1}$ will map the original region onto the interior of the unit circle in the ζ plane, the point (ξ, η) being mapped onto the origin of the ζ plane (cf. Riemann-Weber's "Differentialgleichungen der Physik," vol. 1, p. 546, 1925.)

$$z = c_1 \int \frac{\zeta d\zeta}{\sqrt{\zeta^2 - 1}} + c_2 = c_1\sqrt{\zeta^2 - 1} + c_2. \tag{4}$$

The constants c_1, c_2 may be determined from the requirements that the correspondences assumed above actually are valid numerically. These are that

$$E \to \zeta = 1: \quad b = c_2; \quad D \to \zeta = 0:$$
$$b - id = c_1 i + c_2; \quad C \to \zeta = -1: \quad b = c_2, \tag{5}$$

the condition that $A \to \zeta = -\infty$ being evidently satisfied identically. Equation (4), therefore, becomes

$$z = -d\sqrt{\zeta^2 - 1} + b, \tag{6}[1]$$

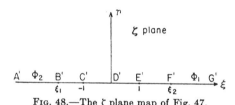

Fig. 48.—The ζ plane map of Fig. 47.

giving a map on the ζ plane as indicated in Fig. 48, B' and F' having the values

$$\xi_1 = -\sqrt{1 + \frac{(c + b)^2}{d^2}},$$

and

$$\xi_2 = +\sqrt{1 + \frac{(c - b)^2}{d^2}},$$

respectively. The problem is now reduced in principle to that of the dam with no sheet piling. For, by shifting the origin of the ζ plane so that it lies midway between B' and F', and changing the scale so as to make the length $B'F'$ equal 2, i.e., by applying the transformation

$$\zeta' = \frac{\zeta - \frac{(\xi_2 + \xi_1)}{2}}{(\xi_2 - \xi_1)/2}; \quad \xi' = \frac{\xi - \frac{(\xi_2 + \xi_1)}{2}}{(\xi_2 - \xi_1)/2}, \tag{7}$$

[1] It should be noted that the negative radical should be taken for $\xi > 1$ and the positive for $\xi < 1$.

the ζ plane becomes transformed into the exact equivalent of the z plane of Fig. 45. Mapping now this ζ' plane onto the λ plane by the transformation,

$$\lambda = \tau + i\theta = \cosh^{-1} \zeta', \tag{8}$$

the semi-infinite strip of Fig. 49 is obtained. The final solution for the potential distribution is then simply

$$\Phi = \frac{\Delta\Phi}{\pi}\theta + \Phi_1, \tag{9}$$

which, in virtue of Eqs. (8), (7), and (6), gives Φ as a function of (x, y) in the original z plane.

For the purpose of studying the uplift pressure on the dam one need consider only values of Φ for which $y = 0$, $|x| < c$.

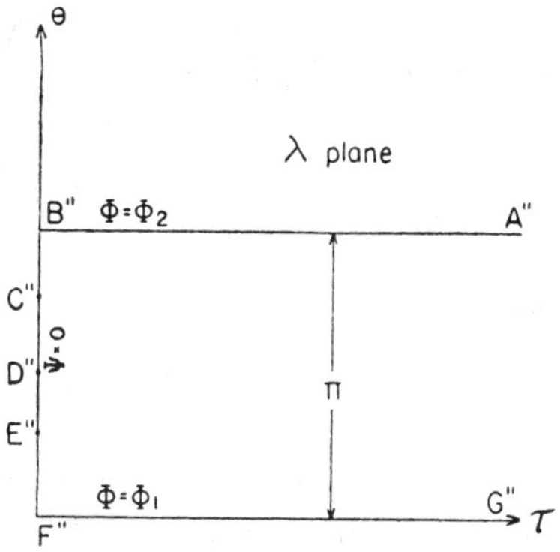

FIG. 49.—The λ-plane map of Fig. 47.

This implies that ζ and ζ' are real and that in fact $|\zeta'| < 1$. Hence for the potential at the base of the dam, Eq. (9) becomes

$$\Phi = \frac{\Delta\Phi}{\pi} \cos^{-1} \zeta' + \Phi_1. \tag{10}$$

Introducing now the notation

$$c = \frac{w}{2}; \qquad b + c = \bar{x}; \qquad x' = x + \frac{w}{2}; \qquad \alpha = \frac{w}{d}; \qquad \beta = \frac{\bar{x}}{d},$$

so that w is the width of the dam, $\bar{x}$ is the distance of the piling from the heel, α is the ratio of the dam width to the depth of the piling, and x' the distance along the base from the heel of the dam, making these substitutions, and returning to the values of the pressures, Eq. (10) may be finally rewritten as

$$p = \frac{\Delta p}{\pi} \cos^{-1} \left[\frac{2\sqrt{1 + \left(\dfrac{x'}{d} - \beta\right)^2} - \sqrt{1 + (\alpha - \beta)^2} + \sqrt{1 + \beta^2}}{\sqrt{1 + (\alpha - \beta)^2} + \sqrt{1 + \beta^2}} \right] + p_1, \tag{11}$$

where the sign of the first radical is to be taken as $+$ for $x' > \bar{x}$ and $-$ for $x' < \bar{x}$.

It is readily seen that for $d = 0$ this equation reduces to that of Eq. 4.10(1) for a dam without piling.

The total uplift force is given by

$$F = p_1 w + \frac{\Delta p}{\pi} \left\{ \int_0^{b_1} \cos^{-1}\left[\frac{-2\sqrt{1 + \left(\frac{x'}{d} - \beta\right)^2} - e + j}{e + j} \right] dx' \right.$$

$$\left. + \int_{b_1}^{w} \cos^{-1}\left[\frac{2\sqrt{1 + \left(\frac{x'}{d} - \beta\right)^2} - e + j}{e + j} \right] dx' \right\}, \quad (12)$$

where

$$e = \sqrt{1 + (\alpha - \beta)^2}; \qquad j = \sqrt{1 + \beta^2}.$$

In general, the evaluation of these integrals involves that of elliptic integrals of the third kind and is simpler to obtain graphically or numerically. In the special cases, however, where the piling is either at the heel or toe of the dam or at its center, F can be expressed in terms of elementary functions. Thus, for

Piling at heel of dam:

$$\beta = 0, \qquad j = 1, \qquad e = \sqrt{1 + \alpha^2}$$
$$F = p_1 w + \frac{d\Delta p(e - 1)}{2}. \qquad (13)$$

Piling at center of base of dam:

$$\beta = \frac{\alpha}{2}, \qquad j = e$$
$$F = p_1 w + \frac{w\Delta p}{2} = \frac{(p_2 + p_1)w}{2}. \qquad (14)$$

Piling at toe of dam:

$$\beta = \alpha, \qquad e = 1, \qquad j = \sqrt{1 + \alpha^2}$$
$$F = p_1 w + d\Delta p\left(\alpha - \frac{(j - 1)}{2}\right). \qquad (15)$$

The total uplift moment is given by

$$M = \frac{p_1 w^2}{2} + \frac{\Delta p}{\pi} d\beta F_0 + \frac{d^2 \Delta p}{8\pi} \Bigg[\pi(3e^2 - j^2 - 2ej - 4)$$
$$+ \{8 + 2je - 3(j^2 + e^2)\} \left\{ \sin^{-1} \sqrt{\frac{j+1}{j+e}} + \sin^{-1} \sqrt{\frac{e+1}{j+e}} \right\}$$
$$- \sqrt{(j+1)(e-1)}(3j - 3e - 2)$$
$$- \sqrt{(e+1)(j-1)}(3e - 3j - 2) \Bigg] \quad (16)$$

where F'_0 is the expression in the bracket in Eq. (12).

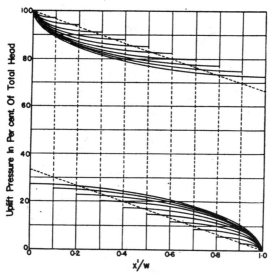

FIG. 50.—The pressure distribution at the base of dams with sheet piling of depth equal to the width of base. x'/w = (distance from heel of dam)/(width of base). The vertical dotted lines give the position of the piling, and their length the pressure drop over the piling. The sloping dotted lines, together with the vertical dotted segments, give the pressure distribution predicted by the "line of creep" theory. (*After Weaver, J. Math. and Physics.*)

Another interesting result is the loss in pressure δp in encircling the piling, which is given by the difference in pressure between the upstream and downstream sides of the piling at the base of the dam. From Eq. (11) this is found to be

$$\delta p = \frac{\Delta p}{\pi} \cos^{-1} \frac{1}{(j+e)^2}[(j-e)^2 - 4 + 4\sqrt{(j^2-1)(e^2-1)}],$$
$$(17)$$

the actual pressure at the bottom of the piling having the value

$$p = \frac{\Delta p}{\pi} \cos^{-1} \frac{j - e}{j + e} + p_1 + \gamma g d. \tag{18}$$

The significance of these results may be most readily understood from the graphical representation of Eqs. (11) to (18). Thus Figs. 50 and 51 show the uplift pressure distribution at the base of the dam for various positions of the piling and for ratios of dam

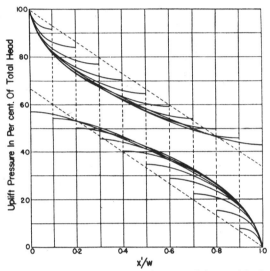

Fig. 51.—The pressure distribution at the base of dams with sheet piling of depth equal to one-fourth of base width. x'/w = (distance from heel of dam)/ (width of base). The vertical dotted lines give the position of the piling, and their length, the pressure drops over the piling. The sloping dotted lines, together with the vertical dotted segments, give the pressure distribution predicted by the "line of creep" theory. (*After Weaver, J. Math. and Physics.*)

width to piling depth of 1 and 4, the ordinates giving the fraction of the total pressure drops corresponding to the $(p - p_1)/\Delta p$ of Eq. (11). The positions of the piling are indicated by the vertical dotted lines whose lengths also give the pressure drop across the piling according to Eq. (17). While these do not vary very markedly with the position of the piling, they decrease rather rapidly with increasing α, as is, of course, to be expected.

The total uplift forces and uplift moments, expressed as fractions of what they would be without sheet piling (and $p_1 = 0$), are given in Figs. 52 and 53 plotted as functions of the position of the piling for different values of α.   It is of interest to note that piling will not affect the total uplift force when set at the center of the

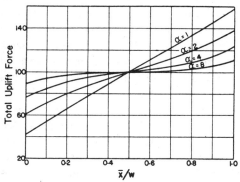

Fig. 52.—The variation of the total uplift force with position of sheet piling under base of dam, expressed as per cent of the force when piling is absent, the downstream fluid head being zero. $\bar{x}/w$ = (distance of piling from heel of dam)/ (width of base). α = (width of base)/(piling depth). (*After Weaver, J. Math. and Physics.*)

base of the dam, but will increase or decrease it as it is placed under the downstream or upstream half of the dam.

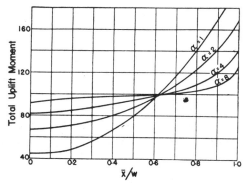

Fig. 53.—The variation of the total uplift moment with respect to heel of dam with position of sheet piling under base of dam, expressed as per cent of the moment when piling is absent, the downstream fluid head being zero. $\bar{x}/w$ = (distance of piling from heel of dam)/(width of base). α = (width of base)/ (piling depth). (*After Weaver, J. Math. and Physics.*)

Although this analysis of the uplift pressures on dams is based upon a rather idealized representation of the problem—the most serious approximations are probably those of the infinitely extended character of the porous material underneath the dam[1] and the lack of penetration of the latter in the porous medium— previous theories have been based upon purely intuitive assumptions. Perhaps the most widespread of such theories[2] has been the "line-of-creep" theory of Bligh, which consists essentially in the assumption that a piling of depth d adds to the effective width of the base of the dam its perimeter of $2d$, the whole pressure drop being distributed linearly over this extended width of $w + 2d$. The resulting pressure distributions at the base of the dam for the cases $\alpha = 1, 4$ are indicated by the sloping dotted lines in Figs. 50 and 51, the vertical dotted lines between the sloping lines again giving the pressure drops across the piling. While these errors in the pressure distribution are as a whole not very large, it may be noted that the pressure drops over the piling in the line-of-creep theory are independent of the position of the piling, whereas the above theory shows that this drop may vary by a factor of 2 as the piling is moved from the heel or toe to the center of the dam.

As to the total uplift force, the line-of-creep theory gives too small values (up to 20 per cent) when the piling is under the upstream half of the dam, and too large values (about 8 per cent) for downstream positions of the piling. The uplift moments predicted by the line-of-creep theory are too low (up to 30 per cent) when the piling is set within 70 per cent of the dam width from the heel, and are slightly too high for piling positions within the 30 per cent of the width from the toe.

While the line-of-creep theory is, therefore, only approximate from a quantitative point of view, it should be mentioned that its fundamental assumption of a streamline following the base of the dam and the sides of the piling is strictly correct. Unfortunately the existence of this streamline is the subject of insistent denial in the discussion by Taylor and Uppal[3] of their otherwise

---

[1] This approximation will give results on the "safe side"—too small pressure drops across the piling and too great uplift forces (*cf.* Fig. 68). The actual errors, however, will be quite small except when the piling extends to within a short distance of the base of the permeable stratum.

[2] Other theories—really nothing more than assumptions—are listed by Weaver (*loc. cit.*).

[3] Taylor, E. M., and H. L. Uppal, Punjab Irrigation Research Institute, *Research Pub.*, 2, nos. 3, 4, 1934. The results of an extensive study of the pressure distributions under an actual dam—the Panjnad Weir—are given

beautiful experimental work on the streamlines in models of dams with and without sheet pilings.   In these experiments the traces of the streamlines in the sand under the dams were formed by introducing at chosen inlet tap funnels a silver nitrate solution, the seepage fluid being a solution of potassium chromate.   Typical photographs of such models are reproduced in Figs. 54, 55, and 56 for dams without piling, with one sheet, and with two sheets. From the fact that no streamlines were observed actually to

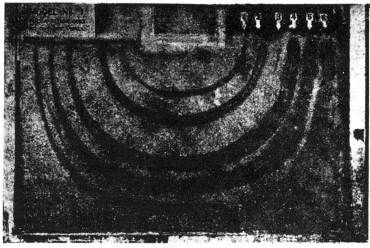

Fig. 54.—Photograph of streamlines under a dam without sheet piling, as obtained by sand-model experiments.   (*After Taylor and Uppal, Research Pub., Punjab Irrigation Research Inst.*)

follow the base of the dam as in Fig. 54, or as in the case of Fig. 55 that the first streamline appears to touch the base of the piling in continuing to the downstream side of the dam, it is concluded that the sides of the piling and base of the dam really do not represent a streamline.   Such a conclusion, however, is fallacious not only because in any potential system an impermeable fixed

---

by A. N. Khosla, *Paper* 162, p. 50, *Proc. Punjab Eng. Cong.*, **21**, 1933.   Cf. also J. B. T. Colman, *Amer. Soc. Civil Eng.*, **80**, 421, 1916, dealing with the pressure distributions in sand models of dams with and without piling, the latter placed either at the heel, toe, or both.   While some of the details reported in these papers cannot be reconciled with the implications of the theory developed here, the general qualitative results confirm the conclusions following from the analytical treatment of the problem.

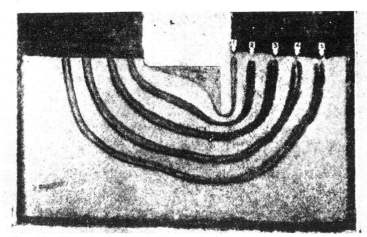

FIG. 55.—Photograph of streamlines under a dam with piling at the heel, as obtained by sand-model experiments. (*After Taylor and Uppal, Research Pub., Punjab Irrigation Research Inst.*)

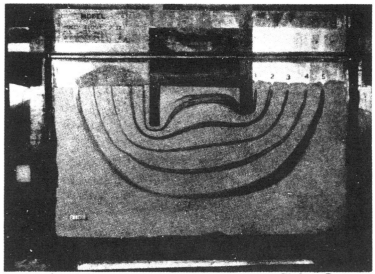

FIG. 56.—Photograph of streamlines under a dam with piling both at the heel and toe, as obtained by sand-model experiments. (*After Taylor and Uppal, Research Pub., Punjab Irrigation Research Inst.*)

boundary must necessarily be a streamline, but also because the streamlines traced in the experimental model seemed to touch the base of the piling only as the result of their finite width, owing to the coarseness of the sand and the natural diffusion of the silver nitrate solution. The crowding of the streamlines about sharp impermeable projections in a flow system necessitates the use of more refined methods if the details of the flow are to be followed with precision. On the other hand, it is to be observed that, except for the case where there is no piling, the limiting streamline following the base of the dam will in general be of lower

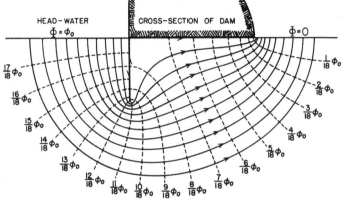

Fig. 57.—Theoretical potential and streamline distributions under a dam with sheet piling at the heel of a depth equal to half of the base width. (*After Weaver, J. Math. and Physics.*)

average velocity than others cutting across the base of the piling (though not actually touching it) and passing directly to the downstream bed. The region between the streamline of highest average velocity and the "creep" streamline following the base of the dam and piling may then appear to be a "dead" zone, although in reality fluid is flowing along all the streamlines of the system.

To see that insofar as the experimental models are refined enough to give details of the motion they do agree with the theory given here, one need only compare Fig. 55 with Fig. 57,[1]

[1] The equipotentials in this figure—taken from Weaver, *loc. cit.*—were originally marked as equipressure curves. These, however, are to be obtained from those shown by the addition to the value of Φ at each point the term $\gamma g y$ (for $k/\mu = 1$). With respect to the detailed variations of the

which gives the computed equipotentials and streamlines for
exactly the same case—a piling depth equal to half of the dam
width. All the significant features are evidently the same. The
persistence in the distortions of the last streamlines of the model
are undoubtedly to be attributed to the finite width of the piling,
finite depth of sand in the piling, and inhomogeneities in the sand.
As a whole, the distortion in the streamlines is seen to extend only
to a distance roughly equal to the depth of the piling. In fact,
these model experiments may be taken as another direct confirma-

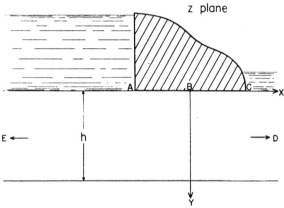

Fig. 58.—*z*-plane representation of a dam without sheet piling setting on a
permeable bed of finite thickness.

tion of the validity of the generalized Darcy's law in the discussion of other than linear-flow systems.

**4.12. The Seepage Flux under Dams of Extended Base Width.
No Sheet Piling. Elliptic-function Transformations.[1]**—In the
last two sections the theory was given of the uplift pressures
underneath dams under the assumption that the underlying sand
was of infinite thickness. While this assumption, as will be seen
below, will in general afford a good approximation with respect to
the pressure distribution and total uplift on the dams, except

---

pressure distribution at the base of the dam, with sheet piling at the heel, a
very satisfactory confirmation of the theory (Figs. 50 and 51) has also been
established by both electrolytic and direct sand-model experiments (*cf.* G.
Ram, V. I. Vaidhianathan, and E. M. Taylor, *Ind. Acad. Sci.*, **2**, 22, 1935).

[1] The results of this and the following two sections are taken from the
paper by M. Muskat, *Physics*, **7**, 116, 1936.

when the sand thickness is small or the piling approaches closely to the base of the permeable bed, it results in a seepage flux of infinite magnitude. To get a physically sensible value for this flux one must, therefore, take into account the finite thickness of the permeable bed. This may be done in exactly the same manner as that outlined in Sec. 4.11, except that the conjugate function transformations of Eq. 4.11(1) or 4.11(8) will have to be replaced by an elliptic-function transformation.

Thus applying the Schwarz-Christoffel theorem—Eq. 4.11(3)—, it is readily seen that the z-plane diagram of the flow system of

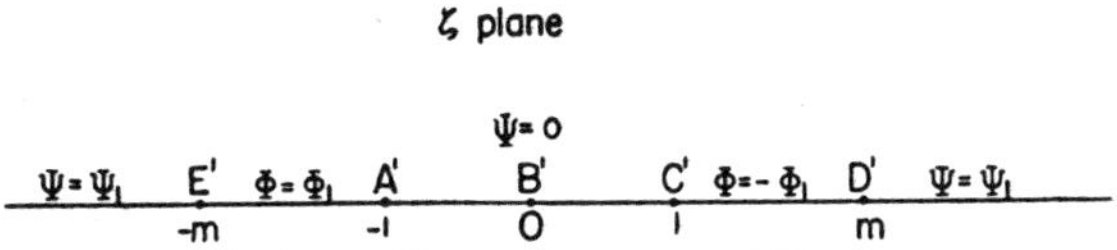

Fig. 59.—The ζ plane map of Fig. 58.

Fig. 58 is transformed onto the upper half of the ζ plane by the transformation function $z(\zeta)$ given by

$$z = c_1 \int \frac{d\zeta}{\zeta^2 - m^2} + c_2. \tag{1}$$

Evaluating the integral and inverting, it is easily found that

$$\zeta = \frac{\tanh \pi z/2h}{\tanh \pi w/4h}, \tag{2}$$

where w is the dam width, h the sand thickness, and the constants have been chosen so that

$$z = 0 \rightarrow \zeta = 0; \qquad z = \pm\frac{w}{2} \rightarrow \zeta = \pm 1. \tag{3}$$

The ζ-plane diagram with the boundary values of the potential and stream function is indicated in Fig. 59, m having the value $\operatorname{ctnh} \pi w/4h$.

Inverting now the point of view outlined in Sec. 4.11 and constructing ab initio the $\omega = \Phi + i\Psi$ plane diagram corresponding to the z plane, it is clear that one will find a rectangular figure as shown in Fig. 60.[1]   Hence if this ω plane is mapped onto the ζ

[1] The value of Ψ_1 is shown as negative as a consequence of the relation $\dfrac{\partial \Psi}{\partial y} = \dfrac{\partial \Phi}{\partial x} < 0$ for the coordinate system of Fig. 58.

plane of Fig. 59, one will have the relation between ω and z, through the intermediate variable ζ, giving the desired potential and streamline distributions in the original system. Applying again the Schwarz-Christoffel theorem to effect this mapping, one finds

$$\omega = \Phi + i\Psi = c_1 \int^{\zeta} \frac{d\zeta}{\sqrt{(\zeta^2 - 1)(\zeta^2 - m^2)}} + c_2, \tag{4}$$

where, as already noted, m, the coordinate of D', must have the value ctnh $\pi w/4h$ in order for Eq. (3) to be satisfied. Setting

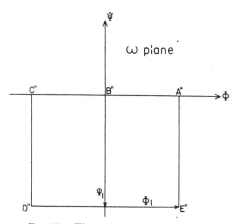

FIG. 60.—The ω-plane map of Fig. 58.

$k^* = 1/m$, $c_2 = 0$, and choosing the limits of the integral so that $\omega = 0$ corresponds to B', Eq. (4) becomes

$$\omega = c_1 \int_0^{\zeta} \frac{d\zeta}{\sqrt{(1 - \zeta^2)(1 - k^{*2}\zeta^2)}} \tag{5}[1]$$

When $\zeta \leqslant 1$, this may also be expressed as

$$\omega = c_1 \int_0^{\sin^{-1}\zeta} \frac{d\theta}{\sqrt{1 - k^{*2} \sin^2 \theta}} = c_1 F(\sin^{-1} \zeta, k^*); \qquad k^* = \tanh\frac{\pi w}{4h}, \tag{6}$$

where F is the elliptic integral of the first kind of "amplitude" $\sin^{-1} \zeta$ and "modulus" k^*.[2]

[1] The constant c_1 is supposed to have absorbed the factor k^*.

[2] The conventional symbol k for the modulus has been replaced by k^* in order to avoid confusion with that used here for the permeability.

Noting that the point C'', at which $\Phi = -\Phi_1$, $\Psi = 0$, corresponds to $\zeta = +1$, it follows that

$$-\Phi_1 = c_1 \int_0^1 \frac{d\zeta}{\sqrt{(1 - \zeta^2)(1 - k^{*2}\zeta^2)}} = c_1 F\left(\frac{\pi}{2}, k^*\right) = c_1 K(k^*), \quad (7)$$

where $K(k^*)$ is the complete elliptic integral of the first kind with modulus k^*. As the total potential drop in the system, $\Delta\Phi$, is $2\Phi_1$, Eq. (5) can be finally rewritten as

$$\omega = \Phi + i\Psi = \frac{-\Delta\Phi}{2K} \int_0^\zeta \frac{d\zeta}{\sqrt{(1 - \zeta^2)(1 - k^{*2}\zeta^2)}}. \quad (8)$$

The seepage flux underneath the dam is evidently given by the value of Ψ_1. This may be found from the value of ω for D'' corresponding to $\zeta = m = 1/k^*$. Thus,

$$-\Phi_1 + i\Psi_1 = \frac{-\Delta\Phi}{2K} \int_0^{\frac{1}{k^*}} \frac{d\zeta}{\sqrt{(1 - \zeta^2)(1 - k^{*2}\zeta^2)}} =$$
$$\frac{-\Delta\Phi}{2K}(K + iK'), \quad (9)[1]$$

where K' is the complete elliptic integral of the first kind with modulus $\sqrt{1 - k^{*2}}$. As $-\Psi_1$ is the total flux, Q, underneath the dam (per unit length of dam), and $\Delta\Phi$ is the total potential drop from upstream to downstream, corresponding to a pressure drop Δp or fluid-head difference ΔH, it follows at once from Eq. (9) that

$$Q = \frac{kK'\Delta p}{2\mu K} = \frac{k\gamma g K'\Delta H}{2\mu K}; \quad k^* = \tanh\frac{\pi w}{4h}. \quad (10)$$

The seepage flux per unit potential drop, *i.e.*, $Q/\Delta\Phi = K'/2K$, is plotted in Fig. 61 as a function of w/h, the ratio of the dam width to the thickness of the permeable stratum.[2] It will be seen that, as is to be expected, the flux, or effective conductivity of the system, decreases with increasing values of w/h from infinitely large values for vanishing w/h and tends to a vanishing flux when w/h becomes infinitely large. In an actual calculation of the

---

[1] WHITTAKER and WATSON, p. 502. A complete discussion of both the elliptic integrals and functions will be found in Chap. XXII of this treatise.

[2] Very complete tables of K, K' and K'/K may be found in "Tafeln der Besselschen, Theta, Kugel- und anderer Funktionen," 1930, by K. Hayashi.

seepage one must, of course, multiply the ordinates of Fig. 61 by $k\gamma g\Delta H/\mu$.[1]

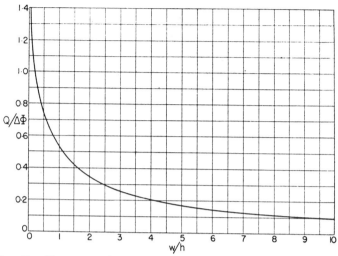

Fig. 61.—The seepage flux under dams with no sheet piling. $Q/\Delta\Phi$ = flux per unit potential drop across the dam, and per unit dam length. w/h = (width of base)/(thickness of permeable stratum below the base). (*From Physics*, **7**, 118, 1936.)

Although Eq. (10) or its equivalent form

$$\frac{Q}{\Delta\Phi} = \frac{K'}{2K} \tag{11}$$

[1] It may be mentioned that a remarkably close approximation to the exact values of Q given by Eq. (10) was derived in 1917 by Ph. Forchheimer (*Wien, Ber.*, **126**, 409, 1917) who obtained his approximate solution by constructing an infinite set of images of the base of the dam so as to attain a vanishing normal flux at $y = h$. The effect of the infinite series of images was summed by introducing a complex variable transformation equivalent to: $X + iY = \cosh \pi\dfrac{(x + iy)}{h}$, which maps an infinite strip, as in Fig. 58, of the (x, y) plane onto the whole of the (X, Y) plane, into the potential function given in the footnote of p. 184. The only error int his analysis is, as pointed out by Forchheimer, that it does not give a strictly uniform potential along the y axis of Fig. 58, as must be required by symmetry (*cf.* Sec. 4.16). The resulting formulas of Forchheimer are

$$\frac{Q}{\Delta\Phi} = \frac{1}{2\pi} \log \frac{\cosh \pi w/8h - 1}{3 \cosh \pi w/8h - 1}, \quad \text{for} \quad \frac{w}{h} < 1 \quad \text{and}$$

$$\frac{Q}{\Delta\Phi} = \frac{1}{\left[0.86 + \dfrac{w}{h}\right]} \quad \text{for} \quad \frac{w}{h} > 1.$$

has been derived here for the special case of the seepage underneath a dam with no sheet piling, it is important to observe that it is a perfectly general result for any system bounded by two segments of uniform potential alternating with two streamline segments. The whole characterization of the particular system of interest with respect to the value of $Q/\Delta\Phi$ lies in the value of the modulus k^* of the elliptic integrals.

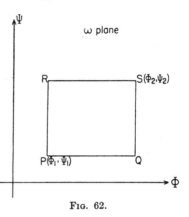

FIG. 62.

Thus assuming that the flow system with an ω-plane (Φ, Ψ) representation, as in Fig. 62, has been mapped onto the upper half of a ζ plane so as to give the representation of Fig. 63, it follows from the Schwartz-Christoffel theorem that the relation between ω and ζ is given by an equation of the form

$$\omega = \Phi + i\Psi = \int \frac{d\zeta}{[(\zeta - r)(\zeta - p)(\zeta - q)(\zeta - s)]^{\frac{1}{2}}}. \quad (12)$$

If now one makes the transformation[1]

$$\lambda = \frac{D\zeta - B}{A - C\zeta} \quad (13)$$

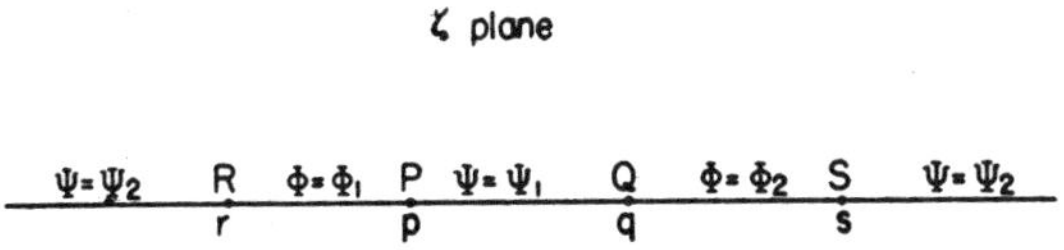

ζ plane

$\Psi = \Psi_2$	R	$\Phi = \Phi_1$	P	$\Psi = \Psi_1$	Q	$\Phi = \Phi_2$	S	$\Psi = \Psi_2$
	r		p		q		s	

FIG. 63.—The ζ-plane map of Fig. 62.

and chooses A, B, C, D so that

$$\left.\begin{array}{ll} A - Cp = B - Dp; & A - Cq = Dq - B \\ A - Cr = k^*(B - Dr); & A - Cs = k^*(Ds - B), \end{array}\right\} \quad (14)$$

it is found that ω assumes the form

$$\omega = \Phi + i\Psi = G \int_0^\lambda \frac{d\lambda}{\sqrt{(1 - \lambda^2)(1 - k^{*2}\lambda^2)}}, \quad (15)$$

[1] *Cf.* BATEMAN, H., "Partial Differential Equations," p. 302, 1932.

where the modulus k^* in Equations (14) and (15) is the root, smaller than unity, of the equation

$$k^{*2}(q - p)(r - s)$$
$$+ k^*[(q - p)^2 + (r - s)^2 - (p + q - r - s)^2]$$
$$+ (q - p)(r - s) = 0 \quad (16)$$

and G is a constant resulting from the transformation.

In addition to reducing Eq. (12) to the standard elliptic integral form of Eq. (15), the transformations of Eqs. (13) and (14) lead to the following correspondences between the ω and λ planes:

$$\lambda(R) = \frac{-1}{k^*}; \quad \lambda(P) = -1; \quad \lambda(Q) = +1; \quad \lambda(S) = \frac{1}{k^*}.$$
$$(17)$$

Hence,

$$\left.\begin{array}{l}
\omega(R) = \Phi_1 + i\Psi_2 = G\displaystyle\int_0^{\frac{-1}{k^*}} \frac{d\lambda}{\sqrt{(1 - \lambda^2)(1 - k^{*2}\lambda^2)}} = \\
\qquad\qquad\qquad\qquad\qquad\qquad -G(K - iK') \\[2ex]
\omega(P) = \Phi_1 + i\Psi_1 = G\displaystyle\int_0^{-1} \frac{d\lambda}{\sqrt{(1 - \lambda^2)(1 - k^{*2}\lambda^2)}} = \\
\qquad\qquad\qquad\qquad\qquad\qquad\quad -GK \\[2ex]
\omega(Q) = \Phi_2 + i\Psi_1 = G\displaystyle\int_0^1 \frac{d\lambda}{\sqrt{(1 - \lambda^2)(1 - k^{*2}\lambda^2)}} = GK \\[2ex]
\omega(S) = \Phi_2 + i\Psi_2 = G\displaystyle\int_0^{\frac{1}{k^*}} \frac{d\lambda}{\sqrt{(1 - \lambda^2)(1 - k^{*2}\lambda^2)}} = \\
\qquad\qquad\qquad\qquad\qquad\qquad G(K + iK')
\end{array}\right\} \quad (18)[1]$$

where K, K' are again the complete elliptic integrals of the first kind with modulus k^*, $\sqrt{1 - k^{*2}}$, respectively.

From these it follows at once that

$$\frac{\Psi_2 - \Psi_1}{\Phi_2 - \Phi_1} = \frac{Q}{\Delta\Phi} = \frac{K'}{2K}, \quad (19)$$

the detailed properties of the system in the z plane entering only in the values of p, q, r, s, which determine the value of k^* by Eq. (16). This result will be found quite useful in the calculations of the

---

[1] *Cf.* Whittaker and Watson, Chap. XXII.

seepage flux underneath dams with sheet piling to be made in the following sections.

It should be further noted that Eq. (10) is not only applicable to the seepage underneath a dam without piling, but it should also afford a good approximation to the seepage around the banks of the dam, if the latter are not directly anchored to impervious rocks. For the part of the dam above the downstream water level, the pressure drop to be used in Eq. (10) may be approximated by the average of that on the upstream side to the depth of the downstream water level. The lateral seepage below the level of the downstream side will correspond to the total fluid-head difference between the two faces of the dam.

Although the main purpose of the analysis of this section has been that of deriving an expression for the seepage flux—Eq. (10)—it is also of interest to compare for this simple case the pressure distribution underneath the dam given here by the exact theory, where the finite thickness of the permeable underlying stratum is taken into account, and that developed in Sec. 4.10 for the case of an infinitely extended permeable stratum. For this purpose it is only necessary to notice that since $\Psi = 0$ along the base of the dam, Eq. (6) gives

$$\frac{p}{\Delta p} = \frac{\Phi}{\Delta \Phi} = -\frac{1}{2K} F(\sin^{-1} \zeta, k^*);$$

$$\zeta = \frac{\tanh \pi x/2h}{\tanh \pi w/4h}; \qquad k^* = \tanh \frac{\pi w}{4h}, \quad (20)$$

where Φ and x are measured from the center of the dam. The resultant pressure distribution for $w/h = 5$ is plotted in Fig. 44 as the dotted curve, that for $w/h = 1$ being indicated by the circles. It will be seen from these that the assumption of an infinitely thick permeable stratum underlying the dam gives a good approximation to the pressure distribution (and hence total uplift) along the base of the dam, even though it does imply an infinite seepage flux.

4.13. The Seepage Flux under Dams of Extended Base Width; Sheet Piling Present.—When the dam is provided with a sheet piling extending to the depth d in the permeable stratum of thickness h, the seepage flux can be again computed by the successive application of two conjugate-function transformations derived from the Schwarz-Christoffel theorem. Thus, the z

plane of Fig. 64 will be mapped onto the upper half of the ζ plane with the correspondences

$$A \to \zeta = -1; \qquad B \to \zeta = 0; \qquad C \to \zeta = +1 \qquad (1)$$

by the transformation function

$$z = c_1 \int \frac{\zeta \, d\zeta}{(\zeta^2 - m^2)\sqrt{\zeta^2 - 1}} + c_2$$

$$= \frac{c_1}{2\sqrt{m^2 - 1}} \log \frac{\sqrt{m^2 - 1} - \sqrt{\zeta^2 - 1}}{\sqrt{m^2 - 1} + \sqrt{\zeta^2 - 1}} + c_2, \qquad (2)$$

where the points D, E are supposed to go into $\zeta = \pm m$, and the· sign of $\sqrt{\zeta^2 - 1}$ should be $-$ for $\zeta < -1$, $+$ for $\zeta > +1$, and

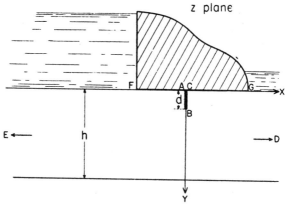

FIG. 64.—The z-plane representation of a dam with sheet piling of depth d setting on a permeable stratum of thickness h.

$+$ imaginary for $-1 \leqslant \zeta \leqslant +1$. Imposing the correspondences of Eq. (1) and the requirement that at $\zeta = \pm m$, z goes over into the line $z = x + ih$, it is readily found that

$$c_1 = -\left(\frac{2h}{\pi}\right) \operatorname{ctn} \frac{\pi d}{2h}, \qquad c_2 = 0, \qquad m = \csc \frac{\pi d}{2h}. \qquad (3)$$

Equation (2) can therefore be written as

$$\zeta^2 = 1 + \operatorname{ctn}^2 \frac{\pi d}{2h} \tanh^2 \frac{\pi z}{2h}. \qquad (4)$$

The ζ-plane diagram is indicated in Fig. 65 with the boundary values of Φ and Ψ, and the extremities of the dam, F', G', denoted by $-a$, b.

To find the seepage flux, it is now only necessary to apply the results of the last section, which give

$$\frac{Q}{\Delta\Phi} = \frac{K'}{2K}, \tag{5}$$

with

$$m(a + b)k^{*2} - 2(ab + m^2)k^* + m(a + b) = 0$$

or

$$\left. \begin{aligned} k^* &= \frac{ab + m^2 - \sqrt{(m^2 - b^2)(m^2 - a^2)}}{m(a + b)}, \\ \text{where} \\ b^2 &= 1 + \operatorname{ctn}^2 \frac{\pi d}{2h} \tanh^2 \frac{\pi w}{2h}\left(1 - \frac{\bar{x}}{w}\right); \\ a^2 &= 1 + \operatorname{ctn}^2 \frac{\pi d}{2h} \tanh^2 \frac{\pi \bar{x}}{2h}, \end{aligned} \right\} \tag{6}$$

$\bar{x}$ being the distance of the piling from the heel of the dam.

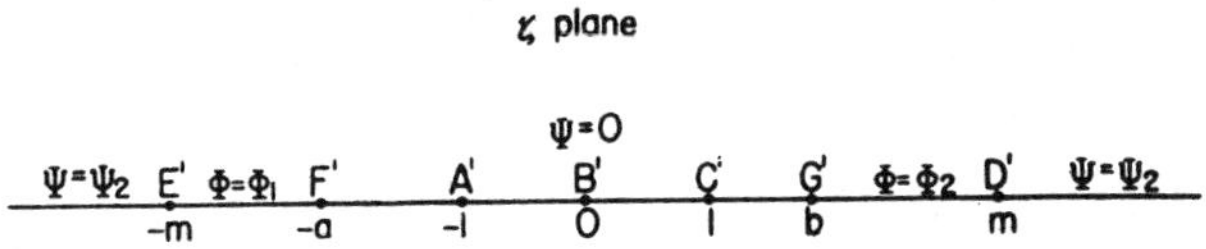

FIG. 65.—The ζ plane map of Fig. 64.

To show the effect of the position of the piling on the seepage flux, $Q/\Delta\Phi$ is plotted in Fig. 66 as a function of $\bar{x}/w$ for $w/h = 1$, $d/h = 0.5$. It will be seen that while the flux is a maximum for the symmetrical position of the piling, the total variation is not large. To see how $Q/\Delta\Phi$ varies with the relative width of the dam w/h and piling depth d/h, it will, therefore, suffice to suppose that the piling is set at the middle of the base. The values so obtained will furthermore give upper limits to the seepage flux, and hence be on the "safe side" when used in calculations of practical flow systems.

The variation of $Q/\Delta\Phi$ with the depth of the piling, d/h, is shown in Fig. 67 for various values of dam width, w/h, and for $\bar{x}/w = \frac{1}{2}$. As is to be expected, the seepage flux decreases with increasing piling depth, and dam width, although for dam widths appreciably larger than the thickness of the permeable stratum the flux is almost independent of the piling depth until the latter extends to the immediate neighborhood of the base of the permeable layer. The fact that the flux rises very sharply as d/h decreases from the value 1 shows that unless the piling is

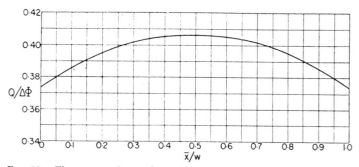

FIG. 66.—The seepage flux under a dam as a function of the position of the sheet piling. $Q/\Delta\Phi$ = flux per unit potential drop across dam and per unit dam length. $\bar{x}/w$ = (distance of piling from heel)/(width of base); (width of base)/(thickness of underlying permeable stratum) = 1; (depth of piling)/(thickness of underlying permeable stratum) = 0.5. *(From Physics, 7, 121, 1936.)*

actually anchored to the impermeable base a considerable fraction of the seepage without piling will still persist even when the piling penetrates 99 per cent of the thickness of the permeable layer.

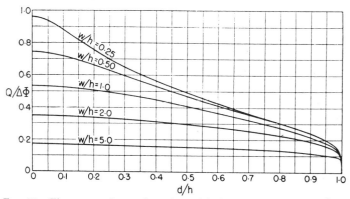

FIG. 67.—The seepage flux under a dam with sheet piling set at the center of the base. $Q/\Delta\Phi$ = flux per unit potential drop across dam and per unit dam length. d = depth of piling; w = width of base; h = thickness of permeable stratum below dam. *(From Physics, 7, 121, 1936.)*

While we shall not enter here into the details of the pressure distribution at the base of dams with sheet piling as given by the analysis developed above, an indication of the accuracy of the more approximate theory of Sec. 4.11 may be obtained by computing the pressure drop over the piling for the special case where the piling is set at the center of the base of the dam. By repeating the analysis leading to Eq. 4.12(5), it is readily found that this pressure drop, δp, is given by

$$\frac{\delta p}{\Delta p} = \frac{1}{K}\int_0^{\frac{1}{b}} \frac{d\zeta}{\sqrt{(1 - \zeta^2)(1 - k^{*2}\zeta^2)}} = \frac{1}{K}F(\sin^{-1}\frac{1}{b}, k^*), \quad (7)$$

where

$$k^* = \frac{b}{m}; \quad b^2 = 1 + \operatorname{ctn}^2\frac{\pi d}{2h}\tanh^2\frac{\pi w}{4h}; \quad m = \operatorname{csc}\frac{\pi d}{2h}, \quad (8)$$

and Δp is the total pressure drop across the dam. Equation (7) is plotted in Fig. 68 for various values of d/h. The curve for

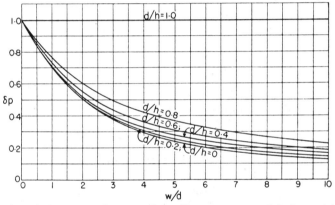

FIG. 68.—The pressure drop over sheet piling set at center of the base of dam. δp = (pressure drop over piling)/(total pressure drop across dam). d = depth of piling; w = width of base; h = thickness of permeable stratum below dam. (*From Physics*, **7**, 122, 1936.)

$d/h = 0$ corresponds to the case of an infinitely thick permeable stratum, where Eq. (7) reduces to the equivalent of Eq. 4.11(17), namely,

$$\frac{\delta p}{\Delta p} = \frac{1}{\pi}\cos^{-1}\frac{(w/2d)^2 - 1}{(w/2d)^2 + 1}. \quad (9)$$

The fact that the deviations from this limiting curve do not become large until $d/h > 0.5$ gives a justification for using the simpler theory of Sec. 4.11 (where $d/h = 0$) in the study of the uplift pressures and moments in dams with sheet piling, at least for smaller values of d/h. In any case, this approximation should be "safe" from the point of view of the practical design of dams with sheet piling, as it will lead to uplift pressures and moments that are higher than those that will actually occur.

As to the numerical values of the seepage flux that correspond to the curves of Figs. 66 and 67, as well as those presented in the following section, it should be observed that the numerical value of the rate of flow $\bar{Q}$ will be given by

$$\bar{Q} = \frac{kL\Delta p}{\mu}\left(\frac{Q}{\Delta\Phi}\right), \tag{10}$$

where k is the permeability of the permeable stratum, μ the viscosity of the water, L the length of the dam, and Δp the pressure differential between the upstream and downstream sides of the dam. If k is in darcys, L in centimeters, Δp in atmospheres, μ in centipoises, $\bar{Q}$ will be expressed in cubic centimeters per second. Thus, taking

$k = 10$ darcys; $\mu = 1$ centipoise; $L = 3,048$ cm. $= 100$ ft.;
$$\Delta p \cong 33.9 \text{ ft. of } H_2O \cong 1 \text{ atm.,}$$

a value of $Q/\Delta\Phi = 0.1$ corresponds to

$$\bar{Q} = 3,048 \text{ cc./sec.} \sim 48.32 \text{ gal./min.} \tag{11}$$

In view of the magnitude of this seepage rate and the fact that, as shown by Fig. 67, values of $Q/\Delta\Phi \sim 0.1$ persist for piling depths as great as 99 per cent of the thickness of the permeable stratum, it becomes clear that real safety from high seepage velocities and the dangers of undermining can only be attained, in the case of dams which are not set themselves directly on an impermeable base, by actually anchoring the piling to the impermeable bed. Unless the piling is so anchored, it will cut down the seepage but little, except in such cases where the width of the dam is small compared to the thickness of the permeable stratum.[1]

[1] For a more detailed discussion of some of the practical aspects of the problem of seepage of water underneath dams, see A. Casagrande, *Proc. Amer. Soc. Civil Eng.*, **61**, 365, 1935; *cf.* also the papers by E. W. Lane, *Proc. Amer. Civil Eng.*, **60**, 929, 1934, in which various data are summarized for more than 250 actual dams, and that by L. F. Harza, p. 967 of the same journal, in which are given the results of the measurements of the uplift

4.14. The Seepage Flux underneath Coffer Dams.—While the results of the last several sections should suffice to give the salient features of the seepage flow of water underneath permanent dam structures, the use of temporary water shut-offs, as coffer dams, during the course of building hydraulic works lends practical interest also to the question of the seepage flux underneath such temporary devices for the shutting off or the reduction of the quantity of seepage water. A typical diagrammatic representation of such a coffer-dam structure is indicated in Fig. 69, where

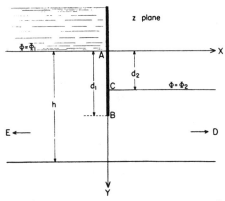

Fig. 69.—The z-plane representation of a coffer dam set to a depth d_1 below the surface with a depth of excavation equal to d_2.

the dam is set to the depth d_1 into the permeable stratum of thickness h, the downstream side having been excavated to the depth d_2, the dam itself being taken as a thin impermeable sheet—AB. Pumps are installed on the downstream side to remove the water seeping into the excavation. A knowledge of the seepage to be expected will be of value not only in estimating the proper capacity of the pumping installation but also in anticipating the dangers of underwashing and undermining the dam itself.

Applying once more Eq. 4.11(3), it is readily found that the z-plane diagram of Fig. 69 will be mapped onto the upper half of the ζ plane by the transformation function defined by

---

pressures under dams, obtained by means of electrolytic models, including a number of cases with 2 sheets of piling and also such where the porous base is stratified.

$$z = c_1 \int \frac{\zeta \, d\zeta}{(\zeta - m_1)(\zeta + m_2)\sqrt{\zeta^2 - 1}} + c_2$$

$$= \frac{c_1}{m_1 + m_2} \left[\frac{-m_1}{\sqrt{m_1^2 - 1}} \log \frac{\sqrt{(m_1 - 1)(\zeta + 1)} + \sqrt{(m_1 + 1)(\zeta - 1)}}{\sqrt{(m_1 - 1)(\zeta + 1)} - \sqrt{(m_1 + 1)(\zeta - 1)}} \right.$$

$$\left. + \frac{m_2}{\sqrt{m_2^2 - 1}} \log \frac{\sqrt{(m_2 + 1)(\zeta + 1)} + \sqrt{(m_2 - 1)(\zeta - 1)}}{\sqrt{(m_2 + 1)(\zeta + 1)} - \sqrt{(m_2 - 1)(\zeta - 1)}} \right] + c_2,$$

$$(1)$$

where the correspondences

$$A \to -1; \quad B \to 0; \quad C \to +1; \quad D \to m_1; \quad E \to -m_2 \quad (2)$$

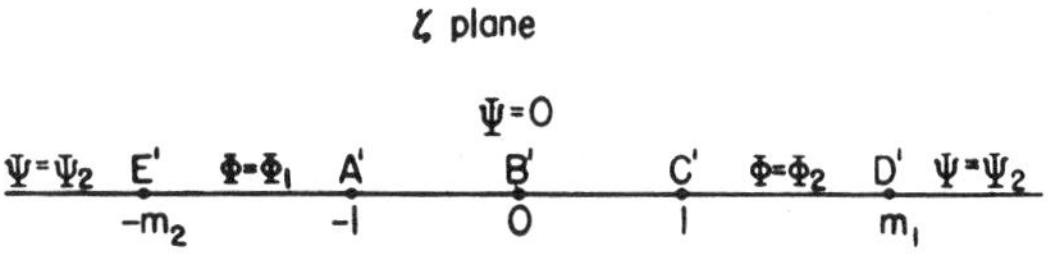

Fig. 70.—The ζ plane map of Fig. 69.

have been presupposed, as indicated in Fig. 70. These conditions give successively

$$\left. \begin{array}{c} 0 = \pi i c_1 \left(\dfrac{-m_1}{\sqrt{m_1^2 - 1}} + \dfrac{m_2}{\sqrt{m_2^2 - 1}} \right) + c_2(m_1 + m_2) \\[2ex] i d_1 = \dfrac{c_1}{m_1 + m_2} \left[\dfrac{-m_1}{\sqrt{m_1^2 - 1}} \log \dfrac{\sqrt{m_1 - 1} + i\sqrt{m_1 + 1}}{\sqrt{m_1 - 1} - i\sqrt{m_1 + 1}} \right. \\[2ex] \left. + \dfrac{m_2}{\sqrt{m_2^2 - 1}} \log \dfrac{\sqrt{m_2 + 1} + i\sqrt{m_2 - 1}}{\sqrt{m_2 + 1} - i\sqrt{m_2 - 1}} \right] + c_2 \\[2ex] i d_2 = c_2. \end{array} \right\} \quad (3)$$

Noting from these that c_1 is real, the condition that $z = x + ih$, for $\zeta > m_1$, gives, on equating the imaginary parts of both sides of Eq. (1) for $\zeta > m_1$

$$ih = \frac{-\pi i m_1 c_1}{(m_1 + m_2)\sqrt{m_1^2 - 1}} + c_2. \quad (4)$$

On eliminating c_2, the resulting three equations for c_1, m_1, m_2 may be written as

$$\left.\begin{array}{c}
\dfrac{\pi m_1 c_1}{\sqrt{m_1^2 - 1}} = -(h - d_2)(m_1 + m_2) \\[3mm]
\dfrac{\pi m_2 c_1}{\sqrt{m_2^2 - 1}} = -h(m_1 + m_2) \\[3mm]
2c_1\left[\dfrac{-m_1}{\sqrt{m_1^2 - 1}}\tan^{-1}\sqrt{\dfrac{m_1 + 1}{m_1 - 1}}\right. \\[3mm]
\left. + \dfrac{m_2}{\sqrt{m_2^2 - 1}}\tan^{-1}\sqrt{\dfrac{m_2 - 1}{m_2 + 1}}\right] = (d_1 - d_2)(m_1 + m_2).
\end{array}\right\} \tag{5}$$

On introducing now the notation

$$m_1 = \sec \alpha_1; \qquad m_2 = \sec \alpha_2 \tag{6}$$

and eliminating c_1, one finally obtains

$$\left.\begin{array}{c}
\dfrac{\sin \alpha_2}{\sin \alpha_1} = 1 - \dfrac{d_2}{h} \\[3mm]
\alpha_1\left(1 - \dfrac{d_2}{h}\right) + \alpha_2 = \pi\left(1 - \dfrac{d_1}{h}\right).
\end{array}\right\} \tag{7}$$

For the special case $d_2 = 0$—no excavation on the downstream side of the dam—Eqs. (7) have the immediate solution

$$\alpha_1 = \alpha_2 = \frac{\pi}{2}\left(1 - \frac{d_1}{h}\right); \qquad m_1 = m_2 = \csc \frac{\pi d_1}{2h}; \qquad d_2 = 0. \quad (8)^1$$

In general, however, Eqs. (7) may be solved graphically without difficulty.

The seepage flux may now be computed by means of Eq. 4.12(19), with k^* given by Eq. 4.12(16). Solving the latter, and using the notation of Eq. (6), it is found that

$$\frac{Q}{\Delta\Phi} = \frac{K'}{2K}; \qquad k^* = \frac{\cos (\alpha_1 + \alpha_2)/2}{\cos (\alpha_1 - \alpha_2)/2}. \tag{9}$$

[1] The solution for this case has also been obtained by Forchheimer (loc. cit.) to a high degree of approximation, by a method similar to that mentioned in the footnote on page 212. His final formulas are:

$$\frac{Q}{\Delta\Phi} = \frac{\pi}{\left[4 \log 2 \operatorname{ctn} \dfrac{\pi(1 - d_1/h)}{4}\right]}, \qquad \text{for} \qquad \frac{d_1}{h} > \frac{1}{2},$$

and

$$\frac{Q}{\Delta\Phi} = \left(\frac{1}{\pi}\right) \log 2 \operatorname{ctn} \frac{\pi d_1}{4h}, \qquad \text{for} \qquad \frac{d_1}{h} < \frac{1}{2}.$$

For the special case where the downstream excavation has reached the depth of the piling, $d_2 = d_1$, the above solution breaks down and the problem must be solved directly with the points C and B taken as coincident from the beginning.   Carrying through such an analysis, in a manner exactly similar to the above, it is found that for this case k^* is given by:

$$k^* = \frac{1}{\left(\dfrac{2h}{d} - 1\right)}; \qquad d_2 = d_1 = d. \tag{10}$$

The values of the seepage flux as given by Eqs. (8), (9), and (10) are shown plotted in Fig. 71 as a function of the fractional dam

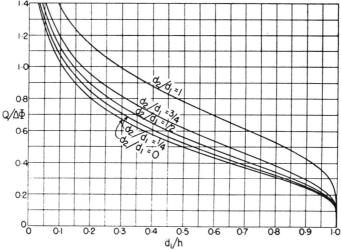

Fig. 71.—The seepage flux under coffer dams. $Q/\Delta\Phi$ = flux per unit potential drop across dam and per unit dam length. d_1 = depth of dam; d_2 = depth of excavation on downstream side; h = thickness of permeable stratum. (*From Physics*, **7**, 125, 1936.)

penetration, d_1/h, for various values of d_2/d_1. Here, too, while the general decrease of $Q/\Delta\Phi$ with increasing d_1/h or decreasing d_2/d_1 is to be expected from physical considerations, one finds the marked persistence of the flux until the dam is set at the very bottom of the permeable layer.   As the physical equivalent of the ordinates is here the same as that given by Eq. 4.13(10), the importance of actually anchoring the coffer dam to the bed rock will be evident.

Although the values of $Q/\Delta\Phi$ derived here and in the last several sections have referred to the total flux seeping into the sand along the whole infinite upstream bed EA and leaving along the whole infinite downstream surface CD, in Fig. 69, the finite limitations of these surfaces in the actual flow systems will not invalidate the analytical results. For it is not difficult to show that the very great part of the flux into or out of the porous medium is concentrated near the singular points A and C. Thus the seepage per unit length out of the surface CD will be given by

$$\frac{d\Psi}{dx} = \frac{d\Psi}{d\zeta} \bigg/ \frac{dx}{d\zeta}. \tag{11}$$

Taking the case where $d_2 = 0$, and $m_1 = m_2 = m$, it is clear that

$$\frac{d\Psi}{d\zeta} = \frac{A}{\sqrt{(\zeta^2 - 1)(m^2 - \zeta^2)}}; \qquad \frac{dx}{d\zeta} = \frac{B\zeta}{(m^2 - \zeta^2)\sqrt{\zeta^2 - 1}}$$

Hence

$$\frac{d\Psi}{dx} = \text{const.} \; \frac{\sqrt{m^2 - \zeta^2}}{\zeta} =$$
$$\text{const.} \; \frac{\text{ctn } \pi d/2h \; \text{sech } \pi x/2h}{\sqrt{1 + \text{ctn}^2 \pi d/2h \; \tanh^2 \pi x/2h}}, \tag{12}$$

as the mapping of the z plane onto the ζ plane is in this case effected by the same function as that defined by Eq. 4.13(4). It is thus seen that $\frac{d\Psi}{dx}$ rapidly approaches a vanishing value as one recedes from the dam by a distance of the order of or twice the thickness of the permeable stratum. In any case, however, the results given here are again on the "safe side," from a practical point of view, in predicting values of the seepage flux that will exceed those in the physical system where EA and CD are actually finite.

4.15. Anisotropic Media.—As has been seen in Chap. II, measurements have shown that quite often the permeability across the bedding plane of a sand is appreciably different from that parallel to the bedding plane. In such cases the sand may be considered as an anisotropic medium with the permeability depending on the direction of flow. While this anisotropy is of no great significance in most practical problems—especially those in which the velocities are essentially confined to planes parallel to the bedding planes— it is of interest to see how the anisotropy may be taken account of when it does have to be considered.

Since the present discussion will be concerned only with the anisotropic character of the medium, it will suffice to assume here that the permeability, though different for directions of flow along the different coordinate axes, is otherwise uniform and independent of the coordinates. Returning then to Sec. 3.3, it is seen that Darcy's law for a homogeneous but anisotropic sand can be written as

$$v_x = -\frac{k_x}{\mu}\frac{\partial p}{\partial x}; \qquad v_y = -\frac{k_y}{\mu}\frac{\partial p}{\partial y}; \qquad v_z = -\frac{k_z}{\mu}\frac{\partial p}{\partial z} + \frac{k_z}{\mu}\gamma g. \qquad (1)$$

Assuming that the individual permeabilities k_x, k_y, k_z are uniform, the equation of continuity gives

$$\frac{\partial v_x}{\partial x} + \frac{\partial v_y}{\partial y} + \frac{\partial v_z}{\partial z} = 0 = -k_x\frac{\partial^2 p}{\partial x^2} - k_y\frac{\partial^2 p}{\partial y^2} - k_z\frac{\partial^2 p}{\partial z^2}.$$

The pressure distribution $p(x, y, z)$ is, therefore, no longer given by Laplace's equation, but rather by the equation

$$k_x\frac{\partial^2 p}{\partial x^2} + k_y\frac{\partial^2 p}{\partial y^2} + k_z\frac{\partial^2 p}{\partial z^2} = 0. \qquad (2)$$

However, a slight change will reduce this to Laplace's equation. For, by transforming the coordinate system to that of $(\bar{x}, \bar{y}, \bar{z})$ defined by

$$\bar{x} = \frac{x}{\sqrt{k_x}}; \qquad \bar{y} = \frac{y}{\sqrt{k_y}}; \qquad \bar{z} = \frac{z}{\sqrt{k_z}}, \qquad (3)$$

it follows at once that

$$\frac{\partial^2 p}{\partial \bar{x}^2} + \frac{\partial^2 p}{\partial \bar{y}^2} + \frac{\partial^2 p}{\partial \bar{z}^2} = 0. \qquad (4)$$

It thus appears that the effect of an anisotropy in the permeability can be replaced by an equivalent shrinking or expansion of the coordinates. Hence, to find the pressure at (x, y, z) according to Eq. (2), it is simply necessary to transform the boundaries by the transformation of Eq. (3), solve Laplace's Eq. (4) for these new boundaries, and then compute the pressure at $(x/\sqrt{k_x},\ y/\sqrt{k_y},\ z/\sqrt{k_z})$. An application of this theory of some practical interest will be made in Sec. 5.5.

With regard to the motion in the original physical system, it may be noted that in general the streamlines will not be normal to the equipotentials. In fact, the angle θ between these directions is readily seen to be given by the equation

$$\cos\theta = \frac{\bar{v}\cdot\nabla p}{|\bar{v}||\nabla p|} = \frac{k_x\left(\dfrac{\partial p}{\partial x}\right)^2 + k_y\left(\dfrac{\partial p}{\partial y}\right)^2 + k_z\left(\dfrac{\partial p}{\partial z}\right)^2}{|\bar{v}||\nabla p|}. \tag{5}$$

The resultant permeability along the streamlines will therefore be

$$k_r = \frac{|\bar{v}|}{|\nabla p|\cos\theta} = \frac{|\bar{v}|^2}{\bar{v}\cdot\nabla p} = \frac{1}{\dfrac{\cos^2\theta_x}{k_x} + \dfrac{\cos^2\theta_y}{k_y} + \dfrac{\cos^2\theta_z}{k_z}}, \tag{6}$$

where $\theta_x,\ \theta_y,\ \theta_z$ are the angles between the vector $\bar{v}$ and the coordinate axes $(x,\ y,\ z)$.[1] On the other hand, the equivalent isotropic permeability in the transformed system will be given by $\sqrt{k_x k_y k_z}$ for three-dimensional systems and $\sqrt{k_x k_y}$ for two-dimensional systems.

4.16. General Potential-theory Results. Green's Reciprocation Theorem.—In the previous sections of this chapter there have been presented the solutions of several two-dimensional flow problems of practical interest which also served to illustrate some of the more powerful analytical methods of potential theory. Since we are primarily interested in the physical interpretation and significance of these problems, only those methods have been illustrated which have immediate application to problems of some practical meaning.[2] However, there are several general results of practical interest which may well be outlined here and which are independent of such detailed specifications as characterize the problems already treated.

---

[1] These results have also been given, with somewhat more complex derivations, by C. G. Vreedenburgh and O. Stevens, *Int. Conf. Soil Mech. and Foundation Eng.*, 1936; *Cf.* also F. *Schaffernak, Die Wasserwirtschaft*, p. 399, 1933.

[2] Perhaps the most frequently used of the methods which have not been outlined in the previous sections is that making use of the Fourier integral. A complete discussion of this method, profusely illustrated with examples, will be found in W. E. Byerly, "Fourier's Series and Spherical Harmonics," Chap. IV.

As the first of these may be mentioned the result that, in general, whatever be the detailed shape of the boundaries, the flow into any set of closed surfaces is always proportional to the pressure difference between the surfaces into which the fluid is flowing and those from which it is flowing, provided only that both the former and latter sets of surfaces are each at a uniform pressure. Although this may be considered as a self-evident consequence of the linearity of Laplace's equation, or derived by making use of the method of the Green's function, the following .demonstration is of interest in illustrating still another principle of potential theory.

For this purpose we make use of the hydrodynamic analogue of the well-known Green's reciprocation theorem usually expressed in electrical phraseology as:[1]

If total charges E_1, E_2, etc., on the separate conductors of a system produce potentials V_1, V_2, etc., and charges E_1', E_2', etc., produce potentials V_1', V_2', etc., then

$$\Sigma E_i V_i' = \Sigma E_i' V_i.$$

Since by a simple extension of the analogies given in Sec. 3.6 it is seen that the charges E and constant potentials V for a set of conductors correspond hydrodynamically to total fluxes Q and uniform potentials Φ, there follows at once the reciprocation theorem for the hydrodynamics of fluid flow through porous media:

If the uniform potentials Φ_1, Φ_2, etc., over several surfaces give rise to fluxes Q_1, Q_2, etc., and if the potentials Φ_1', Φ_2', etc. give rise to fluxes Q_1', Q_2', etc., then

$$\Sigma \Phi_i Q_i' = \Sigma \Phi_i' Q_i. \tag{1}$$

This means physically that if one steady state is characterized by the quantities Φ_i, Q_i, then the Φ_i', Q_i' of any other steady-state flow in the same system must satisfy Eq. (1). This result can be applied to the problem at hand as follows:

Let the drainage surfaces be at the uniform potential Φ_1 and the source surfaces at the uniform potential Φ_2, the corresponding fluxes being Q_1 and Q_2 as indicated in Fig. 72, and let the corresponding values for another steady state be denoted by primes.

---

[1] *Cf.* J. W. Jeans, "Electricity and Magnetism," 5th ed., 1927, p. 92.

Then Eq. (1) implies that

$$\Phi_1 \Sigma Q_1' + \Phi_2 \Sigma Q_2' = \Phi_1' \Sigma Q_1 + \Phi_2' \Sigma Q_2. \qquad (2)$$

As there is no accumulation or destruction of fluid in the system, which is considered as closed, the total fluxes must satisfy the relation

$$\Sigma Q_1 + \Sigma Q_2 = 0 = \Sigma Q_1' + \Sigma Q_2', \qquad (3)$$

which permits Eq. (2) to be rewritten as

$$(\Phi_1' - \Phi_2') \Sigma Q_1 = (\Phi_1 - \Phi_2) \Sigma Q_1',$$

or

$$\frac{\Sigma Q}{\Phi_1 - \Phi_2} = \frac{\Sigma Q'}{\Phi_1' - \Phi_2'} = \cdots = \text{const.} \qquad (4)$$

so that

$$\Sigma Q = \text{const. } \Delta \Phi, \qquad (5)$$

which was the proposition to be proved, since the constant clearly depends only upon the geometry of the system, and is evidently directly proportional to the permeability k.

FIG. 72.

The obvious corollary follows, in addition, that if the potentials be reversed, the fluxes will be reversed, maintaining, however, the same numerical value as before. And finally it may be noted that by the method of Green's function the above result can be shown to hold even for the individual Q's.

Another result of some interest is that quite generally, regardless of the shape of the region drained by a well, the two-dimensional flow into the well may be expressed as

$$Q = \frac{2\pi k \Delta p}{\mu \log c/r_w}, \qquad (6)$$

where Δp is the pressure differential between the external boundary (supposed at uniform pressure) and the well of radius r_w, and c is a constant depending on the shape of the external boundary. To see how this comes about one need only observe that, in general, regardless of the shape of the external boundary, $p(x, y)$ may be expressed as

$$p = q\Phi(x, y) + \text{const.,} \qquad (7)$$

where $\Phi(x, y) \cong \log r$ for small values of r (measured from the center of the well). Setting now

$$\log c = \Phi(r_e), \qquad (8)$$

where $\Phi(r_e)$ represents the uniform value of Φ over the external boundary, applying the boundary conditions to determine q, and computing Q as usual, Eq. (6) is found at once. It is of interest to note that the constant c in particular examples corresponds to reasonable effective average distances of the boundary from the well. Thus in particular:

1. Circular boundary, radius r_e, concentric with well,

$$c = r_e \qquad [cf.\ \text{Eq. } 4.2(10)].$$

2. Circular boundary, radius r_e, with center at distance δ from well,

$$c = \frac{r_e{}^2 - \delta^2}{r_e} \qquad [cf.\ \text{Eq. } 4.6(10)].$$

3. Line source at distance d from well,

$$c = 2d \qquad [cf.\ \text{Eq. } 4.7(8)].$$

4. Rectangular boundary of sides, $2a$, $2b$, with well at center,

$$c = \frac{4\sqrt{ab}}{\pi}.$$

The result of case (4) can be derived without difficulty by using the Green's function for a rectangular region as given in Appendix III. In all cases it will be observed that the values of c do correspond to what might be considered as "effective" average radii of the actual external boundaries. Hence one may reasonably generalize and conclude that for all practical purposes one may compute Q by Eq. (6), and use for c an estimate of an appropriate average of the distance of the external boundary from the well.

In Sec. 4.6 use was made of "Gauss' theorem of the mean" which in terms of the pressure p may be stated thus: The pressure at any point equals the average of the pressure over any circle that does not enclose any sources or sinks and whose center is at the point in question. Here, again, in proving this useful theorem, the method of Fourier series may be used.[1] Thus from Sec. 4.5 it will be clear that the pressure distribution within the circle of radius r_e, about the point of interest, and containing no sources or sinks, may be expressed as

$$p = \sum_0^\infty r^n(a_n \sin n\theta + b_n \cos n\theta), \tag{9}$$

where the a_n, b_n have been so adjusted that p takes on the pre-assigned values over the external boundary $r = r_e$, that is, such that

$$p_e(\theta) = \sum_0^\infty r_e{}^n(a_n \sin n\theta + b_n \cos n\theta). \tag{10}$$

The coefficients c_n, d_n of the corresponding Eq. 4.5(1) must be set equal to 0, if p is to remain finite in the interior of the circle. From Eq. (10) it now follows at once that the average of $p_e(\theta)$ is given by

$$\bar{p}_e = b_0 = p(r = 0), \tag{11}$$

which is the analytic statement of the above theorem.

As another general result of potential theory it will be noted that in any flow problem, two-dimensional with respect to the Cartesian coordinates, one can interchange the equipotential lines and streamlines and still have a system satisfying Laplace's equation. For as was seen in Sec. 4.8, the stream function Ψ not only satisfies Laplace's equation, but moreover the curves $\Psi = $ const. are orthogonal to the equipotential curves $\Phi = $ const. Hence it is clear that one can interchange the roles of Ψ and Φ, considering the former as the potential function and the latter as the stream function. Thus any set of equipotentials and streamlines for a given physical problem may at once be reinter-

---

[1] This result also follows at once by setting $b = 0$ in Poisson's integral, Eq. 4.6(17).

preted as the solution for another physical problem with the role of the equipotentials and streamlines interchanged and the boundary conditions appropriately readjusted. This interchangeability of the equipotentials and streamlines may be illustrated by the specific example of the problem of radial flow in a semicircular system, where the equipotentials are given by the circles

$$\Phi = \frac{k}{\mu} p = \frac{Q}{\pi} \log \frac{\sqrt{x^2 + y^2}}{r_w} + \Phi_w = \text{const.,}^1 \qquad (12)$$

and the streamlines by

$$\Psi = \frac{Q}{\pi} \tan^{-1} \frac{y}{x} = \frac{Q}{\pi} \theta = \text{const.,} \qquad (13)$$

which evidently are the radial lines.

Applying the above results, it is seen that if the streamline $\theta = 0$ is replaced by the equipotential $\Phi = 0$, and the streamline $\theta = \pi$ by the equipotential $\Phi = Q$, the new streamlines will be the circles

$$\Psi = \frac{Q}{\pi} \log \frac{\sqrt{x^2 + y^2}}{r_w} \qquad (14)$$

as indicated in Fig. 73, which corresponds to the flow between two contiguous semi-infinite line sources and sinks, the sources being maintained at a pressure $(k/\mu = 1)Q$ above the pressure of the sinks.

A final theorem of potential theory of practical interest is that establishing a certain symmetry of the potential distribution in a flow system when its geometry has a certain symmetry. For the purposes of this work it may be stated as follows: When a flow system has a plane of symmetry and all the high potential boundaries are at the same uniform potential on one side of the plane and the low potential boundaries are all at a uniform potential symmetrically placed on the other side, the potential distribution within the system will also be symmetrical with respect to the geometrical structure of the streamlines and equipotentials on the two sides of the plane. The numbering of the equipotentials will have the same numerical steps in the increasing sequence

---

[1] The Q here refers to the flux per unit thickness into the semicircular outflow surface, $r = r_w$.

in passing from the plane to the high potential boundaries as in the decreasing sequence in passing to the low potential boundaries. In particular, the plane of symmetry will be at a potential equal to the algebraic average of those at the boundaries.

A simple proof of this almost obvious theorem may be constructed as follows: Supposing the high- and low-potential boundaries S_1, S_0 are at potentials 1 and 0, the potential distribution may be denoted by Φ (x, y, z). Now the potential distribution $\Phi' = 1 - \Phi$ will evidently correspond to the identical state of flow with all the velocities reversed. Furthermore,

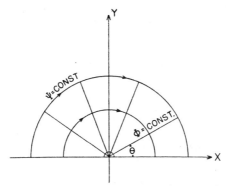

Fig. 73.—The result of interchanging the equipotentials and streamlines in a radial-flow system.

under these conditions, owing to the *geometrical* symmetry of the system, the numbering of the equipotentials beginning with 1 at S_0 will be identical with that in the original system beginning with 1 at S_1, as the Φ' condition can be obtained by a 180-deg. rotation of the flow system about an appropriate axis in the plane of symmetry. Hence if P_1, P_0 are points symmetrical with respect to the plane of symmetry on the sides of S_1 and S_0,

$$\Phi(P_1) = \Phi'(P_0) = 1 - \Phi(P_0), \tag{15}$$

so that

$$\Phi(P_1) + \Phi(P_0) = 1. \tag{16}$$

As P_1 and P_0 approach and reach the plane of symmetry, it is clear that $P_1 = P_0 = \bar{P}$, and hence

$$\Phi(\bar{P}) = \tfrac{1}{2}. \tag{17}$$

Equation (16) can, therefore, be rewritten as

$$\Phi(P_1) - \Phi(\bar{P}) = \Phi(\bar{P}) - \Phi(P_0), \qquad (18)$$

which, together with Eq. (17), is the analytical equivalent of the theorem to be proved.

It may be noted that this theorem is the basis of the conclusions to be drawn in the summary, Sec. 4.18, with respect to the pressure distributions and uplift forces under dams with two sheets of piling placed symmetrically with respect to a vertical plane passing through the center line of the base of the dam.

4.17. Approximate and Nonanalytic Methods of Solving Two-dimensional Flow Problems.—While it has been possible to idealize, without introducing serious errors, the various problems treated in detail in this chapter so that they were amenable to explicit analytical discussion, one cannot, of course, hope to find solutions in analytic form for all practical problems that might arise. For completeness, therefore, we shall briefly mention some of the possible modes of attack that may be applied when one cannot find strictly analytic solutions, although no pretense is made for exhausting the available methods nor for anything more than a bare definition of most of them, as our primary interest here lies in the physical flow problems rather than in the analytical theory of Laplace's equation.

From an analytical point of view, the most interesting of these methods are perhaps the formal procedures of deriving sequences of analytic functions which in the limit approach, individually or in appropriate combinations, the potential function satisfying the prescribed boundary conditions. Among the types of functions involved in some of these successive approximation methods are those which individually satisfy the boundary conditions but not the differential equation (Poincaré's "méthode de balayage"), and those which are individually all solutions of Laplace's equation, but which do not satisfy the boundary conditions (Neumann's "method of the arithmetic mean"). Although these methods afford extremely powerful tools in the study of the formal theory of the solutions of Laplace's equation, they are not particularly suited to the treatment of specific problems. The reader will find a complete review of these methods in the article "Neuere Entwicklung der Potentialtheorie. Konforme Abbildung," by L. Lichtenstein, in

"Encyklopädie der Mathematischen Wissenschaften," vol. II, pt. 3, 1, pp. 177–377, and in "Potential Theory," Chap. XI, 1929, by O. D. Kellogg.

A considerably more practical mode of attack is based on the result of the calculus of variations[1] that the problem of solving Laplace's equation,

$$\frac{\partial^2 \Phi}{\partial x^2} + \frac{\partial^2 \Phi}{\partial y^2} = 0, \tag{1}$$

so as to satisfy preassigned boundary conditions at the boundaries of a given region, is equivalent to that of finding a function Φ satisfying these boundary conditions and such as to make the integral

$$I = \int \int \left[\left(\frac{\partial \Phi}{\partial x}\right)^2 + \left(\frac{\partial \Phi}{\partial y}\right)^2 \right] dx\, dy \tag{2}[2]$$

a minimum. Although the problem of minimizing the integral I is frequently transformed back to that of solving Eq. (1), a direct attempt at an approximate minimization will often be the easier procedure. Among the various direct schemes for minimizing I, mention should be made of that of Ritz[3] and that of Trefftz,[4] which are analogous to the above-mentioned formal analytic methods in that they deal respectively with sequences of functions which satisfy the boundary conditions but not Laplace's equation, and conversely.

The practical formulation of the Ritz method consists essentially in the construction of the function Φ

$$\Phi(x, y) = g(x, y) + \sum_{1}^{n} c_m \Phi_m(x, y), \tag{3}$$

[1] *Cf.* Riemann-Webers, *loc. cit.*, Chaps. V and XX; and Courant-Hilbert, "Methoden der Mathematischen Physik," 1924, Chaps. IV and VI.

[2] This, of course, has the immediate interesting implication that the actual distributions of pressures and velocities in a porous medium carrying a liquid under viscous-flow conditions are such as to make the total macroscopic kinetic energy of the fluid a minimum, when compared with all other distributions consistent with the preassigned boundary conditions. Another implication is that the fluid motion must be such that its total rate of doing work against friction is a minimum, since the integrand of I is proportional to this rate at each point (x, y).

[3] Ritz, W., *J. rein. u. angew. Math.*, **135**, 1, 1909.

[4] Trefftz, E., *Intern. Cong. Applied Mech.*, p. 131, Zürich, 1926.

where $g(x, y)$ is any differentiable function satisfying the boundary conditions, while the Φ_m are a sequence of functions vanishing on the boundary—g may be taken as zero if each of the Φ_m satisfies the boundary conditions—and choosing the constant coefficients c_m in such a way that when Φ is put into Eq. (2), I will be a minimum. By the elementary rules of the calculus this choice of the c_m leads to the set of linear equations in the c_m:

$$\frac{1}{2}\frac{\partial I}{\partial c_m} = \int\int (g_x\Phi_{mx} + g_y\Phi_{my})dxdy + \sum_j c_j \int\int (\Phi_{mx}\Phi_{jx} +$$
$$\Phi_{my}\Phi_{jy})dxdy = 0; \qquad m = 1, 2 \cdots n, \quad (4)$$

where the subscripts x, y indicate differentiation with respect to x or y, or

$$\Sigma c_j \int\int \Phi_m \nabla^2 \Phi_j dxdy = -\int\int \Phi_m \nabla^2 g dxdy; \qquad m = 1, 2 \cdots n, \quad (5)$$

making use of Green's theorem.[1] The accuracy of the resulting function Φ will evidently depend on the choice of the functions Φ_m and the number of elements n taken in the sum. The ease of using this procedure will also obviously depend on the choice of the Φ_m, since the integrals in Eq. (4) or (5) must be evaluated in order to get numerical results. Thus if the Φ_m could be chosen so that their derivatives with respect to x and y form orthogonal sequences, it is clear that the Eqs. (4) would reduce to n separate equations in the single coefficients c_m; otherwise they must be treated as a set of simultaneous equations. In any case, however, it may be shown that unless a finite sum of the chosen Φ_m actually represents the strict solution, the value of the integral of Eq. (2) with any finite series as in Eq. (3) will be greater than its minimal value given by the strict solution.

The procedure given by Trefftz differs from that of Ritz only in the nature of the functions Φ_m. Here the Φ_m are chosen originally as potential functions, and hence they all satisfy Eq. (1), but they do not satisfy the boundary conditions. Taking then the series

$$\chi(x, y) = \sum_1^n c_m \Phi_m(x, y) \qquad (6)$$

and choosing as the equivalent of Eq. (2) the requirement that

$$\int\int |\nabla(\Phi - \chi)|^2 dxdy \qquad (7)$$

[1] *Cf.* E. B. Wilson, "Advanced Calculus," p. 349, 1912.

is a minimum, where Φ is the true solution with boundary values $g(s)$, it is readily found that the c_m are to be obtained from the linear equations

$$\sum_i c_i \int \Phi_i \frac{\partial \Phi_m}{\partial n} ds = \int g(s) \frac{\partial \Phi_m}{\partial n} ds. \tag{8}$$

Here, too, the accuracy of χ will depend on the choice of the Φ_m and the number of terms in the series. However, in contrast to the Ritz method, the value of I in Eq. (2), using the χ of Eq. (6), will here always be less than the true minimal value of I for the correct solution, unless the latter happens to be expressible as a finite sum of the Φ_m.

A different type of procedure for deriving a solution to Laplace's equation corresponding to a definite physical flow problem is that of the graphical integration of the equation. In this method the solution is represented by the geometrical network of equipotentials and streamlines corresponding to the physical problem, this network being obtained by following definite rules which on iteration give successively closer approximations to the network defined by the exact solution. The details of the procedure may be of various kinds, and may be based either upon the principle of so modifying initially arbitrary networks so as to show a behavior at the boundaries corresponding to the preassigned boundary conditions, or upon the extensions of elements of a network, chosen initially so as to satisfy all or part of the boundary conditions, into the interior of the region of interest by rules consistent with the differential equation to be solved. In still another scheme the essential feature is the graphical construction of the Green's function for the region and then the calculation of the final value of the potential by graphical or numerical integration according to Eq. 4.6(1). Detailed descriptions of these methods with complete bibliographies will be found in the article on "Numerische und graphische Integration," by C. Runge and Fr. A. Willers, "Encyklopädie der Mathematischen Wissenschaften," II, 3, 1, pp. 164–171. Examples of potential distributions for flow problems obtained in this way are illustrated in Figs. 74 and 75.

While the above types of graphical integration of the Laplace equation will give immediately only the networks of equipotentials and streamlines, the actual flux in the system can be readily obtained as follows: Thus, supposing that the total potential drop $\Delta\Phi$ in the system, from the inflow to the outflow surfaces, has been divided into n equal parts by $n - 1$ equipotentials lying between the boundaries, it is clear that between each the drop will be $\Delta\Phi/n$. If the separation between two of

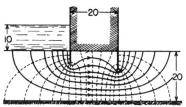

these equipotentials at a given point is s, the flux per unit length along the equipotential will be $\Delta\Phi/ns$. Assuming now that the streamlines have been constructed so as to form a square network with the equipotentials, the flux between two such streamlines will be $\Delta\Phi/n$. Hence, finally, if the square network contains $m - 1$ streamlines between the limiting streamline boundaries, the total flux in the system will be

Fig. 74.—The equipotentials and streamlines under a dam with sheet piling, as obtained by graphical integration. (*After Terzaghi, Die Wasserwirtschaft.*)

$$Q = \frac{m}{n}\Delta\Phi.\qquad(9)$$

It is, therefore, only necessary to count up the numbers of squares in a strip bounded by two streamlines and in one bounded

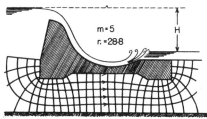

Fig. 75.—The equipotentials and streamlines under a dam with a base of irregular shape, as obtained by graphical integration. m = number of flux units; n = number of equivalent potential units; H = total fluid head. (*After Terzaghi, Die Wasserwirtschaft.*)

by two equipotentials, take their ratio, and then apply Eq. (9) to find the actual flux in the system. The result of such a counting is indicated in Fig. 75, m and n being found to have the values 5 and 28.8.

Still another powerful method for solving Laplace's equation is that of numerical integration. Here, too, a number of special schemes have been developed, all, however, giving the desired solution in the form of the numerical values of the potential or pressure function at a preassigned network of points covering the region of interest.

For practical purposes the only numerical method that need be considered here is that in which the partial differential equation is first replaced by a corresponding difference equation, the numerical treatment being applied to the latter equation.[1] The partial difference equation is constructed by means of the correspondences (for two variables)

$$\left.\begin{array}{l} \delta\dfrac{\partial\Phi}{\partial x} \to \Phi(x+\delta,\,y) - \Phi(x,\,y); \qquad \delta\dfrac{\partial\Phi}{\partial y} \to \Phi(x,\,y+\delta) \\[2mm] \hspace{9cm} -\ \Phi(x,\,y) \\[2mm] \delta^2\dfrac{\partial^2\Phi}{\partial x^2} \to \Phi(x+\delta,\,y) - 2\Phi(x,\,y) + \Phi(x-\delta,\,y) \\[2mm] \delta^2\dfrac{\partial^2\Phi}{\partial y^2} \to \Phi(x,\,y+\delta) - 2\Phi(x,\,y) + \Phi(x,\,y-\delta) \end{array}\right\} \quad (10)$$

for a square mesh of side δ. Taking δ as the unit of length, the difference equation corresponding to Laplace's equation will, therefore, be

$$\Phi(x+1,\,y) + \Phi(x,\,y+1) + \Phi(x-1,\,y) + \Phi(x,\,y-1) - 4\Phi(x,\,y) = 0, \quad (11)$$

x and y being the integral coordinates of the lattice points of the mesh. To solve this equation directly one would write it out successively for each point $(x,\,y)$ within the boundary defining the region of interest, the equations in which $(x,\,y)$ is immediately next to the boundary involving at least one boundary value of Φ for which the preassigned boundary value is to be substituted. There will thus be obtained a set of linear nonhomogeneous equations in the unknown values of Φ at the internal lattice points and equal in number to the number of these points.

---

[1] Mention should also be made of the possibility of solving potential problems by treating numerically the integral equation equivalent of the Laplace partial differential equation (*cf.* Kellogg, *loc. cit.*, page 286), and of the numerical treatment of the Ritz variation approximation method; *cf.* also Chap. IV of *Bull.* 92 of the National Research Council on "Numerical Integration of Differential Equations," 1933.

The solution of these equations by the usual algebraic rules will give the solution to Eq. (11).

This direct solution would obviously be too tedious and laborious for practical use, and approximation methods have been developed for obtaining the desired result even without treating Eq. (11) explicitly as a difference equation. Perhaps the most convenient of these is the "averaging procedure" of Liebmann.[1] Here Eq. (11) is rewritten as

$$\Phi(x, y) = \tfrac{1}{4}[\Phi(x + 1, y) + \Phi(x, y + 1) + \Phi(x - 1, y) + \\ \Phi(x, y - 1)], \quad (12)$$

from which it is seen that the value of Φ at any point is the arithmetic average of those at the four immediate neighbors. Now it may be shown that by taking for the Φ at the boundaries those that are preassigned, and any arbitrary distribution of values in the interior, and applying repeatedly Eq. (12) to obtain "better" values at the interior points, a distribution will be obtained such that further application of Eq. (12) will no longer change them. This limiting distribution will evidently represent the solution to the difference Eq. (11) with the preassigned boundary conditions. A detailed technique for carrying out in a systematic manner these successive approximations obtained by the repeated applications of Eq. (12) as well as a proof of the convergence of the method has been published by Wolf.[2]

It should be observed that aside from the approximation involved in the practical necessity of breaking off the averaging procedure after a finite number of steps, and hence before the limiting solution to the difference equation is actually attained,[3] there is in the above numerical method the approximation inherent in the replacement of the partial differential equation by the partial difference equation. The errors thus introduced will clearly decrease with decreasing size of mesh or increasing number of internal points at which the potential is computed. The size of the mesh to be used in an actual calculation will, therefore, be determined by the accuracy desired in the final results.

[1] LIEBMANN, H., *Munch. Sitzgs. Ber.*, p. 385, 1918.

[2] WOLF, F., *Zeits. Angew. Math. u. Mech.*, **6**, 118, 1926.

[3] It may also be noted that a somewhat different development of the averaging process leads to procedures which, in principle, provide a method for attaining the exact solution of the difference equation in a finite number of steps. *Cf.* L. F. Richardson, *Trans. Roy. Soc., London*, **A-210**, 307, 1910.

A final nonanalytic method of obtaining solutions to specific flow problems, and one of great practical value, is that in which use is made of experimental models of the flow system.[1] Of particular interest are the electrical-model experiments based upon the equivalence of the flow of electric currents in an electrical conduction system to the viscous flow of a homogeneous fluid in a porous medium, as outlined in Sec. 3.6. Thus the equipotentials and lines of current flow in the electrical system correspond to the equipotentials and streamlines in the flow system in the porous medium, and the resistance of the electrical model for unit specific resistance of the model corresponds to the reciprocal of the flux in the flow system for a unit viscosity fluid, a unit permeability medium, and a unit total potential difference.

Perhaps the most flexible type of model is the electrolytic model, in which the porous medium is replaced by an electrolyte, the potential distribution being mapped by probes.   Alternating current must be used to avoid polarization effects at the electrodes representing the inflow and outflow boundaries of the system. Both three- and two-dimensional systems can be studied by means of the electrolytic model, the former in particular being much more readily investigated in this way than in any other, owing to the possibility of probing the interior of the system even when it has no significant symmetry characteristics.   Thus the potential distribution and streamlines in a three-dimensional flow system containing an impermeable barrier can be readily found by the electrolytic model if the barrier is replaced by a geometrically similar nonconducting body.

An electrolytic model which gives a graphic representation of the actual motion of the fluid particles—for two-dimensional systems—is that in which the motion of the ions in the corresponding electrolytic model is made visible by an indicator changing color as the ions advance from the input electrodes. This method has been applied successfully to the problem of water flooding, and is discussed in more detail in Sec. 9.17 in that connection.

An ordinary conduction model suitable for studying three-dimensional systems possessing such symmetry that the significant

[1] Sand-model experiments which have been used by many experimenters are not discussed here as they are really nothing more than small-scale reproductions of the actual flow system.

features are reproduced in a plane composed of stream-lines—such as a radial plane in a system possessing axial symmetry—may be constructed of any homogeneous high-resistance material such as graphite. For most purposes the flow-system boundaries may be simulated either by metallic electrodes—constant-potential boundaries—or by insulated surface elements corresponding to boundaries formed by streamlines. Examples of the use of such models are given in Sec. 8.12 in connection with the analysis of the suppression of water coning in partially penetrating oil wells by shale lenses.

When the system is two-dimensional, plane sheet-conduction models can be used to advantage. The general principles of their use and interpretation are the same as in the other electrical models. Impermeable barriers, as sand lenses, are simulated by cutting out from the conducting sheet figures geometrically similar to the barrier being studied. Some examples of potential distributions obtained with this type of model are given in Sec. 9.21 for various flooding networks.

A particularly convenient feature of the electrical models for two-dimensional systems is that they can be used to give directly both the equipotentials and streamlines of the system. For owing to the mutual orthogonality of these curves they can be interchanged and still represent a possible flow system (*cf.* Sec. 4.16). Hence, by plotting the equipotentials in a model in which boundary streamlines correspond to constant-potential boundary surfaces of the original system and surfaces of uniform potential to streamline surface elements, one will obtain directly the streamlines in the original flow system. In this way one may avoid the rather inaccurate procedure of finding the streamlines by drawing the orthogonal trajectories of the equipotentials.[1]

It should be observed that there is no direct electrical analogue of the effect of gravity in the flow of fluids through porous media. In the case of problems involving vertical velocities, as the seep-

---

[1] One may also note the very interesting application of electrical models recently developed by K. N. E. Bradfield, S. G. Hooker, and R. V. Southwell (*Proc. Roy. Soc.*, **159 A**, 315, 1937). These authors show how electrical models may be used to effect conformal or conjugate-function transformations in two-dimensional potential systems, thus making possible the solution of problems of such geometry where the Schwarz-Christoffel theorem or its generalizations can no longer be applied to practical advantage.

age under dams, it is therefore necessary to construct the model
on the basis of the analogy between the electrical potential and
fluid-velocity potential rather than between the former and the
fluid pressure. Systems involving free surfaces—streamline
surfaces over which the pressure is uniform[1]—on the other hand,
cannot be so easily treated, as the free surfaces, which develop
automatically in the actual gravity-flow systems, will not appear
in the electrical model which will be everywhere cut by lines of
current flow. In such cases one must therefore cut the model so
that it is actually bounded by a curve of the same shape as the
free surface in the physical flow system. This, however, can
only be done by trial and error, as the shape of the free surface
is in general not known initially, and its determination is in fact
one of the objectives of the solution of gravity-flow problems.
The criterion for the correct determination of the shape of the
free surface is that the potential along it should vary linearly
with its vertical height above a horizontal plane, which implies
physically that the pressure is uniform over the free surface, as is
required by the definition of the latter. A trial-and-error
adjustment of the shape of the bounding surface element in a
manner equivalent to the above is described in Sec. 8.10 for the
case of a three-dimensional model used to study the problem of
water coning.

In addition to the trial-and-error adjustment of the boundary
of the electrical model so as to correspond to a free surface, one
must also, in gravity-flow problems such as that of the seepage
through dams, take account of boundary elements of unknown
length constituting "surfaces of seepage."[2] The requirement of
a uniform pressure along such surface segments can be satisfied
by attaching to the model conducting strips along the segment in
question and passing a current through the strip so as to give a
linear potential variation along it. Its length is then adjusted
so as to connect with the free surface which must terminate at
the top of the surface of seepage. The actual application of
this type of model to the problem of the seepage through dams
will be presented in Sec. 6.6.

A final point to be mentioned with respect to model experiments concerns the scales of the models and their numerical

---

[1] *Cf.* Sec. 6.1.

[2] *Loc. cit.*

dimensions. For, although the absolute dimensions are to be determined primarily by convenience and the accuracy desired, it is absolutely necessary for the model and original flow system to be geometrically similar if the results obtained with the model are to be applicable to the physical flow problem. Only if this is so will the potential and streamline distributions in the model be equivalent to those in the actual system. As to the resistance of the model or fluid capacity of the flow system, it is important to note that for a unit specific resistance the total resistance is inversely proportional to one dimension of the system, all the others entering essentially in terms of their ratio to that dimension. Hence, in order to obtain empirical generalizations from model experiments one must study the resistance as a function of the *ratios* of the various dimensions of the model to a fixed dimension, or attempt to express the product of the resistance and the chosen dimension as a function of the ratios of the others to that chosen.

4.18. Summary.—Since water- and oil-bearing sands are often found to possess considerable uniformity in thickness over large areal extents, the problem of fluid flow through such sands involves the analysis of potential problems of a two-dimensional character. This physical approximation is particularly applicable when the wells sunk to drain these horizontal sands of uniform thickness completely penetrate the sands, for then the vertical variations in the potential distribution may be safely neglected. It will be seen in the next chapter that when the well does not completely penetrate the sand the problem takes on a three-dimensional character, which cannot be satisfactorily approximated by two-dimensional simplifications. In view of this restriction of the flow—in such two-dimensional problems— to planes parallel to the horizontal, gravity effectively disappears from the equations. Hence the fluid pressure p, multiplied by k/μ, the ratio of the permeability to the viscosity·of the liquid, can be taken as the equivalent of the velocity potential in the study of horizontal two-dimensional systems.

A detailed analysis is, therefore, made of several physical flow problems governed by the equation

$$\frac{\partial^2 p}{\partial x^2} + \frac{\partial^2 p}{\partial y^2} = 0, \qquad [cf.\ \text{Eq. 4.1(1)}]$$

where x and y represent Cartesian coordinates in a horizontal plane, each chosen to illustrate a situation of practical interest and at the same time one of the general methods of analysis used in two-dimensional-potential theory. In most of these problems the system contains a well of small radius which may be either draining the sand or supplying liquid to it. The differences between these several problems consist in the conditions imposed at the boundaries of the regions surrounding the well, the shape of these regions, and the isotropy of the sand with respect to its permeability. The analysis of these particular problems leads to the following general results.

Each well in a sand into which is flowing a quantity Q of liquid per unit sand thickness[1] per unit time, contributes to the pressure distribution of the system a term such as

$$p = \frac{Q\mu}{2\pi k} \log r, \qquad [cf. \text{ Eq. } 4.2(11)]$$

where k is the sand permeability, μ the viscosity of the liquid, and r is the radial coordinate measured from the center of the well. The resultant pressure distribution of the system is composed of terms of the above type, due to the individual wells in the system, and others depending explicitly on the boundaries of the region and the conditions imposed there.

From an application of the theory of Fourier's series (*cf.* Sec. 4.3) it is found that the flow into a well of radius r_w, at the center of a circular region of radius r_e, may be computed by the equation

$$Q = \frac{2\pi k(\bar{p}_e - \bar{p}_w)}{\mu \log r_e/r_w}, \qquad [cf. \text{ Eq. } 4.5(12)]$$

where $\bar{p}_e$, $\bar{p}_w$ are the averages of the pressures maintained over the external circular boundary and over the well, respectively. Conversely, if the pressure p_w at a given well having a production rate Q is known, one can find the average pressure $\bar{p}_e$ over a circle of radius r_e enclosing the well, by the relation

$$\bar{p}_e = \frac{Q\mu \log r_e/r_w}{2\pi k} + \bar{p}_w. \qquad [cf. \text{ Eq. } 4.5(13)]$$

---

[1] The value of Q both here and in the following three equations refers to the flux per unit sand thickness.

Further, if the pressure is known at any point (x, y), the average of the pressure over any circle about this point as center and enclosing no wells or other fluid sources or sinks, must equal this pressure at the center [cf. Eq. 4.16(11)].

On the other hand, if the external boundary is not circular or is not concentric with the well surface, the fluxes into or out of the well are found by the methods of Green's function and of images and by general considerations to be expressible as

$$Q = \frac{2\pi k \Delta p}{\mu \log c/r_w}, \qquad [cf. \text{ Eq. } 4.16(6)]$$

where Δp is the pressure differential existing between the well of radius r_w and the external boundary, and c is a constant depending on the shape of the external boundary, and which may be approximated by an appropriate average distance of the well from the boundary. Thus if the external boundary is still circular, but the well is off center even by as much as half of the radius of the circular boundary, the error caused by taking c as equal to that radius is less than 5 per cent [cf. Eq. 4.6(11) and Fig. 34]. Further, because of the logarithmic dependence of Q upon these lengths characterizing the dimensions of the system, even rough estimates of the latter will give results of considerable accuracy in the prediction of Q.

If the system contains more than one fluid source or sink and if the sets of sources and sinks are each at a uniform pressure, an application of the so-called Green's reciprocation theorem, or general considerations making use of the concept of the Green's function, shows that the total flux through the system is directly proportional to the pressure differential between the set of sources and the set of sinks, and the permeability of the medium in which they are imbedded [cf. Eq. 4.16(5)].

A practical problem in which the effective external boundary supplying fluid to a well is not even approximately circular is that in which the external boundary is an infinite line source. The analytical idealization of an infinite line source and a single well corresponds to the simplest problem of edge-water encroachment, in which the water advances in the form of a "line drive," displacing the oil into the well near the water-oil interface. An equivalent situation is to be found in the flow of water into an artesian well piercing a sand with its outcrop running into and parallel to a canal or river bed (*cf.* Fig. 38). The solution to this problem—by the method of images—shows that the production rate yielded by such a well is the same as that which would be produced by a well surrounded by a concentric circular boundary —symmetric radial flow—of radius equal to twice the distance of the well from the line source [*cf.* Eq. 4.7(8)].

When the canal or river bed cuts *across* the sand outcrop (*cf.* Fig. 39), the liquid source can no longer be considered as an infinite line, but must be taken as a *finite* line source instead. Such a system may be treated by the method of conjugate functions (*cf.* Sec. 4.8), and leads to a system of confocal ellipses for the equipotentials and confocal hyperbolas for the streamlines (*cf.* Fig. 40). The flow into a well piercing a sand supplied with water by such a finite line source will, of course, be less than if the source were of infinite length, although the difference between the two will become very small if the well is very close to the finite line source. Furthermore, as found on solving this problem by the method of conjugate-function transformations, at any given distance from the source, the production rate will be greatest if the well lies on the perpendicular bisector of the source, and it will decrease as the well is moved toward and beyond the ends of the source.

Another type of two-dimensional problem is that in which the flow takes place in vertical rather than horizontal planes. Such situations arise when the system has uniform dynamical characteristics for extended distances along a horizontal direction, as in the consideration of the seepage under dams of lengths great as compared to their thicknesses. As the force of gravity is equivalent to a uniform vertical pressure gradient, the appropriate dynamical variable for such problems is the velocity potential $\Phi = \frac{k}{\mu}(p - \gamma g y)$,[1] as contrasted with the pressure p for horizontal two-dimensional systems. Furthermore, the streamlines for cases in which the flow is vertical are orthogonal to the curves of constant velocity potential rather than the equipressure curves. Analytically, however, the problems still involve the solution of the two-dimensional Laplace's equation [*cf.* Eq. 4.1(3)].

[1] It is supposed here that $+y$ represents the vertical axis directed downward.

Particularly interesting examples of this type of two-dimensional flow problems are those of the seepage under dams of lengths great as compared to the widths of their bases. These problems involve the two questions of the uplift pressures and moments at the base of the dam and the numerical value of the seepage. For the first it suffices, for most practical purposes, to take the thickness of the permeable stratum underlying the dam as infinite, and thereby simplify the analysis. Thus, when there are no sheet pilings at the base of the dam, the analytical problem becomes formally equivalent to that of the horizontal flow from a finite line source into an infinite sand, with an interchange of the equipotentials and streamlines in the latter system and the rotation of the horizontal into the vertical plane. The pressure at the base of the dam has an arc cosine (cf. Fig. 44) distribution, thus showing large gradients both at the heel and toe of the dam, in contrast to the frequently assumed linear distribution. The total uplift force, however, is the same as on the assumption of a linear pressure distribution, namely, the algebraic average of the pressures at the heel and toe times the base width. On the other hand, the total uplift moment always exceeds that computed by the linear pressure law, and by as much as 11 per cent for vanishing pressures at the toe of the dam.

When there is a single sheet of piling set at the base of the dam, the problem can still be treated analytically by first reducing the geometrical system so as to be equivalent to a dam without piling. This reduction is accomplished by means of the Schwarz-Christoffel theorem, which provides a formula for mapping any polygonal area onto a complex half plane, the boundary of the polygon being transformed into the real axis of the complex plane (cf. Sec. 4.11). The analysis shows that here, too, there are high-pressure gradients both at the heel and toe of the dam. Another feature of practical interest is the pressure drop about the piling. For although this pressure drop decreases rather rapidly as the ratio of the dam width to piling depth increases, it may attain quite large values for small values of this ratio. Thus, when the piling depth is as great as the dam width and is placed either at the heel or toe, the pressure drop over the piling will amount to 72.8 per cent of the total pressure drop from heel to toe (cf. Fig. 50). When the piling is at the center of the base, the pressure drop between its up- and downstream pressures along the base will be 70.5 per cent of the total drop. This slight decrease—for larger ratios of dam width to piling depth the decrease is considerably larger—is in contrast to that implied by the commonly used "line-of-creep" theory, in which the piling is effectively replaced by an additional equivalent base width equal to the perimeter of the piling, the pressure drop along the whole of this extended "line of creep" being taken as linear. In this theory the pressure drop due to the piling is proportional only to its depth and is independent of its position. An immediate consequence of the above high-pressure drops about the piling is that the gradients over the rest of the base are thus necessarily decreased, thereby diminishing the dangers of undermining the base of the dam by the washing away of the sand at the base because of high fluid velocities.

As to the total uplift forces for dams with sheet piling, the theory shows that except when the piling is set just below the center of the dam, very appreciable changes may be introduced by the piling. Thus for piling of depth equal to the dam width, the uplift force will be decreased by 59 per cent if the piling is at the heel, and increased by the same amount when the piling is set at the toe. Of course, this effect decreases as the ratio of the dam width to the piling depth increases. Here again, the line-of-creep theory gives only approximate results, its predictions of the total uplift force being by as much as 20 per cent too small for upstream piling, and as much as 8 per cent too high for down-stream piling.

The effect of sheet piling on the total uplift moment (about the heel of the dam), however, is not symmetrical about the center position of the piling, the piling decreasing the uplift moments when placed within about 62 per cent of the width from the heel and increasing the moments when set within the 38 per cent of the width nearest the toe. The magnitude of the changes is, however, very considerable, especially for small ratios of dam width to piling depth. Thus when this ratio is 1, piling at the heel will cut the total uplift moment to 46 per cent of the value for a dam without piling, while if it is set at the toe the uplift moment will be 202.3 per cent of that without piling. Here, too, the line-of-creep-theory predictions differ from those of the correct theory, the predictions of the former being too low for piling set within 70 per cent of the heel and slightly too high for piling positions within 30 per cent of the toe of the dam.

Although the line-of-creep theory is actually incorrect in its quantitative predictions, the line of creep formed by the base of the dam and the perimeter of the piling does represent a limiting streamline of the flow system, and in fact is the streamline of maximum velocity if the base of the dam is always convex downward. However, for dams with piling, the streamlines of maximum velocity are those following the shorter paths cutting across the sand between the lowest points reached by the piling and passing directly to the toe of the dam.

While the analytical theory has been developed here[1] quantitatively only for dams with one set of piling, the effect of additional sheets of piling can be readily inferred from the simpler problem. Thus, if there are two sets of piling of equal depth, one at the heel and one at the toe, or placed at any positions symmetrical with respect to the base of the dam, the pressure distribution underneath the dam will possess symmetry about the vertical plane passing through the center line of the base of the dam (*cf.* Sec. 4.16). In particular, this central line will be the equipotential of magnitude equal to the average of the potentials at the upstream and downstream beds. The potential difference between this line and two points symmetrically placed about it will be equal. Likewise the pressure drops about the two pilings will be equal. Furthermore, the total uplift force will be unaffected by two such symmetrically placed pilings. The absolute magnitude of the pressure drops about the pilings will be less than if each piling were alone under the dam, although of course, the resultant drop will be greater than for a single piling of the same depth. Intermediate pilings, however, will in general induce much smaller pressure drops, owing to the relatively small gradients along the floor of the dam if the pilings near the heel and toe are of appreciable depth. But in all

---

[1] A numerical treatment of the problem of a dam with similar piling at both the heel and toe, based upon a procedure of correcting Forchheimer's method of solution (*cf.* footnote on p. 212) by the addition of Fourier series potentials to those of Forchheimer, has been given by R. Hoffmann, *Die Wasserwirtschaft*, **1**, 108, 1934. In this paper will also be found a description of sand-model experiments which strikingly confirm both the streamline and potential distributions computed theoretically.

cases whether the total uplift is increased or decreased, the piling will decrease the pressure gradients and velocities near the heel and toe of the dam and thus lessen the danger of washing away the sand and undermining the dam.

In studying the question of the seepage underneath dams, the finite thickness of the underlying permeable strata must be taken into account, as the seepage will be infinite even for dams with sheet piling, if the underlying permeable stratum is of infinite thickness. While the general analytic method of the conjugate-function transformations suffices here as well as for the systems with infinite permeable strata, the finiteness of the latter leads to elliptic-function transformations where before the functions were elementary (*cf.* Sec. 4.12). Among incidental results of this analysis may be mentioned, first, the general confirmation of the pressure distributions at the base of the dam found by the previous simpler theory. Thus for a dam without sheet piling, the finite thickness of the permeable stratum will not have a significant effect on the pressure distribution unless the thickness of the stratum is several folds smaller in magnitude than the width of the dam (*cf.* Fig. 44). And in the case of dams with sheet piling, the pressure drop over the piling will not be appreciably affected by the finite thickness of the permeable stratum unless the piling penetrates through more than 50 per cent of the thickness of the stratum (*cf.* Fig. 68).

The seepage flux, however, is largely determined by the thickness of the permeable stratum, or, more specifically, by the ratio of the dam width to the thickness of the stratum, the flux decreasing continuously from infinitely high to vanishing values as this ratio increases from zero to infinite values (*cf.* Fig. 61). The position of the piling affects the seepage flux but slightly (*cf.* Fig. 66), the flux decreasing symmetrically from a maximum for piling set at the center of the base to minimal values, about 9 per cent lower, for piling set either at the heel or toe.

As is to be expected, the seepage flux decreases with increasing depth of piling. However, the variation is rather small unless the thickness of the underlying permeable stratum is of the order of or greater than the dam width. But in all cases, the calculations show the rather surprising result that the flux will persist with relatively high values as the piling depth is increased until it approaches the immediate vicinity of the bottom of the permeable stratum (cf. Fig. 67). Thus, when the dam width equals the thickness of the permeable stratum and the piling (at the center of the base) penetrates 99 per cent of the stratum, the seepage flux will still be 25.3 per cent of what it would be if there were no piling at all; and if the dam width equals five times the thickness of the permeable stratum, the seepage flux with piling of a 99 per cent penetration will be 60.6 per cent of that under the same dam with no piling. It is seen, therefore, that while pilings of even moderate depth do very materially affect the uplift pressures on a dam when set near the heel or toe, their effect on the seepage flux will not be large unless they are practically anchored to the bottom of the permeable stratum. Even a 3-in. opening between the bottom of the piling and that of a permeable stratum of 25 ft. thickness will permit a seepage that may be as high as 60 per cent of that which would flow under the dam if it had no piling whatever. Furthermore, this large seepage passing through such restricted channels will evidently involve very high fluid velocities and hence may lead to serious sand-flow difficulties.

The same type of analysis which leads to a solution of the problem of the seepage flux under dams of extended base width can be used for the calculation of the seepage flux under coffer dams, or other temporary water shut-offs of small width, for varying depths of excavation on the downstream side of the dams (Sec. 4.14). As is to be expected, such calculations show that the seepage flux will increase as the excavation on the downstream side approaches in depth that of the dam (cf. Fig. 71). However, the total increase from the condition at the beginning of the downstream excavation to that when it reaches to the bottom of the dam or water shut-off is not very large, amounting, for example, to but 56 per cent when the dam penetration into the permeable stratum is 50 per cent. The decrease of the flux with increasing penetration of the dam or shut-off sheet into the permeable stratum is here more marked than for the corresponding variation with the piling depth in the case of dams of extended base width. On the other hand, the seepage flux persists here, too, with values of appreciable magnitude until the dam or water shut-off is in effect actually anchored to the impermeable bedrock.

The values of the seepage fluxes as given directly by the analyses, $Q/\Delta\Phi$, are those per unit potential drop between the upstream and downstream faces of the dam and per unit length of the dam. When translated into terms of more practical meaning, one finds that a value of $Q/\Delta\Phi = 0.1$ is equivalent to a seepage flux of 48.3 gal./min. of water under a dam of 100 ft. length with an upstream-downstream fluid-head difference of 34 ft. When it is noted that the above-mentioned persistence of the flux with increasing depth of piling or water shut-off involves values of $Q/\Delta\Phi$ equal to, or considerably exceeding, 0.1 for depths of coffer dam or piling as high as 99 per cent of the thickness of the permeable stratum, the practical importance of actually anchoring the piling or water shut-off onto bedrock becomes obvious.

While these considerations and results have referred to systems in which the media have been taken as homogeneous and isotropic, the analysis can also be carried through if the sand is uniform but anisotropic—the permeability is different in different directions—as it is usually found to be when vertical flows across the bedding planes of a sand are compared with those parallel to the bedding planes. Such problems can also be solved by potential-theory methods, provided one first shrinks or expands the coordinates properly [*cf.* Eq. 4.15(3)] so as to reduce the equation for the pressure distribution to the Laplacian form.

Another feature of two-dimensional problems of liquid flow through porous media worthy of mention is the interchangeability of the equipotential curves and streamlines [*cf.* Eq. 4.8(4)], which are the curves along which the fluid particles flow [*cf.* Eq. 4.8(7)]. This relation follows from the facts that the equipotential curves and streamlines form a mutually orthogonal network [*cf.* Eq. 4.8(5)] and that the stream functions also satisfy Laplace's equation [*cf.* Eq. 4.8(4)]. Hence each solution of Laplace's equation in two dimensions represents the solution to two distinct physical problems in which the roles of potential and stream function are mutually interchanged.

A final theorem of potential theory of interest in the discussion of certain types of flow problems is that when the flow system has a *geometrical* plane of symmetry and the boundary conditions are symmetrical—in type rather than numerical values—about this plane, the distributions of potential and streamlines within the system are also symmetrical about the plane, provided one numbers the equipotentials by the absolute value of their difference from the potential of the plane of symmetry.

While these various analytic solutions outlined above do cover the more important two-dimensional flow problems of practical interest, it is to be observed that even small changes in the geometries of the various flow systems may not only invalidate the original analytic solutions but may even lead to unsurmountable mathematical difficulties in the derivation of the new correct solutions. In such cases one must resort to either approximate analytic methods or even nonanalytic or empirical methods. Formal approximate analytic procedures for solving potential problems have been developed in the course of the derivation of proofs of the existence of solutions to Laplace's equation with prescribed boundary conditions. These are, however, unsuitable for application to specific problems.

A powerful method of considerable generality is based upon the observation that the problem of solving Laplace's differential equation with given boundary conditions is analytically equivalent to that of finding the velocity potential satisfying these boundary conditions which will make a minimum of the total kinetic energy of the fluid in the system [*cf.* Eq. 4.17(2)]. This latter problem may then be approximately solved by choosing for the potential function a linear combination of individual functions which either satisfy the boundary conditions but not Laplace's equation, *i.e.*, the Ritz method, or Laplace's equation but not the boundary conditions, *i.e.*, the method of Trefftz. Having chosen such sets of individual functions, the actual analytical processes required to give the approximate solutions reduce to those of solving systems of simultaneous linear algebraic equations for the constant coefficients in the linear combinations of the individual functions. The coefficients in these algebraic equations are integrals involving both the individual functions and the preassigned boundary values which the rigorous solution must assume at the boundaries.

A less formal procedure for solving flow problems which are intractable by rigorous analysis is that of constructing by graphical means the potential and streamline distributions in the system. Networks of these distributions may be obtained with successively increasing accuracy by following definite rules in their construction which are consequences of the differential equation being solved. When such a graphical integration of the Laplace equation has been completed in the form of a square network of equipotentials and streamlines, the flux through the system, for a unit total potential drop, will be given by the ratio of the number of squares lying between two neighboring equipotentials, extending from one bounding streamline surface to the other, divided by the number of squares lying between two neighboring streamlines extending from the high- to the low-potential boundaries [*cf.* Eq. 4.17(9)].

Strictly numerical procedures may also be applied to the solution of flow problems. These are based upon the replacement of the Laplace partial differential equation by a corresponding difference equation [*cf.* Eq. 4.17(10)]. This latter—Eq. 4.17(11)—may be solved in principle by algebraic means. But techniques have been developed for deriving its solution by strictly iterative numerical processes, which successively give increasingly accurate values for the potential at the lattice points of a square grid covering the interior of the flow system.

Finally one may avoid entirely all analytic processes and study specific flow problems by means of model experiments. While sand-model experiments have been frequently performed to give a direct picture of the conditions of flow in specific cases, these are really nothing more than small-scale reproductions of the actual flow systems, and can hardly be considered as representing an application of the fundamental laws of flow in the generalization of the elementary basic experiments establishing Darcy's law.

A very powerful empirical method, and one that does not involve a reversion to the use of sand models, is that based upon the analogy between the flow of current in an electrical-conduction system and that of a liquid in a porous medium (*cf.* Sec. 3.6). In this method one may use either electrolytic-conduction models or solid models consisting of semiconductors or metallic sheets.

The main advantage of electrolytic models lies in the fact that they permit the direct determination of the potential distribution in the interior of three-dimensional systems which do not have axial symmetry. When the three-dimensional system does have axial symmetry, such as that of the partially penetrating well, the interior potential distribution may be determined by mapping that on the axial planes of solid-sector models made of a high-resisting material as graphite (*cf.* Sec. 8.10). The constant-potential surfaces in all electrical models are to be simulated by metallic electrodes.

A modified analogue of sand-model experiments, in which the streamlines are traced by the injection of dyes at various points in the sand, may be obtained electrically in the case of two-dimensional systems by means of electrolytic-sheet models. These may be provided with indicator solutions so as to change color as the ions, simulating the inflow liquid, advance from the input electrodes. Such models are particularly adapted to show graphically the progress of the injected water in various flooding operations (*cf.* Sec. 9.17).

An especially convenient type of model experiment for the study of two-dimensional flow systems is that of the plane-conduction-sheet model (*cf.* Sec. 9.21). Impermeable barriers, as very tight sand lenses or sheet piling or other water shut-offs, are easily simulated by cutting out from the conduction sheets figures which are geometrically similar to the barriers in question. By interchanging the equipotential and streamline surfaces in such models, the probing and determination of the equipotentials in the new system will give at once the streamlines in the original system.

While there is no direct electrical analogue of the effect of gravity, it is formally taken into account by constructing the model so that the electrical potential is the analogue of the velocity potential rather than the fluid pressure. This, however, is not entirely sufficient in the case of gravity-flow systems (*cf.* Chap. VI), where the liquid does not completely fill out the porous medium but only occupies a region which is bounded at its points of maximum vertical height by a streamline surface along which the pressure is uniform—a "free surface." This boundary surface is not known a priori, but must be found by a trial-and-error cutting of the model in such a way that it will correspond to a streamline surface of uniform pressure (*cf.* Sec. 6.6). In addition, in the study of such problems as the seepage of water through dams, provision must be made to simulate the "surface of seepage," which represents the part of the outflow surface along which the pressure rather than the potential is uniform.

As the uniformity of the pressure implies a linear variation of the potential, the proper boundary conditions at the surface of seepage can be obtained by attaching a conducting strip to the surface and passing a current through it, thus giving a linear variation of the potential. The length of this surface segment, however, must be adjusted by trial and error so as to connect at its upper extremity with the free surface.

A general requirement to be observed in the use of any type of model experiment is that the model be strictly geometrically similar to the physical flow system. The potential and streamline distributions depend only on the shape of the model and not upon its absolute dimensions, which may be chosen on the basis of convenience and accuracy. The total flux through the model or actual flow system will be proportional to one of the absolute dimensions, but all the others will enter only in ratio form. In fact, the significant variables to be used in the study of systems of the same type but with gradually changing geometry, such as the seepage flux under dams with varying depth of sheet piling, should always be expressed as ratios of two of the dimensions, as the piling depth divided by the dam width or thickness of the permeable stratum (*cf.* Sec. 4.13).

CHAPTER V

THREE-DIMENSIONAL PROBLEMS

5.1. Introduction.—Although many problems of the flow of fluids through porous media which are of practical interest may be fairly well approximated by one of the several two-dimensional systems discussed in the previous chapter, there are still others of considerable importance that are definitely of a three-dimensional character. Thus, if the well penetrating a sand does not pass through it completely, the flow in the region of the sand not penetrated by the well will have an upward component tending to bring the fluid into the well, while the flow in the upper portions of the sand will still be largely radial and will have but a small vertical component. The pressure distribution in the sand body will, therefore, vary with the vertical coordinate, and the problem will be a three-dimensional one.

With regard to general methods of treating three-dimensional problems, it may be mentioned that all but one of the methods discussed in Chap. IV as applying to two-dimensional systems have their analogues when the third coordinate is included. Only the method of conjugate functions has no analogue in the case of the three-dimensional Laplace's equation. For the problems of practical interest, however, we shall find that the methods available will suffice to lead to the desired solutions.

Since the addition of the vertical coordinate introduces the possibility for gravity to play its role in affecting the pressure distribution, we shall deal in this chapter with the potential function

$$\Phi = \frac{k}{\mu}(p - \gamma g z) \qquad [cf. \text{ Eq. } 3.3(3)] \qquad (1)$$

rather than with the pressure p, when the porous medium is isotropic. The basic differential equation will, therefore, be Laplace's equation in Φ,

$$\nabla^2\Phi = \frac{\partial^2\Phi}{\partial x^2} + \frac{\partial^2\Phi}{\partial y^2} + \frac{\partial^2\Phi}{\partial z^2} = 0. \tag{2}$$

Equation (2) will now be solved for several special cases of practical interest.[1]

5.2. Spherical Flow.—The analogue of the two-dimensional problem of radial flow (*cf.* Sec. 4.2) is clearly that in which the potential and velocity distributions depend only on the radius r of a spherical-coordinate system. Since the general form of Laplace's equation in the spherical coordinates (r, θ, χ)

$$\frac{1}{r^2}\frac{\partial}{\partial r}\left(r^2\frac{\partial\Phi}{\partial r}\right) + \frac{1}{r^2\sin\theta}\frac{\partial}{\partial\theta}\left(\sin\theta\frac{\partial\Phi}{\partial\theta}\right) + \frac{1}{r^2\sin^2\theta}\frac{\partial^2\Phi}{\partial\chi^2} = 0$$

[*cf.* Eq. 3.7(6)]   (1)

reduces in this case of "spherical flow" to

$$\frac{1}{r^2}\frac{\partial}{\partial r}\left(r^2\frac{\partial\Phi}{\partial r}\right) = 0, \tag{2}$$

it follows at once that

$$r^2\frac{\partial\Phi}{\partial r} = \text{const.} = c_1. \tag{3}$$

Hence,

$$\Phi = -\frac{c_1}{r} + c_2. \tag{4}$$

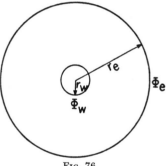

Fᵢɢ. 76.

This is the general distribution function for the potential Φ in a spherical-flow system. Its salient feature is that Φ varies inversely with the radius r. This is evidently a more rapid variation than the logarithmic dependence on r characteristic of the two-dimensional radial-flow system [*cf.* Eq. 4.2(6)]. To see the physical meaning of the two constants c_1, c_2, one need only apply Eq. (4) to a specific case defined by the boundary conditions (*cf.* Fig. 76)

$$\left.\begin{array}{ll}\Phi = \Phi_w: & r = r_w \\ \Phi = \Phi_e: & r = r_e.\end{array}\right\} \tag{5}$$

[1] Still other three-dimensional-flow problems involving sands of non-uniform permeability or two-fluid systems will be treated in Secs. 7.9, 7.10, and 8.10.

These give

$$\Phi_w = -\frac{c_1}{r_w} + c_2; \qquad \Phi_e = -\frac{c_1}{r_e} + c_2,$$

so that

$$\Phi = \frac{\Phi_e - \Phi_w}{\dfrac{1}{r_e} - \dfrac{1}{r_w}}\left(\frac{1}{r} - \frac{1}{r_w}\right) + \Phi_w. \tag{6}$$

It is now seen that Φ is proportional to the potential difference $\Phi_e - \Phi_w$ between the spherical boundaries at r_e, r_w.

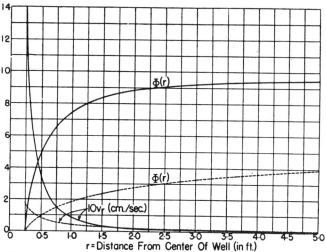

Fig. 77.—Velocity (v_r) and potential (Φ) distributions in spherical (solid curves) and radial-flow (dotted curves) systems. $\Phi(r)$ is taken as 0 at $r = \frac{1}{4}$ ft. and 10 at $r = 500$ ft.

The velocity in the system is, as usual, obtained by differentiation,

$$v_r = -\frac{\partial \Phi}{\partial r} = \frac{\Phi_e - \Phi_w}{\dfrac{1}{r_e} - \dfrac{1}{r_w}}\frac{1}{r^2}, \tag{7}$$

and the total flow through the system is given by

$$Q = -\int_0^{2\pi} d\chi \int_0^{\pi} r^2 \sin\theta v_r d\theta = \frac{4\pi(\Phi_e - \Phi_w)}{\dfrac{1}{r_w} - \dfrac{1}{r_e}}. \tag{8}$$

Hence Φ and v_r can be rewritten as

$$\Phi = \frac{Q}{4\pi}\left(\frac{1}{r_w} - \frac{1}{r}\right) + \Phi_w, \qquad (9)$$

$$v_r = -\frac{Q}{4\pi r^2}. \qquad (10)[1]$$

It is seen from these that both Φ and v_r vary directly as the flux Q, as they do in the two-dimensional case [*cf.* Eq. 4.2(11), 4.2(12)], although their variation with r is much more rapid here. This may be seen more clearly from the curves for Φ and v_r in Fig. 77 for the case where

$$p_w = 0; \qquad p_e = 10 \text{ atm. at } z = 0; \qquad \text{and} \qquad \frac{k}{\mu} = 1,$$

so that[2]

$$\Phi_w = 0; \qquad \Phi_e = 10;$$
$$r_w = \tfrac{1}{4}' = 7.62 \text{ cm.}; \qquad r_e = 500' = 15{,}240 \text{ cm.}$$

Hence,

$$\Phi = 10.005 - \frac{76.24}{r},$$

$$v_r = -\frac{76.24}{r^2} \text{ cm./sec.}; \qquad Q = 958.0 \text{ cc./sec.}$$

A comparison with the dotted curves, which refer to the radial-flow case for the same boundary conditions, will show at once

[1] Just as in the case of radial flow, Eq. (10) could have been written down at once from the integrated form of the equation of continuity, namely, that $Q = -4\pi r^2 v_r = \text{const.}$ A single integration, after replacing v_r by $-\dfrac{\partial \Phi}{\partial r}$, will then give Eq. (9) at once.

[2] It should be observed that the apparently indirect specification of the boundary conditions for Φ by a preliminary assignment of the values of the pressures is not a complication inherent in the use of the potential function Φ, but arises rather from the almost universal habit of thinking of the pressure as the quantity of fundamental physical significance even in three-dimensional systems. In reality, however, the potential function is of fundamental significance, when gravity is taken into account, although both the pressure and Φ satisfy Laplace's equation. For if, in the above system, the pressure had been specified as constant over $r = r_w, r_e$, its distribution would be of the same form as Eq. (6) and hence spherically symmetrical. The velocity distribution, however, would no longer be radial, and hence the system as a whole would really not be spherically symmetrical.

that the high potential gradients in the case of spherical flow are localized near the small radius boundary with a much greater concentration than the already highly concentrated large gradient region in the radial-flow system. Another important difference lies in the value of Q. Thus for practical cases where $r_w \ll r_e$, Eq. (8) gives

$$Q \cong 4\pi(\Phi_e - \Phi_w)r_w, \tag{11}$$

with a flux varying as the radius r_w, whereas in the case of radial flow, Q varies in the much slower logarithmic manner [cf. Eq. 4.2(10)]. Further, Eq. (11) shows that the flux in the case of spherical flow is independent of the external boundary radius as long as it is large compared to r_w. In the radial-flow case, it will be recalled, Q varies logarithmically with both the external and well radius.

The practical significance of the problem of spherical flow is that it corresponds to a well (of small radius) just tapping a relatively thick sand. This correspondence will appear as a special case of the analysis of the next section (cf. Fig. 78). Of course, Eq. (11) itself shows Q to be independent of the radius of the external boundary and hence of its exact shape, provided the external boundary radius is large as compared to the well radius.[1] In other words, if the area of the surface representing the well is small compared to that of the external boundary surface, no part of which lies close to the well surface, the well may be replaced by a small spherical cavity, and the flow into it may be taken as spherical regardless of the detailed shape of either the well or external bounding surfaces.

It should be mentioned that in a practical case where the well just taps the sand the well surface is really a hemisphere. The flux will then be half of that in Eqs. (8) and (11), and in Eqs. (9) and (10) the $Q/4\pi$ should be changed to $Q/2\pi$.

---

[1] The assumption of a uniform external boundary potential Φ_e can also be dropped if Φ_e in Eqs. (8) and (11) is replaced by the average of the actual potential over the external boundary. This result can be derived in a manner entirely analogous to that developed in Sec. 4.5 for two-dimensional-flow systems, the Fourier series there being replaced here by the corresponding functions—the surface spherical harmonics—of the polar and azimuthal angles θ and χ. For details see Byerly, "Fourier's Series and Spherical Harmonics," Chap. VI.

Finally it is of interest to compare the "effectiveness" of a well producing by spherical flow with one producing by radial flow, with the same potential drop, $\Delta\Phi$. Denoting the production capacity of the former by Q_s and that of the latter by Q_r, the above results show that

$$\frac{Q_s}{Q_r} = \frac{r_w}{h} \log \frac{r_e}{r_w}, \tag{12}$$

where h is the sand thickness in the case of radial flow. For the numerical example considered above it follows that

$$\frac{Q_s}{Q_r} = \frac{1.9}{h}, \qquad h \text{ in feet}, \tag{13}$$

so that for a sand thickness of 50 ft. the radial-flow system will produce at a rate more than 25 times as high as the spherical-flow system when flowing under the same potential drop. In view of these radically different production capacities of radial-flow systems—completely penetrating wells—and spherical-flow systems—"nonpenetrating" wells—it is clear that the only conditions under which one should deliberately plan to produce from the latter type of flow system would be those where bottom waters underly an oil zone and the difficulties of water coning would ensue from high well penetrations (*cf.* Sec. 8.10).

5.3. Partially Penetrating Wells. Potential Distributions.— It will be apparent from the discussion of the last section that neither the limit of radial flow, corresponding to a well completely penetrating a sand, nor the limit of spherical flow, representing the case where the well just taps the top surface of the sand, can be used satisfactorily to describe the situation when the well penetration is neither complete nor vanishing. Since this latter condition of partial penetration is encountered in practice very frequently, we shall outline the analysis of this problem, omitting, however, the numerical details[1] of the solution.

It will be supposed, as usual, that the well is symmetrically placed with respect to the surrounding sand, at the boundary of

[1] MUSKAT, M., *Physics*, **2**, 329, 1932. The treatment given there is in terms of the analogous electrical problem of an electrode partially penetrating a large cylindrical disk; the analogy of this problem to that considered here follows at once from the equivalence of the boundary conditions [*cf.* Eqs. (2) below] in the two systems (*cf.* also A. F. Samsioe, *Zeits. angew. Math. u. Mech.*, **11**, 124, 1931).

which the potential is maintained at a uniform value. The system will then be radially symmetrical and the natural coordinates will be those of the cylindrical system. Specifically, it will be supposed that a well of radius r_w penetrates a sand of thickness h to a depth b. The sand is bounded externally by the circle $r = r_e$ concentric with the well, and is bounded at the top and bottom by impermeable sands (cf. Fig. 78).

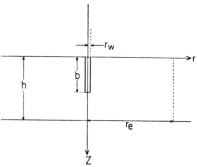

FIG. 78.—Diagrammatic representation of a partially penetrating well.

Analytically this problem may be restated as that of finding a solution Φ to the system of equations

$$\frac{1}{r}\frac{\partial}{\partial r}\left(r\frac{\partial\Phi}{\partial r}\right) + \frac{\partial^2\Phi}{\partial z^2} = 0, \tag{1}[1]$$

$$\left.\begin{array}{l} v_z = -\dfrac{\partial\Phi}{\partial z} = 0: \quad z = 0,\, h \\[2mm] \Phi = \text{const.} = \Phi_w: \quad r = r_w, \quad z \leqslant b \\[2mm] \Phi = \text{const.} = \Phi_e: \quad r = r_e. \end{array}\right\} \tag{2}$$

The first of Eqs. (2) is equivalent to the requirement that no fluid pass across the sand faces, since it is bounded by impermeable sands. The second boundray condition means that the well surface is kept at a uniform potential, as it will be if the well bore is full of liquid at least to the top of the sand. The third condition repeats the assumption that the external boundary of the sand is at a uniform potential.

[1] It is to be observed that although Eq. (1) is an equation in only two independent variables (r, z), it cannot be considered as a Laplacian equation in two "dimensions" unless it can be transformed into the fundamental form of Eq. 4.1(1). Compare Eq. (1) with Eq. 4.2(1) which is a two-dimensional Laplace's equation.

The starting point of the analysis may be chosen in two ways. Thus one may obtain a formal solution of Eq. (1) directly as a series or integral over Bessel functions with coefficients adjusted so that the Eqs. (2) are satisfied. Or, one may begin with the less elegant method of images, which here, however, is more suitable for numerical discussion. To facilitate such a discussion the image method will be presented here and the other will be used in the more difficult problem of stratified horizons (*cf.* Sec. 7.9).

Before proceeding with the analysis it will be convenient to introduce as a unit of length one which is equal to twice the sand thickness, h. The variables may then be redefined as:

$$\rho = \frac{r}{2h}; \qquad w = \frac{z}{2h}; \qquad x = \frac{b}{2h}; \qquad \rho_w = \frac{r_w}{2h}; \qquad \rho_e = \frac{r_e}{2h}. \qquad (3)$$

Equation (1) is evidently unchanged by this transformation.

To apply now the method of images one notes first that a particular solution of Eq. (1) is the function

$$d\Phi = \frac{q\,d\alpha}{\sqrt{\rho^2 + (w - \alpha)^2}}, \qquad (4)$$

and that it represents, physically, a fluid-source element[1] of strength $q\,d\alpha$ placed along the w axis at the distance α from the top of the sand. Recalling now the discussions of the last chapter in which the wells in two-dimensional problems were replaced by point sources or sinks located at the well center, it is seen that the natural extension in the case of three dimensions is that of replacing a well by a continuous distribution of sources or sinks along the well axis from the top of the sand to the well extremity.

Although in view of this method of representation one might now attempt to solve the problem by integrating Eq. (4) with respect to α from 0 to x, and choosing q as a function of α so that the boundary conditions, Eqs. (2), are satisfied, it will be more convenient to develop solutions step by step which successively satisfy the various boundary conditions.

---

[1] From a physical point of view an element of a well draining a sand corresponds to a sink rather than a source. We are using the latter representation, however, because it avoids the repeated use of − signs as coefficients of the flux densities q.

Thus we shall first develop a solution which satisfies the first of the boundary conditions of Eqs. (2), *i.e.*, that there should be no fluid flow across the sand faces. To do this it is convenient to make use of the method of images.

By following a procedure similar to that outlined in Sec. 4.9, it is readily found that the condition of zero flux across the bounding planes, $w = 0$, $\frac{1}{2}$, may be satisfied by superposing an infinite array of flux elements placed at the points $(0, \pm n \pm \alpha)$, where n goes from 0 to ∞ (*cf.* Fig. 79). The resultant potential

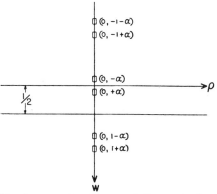

Fig. 79.—Image system for a partially penetrating well.

due to this array is simply the sum of the potentials due to the individual elements, and hence is equal to

$$d\Phi = q \, d\alpha \Bigg\{ \frac{1}{[\rho^2 + (w - \alpha)^2]^{1/2}} + \frac{1}{[\rho^2 + (w + \alpha)^2]^{1/2}}$$

$$+ \sum_{1}^{\infty} \Bigg[\frac{1}{\{\rho^2 + (n + w - \alpha)^2\}^{1/2}} + \frac{1}{\{\rho^2 + (n + w + \alpha)^2\}^{1/2}}$$

$$+ \frac{1}{\{\rho^2 + (n - w - \alpha)^2\}^{1/2}} + \frac{1}{\{\rho^2 + (n - w + \alpha)^2\}^{1/2}} \Bigg] \Bigg\}. \quad (5)$$

Formally, the first stage of the problem is now solved. However, Eq. (5), as it stands, is quite unsuited for numerical computation, and must be transformed into forms appropriate for use in the region of interest. Thus to compute the $d\Phi$, as given by Eq. (5), for small values of ρ, *i.e.*, in the region close to the well, one may expand each term in Eq. (5) as a power series in ρ. After a little manipulation it is found that

$$
d\Phi = qd\alpha \left\{ \frac{1}{[\rho^2 + (w - \alpha)^2]^{1/2}} + \frac{1}{[\rho^2 + (w + \alpha)^2]^{1/2}} \right.
$$

$$
\begin{aligned}
&- \psi(1 - w - \alpha) - \psi(1 - w + \alpha) - \psi(1 + w + \alpha) \\
&- \psi(1 + w - \alpha) - \tfrac{1}{2}\rho^2[\zeta(3, 1 - w - \alpha) \\
&+ \zeta(3, 1 + w + \alpha) + \zeta(3, 1 - w + \alpha) \\
&+ \zeta(3, 1 + w - \alpha)] + \tfrac{3}{8}\rho^4[\zeta(5, 1 - w - \alpha) \\
&+ \zeta(5, 1 + w + \alpha) + \zeta(5, 1 - w + \alpha) \\
&\left. + \zeta(5, 1 + w - \alpha)] \cdots \right\}, \quad (6)
\end{aligned}
$$

where ψ is a function defined in terms of the Γ function[1] by

$$
\psi(y) = \frac{\Gamma'(y)}{\Gamma(y)} = -0.5772 - \frac{1}{y} + \lim_{n \to \infty} \sum_{1}^{n}\left(\frac{1}{m} - \frac{1}{y + m}\right),
$$

and

$$
\zeta(s, y) = \sum_{0}^{\infty} \frac{1}{(n + y)^s}.
$$

For large values of ρ, *i.e.*, values of the order of 1, Eq. (6) converges very slowly. However, a direct solution of Laplace' equation for a system composed of the set of images $(0, \pm n \pm \alpha)$ is found to be particularly suited for this purpose. It is given by

$$
d\Phi = 4qd\alpha\left[2\sum_{1}^{\infty} K_0(2n\pi\rho) \cos 2n\pi w \cos 2n\pi\alpha + \log \frac{2}{\rho}\right], \quad (7)
$$

where K_0 is the Hankel function[2] of order zero. Since this function decreases exponentially for large arguments, one or two terms of the series in Eq. (7) suffice for all purposes, and even for ρ as small as 0.5.[3]

Now the series of Eqs. (6) and (7) are potential functions representing the effect of a single flux element $qd\alpha$ at $(0, \alpha)$ and its images. To get the potential due to a whole well of depth b one must distribute such elements over the whole length of the

[1] *Cf.* Whittaker and Watson, Chap. XII.

[2] *Cf.* Whittaker and Watson, p. 373.

[3] The difficulty with this series for small ρ is that the K_0 become logarithmically infinite as ρ approaches zero.

well; *i.e.*, Eqs. (6) and (7) must be integrated with respect to α, from $\alpha = 0$ to $\alpha = x = b/2h$. Assuming that the source strength is uniform along the well, the results may be expressed as:

For small values of ρ,

$$
\begin{aligned}
\Phi = q\Bigg\{ &-\log \frac{\Gamma(1 + w + x)\Gamma(1 - w + x)}{\Gamma(1 - w - x)\Gamma(1 + w - x)} \\
&+ \log \frac{w + x + [\rho^2 + (w + x)^2]^{1/2}}{w - x + [\rho^2 + (w - x)^2]^{1/2}} \\
&\quad -\tfrac{1}{4}\rho^2[\zeta(2, 1 - w - x) - \zeta(2, 1 - w + x) \\
&\qquad\quad +\zeta(2, 1 + w - x) - \zeta(2, 1 + w + x)] \\
&\qquad\qquad\qquad\qquad\qquad\qquad\qquad + 0(\rho^4)\Bigg\}; \quad (8)
\end{aligned}
$$

and for large values of ρ,

$$
\Phi = 4q\left[\frac{1}{\pi}\sum_{1}^{\infty}\frac{1}{n}K_0(2n\pi\rho)\,\cos 2n\pi w\,\sin 2n\pi x + x\log\frac{2}{\rho} \right]. \quad (9)
$$

In these developments thus far the purpose has been to obtain solutions giving no flux across the sand faces. To see if these solutions are the final answer to the problem of Eqs. (1) and (2), one must test them to determine whether they satisfy the last boundary conditions of Eq. (2). Considering first the requirement that Φ be constant at $r = r_e$, it is seen from Eq. (9) that though this is not satisfied exactly,[1] the variation of Φ with w or z for $\rho \geqslant 1$ may be safely neglected for all practical purposes. This is, of course, due to the fact that for $\rho \geqslant 1$, $K_0(2n\pi\rho)$ is so much smaller than $x\log\frac{2}{\rho}$ (except for the special values $\rho \sim 2$) that the whole trigonometric series may be dropped.

---

[1] The strict constancy of Φ at the external radius $\rho = \rho_e$ could be obtained simply by replacing the terms $K_0(2n\pi\rho)$ in Eqs. (7) and (9) by

$$
K_0(2n\pi\rho) - K_0(2n\pi\rho_e)I_0(2n\pi\rho)/I_0(2n\pi\rho_e)
$$

where I_0 is the zero-order Bessel function of the third kind which, in contrast to K_0, becomes exponentially large for large arguments and equals unity for zero argument (Whittaker and Watson, *loc. cit.*). However, the added term will be of insignificant numerical magnitude for cases of practial interest.

It remains to be seen whether Eqs. (8) and (9) satisfy the final boundary condition that Φ be uniform over the well surface, *i.e.*, over $\rho = \rho_w$, $w \leqslant x$. A calculation by means of Eq. (8) gives results such as are shown by the dotted curves plotted in Fig. 80. Evidently this condition is not satisfied, and it will be necessary to modify the above solution. In fact, a little reflection shows that the uniform flux density $q(\alpha)$ assumed in the derivation of Eqs. (8) and (9) cannot be expected to give an exact representation of the problem under consideration. For

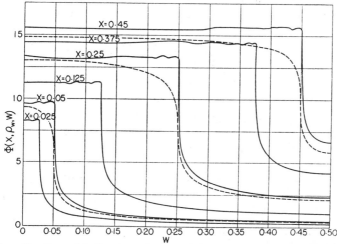

Fig. 80.—Potential distributions over the well surface ($\rho_w = 0.001$) as a function of the depth w after adjustment of the flux-density distribution. $x =$ (well penetration)/(2 · sand thickness); dotted curves give unadjusted distributions. (*From Physics*, **2**, 355, 1932.)

in addition to the ordinary flux contribution of a radial character entering each unit of length along the well surface, the lower parts of the well will receive most of the flux coming from the sand not penetrated by the well. This additional flux will not simply converge at the well extremity, but will be distributed all along the lower parts of the well, though, of course, with a greater concentration toward the actual extremity.

From this point of view the remedy for the lack of uniformity in the potential distribution over the well surface evidently appears to lie in abandoning the assumption of the uniformity of the flux density $q(\alpha)$ and in choosing a flux distribution such that the well potential is uniform. The exact choice of this

$q(\alpha)$, however, involves analytical difficulties for the wells of finite radius, although when the well radius becomes vanishingly small it can be shown that $q(\alpha)$, in general, should be proportional to the well potential at α, except at the extremity, where it must be doubled. For wells of finite radius, the $q(\alpha)$ can be determined by breaking it into discontinuous elements and adjusting their strengths until the potential at the well surface is found to be uniform, to as high an accuracy as desired. The solid curves of Fig. 80 show the results of such adjustments, the final potentials in these cases having a maximum variation of 2 per cent from the average, the dotted curves representing

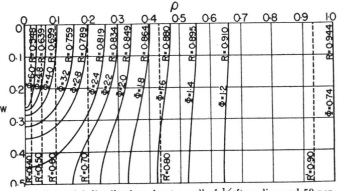

Fig. 81.—Potential distribution about a well of ¼ ft. radius and 50 per cent penetration in a 125-ft. sand. R = fraction of total potential drop across a 500-ft.-radius reservoir. Dotted curves and R' correspond to a strictly radial-flow—complete well penetration—system. Unit of distance = 2 · sand thickness. (*From Physics,* **2**, 359, 1932.)

the unadjusted potentials, *i.e.*, those for uniform flux-density distributions, as indicated above. The flux densities required for these adjustments increase markedly—by as much as 45 per cent—as the well extremities are approached, and the excess flowing into the actual bottoms of the wells may be represented by point sources placed there.

The resultant potential distribution in the sand for the case of a 50 per cent penetrating well, of radius equal to ⅟₅₀₀ of the sand thickness, is shown in Fig. 81.

The concentrated character of the potential drop about the well is indicated by the values of R which give the fraction of the total drop across the sand. By comparison with the corresponding values for strict radial flow, indicated by the dotted equipotentials and fractions R', the markedly higher concentrations of the partially penetrating well distributions become evident. It is to be noted, however, that the equipotentials for the partially penetrating case rapidly change to a radial type and can hardly be distinguished from those for a radial system at a distance from the well equal to only twice the sand thickness.[1] Obviously, for increasing penetrations this change to a radial character will take place even more rapidly.

For the other extreme, when the well just taps the top of the sand—represented by the case when $x \sim \rho_w$ in the above equations—the corresponding system of equipotentials is given in Fig. 82. It will be noted that in this case the equipotentials

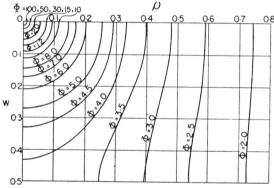

Fig. 82.—Potential distribution about a "nonpenetrating" well. (*From Physics*, **2**, 336, 1932.)

near the well are closely spherical, as one should expect. But here, again, at receding distances from the well the equipotentials flatten and assume a radial character, although, of course, not as rapidly as when the wells actually penetrate the sand for some considerable fraction of its thickness.

5.4. The Production Capacities of Partially Penetrating Wells. In the last section was discussed the nature of the potential distributions existing about wells penetrating a sand only partially. Important as are these qualitative characteristics for an understanding of the mechanism of flow in a partially penetrating well system, one is usually faced, in practical cases, with the

[1] This fact verifies the above-mentioned remark that the solution given by Eq. (9) automatically satisfies the last of Eqs. (2), although no explicit attempt was made to effect this rigorously.

question of the numerical values of the production rates to be expected from such systems. The analysis of the last section will, therefore, be applied to this problem.

As a first step in the discussion of the production capacities, it is useful to derive a more exact interpretation of the flux density q appearing in the equations of the last section. This is found by simply computing the actual flux in a partially penetrating well with a uniform value of q. Thus using Eq. 5.3(9) and recalling the definitions of ρ and w of Eq. 5.3(3), it follows that

$$Q = -4\pi h \int_0^{\frac{1}{2}} \rho \frac{\partial \Phi}{\partial \rho} dw = 8\pi hqx = 4\pi qb, \tag{1}$$

since the series terms in Eq. 5.3(9) vanish on integration. q is therefore seen to be $1/4\pi$ of the actual flow into[1] the well per unit of its length exposed to the producing sand.

It has been seen, however, that to obtain a uniform potential over the well surface, the flux density q must be taken as variable, and may be approximated by a superposition of separate flux elements of strengths q_m and extending to depths[2] x_m. In this case, then, Q will evidently be given by

$$Q = 8\pi h \Sigma q_m x_m, \tag{2}$$

since the potential will be given by a sum of terms as in Eq. 5.3(9).

Now, as already pointed out, at distances from the well of the order of the sand thickness or larger, one may, to a very close approximation, drop the series in Eq. 5.3(9) and rewrite it, with q taken as a positive quantity, as

$$\Phi = -4qx \log \frac{2}{\rho}.$$

Hence in the case of the superposed system of flux elements, the potential for $\rho \gtrless 1$ may be written as

---

[1] It has been assumed that the sign of q in Eq. 5.3(9) has been changed so as to make the well an outflow surface for the sand.

[2] It is to be noted that in this representation one really places first a segment of the maximum flux density extending to the bottom of the well ($x_m = x$), and then successively superposes shorter elements with *opposite* flux densities until distributions of the type shown in Fig. 80 are obtained.

$$\Phi = -4 \sum q_m x_m \log \frac{2}{\rho} = -\frac{Q}{2\pi h} \log \frac{2}{\rho}, \qquad (3)$$

by Eq. (2). Denoting explicitly the potential over the well surface by Φ_w, the difference in Φ between the well and ρ which is ~ 1 becomes

$$\Delta\Phi = \frac{Q}{2\pi h}\left(\frac{-\Phi_w}{4\Sigma q_m x_m} - \log \frac{2}{\rho}\right). \qquad (4)[1]$$

Finally, by specifying that $\Delta\Phi$ refer to the total potential drop between the well of radius ρ_w and the external sand boundary, of radius ρ_e, and reverting to the original units of length, and denoting the sand thickness by h, the value of the production rate per unit of time takes the form:

$$Q = \frac{-2\pi h \Delta\Phi}{\dfrac{\Phi_w(r_w)}{4\Sigma q_m x_m} + \log \dfrac{4h}{r_e}}. \qquad (5)$$

To use this formula one must know, in addition to the pre-assigned physical constants, r_w, r_e, and h, the values of Φ_w and $\Sigma q_m x_m$. However, as mentioned in the last section, this requires a trial-and-error adjustment of the flux elements, and this is quite laborious. Nevertheless, all the details and the complete procedures have been carried through for the two cases where $\rho_w = \frac{1}{600}$ and $\rho_w = \frac{1}{1000}$, i.e., for sand thicknesses of 75 and 125 ft., taking the well radius as $\frac{1}{4}$ ft. With these results as criteria, it has been found that the following approximation will give values accurate to 0.5 per cent, namely, one may assume the flux density over the well to be uniform and then take as the potential Φ_w an "effective average," which turns out to be that at three-fourths of the distance from the top of the sand to the bottom of the well.

To get the approximate formula, one may replace in Eq. (5) $\Sigma q_m x_m$ by qx, and Φ_w by the value of Φ in Eq. 5.3(8) at $w = \frac{3}{4}x$. On dropping terms of the order of ρ^2, it is found in this way that

---

[1] This expression may be compared with the corresponding result for strict radial flow which by Eq. 4.2(11) is equivalent to

$$\Delta\Phi = \frac{Q}{2\pi h}\left(\log \frac{2}{\rho_e} - \log \frac{2}{\rho}\right).$$

$$Q = \frac{2\pi kh\Delta p/\mu}{\frac{1}{2\bar{h}}\left\{2 \log \frac{4h}{r_{w.}} - \log \frac{\Gamma(0.875\bar{h})\Gamma(0.125\bar{h})}{\Gamma(1 - 0.875\bar{h})\Gamma(1 - 0.125\bar{h})}\right\} - \log \frac{4h}{r_e}},$$

$$(6)^1$$

where $\bar{h}$ is the well penetration given as a fraction of the sand thickness h, k is the sand permeability, and μ is the liquid viscosity.

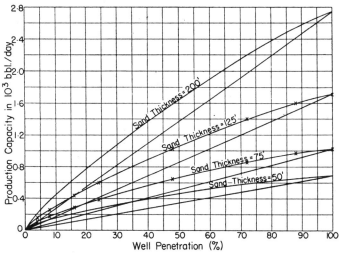

Fig. 83.—Production capacities of partially penetrating wells as functions of the well penetration. Straight lines give production capacities if the flow is strictly radial. Crosses give values computed by Eq. 5.4(6), whereas the continuous curves for sand thicknesses of 75 ft. and 125 ft. were computed by the direct analytical procedure. Total potential drop is taken as unity (pressure differential = 1 atm. if $k/\mu = 1$); well radius = $\frac{1}{4}$ ft.; external-boundary radius = 500 ft.

In Fig. 83, Q is plotted as a function of the per cent well penetration ($100\bar{h}$) for various sand thicknesses, r_e/r_w being taken as 2,000, and $k\Delta p/\mu$ as unity. The solid curves for $h = 75$ and 125 ft. were computed by the exact method implied in the use of Eq. (5). The crosses on these curves are values computed by Eq. (6). It will be evident from these that Eq. (6) represents an

---

[1] It may be noted that Kozeny (*Wasserkraft u. Wasserwirtschaft*, **28**, 101, 1933) has found that the resulting fluxes plotted in Fig. 83 can be represented by the still simpler formula

$$Q = \frac{2\pi kh\bar{h}\Delta p/\mu}{\log r_e/r_w}\left(1 + 7\sqrt{\frac{r_w}{2\bar{h}h}} \cos \frac{\pi\bar{h}}{2}\right).$$

approximation which is certainly satisfactory for all practical purposes. The remaining curves of Fig. 83 were then computed by means of Eq. (6).

The straight lines in Fig. 83 represent the production capacities that would be obtained if the flow into the partially penetrating wells were strictly radial. It is seen from these that as the penetration decreases the excess of the actual production capacity over that for strictly radial flow continually increases until at penetrations of about 20 per cent the excess may even

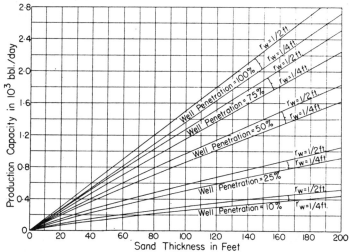

Fig. 84.—Production capacities of partially penetrating wells as functions of the sand thickness. Total potential drop is taken as unity. r_w = well radius; external-boundary radius = 500 ft.

exceed 50 per cent of the strictly radial flow. The approximation of strictly radial flow will, therefore, lead to large errors. Further, these results also show that one cannot consider the actual system as being equivalent to a simple superposition of a radial flow into the well proper and a semipherical flow into the well extremity, since the contribution of the latter is only of the order of 2 to 3 per cent of the radial flow.

In Fig. 84 the results are presented in a somewhat different manner. Here the production capacities are plotted against the sand thickness for various penetrations and two well radii. It is of interest to note that for sand thicknesses greater than about 50 ft. the variation of the production capacity with the sand thickness is almost exactly linear even for the partially penetrating wells. This fact makes it possible to both extrapolate and interpolate on the curves with considerable accuracy.

Considering essentially the same results from still another point of view, the production capacities have been plotted in Fig. 85 as a function of the sand thickness for various fixed values of the actual well penetration in feet. These curves show directly the effect of adding on additional sand thickness below the bottom of the well bore. In particular, it will be seen that beyond the first few feet of sand below the well bottom the additional layers of sand give successively decreasing contributions

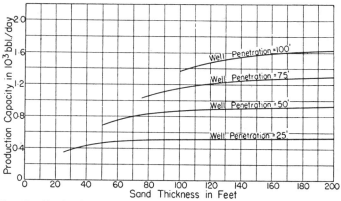

Fig. 85.—Production capacities of partially penetrating wells of fixed penetration as functions of the sand thickness. Total potential drop is taken as unity; well radius = ¼ ft.; external-boundary radius = 500 ft.

to the production from the well. Thus for a 25-ft. well penetration, increasing the sand thickness from 125 to 200 ft. will increase the production capacity by less than 2 per cent, whereas the first 75 ft. below the well bottom (from 25 to 100 ft.) increases the production capacity by 45.6 per cent.

Finally, as to the variation of the production capacities with the well radius, the curves of Fig. 84 show that for the large penetrations the production capacities vary logarithmically with the well radius (as they do exactly for the radial-flow case), and then increase in their rate of variation with decreasing penetration until in the limit of the nonpenetrating well the production capacities vary directly as the well radius, as in the case of spherical flow [*cf.* Eq. 5.2(11)].

5.5. Partially Penetrating Wells in Anisotropic Sands.—In the discussions of Sec. 5.3 and 5.4 it has been explicitly supposed that the sand is perfectly isotropic. If often happens, however, that the permeability along the direction perpendicular to the bedding plane is considerably less than that parallel to the bedding plane.[1] It is, therefore, of interest to see what effect this anisotropy will have upon the production capacity of a well which only partially penetrates the sand.

In a general qualitative way, when the vertical permeability is less than the horizontal permeability, the effect of the anisotropy will be essentially that of diminishing the vertical velocities in the system so as to make the flow more exactly radial in character. Thus the contribution to the production rate of the well which is due to the part of the sand not actually penetrated by the well will be decreased, and with it the resultant total production capacity. Since this contribution decreases uniformly as the well penetration is increased, the effect of the anisotropic character of the sand should increase from a vanishing value for a completely penetrating well to a maximum for wells just tapping the sand.

First, however, it will be convenient to separate the permeability k from the potential function Φ, and use in its place the simplified function

$$\varphi = \frac{1}{\mu}(p - \gamma g z). \qquad (1)$$

Analytically, the problem may be treated by the method suggested in Sec. 4.15. Thus, taking for convenience the permeability parallel to the bedding plane as unity and that perpendicular to it (parallel to the z axis[2]) as k_z, we may write the differential equation for φ in cylindrical coordinates as

$$\frac{1}{r}\frac{\partial}{\partial r}\left(\frac{r\partial\varphi}{\partial r}\right) + k_z\frac{\partial^2\varphi}{\partial z^2} = 0. \qquad (2)$$

---

[1] *Cf.* Sec. 2.12. Although, as was seen there, the reverse situation where the permeability perpendicular to the bedding planes exceeds that parallel to them will also be found with some frequency, only the more commonly occurring inverse case will be treated here numerically. The analysis, however, will be applicable to both types of flow system.

[2] It is assumed here that the sand strata are horizontal.

This equation may now be reduced to a form identical with that for an isotropic system by making either of the changes of variables

$$r = \frac{r'}{\sqrt{k_z}}: \quad z = z'; \quad \text{or} \quad r = r': \quad z = z'\sqrt{k_z}. \quad (3)$$

In either case Eq. (2) becomes

$$\frac{1}{r'}\frac{\partial}{\partial r'}\left(r'\frac{\partial \varphi}{\partial r'}\right) + \frac{\partial^2\varphi}{\partial z'^2} = 0. \quad (4)$$

Introducing again the dimensionless variables

$$\rho' = \frac{r'}{2h'}; \quad w' = \frac{z'}{2h'}, \quad (5)$$

it is clear that the boundary conditions to be applied for Eq. (4)

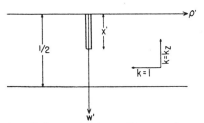

Fig. 86.—A partially penetrating well in an anisotropic sand.

are (cf. Fig. 86)

$$\left.\begin{array}{l} \dfrac{\partial \varphi}{\partial w'} = 0: \quad w' = 0, \dfrac{1}{2}, \\[2mm] \varphi = \text{const.}: \rho_w' = \sqrt{k_z}\rho_w: \quad w' \leqslant x' = x, \\[2mm] \varphi = \text{const.}: \rho_e' = \sqrt{k_z}\rho_e. \end{array}\right\} \quad (6)$$

It follows that the potential distribution in terms of the variables (w', ρ') corresponds to that in an *isotropic* sand with the *same* fractional well penetration and with well and external radii equal to $\sqrt{k_z}$ times those in the actual physical problem. The potential drop between the well and external boundary will, therefore, be given by

$$\Delta\varphi = \varphi_e(\sqrt{k_z}\rho_e, x) - \varphi_w(\sqrt{k_z}\rho_w, x). \quad (7)$$

Now $\varphi_w(\sqrt{k_z}\rho_w, x)$ can evidently be obtained from Eq. 5.3(8) with the value of $w = \frac{3}{4}x$, as suggested on page 273. Hence, assuming a unit flux density along the well

$$\varphi_w(\sqrt{k_z}\rho_w, x) = -2 \log \frac{2}{\sqrt{k_z}\rho_w} +$$
$$\log \frac{\Gamma(1.75x)\Gamma(0.25x)}{\Gamma(1 - 1.75x)\Gamma(1 - 0.25x)}. \quad (8)$$

With regard to φ_e, one may again return to Sec. 5.3 and use the equivalent of Eq. (8) suitable for large values of ρ, *i.e.*, Eq. 5.3(9), which may be rewritten, for $w = 0$, as

$$\varphi_e(\sqrt{k_z}\rho_e, x) =$$

$$-4\left[\frac{1}{\pi}\sum_{1}^{\infty}\frac{1}{n}K_0(2n\pi\sqrt{k_z}\rho_e) \sin 2n\pi x + x \log \frac{2}{\sqrt{k_z}\rho_e}\right]. \quad (9)$$

Although in Secs. 5.3 and 5.4 the series was dropped in applying Eq. 5.3(9) because of the very rapid decrease of K_0 with increasing argument, it is to be noted that when k_z is so small that $\sqrt{k_z}\rho_e$ is appreciably less than 1, at least the first few terms of the series must be taken into account. And when k_z approaches 0, so many terms of the series would have to be used that it becomes more convenient to use for φ_e an expression of the form of Eq. (8), so that

$$\Delta\varphi(k_z \sim 0) = 2 \log \rho_e/\rho_w$$

as indeed should be expected.

With $\Delta\varphi$ essentially determined, one need only find the production rate or flux, Q, to which the $\Delta\varphi$ corresponds. Applying Eq. 5.4(1) to Eq. (9), it is found that

$$Q = 2\pi \int_0^h \rho_e \frac{\partial \varphi_e}{\partial \rho_e} dz = 8\pi hx. \quad (10)$$

The quantity $Q/h\Delta\varphi$, or the production capacity per unit potential drop and per unit sand thickness, can now be computed. Figure 87 shows the values of $Q/h\Delta\varphi$ plotted as a function of k_z for various values of the well penetration for a sand of 125-ft. thickness. The extreme values of $k_z = 1$ and $k_z = 0$ correspond, respectively, to the case of an isotropic sand, as treated in Secs. 5.3 and 5.4, and to the case of strict radial flow confined to the part of the sand actually penetrated by the well.   Although the effect of the variation in the vertical permeability may not

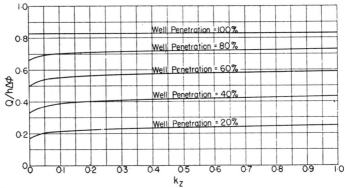

FIG. 87.—Production capacities of partially penetrating wells as functions of the vertical permeability k_z. $Q/h\Delta\varphi$ = production rate per unit pressure differential and per unit sand thickness.   Total sand thickness = 125 ft.; well radius = ¼ ft.; external-boundary radius = 500 ft.   Horizontal permeability = 1; liquid viscosity = 1.0.

appear large from Fig. 87, a plot of the ratios of the extreme values of the production capacities, *i.e.*, $\dfrac{Q(k_z = 1)}{Q(k_z = 0)}$, as is given

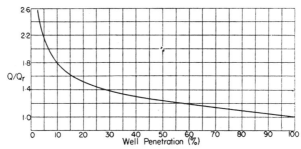

FIG. 88.—Ratios of production capacities of partially penetrating wells to those for strict radial-flow systems. Q = total production capacity; Q_r = strictly radial component of Q; sand thickness = 125 ft.; well radius = ¼ ft.; external-boundary radius = 500 ft.

in Fig. 88, shows that the addition to the flux due to the vertical flow is after all quite considerable, especially for the smaller well penetrations.   Thus for a well penetration of 20 per cent the strictly radial flow is increased by 50 per cent on introducing a vertical permeability equal to the horizontal permeability; and for a 10 per cent well penetration the increase is more than 75 per cent.

It is to be noted, however, that the increase in Q due to the vertical permeability is not simply proportional to k_z, for if that were so, the curves of Fig. 87 would be straight lines connecting the points for $k_z = 0$ and $k_z = 1$. Rather, the increase in Q due to k_z sets in rapidly even for small k_z, and changes but slowly for $k_z > 0.1$. This result is of practical significance in that it shows that, as long as k_z is not excessively small, it need not be determined with any great precision, while still permitting a reasonably accurate estimate of the value of the flux into the well. On the other hand, unless k_z is an appreciable fraction of the horizontal permeability, the above analysis shows that the anisotropy of the sand may cause a very appreciable diminution in the production rate that will enter a well which only partially penetrates the anisotropic sand, and hence should be taken into account whenever possible.

5.6. Summary.—It is a frequent occurrence that wells drilled into fluid-bearing sands do not completely penetrate them. This failure to drill the well through the whole thickness of the producing horizon may be due, in the case of artesian wells, to the lack of previous knowledge as to the actual thickness of the producing sand which, however, is subsequently obtained from the records of neighboring wells, or, as in the case of oil wells, it may be the result of the desire to avoid underground waters which may be known or suspected to underlie the oil sand. For the difficulties created by the entry of water into the well bore of an oil well are often so serious that it is a common practice, when bottom waters are suspected, to stop the drilling before they are reached; or, if they have been penetrated by mistake, the usual procedure is to plug the well back if at all possible.

In these cases where the well does not extend through the whole thickness of the producing sand the flow will not be radial, as it is when the well penetration is complete, and which has been discussed in the last chapter. For in addition to the fluid lying in the sand above the depth penetrated by the well, the fluid in the sand below this depth will also move toward the well. Evidently this latter fluid must move upward as well as radially in order to reach the well. In this way the flow in the system assumes a three-dimensional character, as both horizontal and vertical velocities are necessarily involved in a complete description of the dynamics of the flow.

The methods of treatment of three-dimensional problems are quite analogous to those that were used in the last chapter for two-dimensional flow systems. One only begins with the more general three-dimensional Laplace equation [cf. Eqs. 5.1(2) and 5.2(1)].

Perhaps the simplest of such problems is that of spherical flow—the exact analogue of the strictly radial two-dimensional flow system. Physically, it corresponds to a well just tapping a uniform sand of great thickness, the well itself being represented by a hemispherical surface of radius equal to the well radius.[1] The equipotential surfaces of the system are hemispheres concentric with the well surface, and the streamlines are the radii entering the center of the well. The potential [cf. Eq. 5.1(1)] varies inversely as the radial distance [cf. Eq. 5.2(6)] from the center of the well, and hence the gradients are much more concentrated about the well than in the case of radial flow. The velocity decreases inversely as the square of the radial distance [cf. Eq. 5.2(7)], which again gives a much more rapid decrease than in the radial-flow case. These high concentrations of the pressure gradients near the well and their almost vanishing value at larger distances lead to the result that the production capacity of such a spherical-flow system is practically independent of the external radius at which the high potential is applied. On the other hand, it is sensitive to the radius of the well, being in fact approximately proportional to it [cf. Eq. 5.2(11)]. In absolute value, however, a well producing by spherical flow has a much lower capacity than a well completely penetrating a sand, and hence producing by radial flow, the ratio being 1 to 26 in the case of a 50-ft. sand and a well radius of $\frac{1}{4}$ ft. [cf. Eq. 5.2(13)].

In the more practical problem where the sand is of finite thickness and is partially penetrated by the well, the analysis becomes considerably more complicated. Here, a potential distribution must be developed which gives no flux at the top and bottom of the sand—corresponding to a sand lying between impermeable shales—and a uniform potential over the well surface. To take

[1] Such wells, when drilled into water-bearing sands, are commonly termed "open-bottom" wells.

care of the first requirement it is convenient to apply the method of images, already discussed in Chap. IV, and successively reflect the well surface in the upper and lower sand faces, finally forming an infinite series of such images. To make the potential function assume a uniform value over the well surface is, however, more difficult. For in the analytical treatment of the problem one begins with the potential due to an element of flux placed at some point along the well axis. The real problem, then, is to find such a distribution of the flux elements along the well axis that the potential over the well surface will be uniform. Thus when the well completely penetrates the sand, equal amounts of flux evidently enter in each unit depth along the well surface. But when the well penetration is not complete, it is clear that the flow coming from the portions of the sand lying below the depth of the well will be mostly concentrated near the bottom of the well, thus causing an increase in the flux density entering the well surface as the bottom is approached.

After determining the flux distribution which will give a uniform potential over the well surface, the problem is essentially solved. For then it is possible to derive not only the potential distribution within the sand body, but the relation between the production rates and the potential drop across the sand can also be found at once. As already indicated, the flux density into the well surface increases uniformly as the bottom of the well is reached, owing to the fluid entering the lower portion of the well surface from the part of the sand lying below the depth of penetration. The integrated sum of these flux densities over the well surface then gives the total production capacity.

Summing, also, the contributions to the potential due to these individual flux elements, the resultant potential distribution is obtained. The striking feature of these distributions is that they assume an almost exact radial character as soon as one recedes from the well by a distance of the order of twice the sand thickness, and actually vary logarithmically with the radial distance [*cf.* Eq. 5.3(9)]. Near the well surfaces, however, the equipotential surfaces follow closely the contour of the well, being closely spherical for a nonpenetrating well (*cf.* Fig. 82) and markedly cylindrical for a partially penetrating well (*cf.* Fig. 81).

Of more practical interest, however, are the production rates that correspond to the pressure distributions. The analysis shows that the total production capacity of a well partially penetrating a sand of fixed thickness increases more rapidly than the penetration for small penetrations, but less rapidly as the penetration approaches 100 per cent (*cf.* Fig. 83). The interpretation is again to be found in the consideration of the nonradial flux that enters the well from the part of the sand lying below the depth of penetration. For when the well penetration is small, this nonradial flux will increase with the well penetration, owing to the fact that more well surface is available into which the nonradial flux can flow. Since the strictly radial part of the flow increases directly with penetration, the sum will increase more rapidly. As the penetration is increased beyond a certain point, however, the fact that the thickness of the nonpenetrated part of the sand, from which the nonradial flow comes, decreases, counterbalances the effect of an increasing well surface, with a resulting variation in the total flux that is less rapid than the penetration.

With regard to the magnitude of the nonradial part of the flux rate as compared to the strictly radial contribution, it is found that it increases uniformly from a vanishing value at 100 per cent penetration—where the flow is entirely radial—to a magnitude of 50 per cent of the radial flow at a penetration of 20 per cent, and finally exceeds the radial-flow contribution at penetrations of 6 per cent or less. This shows, incidentally, that the actual flow system of a partially penetrating well cannot be even roughly approximated by a simple superposition of a radial flow into the well proper and a semispherical flow into the well extremity.

Considering the effect of the sand thickness on the production capacity of a well of a given fractional penetration, the analysis shows that for sand thicknesses greater than 50 ft. the variation of the production capacity with the sand thickness is almost exactly linear, the slopes of the curves increasing with the value of the fractional well penetration (*cf.* Fig. 84). For smaller thicknesses the production capacities increase with the sand thickness somewhat more rapidly, though this effect is appreciable only for the smaller penetrations.

As might be expected, the variation of the production capacity of a partially penetrating well with the well radius is intermediate between the variations in the case of strict spherical flow and exact radial flow. That is, for large penetrations the flow approximates the radial type and the production capacity varies logarithmically with the well radius. But this rate of variation increases with decreasing penetration until in the limit of the nonpenetrating well—which corresponds to a spherical-flow system—the production capacities vary directly as the well radius.

A modification of considerable practical interest of the problem of the production from partially penetrating wells as just outlined is that in which the effect of an anisotropy in the sand permeability is taken into account. When it is noted that most measurements of permeability, on single samples of consolidated sands, that have been made both parallel and perpendicular to the bedding planes have shown appreciable differences between the two permeabilities, the matter of anisotropy assumes more than academic interest. Fortunately, however, the analytical treatment of an anisotropic sand can be carried through for the problem of the partially penetrating wells by making only slight formal changes in the analysis developed for the same problem in isotropic porous media.

For this purpose it is convenient to take the permeability parallel to the bedding plane as unity and denote that perpendicular to the bedding plane by k_z, so that k_z really represents the ratio of the two permeabilities. It then turns out that by the simple expedient of either multiplying the radial coordinates by the factor $\sqrt{k_z}$ or the vertical coordinates by the factor $1/\sqrt{k_z}$ the equations in the *new* coordinates become exactly the same as those for an isotropic sand [*cf.* Eq. 5.5(4)]. Furthermore, the physical equivalent of the transformed system is that of a partially penetrating well in an *isotropic sand* of the *same* fractional penetration as that in the real anisotropic sand, and with "effective" (numerical radii divided by twice the sand thickness) well and external radii equal to those in the actual system multiplied by $\sqrt{k_z}$ [*cf.* Eq. 5.5(6)]. With these changes in the numerical values of the apparent physical dimensions of the system, the analysis can be carried through without taking any further explicit account of the physical anisotropy of the system, *i.e.*, by the same analysis as that used for isotropic sands.

As might be anticipated, the effect of the anisotropy on the production capacities is most pronounced for the small well penetrations. Furthermore, the production capacities increase continuously with increasing vertical permeability k_z, from a value of $k_z = 0$ corresponding to strictly radial flow, to the value of $k_z = 1$ corresponding to an isotropic sand (cf. Fig. 87). The increase, however, is not linear, but is most rapid for small values of k_z. This has the result that where k_z does not differ greatly from 1 the anisotropy will not be of particular consequence except for very small penetrations. On the other hand, when $k_z < 0.1$ and the well penetrations are not larger than 20 per cent, the anisotropic sand will yield less than 80 per cent of the production rate from the same well in an isotropic sand. And, if the vertical permeability should actually vanish, the production capacity for well penetrations of 20 per cent will fall to 65 per cent of that for the isotropic sand, and for 5 per cent well penetration to less than 47 per cent of the production capacity of an isotropic sand.

Another interesting problem arising in the study of partially penetrating wells is that where the producing horizon is stratified rather than composed of a single homogeneous layer. While the flow in such cases is also three-dimensional, the effect of the nonuniformity of the sand requires a somewhat different type of analysis. Its discussion will, therefore, be deferred to Chap. VII where the general methods of treating such systems of nonuniform permeability will be developed. In this same chapter will also be treated the three-dimensional problem of the well with a sanded liner, as it too involves regions of different permeability in a single connected porous medium.[1] A final three-dimensional flow problem of practical interest to be treated ·in this work is that of water coning. Here, however, the flow system involves two homogeneous fluids and must again be given a type of treatment different from that used for systems carrying only a single homogeneous fluid. Its discussion is, therefore, also deferred to a later chapter (Chap. VIII) which is devoted entirely to the analysis of this class of problems.

---

[1] A somewhat similar type of analysis is given by Tolke in the discussion of his method for determining the permeability of consolidated porous media by nonlinear-flow experiments (cf. footnote on p. 75). However, as it is unlikely that this method will find wide use in actual permeability measurements, his theory is not reproduced here.

CHAPTER VI

GRAVITY-FLOW SYSTEMS

6.1. Introduction.—It has already been seen that in a general three-dimensional flow system and in two-dimensional systems involving flow in vertical planes the effect of gravity may be formally taken into account by the introduction of the velocity potential

$$\Phi = \frac{k}{\mu}(p \pm \gamma g z) \tag{1}$$

into the equation of continuity, where the $\pm$ signs correspond to the upward or downward direction of the vertical coordinate, $+z$. This procedure leads at once to Laplace's equation for Φ,

$$\nabla^2\Phi = 0, \tag{2}$$

which may be taken as the basis for describing the details of all steady-state three-dimensional liquid-flow systems. This formulation was used in the last chapter, and solutions were given to several three-dimensional problems of practical interest. And, strictly speaking, this same formulation suffices to describe all possible steady-state incompressible-liquid flow systems in a homogeneous medium in which the effect of gravity must be taken into account.

As already indicated in previous discussions, however, the particular methods to be used for the solution of any specific problem will depend largely upon the nature of the boundaries of the fluid system and upon the conditions specified at these boundaries. The class of problems to be discussed in this chapter will be characterized by the fact that a part of the boundary defining the fluid system is a "free surface." As the term suggests, a free surface is a fluid surface in equilibrium with the atmosphere rather than with a rigid impermeable boundary. From a strictly hydrodynamic point of view, it may be defined as a streamline along which the pressure is uniform.

287

The practical significance of such systems whose boundaries include free surfaces is that every flow system in which gravity is the essential driving force gives rise to a free surface. Thus the water leaking through a dam to a lower level possesses a free surface within the dam. Likewise, the water seeping out from an irrigation ditch possesses a free surface in the soil surrounding the ditch. Finally, the water in a sand feeding a pumping well, in which the fluid level is kept below the top of the sand, flows with a free surface in equilibrium with a uniform pressure existing over the fluid body in the sand and equal to that above the fluid level in the well bore.

Supposing that the sand carrying the water into an artesian well is covered by an impermeable boundary, it is clear that if the water level at the well surface be raised to equal the sand height the flow will stop unless there is an external pressure head exerted at the outer boundary of the system. In the latter case there would be no free surface, and the flow could be described by the methods of the previous and following chapters. However, when a free surface is present, the mathematical difficulties of the problem immediately become very great, and indeed practically insurmountable in the case of three-dimensional systems. The reason is that not only are the boundaries no longer of simple geometrical form, but the exact shape of the free surface is actually unknown. Rather, the shape of the free surface must be determined *simultaneously* with the pressure distribution within the system. The possibility of a solution to such a problem is indeed analytically assured, because of the double requirement that the free surface be both a streamline and constant-pressure surface. But unfortunately there are no analytical tools available which are sufficiently powerful for finding explicit solutions to such problems, except in the case of two-dimensional systems, where the method of conjugate-function transformations will, in principle, lead to the required results. On the other hand, as will be seen presently, even this powerful method entails a considerably more complex analysis than is involved in its application to systems without free surfaces, such as those treated in Chap. IV.

An additional complication in the study of general gravity-flow systems arises from the occurrence in most of such systems of "surfaces of seepage."[1] As in the case of free surfaces, the pressure along these surface elements is uniform. However, they do not represent streamlines but are simply parts of the boundary of the porous medium where the liquid leaving the system enters a region free of both liquid and porous medium. On the other hand, the length of these surface elements is not known a priori, as their upper terminal always joins with a free surface, which is also unknown initially. Although the existence of such surface elements has been recognized for some time, it is only recently (*cf.* Sec. 6.3) that an analysis has been developed in which they can be explicitly taken into account.

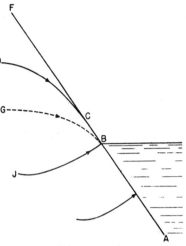

The physical basis for the occurrence of these surfaces of seepage lies in the observation that not only must a free surface always have a downward slope in approaching an outflow surface, but that furthermore the velocity along the free surface, being the component of the gradient of gravity along the surface,

Fig. 89.—Diagrammatic representation of the trend of the streamlines at an inclined outflow face of a gravity-flow system.

must always be finite.[2] Thus supposing that AF is the physical outflow boundary of the gravity-flow system with an outflow fluid level extending to B (*cf.* Fig. 89), the limiting streamline of those terminating along AB must meet AF at B at right angles, as JB. If now it be assumed that the free surface also terminates at B, it must form a finite angle with JB, as GB. However, the resulting convergence of the two streamlines to a point must lead to infinite velocities at B, both along JB and GB. Because of

---

[1] The literal translation of the German "Hang-quelle" hardly seems as expressive as the term used here.

[2] In fact it may be readily shown that the total velocity along the free surface can never exceed the vertical free fall velocity due to gravity, namely, $k\gamma g/\mu$ [*cf.* Eq. 6.3(5)].

the above-mentioned limitation of the velocities along free surfaces, this is impossible, and the free surface must, therefore, terminate *above B*, as at *C*. The intervening segment *BC* is then a surface of seepage, as it is not a streamline but is exposed to a uniform pressure.

In case the outflow surface *AF* is vertical, the above argument

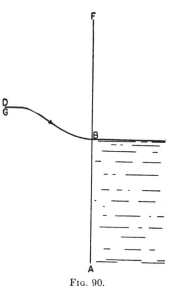

Fig. 90.

breaks down, as one might suppose the free surface *DB* to coincide with the last streamline *GB* of those cutting *AB* at right angles (*cf.* Fig. 90). Such a condition is not possible, however, as the nonvanishing horizontal velocities at *AB* require that the fluid pressure has a nonvanishing horizontal derivative along *AB*. As the slope of *DB* is zero at *B*, the horizontal increase in pressure in leaving *AB* must be due to the integrated effect of vertical velocities near *AB*, directed *upward*.[1] But as the potential along *AB* is uniform and as the excess flow into *AB* beyond that corresponding to the strictly linear flow is that coming from a level above *B*, *upward* vertical velocities are physically inadmissable.

When the inclination of the outflow surface *AF* is less than 90 deg., the necessity for separating the top of the outflow-water level from the terminus of the free surface by a surface of seepage does not follow from either of the above arguments. Its existence can, however, be inferred from the fact that such surfaces of seepage are obviously necessary in all cases for zero outflow-fluid head, and it is physically unreasonable that this should suddenly collapse as a nonvanishing outflow-fluid head is added to the system. On the other hand, a study[2] of the hodographs (*cf.* Sec.

---

[1] As the upward velocities would make the streamlines convex upward near *AB*, the coincidence of *DB* with *GB* would have the further consequence that the streamlines would converge into a cusp at *B*.

[2] *Cf.* B. Davison, *Phil. Mag.*, **21**, 904, 1936.

6.3) for such systems indicates that under certain special conditions an outflow face with an inclination less than 90 deg. might be devoid of a surface of seepage, provided the outflow-fluid level is nonvanishing.

Because of the requirement that the tangential velocity along the outflow surface must be continuous in passing from the terminus of the free surface onto the surface of seepage, it readily follows that for inclinations of the outflow surface which are equal to or greater than 90 deg., the free surface must meet the surface of seepage, or outflow surface, tangentially. When the outflow-surface inclination is less than 90 deg., this same condition leads to the result that the free surface will no longer be tangential to the outflow surface, but will rather cut it vertically.

With regard to the behavior of the free surface at an inflow surface, a little consideration shows that for inflow faces of inclinations $\geqslant$ 90 deg. the free surface will enter the sand horizontally, whereas if the inflow face of the porous medium has an inclination $\leqslant$ 90 deg. the free surface will enter at right angles to the face.

Finally, it may be noted that in all cases the velocity at the lower terminus of the surface of seepage will be theoretically[1] infinite, owing to the discontinuity in the tangential velocity above and below the terminus of the surface of seepage. And when the inclination of the outflow surface is less than 90 deg., the velocity at its base (horizontal) will also be infinite, as it will be a point of convergence for streamlines which will enter this point with a nonvanishing angular separation. When, however, the outflow or inflow equipotential surfaces make acute angles with a boundary streamline, as at the heel or toe of a dam with faces sloping toward each other, the velocities at these corners[2] will be zero. These latter properties are, of course, applicable to all flow systems and have nothing to do with the gravity component of the flow.

Although it is possible to give a rather complete qualitative discussion of gravity-flow systems, their quantitative treatment

[1] In the physical system, of course, no such infinities will occur, owing to the breakdown of Darcy's law after the critical Reynolds number for the system is exceeded.

[2] In the case of the outflow surface, this assumes that there is a non-vanishing outflow-fluid level. When the latter vanishes, the velocity will be infinite at the toe.

is, as a whole, in a rather unsatisfactory state. For while the method of hodographs does, in principle, provide a means for treating practically any two-dimensional system, the labor of applying it numerically is so formidable that it has been carried through in detail only for a small number of specific cases. And in the case of three-dimensional systems, there is not a single problem for which there is available an exact analysis. One must, therefore, resort to approximate methods, even though their accuracy can in some cases only be estimated by physical intuition.

In view of this situation, no attempt will be made here to classify the various gravity-flow problems and to treat each class systematically. Rather, we shall first present several of the available rigorous solutions for two-dimensional problems as obtained analytically and by electrical models, then give some approximate treatments of other problems of practical interest, next review some empirical studies of the three-dimensional radial-flow problem, and finally outline an approximate theory for the calculation of the flux through gravity-flow systems. The order of presentation will thus be determined largely by the type of the method of treatment rather than by the physical interrelationships of the various problems.

6.2. The Drainage of a Sloping Sand.[1] **The Problem of Hopf and Trefftz.**—The problem of Hopf and Trefftz is essentially that of the drainage of an inclined water sand by a ditch dug at the top of the sand. Although the solution of this problem to be presented here is not as satisfying analytically as those given by the method of hodographs (*cf.* Sec. 6.3), since it leaves out of account the surface of seepage and introduces the shape of the drainage ditch only indirectly, it does possess the advantages of both analytical and numerical simplicity, as compared to the more exact method. In fact, in view of the very formidable numerical difficulties that the latter method would involve in treating general two-dimensional gravity-flow systems, a detailed presentation of the solution of Hopf and Trefftz appears to be fully justified, in spite of its limitations.

---

[1] Hopf, L., and E. Trefftz, *Zeits. angew. Math. u. Mech.*, **1**, 290, 1921. Here and in all the analytical developments to follow no account will be taken of the capillary zone of saturation overlying the free surfaces of the system.

The flow system may be represented as in Fig. 91. The water flowing down slope from A is partially caught by the ditch, the remainder continuing downward toward D. It is assumed that the impermeable bed is inclined at an angle α to the horizontal, and that far up on the upper side of the ditch the water fills the sand to a height h_1 above the bed. On the lower side of the ditch the water level falls to a height h_2 above the bed. Finally, the system will be taken as two-dimensional, extending to infinity in both directions normal to the plane of the paper.

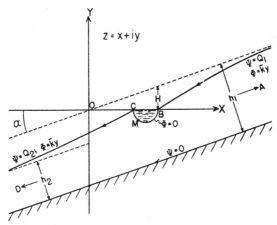

Fig. 91.—The z-plane representation of a ditch lying at the top of a sloping sand. (*After Hopf and Trefftz, Zeits. angew. Math. u. Mech.*)

Evidently the impermeable bed itself corresponds to a streamline; this may be taken as $\Psi = 0$. Likewise, the uppermost surfaces of the water both above and below the ditch will be streamlines, which may be denoted by $\Psi = Q_1$, and $\Psi = Q_2$, respectively, where $Q_1 - Q_2$ is the flux taken away by the ditch.[1] Besides being streamlines these surfaces $\Psi = Q_1$, Q_2 have the

---

[1] As pointed out by Hamel (*Zeits. angew. Math. u. Mech.*, **14**, 129, 1934), the streamline $\Psi = Q_1$ will in general enter the outflow surface above the fluid level in the latter, thus giving rise to a surface of seepage (*cf.* Sec. 6.1). However, as the analysis for the more rigorous treatment, in which the surface of seepage is not neglected, has not yet been carried through, and as the Hopf and Trefftz theory should in any case give the essential features of the solution, it will be presented here in detail as an example of a method that is essentially different from that to be presented in the following three sections.

still more important characteristic that they are "free" surfaces in equilibrium with the atmosphere. As already pointed out in Sec. 6.1, this implies that the pressure is uniform over the free surface, and hence may be taken as 0. The potentials Φ over these surfaces therefore have the value

$$\Phi = \frac{k\gamma g}{\mu} y = \bar{k}y, \tag{1}[1]$$

as indicated in Fig. 91, where y is taken as the vertical coordinate. It is, of course, just this condition at one of the boundaries which characterizes all gravity-flow systems, and which on the one hand gives rise to the difficulties of the problem, and on the other hand makes the solution analytically possible, since it is a "compensation" for the unknown shapes of the free surfaces. In the particular method of solution used here, however, the specification of the shape of the free surfaces is avoided by an appropriate transformation of coordinates [Eq. (4)], and the chief difficulty is the determination of the exact shape of the drainage ditch, as will be seen below.

As a preliminary to the main part of the analysis of the present problem, it may be observed that the complex potential

$$\omega_0 = \Phi + i\Psi = z_0\bar{k} \sin \alpha e^{-i\alpha} + iQ_1; \qquad \frac{Q_1}{\bar{k} \sin \alpha} = h_1 \tag{2}$$

gives a complete representation of the inclined-flow system with the ditch removed (*cf.* Fig. 92). Thus equating the real and imaginary parts of Eq. (2), it is seen that

$$\left.\begin{array}{l} \Phi = \bar{k} \sin \alpha(x_0 \cos \alpha + y_0 \sin \alpha) \\ \Psi = Q_1 - \bar{k} \sin \alpha(x_0 \sin \alpha - y_0 \cos \alpha). \end{array}\right\} \tag{3}$$

---

[1] The quantity $k\gamma g/\mu$ has been denoted by $\bar{k}$ since it represents an "effective" permeability for systems subjected entirely to gravity driving agents and which deal exclusively with a single liquid, as water. Although it does not express the meaning of the permeability constant from a fundamental physical point of view, it is convenient from a practical stand-point to use this special form of an "effective" permeability is such fields as civil and agricultural engineering where the exclusive use of hydraulic gradients in terms of water makes it unlikely that there will be an erroneous extrapolation to other types of fluid flow through porous media. Of course, the units of $\bar{k}$ are those of a velocity, that is, cm/sec. *Cf.* also Sec. 2.3.

Evidently the streamline $\Psi = 0$ coincides with the impermeable base of Fig. 92, and the upper surface $y_0 = x_0 \tan \alpha$ clearly gives: $\Psi = Q_1$, $\Phi = \bar{k}y_0$.

The variable z is now transformed to z_1 by the relation

$$z = z_0(\omega) + z_1(\omega),$$

where

$$z_0(\omega) = \frac{(\omega - iQ_1)e^{i\alpha}}{\bar{k} \sin \alpha}; \qquad \omega = \Phi + i\Psi, \tag{4}$$

so that $z_1(\omega)$ represents the particular effect of the ditch.

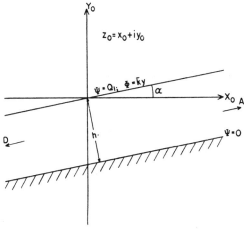

FIG. 92.—The inclined linear-flow system of Fig. 91 without the ditch.

Taking into account the boundary conditions indicated in Fig. 91, it is readily seen that the z plane is mapped by Eq. (4) onto the z_1 plane according to the correspondences

$$
\begin{aligned}
BA\rightarrow: \quad & y_1 = 0; \quad x_1(B) = x(B); \quad x_1(A) = 0 \\
AD\rightarrow: \quad & x_1 \sin \alpha - y_1 \cos \alpha = 0 \\
DC\rightarrow: \quad & y_1 = \frac{Q_1 - Q_2}{\bar{k}} \operatorname{ctn} \alpha; \quad x_1(D) = y_1 \operatorname{ctn} \alpha = \\
& \qquad\qquad\qquad\qquad \frac{Q_1 - Q_2}{\bar{k}} \operatorname{ctn}^2 \alpha \\
x_1(C) = x(C) & - \frac{Q_1 - Q_2}{\bar{k}}; \quad x_1(B) - x_1(C) = b + \\
& \qquad\qquad\qquad\qquad \frac{Q_1 - Q_2}{\bar{k}},
\end{aligned}
\tag{5}
$$

where $b = x(B) - x(C)$ = breadth of ditch, and $h_2 = Q_2/\bar{k} \sin \alpha$.

The section $C'B'$ is undetermined until the shape of the ditch is chosen. However, it will be easier to *assume* the shape of $C'B'$ (*i.e.*, Q_1, Q_2, Q_m) and see later the form of the ditch so implied, rather than to put in directly the shape of the ditch. The z_1-plane representation of Fig. 91 is shown in Fig. 93.

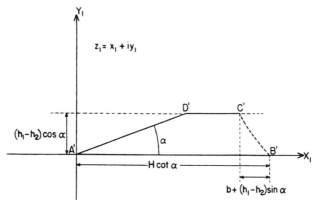

FIG. 93.—The z_1-plane map of Fig. 91.

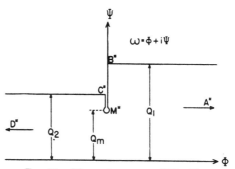

FIG. 94.—The ω-plane map of Fig. 91.

The system of Fig. 91 is still more simply mapped onto the ω plane, the result being indicated in Fig. 94. In this representation, the cut $B''M''C''$ corresponds to the ditch, over which $\Phi = 0$, the value of Ψ decreasing from Q_1 to a minimum, Q_m, and then rising to Q_2.

The next step in the solution of the problem is to determine z_1 as a function of ω such that along the Φ axis,

$$x_1 \sin \alpha - y_1 \cos \alpha = 0;$$

along $\Psi = Q_1$: $y_1 = 0$; along $\Psi = Q_2$: $y_1 = \dfrac{Q_1 - Q_2}{k}$ ctn α, and such that the cut $B''M''C''$ goes into the ditch $y_1 = F_1(x_1)$. Instead, however, of attempting to map directly the ω onto the z_1 plane, it is simpler to follow the more indirect method of mapping both planes on the first quadrant of a $\zeta = \xi + i\eta$ plane (*cf.* Fig. 95). In this mapping we shall make the correspondences:

$$C'' \to \zeta = 0; \qquad B'' \to \zeta = \infty; \qquad A'' \to \zeta = a;$$
$$D'' \to \zeta = d; \qquad M'' \to \zeta = im.$$

By a slight modification[1] of the Schwarz-Christoffel theorem of Sec. 4.11 it is easily seen that the appropriate transformation function is

$$\omega = \frac{Q_1}{\pi} \log \frac{a + \zeta}{a - \zeta} - \frac{Q_2}{\pi} \log \frac{\zeta + d}{\zeta - d}. \tag{6}$$

For the z_1 plane, the mapping function may be found by the generalization of the Schwarz-Christoffel transformation

$$z_1 = \int_a^\zeta \frac{f(\zeta)d\zeta}{(\zeta - d)^{\frac{\alpha}{\pi}}(\zeta - a)^{1 - \frac{\alpha}{\pi}}}, \tag{7}$$

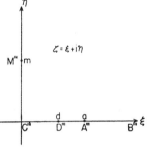

FIG. 95.—The ζ plane map of the z_1 and ω planes of Figs. 93 and 94.

where $f(\zeta)$ has no singularities in the first quadrant, is positive on the $+$ side of the real (ξ) axis, and vanishes at infinity. Its form determines the shape of the ditch. It is readily verified that the transformation of Eq. (7) maps the boundary $C'D'A'B'$ on the positive real (ξ) axis of the ζ plane and the ditch BC onto the positive imaginary (η) axis.

The relation implied by Eqs. (6) and (7) between ω and z_1 (and hence z) constitutes the formal solution to the problem. However, it cannot be explicitly derived until a specific assumption is made with regard to the function $f(\zeta)$, *i.e.*, the shape of the ditch. But before specializing the form of $f(\zeta)$ it will be of

---

[1] One need simply combine Eq. 4.11(3) with the further transformation $t = \zeta^2$, which will map the upper half of the t plane onto the first quadrant of the ζ plane.

interest to derive first the relation between the flux caught by the ditch, $Q_1 - Q_2$, and the other constants of the system.

Thus since the point B corresponds to $\zeta = \infty$, and

$$x_1(B) = x(B),$$

it follows that

$$x_1(B) = \int_a^\infty \frac{f(\zeta)d\zeta}{(\zeta - d)^{\frac{\alpha}{\pi}}(\zeta - a)^{1 - \frac{\alpha}{\pi}}}.$$

For the cases of practical interest, α/π is very small. One may, therefore, make the approximations that except in the vicinity of $\zeta = a, d$

$$(\zeta - d)^{\frac{\alpha}{\pi}} \cong 1; \qquad (\zeta - a)^{1 - \frac{\alpha}{\pi}} \cong \zeta - a.$$

It follows then that the drawdown H of the upper edge of the ditch from the unperturbed level is given by

$$H = x_1(B) \tan \alpha = \pi f(a) + 0\left(\frac{\alpha}{\pi}\right).$$

Further, since D corresponds to $\zeta = d$,

$$z_1(D) = \int_a^d \frac{f(\zeta)d\zeta}{(\zeta - d)^{\frac{\alpha}{\pi}}(\zeta - a)^{1 - \frac{\alpha}{\pi}}},$$

so that

$$y_1(D) = \sin \alpha \int_d^a \frac{f(\zeta)d\zeta}{(\zeta - d)^{\frac{\alpha}{\pi}}(a - \zeta)^{1 - \frac{\alpha}{\pi}}} = \frac{\pi}{\alpha} f(a) \sin \alpha + 0(a) \cong$$

$$\pi f(a).$$

Combining these results with Eq. (5), it follows that to the order of α/π

$$H = y_1(D) = \frac{Q_1 - Q_2}{\bar{k} \sin \alpha},$$

so that

$$\frac{Q_1 - Q_2}{Q_1} = \frac{H}{h_1}. \tag{8}$$

Hence to the order of α/π the fraction of the original flux in the sand that is caught by the ditch is equal to the ratio of the drawdown of the free surface at the ditch to the thickness of the undisturbed layer of water-saturated sand. Thus whereas the previous formulas that have been used for this problem have involved the depth of the water in the ditch or its width, Eq. (8) shows that the ratio $(Q_1 - Q_2)/Q_1$ is determined by the drawdown H and the initial fluid depth h_1 alone.

Without going into the details, which may be found in the original paper, we shall present the results of a specific numerical calculation carried out by Hopf and Trefftz. The assumptions made were the following:

$$\alpha = 0.003;$$
$$Q_2 = 0.5Q_1;$$
$$Q_m = 0.45Q_1,$$

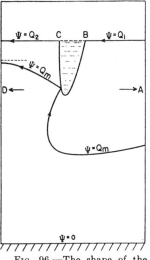

so that $0.55Q_1$ enters the ditch, and $0.05Q_1$ flows back out. In units of the depth h_1 of the undisturbed flow, it follows that

$$h_2 = \frac{Q_2}{k \sin \alpha} = 0.5;$$

$$h_m = \frac{Q_m}{k \sin \alpha} = 0.45;$$

$$H = 0.5.$$

FIG. 96.—The shape of the ditch and limiting streamlines corresponding to Eq. 6.2(9). (*After Hopf and Trefftz, Zeits. angew. Math. u. Mech.*)

The shape of the ditch was determined after choosing for $f(\zeta)$ the expression

$$f(\zeta) = \frac{A_1}{\zeta + l} + \frac{A_2}{(\zeta + l)^2}. \tag{9}$$

l was taken as 7.25, and A_2 was chosen so that the breadth b of the ditch was $\frac{1}{15}$. The shape of the ditch and the position of the minimal streamline $\Psi = Q_m$ that reaches the ditch are indicated in Fig. 96. The general appearance of the flow system is shown in Fig. 97. Although the shape of the ditch as shown in Fig. 96 may seem artificial, one may without great difficulty adjust the constants A and l or change the form of $f(\zeta)$ so that it will at least roughly approximate any preassigned shape. Such changes, however, will not materially affect the drainage capacity of the ditch or the general shape of the free surface of the system except in the immediate vicinity of the ditch.[1]

6.3. The Treatment of Two-dimensional Gravity-flow Systems by the Method of Hodographs.   The Mapping of the Boundaries. A very powerful though rather difficult method of treatment of two-dimensional systems containing simultaneously impermeable boundaries, constant-potential surfaces, and surfaces of seepage,

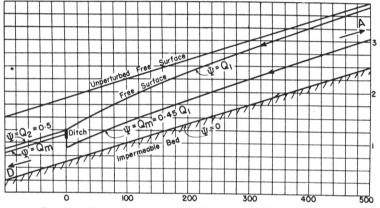

Fig. 97.—(*From Hopf and Trefftz, Zeits. angew. Math. u. Mech.*)

is the method of hodographs, which has been developed in considerable detail by Hamel,[2] and whose treatment will be followed

---

[1] The neglect of the surface of seepage at B does, however, make the values of the flux removed by the ditch, Q_1, $- Q_2$, for a given drawdown H, as given by Eq. (8), somewhat less than what they will be with a surface of seepage at B.

[2] Hamel, G., *Zeits. angew. Math. u. Mech.*, **14**, 129, 1934.   A brief qualitative outline of this theory was given even earlier by P. Nemenyi in "Wasserbauliche Strömungslehre," p. 204, 1933.   This method is quite similar to that developed by Helmholtz and Kirchhoff for the study of discontinuous fluid motions in the classical hydrodynamics, the detailed analysis here, however, being more difficult because of the necessity of dealing with the circular segments in the hodograph plane (*cf.* also M. Muskat, *Physics*, **6**, 402, 1935). An essentially equivalent development, but one phrased in a more formal analytic terminology, has been published by B. Davison, *Phil. Mag.*, **21**, 881, 904, 1936, expanding that author's earlier Russian paper in the *Mem. l'Inst. Hydr.*, **6**, 121, 1932, Leningrad, which was apparently unnoticed by Hamel. While Davison also applies the general theory to the problem of the dam with vertical faces (*cf.* Sec. 6.4), he gives no numerical discussion.   It may also be noted that the combination of a reciprocal hodograph transformation with the method developed in Secs. 6.8 and 6.9 has been applied by V. V.

here. A "hodograph" is a representation of a dynamical system in which the coordinates are the velocity components of the particles of the system. In two-dimensional problems the original geometrical representation of the system may be considered as being in the z plane, while the hodograph is in the hodograph or (u, v) plane, where u, v are the velocity components along the original x and y axes. The particular advantage of this (u, v) plane representation is that, whereas in the original z plane the shapes of free surfaces are in principle unknown until the whole dynamical problem is solved, their hodographs are simply circles of definite and fixed parameters. Furthermore, the surfaces of seepage which cannot be fixed in the z plane until the exact shape of the free surface is known, can also be given unique hodograph representations preliminary to the analysis of the problem as a whole. Once the boundaries of the system are fixed—in the (u, v) plane—the methods of conjugate-function theory can be applied to the final solution of the problem, although the transformations of the circular segments representing the free surfaces involve[1] the theory of modular elliptic functions, which has not heretofore been widely used in the discussion of physical problems.

To construct the hodograph of a system in the z plane one need only map the various boundary segments onto the (u, v) plane. This may be done as follows

a. Impermeable Boundaries.—Denoting the length along the impermeable boundary by s, the velocity components at the boundary will evidently be given by

---

Wedernikow to the solution of a specific case of the seepage of water into a canal (*C.R.*, **202**, 1155, 1936), and to the seepage of water out of ditches of rectilinear section (*Zeits. angew. Math. u. Mech.*, **17**, 155, 1937).

[1] When, however, the rectilinear segments all pass through a point lying on the circular segment, an inversion transformation will make it unnecessary to use the modular elliptic functions (*cf.* Wedernikow, *loc. cit.*). Such cases will arise either when all the constant potential surfaces are horizontal, as in the seepage into a dry ditch, or when there are no surfaces of seepage, as in the problem of the seepage out of ditches. Moreover, for a large class of the latter type of problem, the method of hodographs can be entirely avoided. (*Cf.* Secs. 6.8, 6.9). On the other hand, in certain special cases, as the seepage of water over a vertical cut-off, the hodographs may be so simple that the modular functions degenerate into elementary functions which can be constructed by inspection (*cf.* R. Dachler, "Grundwasserstromung," p. 95, 1936).

$$u = -\frac{\partial \Phi}{\partial s} \cos \alpha; \qquad v = -\frac{\partial \Phi}{\partial s} \sin \alpha; \qquad \frac{v}{u} = \tan \alpha, \qquad (1)$$

where Φ is the velocity potential and α is the inclination to the x axis of the boundary element at s. Hence for a rectilinear impermeable boundary α will be constant, and its representation in the hodograph plane will be a straight line through the origin—of the (u, v) plane—*parallel* to the boundary.

b. *Constant-potential Surface—Stationary Body of Liquid.*—Here, by definition, $\Phi = $ const., and the velocity components at the surface are

$$u = -\frac{\partial \Phi}{\partial n} \sin \alpha; \qquad v = +\frac{\partial \Phi}{\partial n} \cos \alpha; \qquad \frac{v}{u} = -\text{ctn } \alpha, \qquad (2)$$

where n is the normal to the surface. It follows, therefore, that a rectilinear constant potential boundary gives a line through the origin in the (u, v) plane *normal* to the real boundary.

c. *Free Surfaces.*—As by definition the pressure is constant along a free surface, Φ is given by

$$\Phi - \bar{k}y = C; \qquad \bar{k} = \frac{k\gamma g}{\mu}, \qquad (3)$$

where the vertical coordinate, directed upward, has been taken as $+ y$. The "effective" permeability $\bar{k}$ is evidently equal to the vertical-free-fall velocity of the liquid due to gravity. Hence, as $\frac{\partial y}{\partial s} = \sin \alpha$,

$$-\bar{k} \sin \alpha + \frac{\partial \Phi}{\partial s} = 0. \qquad (4)$$

Since the free surface must also be a streamline of the system, $\frac{\partial \Phi}{\partial s}$ gives the total velocity at the surface. Multiplying through Eq. (4) by $\frac{\partial \Phi}{\partial s}$, it follows therefore that

$$v^2 + u^2 + \bar{k}v = 0, \qquad (5)$$

which is evidently a circle passing through the origin, of radius $\bar{k}/2$ and center at $(0, -\bar{k}/2)$, in the (u, v) plane.

d. *Surfaces of Seepage.*—At a surface of seepage, through which the liquid passes and enters a region free of both porous medium

and liquid, the pressure is constant. Equation (4) must, therefore, apply. However, as this type of surface is not a streamline
surface, the velocity along it is given by

$$-\frac{\partial \Phi}{\partial s} = u \cos \alpha + v \sin \alpha, \tag{6}$$

so that

$$\bar{k} \sin \alpha + u \cos \alpha + v \sin \alpha = 0. \tag{7}$$

Hence if the surface is plane, its representation in the (u, v)
plane is a straight line normal to the surface and passing through
the point $(0, -\bar{k})$, which is the lowest point of the circle of Eq.
(5) representing the free surface.

6.4. The Seepage through a Dam with Vertical Faces; Analytical Theory.—Having seen how to construct the hodograph of a

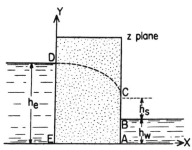

FIG. 98.—The z- plane diagram for the seepage through a dam with vertical faces.

given dynamical system we are ready to begin the analysis of a
practical problem. Perhaps the simplest problem, which at the
same time contains boundary segments satisfying all types of
boundary conditions, is that of the seepage through a vertical
earthen dam whose length is great compared to its thickness, so
as to be equivalent to a two-dimensional system. It will be
supposed that the capacities of both the upstream and downstream sides of the dam are so large that in spite of the seepage
through the dam the system is in hydrostatic equilibrium except
within the interior of the dam.

To construct now the hodograph of the system in the z plane,
(Fig. 98), it is to be noted first that since the region to the left
of DE is in hydrostatic equilibrium, the segment DE is at a
uniform potential and hence gives rise to a straight segment in

the (u, v) plane normal to DE, i.e., parallel to the u axis.   In fact, it lies on the u axis, as it must pass through the origin [cf. Eq. 6.3(2)].   Further, EA is an impermeable boundary along which also $v = 0$, so that it, too, is mapped on the u axis. Along AB the potential is again constant with vanishing vertical velocities, and is, therefore, mapped onto $A'B'$ in the (u, v) plane. BC is a surface of seepage parallel to the y axis.   Hence it must be mapped on the parallel to the u axis:

$$ v = -\bar{k} = \frac{-k\gamma g}{\mu}. $$

Finally, the free surface $C\acute{D}$ completes the diagram as a semicircle passing through D' and C'.[1]   The complete hodograph is,

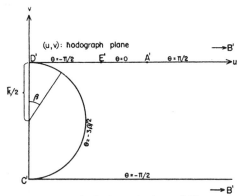

Fig. 99.—The hodograph plane of Fig. 98.

therefore, as shown in Fig. 99.   It is to be noted that, whereas in the z plane the exact form of the boundary segment DC and hence also the upper terminus of BC are in principle unknown until the problem as a whole has been solved, the boundaries of the system in the (u, v) plane are all known except for the numerical values of the abscissas of E' and A'.

The next step is that of finding another function of (u, v) whose actual values on the boundary segments can be preassigned.

---

[1] While the boundary conditions alone do not uniquely fix the exact position of the hodograph of the free surface, its position as given in Fig. 99 may be shown to follow from the requirement that there be no sources or sinks within the flow system.

Such a function can be obtained as follows: Introducing first the complex potential

$$f(z) = \Phi + i\Psi, \tag{1}^1$$

where Φ is the velocity potential and Ψ is the stream function, it follows at once that

$$u - iv = -f'(z); \qquad u + iv = -\overline{f'(z)}, \tag{2}$$

where the bar indicates the complex conjugate. From this function one proceeds to the intermediate potentials τ, θ, defined by

$$\tau + i\theta = -\log\left[-f''(z)\right], \tag{3}$$

of which θ can actually be specified over the boundaries of the region in the (u, v) plane. For,

$$-f''(z) = \frac{-(d/dt)f'(z)}{(dz/dt)} = \frac{\dot{u} - i\dot{v}}{u + iv} = \frac{be^{-i\chi}}{ne^{i\nu}}, \tag{4}$$

where n, b are the absolute values of the velocity and acceleration, and χ, ν are the inclinations of their directions (the dots indicating time differentiation). Hence

$$\tau = \log\frac{n}{b}; \qquad \theta = \nu + \chi. \tag{5}$$

The boundary values of θ may now be determined in the following manner. For the impermeable base, EA:

$$u = -\frac{\partial\Phi}{\partial x}; \qquad v = 0; \qquad -f''(z) = \frac{du - idv}{dx + idy} = \frac{du}{dx}.$$

Hence $f''(z)$ is real and

$$\theta_{EA} = 2m\pi = 0, \tag{6}$$

as the velocity and acceleration are evidently both parallel to the x axis.

For the constant-potential boundary AB,

$$u = -\frac{\partial\Phi}{\partial x}; \qquad v = 0; \qquad -f''(z) = \frac{du - idv}{dx + idy} = \frac{1}{i}\frac{du}{dy}.$$

[1] The use of Φ and Ψ directly o as to map the z plane on the (u, v) plane is not convenient, as neither the value nor variation of $\frac{d\Psi}{d\Phi} = -\frac{u}{k}$ on the surface of seepage BC is known a priori.

As $\dfrac{du}{dy} > 0$,

$$\theta = \frac{\pi}{2} + 2m'\pi.$$

Since $\nu = 0$ along AB, $\theta = \chi$. Now as the streamlines are convex upward as they leave DE, and must cut AB at right angles, the acceleration vector near AB must be directed upward, so that $\chi > 0$. Hence $m' = 0$, and

$$\theta_{AB} = \frac{\pi}{2}. \tag{7}$$

For the surface of seepage, BC,

$$v = -\bar{k}; \qquad -f''(z) = \frac{1}{i}\frac{du}{dy}.$$

Here $\dfrac{du}{dy} < 0$, so that

$$\theta = -\frac{\pi}{2} + 2m''\pi.$$

At B, $\nu = 0$, while at C, $\nu = -\pi/2$. Hence, as

$$|\theta - \nu| = |\chi| \leqq 2\pi,$$

the only values possible for m'' are 0, 1. If $m'' = 1$, χ will equal $(3/2)\pi$ at B and immediately above, while immediately below B, on AB, $\chi = \pi/2$, so that because of the continuity of χ within the system there will be points in the neighborhood of B where $\chi = \pi$. This, however, is physically unreasonable, as it would give a retardation of the fluid particles as they approach the outlet surface. It follows that $m'' = 0$, and

$$\theta_{BC} = -\frac{\pi}{2}. \tag{8}$$

Along the free surface CD,

$$n^2 = u^2 + v^2 = -\bar{k}v,$$

so that

$$\frac{du}{dv} = -\left[\frac{v + (\bar{k}/2)}{u}\right]; \qquad \frac{dy}{dx} = \frac{v}{u};$$

$$-f''(z) = \frac{du - idv}{dx + idy} = \frac{2uv + \dfrac{\bar{k}u}{2} + i\left(u^2 - v^2 - \dfrac{\bar{k}v}{2}\right)}{\bar{k}v}\frac{dv}{dx}.$$

Introducing the polar angle β (*cf.* Fig. 99), u and v may be expressed as

$$u = \bar{k} \sin \frac{\beta}{2} \cos \frac{\beta}{2}; \qquad v = -\bar{k} \sin^2 \frac{\beta}{2}, \tag{9}$$

so that $f''(z)$ takes the form

$$-f''(z) = -\frac{e^{3i\beta/2}}{2 \sin \beta/2} \frac{dv}{dx}. \tag{10}$$

Hence

$$\theta = -\frac{3\beta}{2} + 2m'''\pi,$$

as $\dfrac{dv}{dx} < 0$. Since $\beta = 0$, $\nu = 0$, and $\chi = 0$, at D, it follows that $m''' = 0$, or

$$\theta_{CD} = -\frac{3\beta}{2}. \tag{11}$$

Finally, along the constant-potential surface DE,

$$v = 0, \qquad u = -\frac{\partial \Phi}{\partial x}; \qquad -f''(z) = \frac{1}{i} \frac{du}{dy}.$$

Here, $\dfrac{du}{dy} < 0$, so that

$$\theta = -\frac{\pi}{2} + 2m^{IV}\pi.$$

Along DE, $\nu = 0$ and the particles are accelerated downward; hence $\chi = -\pi/2$, and $m^{IV} = 0$, or

$$\theta_{DE} = -\frac{\pi}{2}. \tag{12}$$

Now that the boundary values of θ have been determined, one can proceed with the mapping of the hodograph in the (u, v) plane onto the infinite half plane. First, however, it is convenient to transform the (u, v) plane hodograph into the more standard form of Fig. 100, obtained by a reflection of the hodograph in the u axis, multiplication by $1/\bar{k}$ to make the length $D'C'$ unity, then a positive rotation of 90 deg., and finally a translation of unity. This transformed or q plane is, therefore, defined by the variable

$$q = \frac{i}{k}(u - iv) + 1 = 1 + \frac{v}{k} + \frac{iu}{k}. \tag{13}$$

The region $B'' C'' D'' E'' A'' B''$ in the q plane can now be mapped onto the upper half of the λ plane by means of the modular elliptic function[1] $\lambda(q)$, the function being so chosen that

$$\lambda(0) = 0; \quad \lambda(1) = 1; \quad \lambda(i\infty) = -\infty; \quad \lambda(1 + i\infty) = +\alpha \quad (14)$$

The constant-boundary values of θ will evidently be retained in the λ plane. The angle β will also be retained for the segment

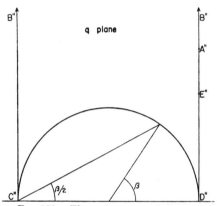

Fig. 100.—The q-plane map of Fig. 99.

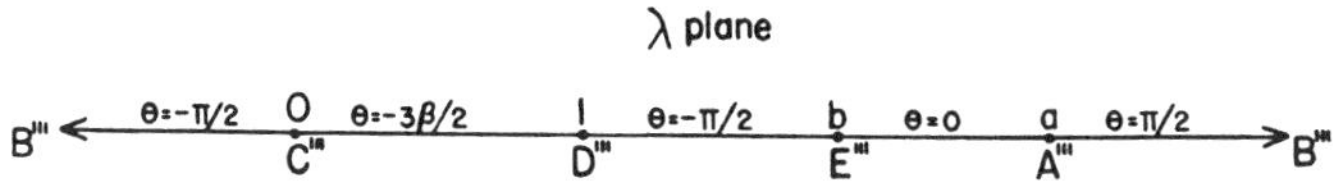

Fig. 101.—The λ plane map of Fig. 100.

$0 \leq \lambda \leq 1$, the λ—β correspondence being given by:

$$\lambda = \lambda[\tfrac{1}{2}(1 + e^{i\beta})].$$

The representation in the λ plane is indicated in Fig. 101.

As θ is known on the real axis of the λ plane, the complex function $\theta - i\tau$ may be expressed in the whole upper λ plane by the generalized Poisson integral as

$$\theta - i\tau = -i\tau_0 + \frac{1}{\pi i}\int_{-\infty}^{+\infty} \frac{\theta(\epsilon)(\lambda\epsilon + 1)d\epsilon}{(\epsilon - \lambda)(1 + \epsilon^2)}, \quad (15)^2$$

[1] A discussion giving the useful formulas relating to the modular elliptic function may be found in H. A. Schwarz "Formeln und Lehrsätze zum Gebrauche der elliptischen Funktionen." A summary of the results needed for the present discussion is given in Appendix IV.

[2] *Cf.* Appendix V.

where τ_0 is an arbitrary constant. If now the values of λ corresponding to E''' and A''' be denoted by b and a, Eq. (15) becomes, on putting in the boundary values of θ,

$$\theta - i\tau = -i\tau_0 + \frac{1}{\pi i}\left[\frac{-3}{2}\int_0^1 \frac{\beta(\epsilon)d\epsilon}{\epsilon - \lambda} + \frac{\pi}{2}\log\frac{\lambda - 1}{(a - \lambda)(\lambda - b)\lambda}\right],$$

(16)

where certain constants have been absorbed in τ_0.

Recalling the definition of θ, τ from Eqs. (4) and (5), it follows that

$$-f''(z) = \sqrt{\frac{(\lambda - b)(a - \lambda)\lambda}{\lambda - 1}}e^{-\tau_0 + \frac{3}{2\pi}\int_0^1 \frac{\beta(\epsilon)d\epsilon}{\epsilon - \lambda}}$$

(17)

As $\lambda = \lambda(q)$ is a known function of $u - iv = -f'(z)$, it follows that $f''(z) = F[f'(z)]$, and

$$z = -\int \frac{d(u - iv)}{F(u - iv)} = \int e^{\tau + i\theta}d(u - iv) =$$
$$C\int\sqrt{\frac{\lambda - 1}{\lambda(a - \lambda)(\lambda - b)}}e^{-\frac{3}{2\pi}\int_0^1 \frac{\beta(\epsilon)}{\epsilon - \lambda}d\epsilon}d(u - iv), \quad (18)$$

where $C = e^{\tau_0}$.

The sign of the radical for real λ is the same as that for $e^{-i\theta}$. Thus, in principle, the problem is solved, as by Eq. (18) z is given as a function of $(u - iv)$, and hence the velocity components as functions of z.

6.5. Numerical Applications.[1]—The physical content of the analytic solution contained in Eq. 6.4(18) may be obtained by the evaluation of the geometrical dimensions of the flow system, as implied by the constants a and b, the determination of the shape of the free surface, the computation of the velocity distributions along the boundaries, the calculation of the total flux through the system, and the mapping of the potential and streamline distributions in the system. The specific expressions giving these results are readily seen to be the following equations, taking note of the values of θ indicated in Fig. 101, and setting the constant C equal to unity.

a. Velocity distribution $[x = x(u)]$ *along* EA, *its length* L, *and the fluid-head distribution* $h(x)$:

[1] MUSKAT, *loc. cit.*

$$x(EA) = \int_{u(E)}^{u} e^{\tau} du; \quad L = \int_{u(E)}^{u(A)} e^{\tau} du, \quad \text{(1)}$$

$$h(x) = h_e - \int_0^x \left(\frac{u}{\bar{k}}\right) dx = h_e - \frac{1}{\bar{k}} \int_{u(E)}^u u e^{\tau} du. \quad \text{(2)}$$

b. Velocity distribution $[y_w = y_w(u)]$ along the outlet face AB, and the outlet-fluid height h_w:

$$y_w = \int_{u(A)}^u e^{\tau} du; \quad h_w = \int_{u(A)}^\infty e^{\tau} du. \quad \text{(3)}$$

c. Velocity distribution $[y_s = y_s(u)]$ along the surface of seepage BC, and the length h_s of BC:

$$y_s = -\int_{u(B)}^u e^{\tau} du; \quad h_s = -\int_\infty^0 e^{\tau} du. \quad \text{(4)}$$

d. The shape of and velocity distribution

$$[x = x(u, v); \quad y = y(u, v)]$$

along the free surface CD:

$$z - z_c = x - L + i(y - h_w - h_s) = \int e^{\tau - \frac{3i\beta}{2}} d(u - iv) =$$
$$\frac{\bar{k}}{2} \int_\pi^\beta e^{\tau - \frac{i\beta}{2}} d\beta. \quad \text{(5)}$$

e. Velocity distribution $[y_e = y_e(u)]$ along the inflow surface, and the inflow-fluid height h_e:

$$y_e = -\int_{u(E)}^u e^{\tau} du; \quad h_e = -\int_{u(E)}^0 e^{\tau} du. \quad \text{(6)}$$

A convenient check between the computation of h_e and h_w is obtained from Eq. (2) in the relation

$$h_e - h_w = \int_0^L \left(\frac{u}{\bar{k}}\right) dx = \frac{1}{\bar{k}} \int_{u(E)}^{u(A)} u e^{\tau} du. \quad \text{(7)}$$

f. Total flux Q through the system:

$$Q = \int_0^{h_e} u\, dy_e = -\int_{u(E)}^0 u e^{\tau} du$$
$$= \int_0^{h_w} u\, dy_w + \int_0^{h_s} u\, dy_s = \int_{u(A)}^\infty u e^{\tau} du - \int_\infty^0 u e^{\tau} du. \left.\right\} \quad \text{(8)}$$

g. To find the potential and streamline distributions in the interior of the system analytically would require the general evaluation of Eq. 6.4(18), so as to first get the internal velocity distribution, and then finding $\Phi(x, y)$ by integration. A more rapid and convenient procedure would be to find $\Phi(x, y)$ numerically or graphically by the methods outlined in Sec. 4.17, after having determined the shape of the free surface by Eq. (5) and the values of Φ along EA by Eq. (2). Or, if an electrical model is available, the equipotentials can be quickly mapped with the model after making the adjustments in the boundary conditions described in Sec. 4.17.[1]

Although it is important to get a clear physical picture of all the essential characteristics of the flow system, in order to understand the significance of the approximations that have been made in earlier theories of gravity-flow systems—which will be discussed below—the most important feature of the solution from the hydrodynamic point of view is the relation between the total flux Q through the dam and the inlet and outflow-fluid heads h_e, h_w.[2] We shall, therefore, stress mainly this phase of the solution and shall give the other results only as they are derived incidentally to the calculation of Q.

In order, however, to get any numerical results at all, it is necessary to determine the value of the integrand of Eqs. (1) to (8):

$$e^{\tau} = \sqrt{\frac{\lambda - 1}{\lambda(a - \lambda)(\lambda - b)}} e^{-\frac{3}{2\pi}\int_{0_\epsilon}^{1} \frac{\beta(\epsilon)}{\epsilon - \lambda} d\epsilon} \equiv \sqrt{\frac{\lambda - 1}{\lambda(a - \lambda)(\lambda - b)}} e^{\alpha}. \quad (9)$$

In doing this it is necessary first to notice that whereas the modular elliptic function $\lambda(q)$ has been defined here by Eq. 6.4(14), its properties are usually given in terms of the function $k^{*2} = \Theta(q)^3$ defined by the conditions

[1] Some potential distributions obtained by such models will be presented in Sec. 6.6.

[2] It should be observed, however, that from a strictly engineering point of view the erosion of the outflow face and its effect upon the stability of the dam are of even greater interest. For the study of this phase of the problem attention must be focused upon the point of emergence of the free surface and the velocity distribution along the outflow face. The consideration of these questions is, however, beyond the scope of this work, and the reader who is interested is referred to such works as "Erdbaumechanik," by K. Terzaghi, and a recent paper by A. Casagrande, *J. Boston Soc. Civil Eng.*, **23**, 13, 1936.

[3] The modulus of the elliptic integrals is here again taken as k^* to avoid confusion with the notation k for the permeability constant.

$$\Theta(0) = 1; \qquad \Theta(1) = \infty; \qquad \Theta(i\infty) = 0;$$
$$\Theta(1 + i\infty) = 0. \tag{10}$$

Comparing these with Eq. 6.4(14), it is evident that

$$\lambda(q) = 1 - \frac{1}{\Theta(q)} = 1 - \frac{1}{k^{*2}}. \tag{11}$$

Hence, making use of the formulas quoted in Appendix IV,

$$\lambda(q) = \Theta\left(1 - \frac{1}{q}\right), \tag{12}$$

which can now be expanded by the rules already established for the function Θ. This relation affords the link between the variable λ entering in the integrand e^{τ} of Eq. 6.4(18) and the variable of integration $(u - iv) = -\bar{k}i(q - 1)$, and thus also permits changing the variable of integration from $u - iv$ to λ.

Finally one must evaluate the integral

$$\alpha(\lambda) = -\frac{3}{2\pi} \int_0^1 \frac{\beta(\epsilon) d\epsilon}{\epsilon - \lambda}. \tag{13}$$

This is effected by noting that the variable ϵ in Eq. 6.4(15) really represents the real axis of the λ plane on which the q plane of Fig. 100 is mapped by the function $\lambda(q)$. Hence,

$$\epsilon = \lambda(q) = \lambda\left(\frac{1}{2} + \frac{e^{i\beta}}{2}\right) = \Theta\left(1 - \frac{1}{q}\right) = \Theta\left(i \tan \frac{\beta}{2}\right) = k^{*2}, \tag{14}$$

from which β can be computed as a function of ϵ. The result of the calculations carried out by means of this relation is shown in Fig. 102,[1] and may be used in evaluating the e^{τ} of Eq. (9) for all values of a or b.

e^{α} being known as a function of λ, e^{τ} can then be computed by Eq. (9) after choosing a and b. The connection between the

---

[1] The author is indebted to Dr. G. Hamel for sending him before publication the numerical calculations of Mr. Günther in which the values of e^{α} are given for $0.01 \leqq \lambda \leqq +20$ and $-20 \leqq \lambda \leqq 0$. For the extension of these results to $\lambda < 0.01$ and $|\lambda| > 20$ as well as a discussion of five numerical cases other than that of Günther ($b = 5$; $a = 10$), cf. M. Muskat, *loc. cit.* The calculations of Günther have since been published in the *Zeits. angew. Math. u. Mech.*, **15**, 255, 1935.

variable λ and q, or u in Eqs. (1) to (8), is then made by means of the inversion formula, stating that if

$$\lambda(q) = \Theta(\xi) = k^{*2},$$

then

$$\pi i \xi = -\pi \frac{K'}{K} = G(k^{*2}) = \log \frac{k^{*2}}{16} + \frac{k^{*2}}{2} + \frac{13 k^{*4}}{64}$$
$$+ \frac{23 k^{*6}}{192} + \cdots , \quad (15)$$

where K' and K are the complete elliptic integrals of the first kind with modulus $\sqrt{1 - k^{*2}}$ and k^*, respectively, and are listed

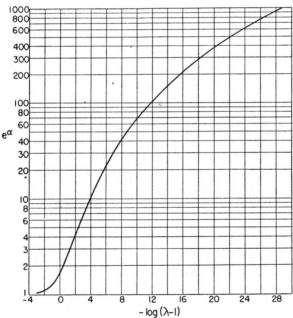

Fig. 102.—Curve giving $\alpha(\lambda)$ as defined by Eq. 6.5(13). (*From Physics*, **6**, 409, 1935.)

individually and in the ratio K'/K as a function of k^{*2} in "Tafeln der Besselschen, Theta-, Kugel- und anderer Funktionen," by K. Hayashi. The final integrations can then be carried out graphically, in the variable u for finite u, and by analytic approximations in λ for infinite and near singular values of $\lambda(a, b)$.

Figure 103 gives the velocity distributions along the inflow and outflow faces, as well as the fluid-head distribution along the base of a dam corresponding to $a = 10$, $b = 5$. The dimensions for

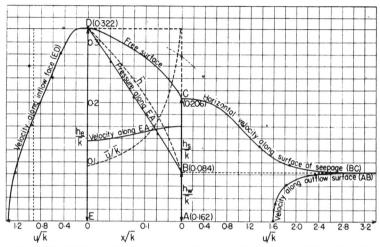

FIG. 103.—Velocity distributions over faces and base of dam of Case I, Table 14. h_e, h_w, h_s are inflow head, outflow head, and length of surface of seepage. $\bar{h}$, $\bar{u}$ are free surface and base velocities according to the Dupuit-Forchheimer theory. Circles are values calculated by theory of Sec. 6.20. (*From Physics*, **6**, 411, 1935.)

TABLE 14

Case	I	II	III	IV	V	VI
*b	5	2	10	5	2	1.2
a	10	5	∞	∞	∞	∞
†$h_e/\bar{k}$	0.322	0.670	0.672	0.872	1.286	1.823
$h_w/\bar{k}$	0.084	0.158	0	0	0	0
$h_s/\bar{k}$	0.122	0.202	0.430	0.519	0.646	0.719
$L/\bar{k}$	0.162	0.444	0.329	0.484	0.906	1.692
$Q/\bar{k}^2$	0.299	0.480	0.687	0.783	0.913	0.983
$\bar{Q}/\bar{k}^2$	0.298	0.477	0.686	0.786	0.913	0.982

* It may be noted that the case $b = 1$ corresponds to the seepage out of a sand or dam of infinite thickness, *i.e.*, to $L = ∞$.

† It should be remembered that all of the quantities h_e, h_w, h_s, L, and Q may be multiplied by a constant and yet correspond to the same parameters a, b, since the constant C of Eq. 6.4(18), which has here been taken as 1, may be chosen arbitrarily. The expression of the lengths $h_e - L$ as ratios with respect to $\bar{k}$ arises from the fact that the tabulated values of K'/K, of Eq. (15), which is used as the variable of integration in Eqs. (1) to (8), correspond to $u/\bar{k}$ rather than to u itself.

this case were computed by Günther,[1] the lengths of EA and DE being found from the terminus D of the free surface with respect to C, as given by Eq. (5).

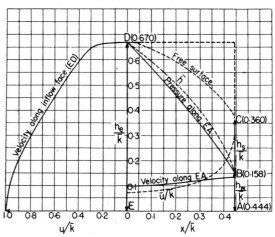

Fig. 104.—Velocity distributions over faces and base of dam of Case II, Table 14. (*From Physics*, **6**, 411, 1935.)

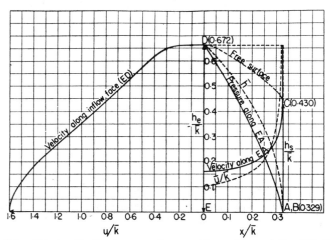

Fig. 105.—Velocity distributions over faces and base of dam of Case III, Table 14.   (*From Physics*, **5**, 412, 1935.)

Table 14 gives the numerical results for five other cases, including the values of the significant dimensions of the systems, the

[1] Günther, *loc. cit.*

correct fluxes $Q/\bar{k}^2$, and those, $\bar{Q}/\bar{k}^2$, calculated by means of the equation

$$\bar{Q} = \bar{k}\frac{(h_e{}^2 - h_w{}^2)}{2L}, \tag{16}$$

which may be obtained by various approximate treatments of the problem (*cf.* Secs. 6.17 and 6.20). The calculated pressures and

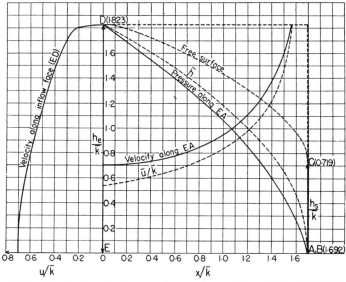

Fig. 106.—Velocity distributions over faces and base of dam of Case VI, Table 14.
(*From Physics*, **6**, 412, 1935.)

velocity distributions for cases II, III, and VI are shown in Figs. 104 to 106. The free surfaces for these cases were drawn in without calculation, following the general features indicated by that for case I and so adjusted as to pass through the points C, $(h_s + h_w)$, given by the calculations.

The curves $\bar{h}$ represent the free surfaces given by the Dupuit-Forchheimer theory (*cf.* Sec. 6.17), *i.e.*, by the equation

$$\bar{h}^2 = -(h_e{}^2 - h_w{}^2)\frac{x'}{L} + h_e{}^2, \tag{17}$$

and the velocitites $\bar{u}/\bar{k}$ are those corresponding to $\bar{h}$, on the Dupuit assumption that the horizontal velocities are proportional to the hydraulic gradient of the free surface.

An especially interesting result of these calculations, from a practical point of view, is the fact that the correct fluxes Q are so closely reproduced by Eq. (16).[1] For practical purposes, therefore, it will suffice to compute the flux by the simple formula of Eq. (16), and thus avoid the tedious calculations of the exact theory. The success of Eq. (16) alone must not, however, be considered as justifying the approximations involved in the assumptions underlying its derivation. Indeed, as will be seen in Sec. 6.17, the Dupuit-Forchheimer theory, which also leads to Eq. (16), is based upon such assumptions that even its approximate validity cannot be justifiably anticipated, and its implication of Eq. (16) must be considered as wholly fortuitous. Thus, as already mentioned above, this theory also predicts that the actual free surface within the dams should follow Eq. (17), or the curves $\bar{h}$ in Figs. 103 to 106. But here even for the cases I and II, where the fluid heads h_w are nonvanishing, the discrepancies between the correct free surfaces and the curves $\bar{h}$ are so great that the only semblance of agreement is in the monotonic decline of both sets of curves as the outflow surfaces are approached.

These curves marked $\bar{h}$ are of interest not only in demonstrating the failure of the Dupuit-Forchheimer theory in predicting the free surfaces in gravity-flow systems, but they furthermore show the caution that must be used in extrapolating the results obtained analytically or empirically for systems of one geometry to those of another. Thus whereas the equivalent of Eq. (17) for a radial gravity-flow system has been found to reproduce quite accurately the actual pressure or fluid-head distribution at the base of radial sand models (*cf.* Sec. 6.18), it evidently gives only a rough approximation of the correct fluid-head distribution in the linear system of the gravity seepage through a dam. These discrepancies become still more marked when the actual velocity distributions along the bases of the linear systems are compared with those corresponding to Eq. (17), and denoted by the curves $\bar{u}/\bar{k}$, assuming that it gives either directly the fluid-head distribution at the base or, as in the Dupuit-Forchheimer theory, the free surfaces, whose slope is proportional to the horizontal velocity.

[1] In fact, the differences between $\bar{Q}$ and Q are no greater than the uncertainties in the latter owing to the errors involved in the graphical evaluation of the integrals of Eq. (8).

While this disparity appears more pronounced in Figs. 103 and 104 where $h_w \neq 0$, the ratio of $\bar{u}/\bar{k}$ to the correct velocity will even become infinitely large as the outflow surface is approached along the base, for $h_w = 0$, as the latter becomes infinite logarithmically, whereas $\bar{u}/\bar{k}$ becomes infinite as $1/\sqrt{x}$ for $x \to 0$, x being the distance from the point A (Fig. 98).

In spite of these failures of the other features of the Dupuit-Forchheimer theory to reproduce even approximately the interior behavior of the linear gravity-flow system, the important fact remains that the resultant flux is given, to an accuracy sufficient for all practical purposes, by the simple formula of Eq. (16), which was originally derived on the basis of the Dupuit-Forchheimer theory. This paradoxical situation with respect to Eq. (16) will be cleared up in Sec. 6.20, where it will be seen that Eq. (16) may also be derived from a physically reasonable approximation theory which at the same time gives a close approximation to the correct pressure distribution at the base of the flow system. It is this theory which gives a physical significance to Eq. (16), and not the Dupuit-Forchheimer theory, the prediction by which of Eq. (16) must be considered as nothing more than a coincidence.

6.6. The Study of Gravity-flow Systems by Means of Electrical Models.—As already pointed out in Sec. 4.17, there is no immediate electrical analogue of the effect of gravity in a flow system. While this does not destroy the equivalence between electrical models and nongravity-flow systems of the same geometry involving vertical velocities, if the boundary conditions for the latter are expressed in terms of the velocity potential, electrical models of gravity-flow systems are not so easily constructed. For the lack of an a priori knowledge of the shape of the free surface leaves the corresponding boundary of the electrical model undetermined. Furthermore, as the electric current cannot be subjected to a uniform external "body force" such as gravity, it will traverse the whole of every conducting body in which it is flowing, and the bounding streamlines will always coincide with the physical boundaries of the model. The free surface in the electrical model must, therefore, be introduced artificially by varying the physical boundaries of the conducting model. Although its exact shape is not known a priori, its definition implies that the potential along it should vary linearly with the coordinate corresponding to the vertical coordinate in the flow system.   Hence if the shape of the free surface be cut—by trial and error—in such a way that this condition is fulfilled, and the other boundary conditions are also satisfied, the internal potential and streamline distribution will be exactly the same as if the free surface had developed automatically as the result of a real body force such as gravity.

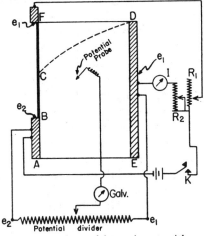

Fig. 107.—The electrical circuit for model experiments with gravity-flow systems. (*After Wyckoff and Reed, Physics.*)

The electrical circuit for such models, as constructed[1] for a study of the seepage of water through a dam with vertical faces is shown in Fig. 107.   *AEDF* is a high-resistance sheet equivalent to the permeable cross section of the dam.   A high resistance is necessary in order to minimize the current through it and hence the perturbation of the linear potential distribution along *BF*, which simulates the part of the outflow face above the outflow-

[1] *Cf.* R. D. Wyckoff and D. W. Reed, *Physics*, **6**, 395, 1935.   Figures 108 to 111 are also taken from this paper, which contains complete details of the experimental technique to be used with these electrical models.   An even earlier study of gravity-flow systems based on the electrical analogy was made by C. G. Vreedenburgh and O. Stevens (*De Ingenieur*, **48**, 187, 1933) who used electrolytic models.   However, they had to assume that the shapes of the free surfaces had been predetermined by preliminary sand-model experiments.   Furthermore the potential variation along the surfaces of seepage were taken into account only approximately.

fluid level. Furthermore, it must be of such material as can be easily cut for the purpose of finding the free surface DC. These requirements are both met by making the sheet of Bristol board or heavy firm paper and spraying it with a number (12-20) of coats of a graphite colloid, as aquadag. The homogeneity of the coating is readily tested electrically by mapping the potential distribution between two parallel electrodes of uniform potential and equal length placed along the opposite edges of the original rectangular sheet.

The appropriate boundary conditions are simulated as follows. The inflow constant-potential face is represented by the highly conducting electrode ED maintained at the potential e_1, and the uniform potential outflow surface by the electrode AB, at potential e_2, the ratio AB/ED being equal to h_w/h_e, and AE/ED having the value L/h_e. A resistance strip is placed along the remainder, BCF, of the outflow face, the terminals being kept at potentials e_2 and e_1. As the resistance of the strip along BCF is much smaller than that of the sheet $AEDF$, the current in the former greatly exceeds that in the latter, so that the variation of the potential along BF remains closely linear, thus corresponding to the uniform pressure above AB. The final step in the construction of the model consists in cutting away the upper left corner DFC in such a way that the potential at any point along DC will vary linearly with the distance of the point above AE. This is most conveniently effected by adjusting the shape of CD so that the terminae on CD of equally spaced equipotentials have equidistant projections on CF. A moderately sharp needle point makes a satisfactory potential probe.

Once the shape of CD has been found, the potential distribution may be mapped within $ABCDE$ exactly as with any other electrical model. The streamlines may be constructed as the orthogonal trajectories of the equipotentials. The flux through the system may be determined by measuring the current through the meter with the potential-divider circuit open. If this current is I, for the potential drop $e_2 - e_1$, and if the specific resistance of the sheet is σ, the flux in the corresponding dam of permeability k will be

$$Q = \frac{k\gamma g(h_e - h_w)}{\mu} \frac{\sigma I}{e_1 - e_2}. \tag{1}$$

The potential and streamline distributions obtained with electrical models of dams with both vertical and sloping faces in the manner described above are shown in Figs. 108 to 111. Figure 108, representing a dam chosen with relative dimensions for AB, AE, and ED to correspond to that of case I treated analytically in Sec. 6.5, gives a good verification of the applicability of the electrical model to these problems, as the shape of

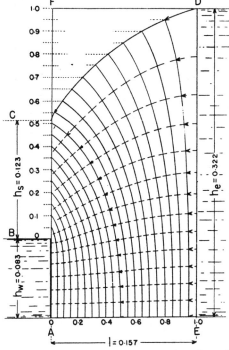

FIG. 108.—Potential and streamline distributions in a dam with vertical faces corresponding to the case of Fig. 103, as obtained by electrical-model experiments. (*After Wyckoff and Reed, Physics.*)

the free surface and the height BC are in excellent agreement with those computed in Sec. 6.5 (*cf.* Fig. 103). Moreover, the velocity distributions along ED and AC and the potential distribution along EA given by Fig. 108 check closely those shown in Fig. 103.

Figure 109 gives the potential and streamline distributions for a dam with vertical faces and vanishing outflow-fluid level. The dimensions are approximately proportional to those of case VI, Sec. 6.5.

The potential and streamline distributions for dams with faces sloping at angles of 30 and 45 deg. are shown in Figs. 110 and 111. The low gradients at the heels, E, and the inflexions in the free surfaces are to be noted. While the theoretically predicted

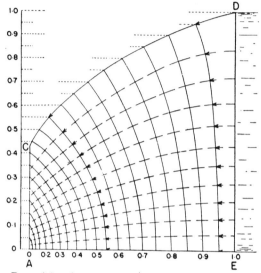

FIG. 109.—Potential and streamline distributions in a dam with vertical faces and zero outflow fluid level, as obtained by electrical model experiments. (*After Wyckoff and Reed, Physics.*)

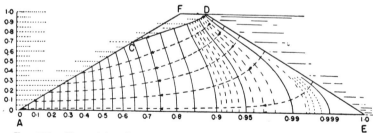

FIG. 110.—Potential and streamline distributions in a dam whose faces have an inclination of 30 deg., as obtained by electrical model experiments. (*After Wyckoff and Reed, Physics.*)

values of zero velocity at E (*cf.* Sec. 6.1) are thus obviously verified, the infinitely high velocities to be expected at the toe, A, must evidently be highly concentrated about A, as even the last of the equipotentials drawn in Figs. 110 and 111 give practically no suggestion of the attainment of infinitely high velocities at A.

Although the illustrations given here of the application of the method of electrical models in the study of gravity-flow systems have involved rather idealized problems of the seepage of water through dams, it may be noted that it is practically unlimited in its scope. Thus models cannot only be constructed to simulate dams with central impermeable cores, but two-dimensional systems consisting of regions of different permeability can be treated as well. Impermeable regions can be simulated by cutting out from the conducting-sheet figures geometrically similar to the impermeable regions, while the effect of variations in permeability can be studied by varying the number of coatings of graphite sprayed over the different parts of the model.

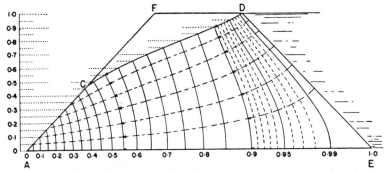

FIG. 111.—Potential and streamline distributions in a dam whose faces have an inclination of 45 deg., as obtained by electrical-model experiments. (*After Wyckoff and Reed, Physics.*)

6.7. Some Exact Solutions of Laplace's Equation Appropriate to Gravity-flow Systems.

—Although the methods of Hopf and Trefftz and of Hamel do attack directly problems of gravity flow in leading to solutions for systems of preassigned geometry, the difficulty of carrying through completely[1] the necessary analysis is a serious limitation to their general applicability to a variety of specific problems. There seems to be some justification, therefore, in the consideration of not only approximate methods of treatment, but even in the more primitive device of constructing

[1] It should be noted, however, that the difficulties of mapping the hodograph planes are entirely of an analytical nature. From a practical point of view, therefore, the essential features of the solution may be obtained by effecting the transformations graphically (*cf.* F. Weinig and A. Shields, *Wasserkraft u. Wasserwirtschaft*, **31**, 233, 1937).

ab initio typical potential distributions and then seeing if they correspond to systems of practical interest. As this latter procedure has the merit of at least providing an exact solution to the potential problem, even though the geometry of the corresponding physical system is not identical with that in question, typical examples of such solutions[1] will be presented first.

As the solutions to be given here will all be conjugate functions resulting from the separation of complex variable functions into their real and imaginery parts, they will necessarily refer to two-

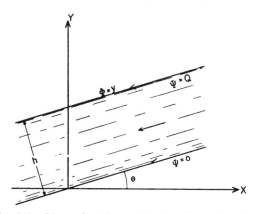

FIG. 112.—Linear-sheet flow under the action of gravity.

dimensional systems (*cf.* Sec. 4.8). It will also be convenient to take the pressure over the free surfaces as zero and $\bar{k}$ or $k\gamma g/\mu$ as 1, so that the condition to be satisfied by Φ along a free surface is that $\Phi = y$.

a. Linear Sheet Flow.—The simplest case of a complex-variable representation of a gravity-flow system is that describing the flow in an infinitely extended inclined stratum undisturbed by either sources or sinks.[2] In the notation of Fig. 112 the appropriate relation is

$$\omega = \Phi + i\Psi = z \sin \theta e^{-i\theta} + h \cos \theta, \qquad (1)$$

[1] The solutions given here are taken from the paper by J. Kozeny in *Wasserkraft u. Wasserwirtschaft*, **26**, 28, 1931; *cf.* also V. V. Wedernikow, *ibid.*, **29**, 128, 1934.

[2] This has already been used in the treatment of the problem of Hopf and Trefftz [*cf.* Eq. 6.2(2)].

so that

$$\left.\begin{array}{l}\Phi = (x\cos\theta + y\sin\theta)\sin\theta + h\cos\theta \\ \Psi = (-x\sin\theta + y\cos\theta)\sin\theta.\end{array}\right\} \qquad (2)$$

Evidently, then, the streamline

$$\Psi = 0; \qquad y = x\tan\theta, \qquad (3)$$

represents the impermeable bed, and

$$\Psi = Q = h\sin\theta; \qquad y = x\tan\theta + h\sec\theta, \qquad (4)$$

the upper surface. That this is a *free* surface, follows at once from the substitution of Eq. (4) into Eq. (2), which gives

$$\Phi = y, \qquad (5)$$

as it should.

b. *Parabolic Free Surfaces.*—Setting

$$x + iy = z = \frac{1}{2Q}\omega^2, \qquad (6)$$

and separating the real and imaginary parts, it follows that

$$x = \frac{1}{2Q}(\Phi^2 - \Psi^2); \qquad y = \frac{\Phi\Psi}{Q}. \qquad (7)$$

The streamline $\Psi = Q$ is, therefore, the free surface

$$y^2 - Q^2 - 2Qx = 0, \qquad (8)$$

which is a parabola with focus at the origin. The positive x axis corresponds to the impermeable base (*cf.* Fig. 113), whereas that from the origin to $x = -Q/2$ is a zero-potential surface. Except for the porous medium to the right of the y axis, this system should approximate that of the seepage through a very thick dam, the presence of the medium to the right of the y axis evidently diminishing the total flux Q from that passing through the corresponding dam. In fact, supposing that at a great distance L the height of the free surface, and fluid head, is h_e, it follows from Eq. (8) that Q may be expressed as

$$Q = L\left(\sqrt{1 + \frac{h_e^2}{L^2}} - 1\right), \qquad (9)$$

which is always less than $h_e^2/2L$, an expression found in Sec. 6.5 to reproduce quite accurately the seepage through a dam with zero outflow-fluid head. For small values of h_e/L, however, Eq. (9) does approximate Eq. 6.5(16) (with $h_w = 0$) quite closely, as should indeed be expected, since the excess porous medium at the outflow surface in Fig. 113 should become of diminishing significance in comparison to the main body of the dam as its thickness increases. It should be noted that any parabolic streamline: $0 < \Psi < Q$ can be considered as the base of the flow system without changing the free surface given by Eq. (8).

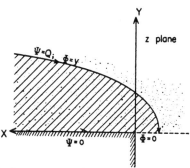

Fig. 113.—A gravity-flow system with a parabolic free surface. (*After Kozeny, Wasserkraft u. Wasserwirtschaft.*)

c. *The Drainage from Ditches.*—A different type of complex variable relation leading to systems with free surfaces corresponding to the seepage of water out of ditches is given by

$$z = -He^{\frac{\pi\omega}{Q}} - i\omega + \frac{Q}{2},$$ (10)

so that

$$\left. \begin{array}{l} x = -He^{\frac{\pi\Phi}{Q}} \cos \frac{\pi\Psi}{Q} + \Psi + \frac{Q}{2} \\[2mm] y = -He^{\frac{\pi\Phi}{Q}} \sin \frac{\pi\Psi}{Q} - \Phi, \end{array} \right\}$$ (11)

where, since $+y$ is taken here as downward (*cf.* Fig. 114), the potential is $(k/\mu)(p - \gamma gy)$, and the condition for a streamline $\Psi = $ const. to be a free surface is $\Phi = -y$ (with $k\gamma g/\mu = 1$).

The two streamlines, $\Psi = 0, -Q$ are evidently free-surface streamlines, and are specifically defined by the curves

$$x = -He^{-\frac{\pi y}{Q}} + \frac{Q}{2}; \quad x = +He^{-\frac{\pi y}{Q}} - \frac{Q}{2}$$ (12)

which are symmetrical about the y axis.   The shape of the ditch, along which $\Phi = 0$, is given by

$$\left(x - \frac{Q}{\pi} \cos^{-1} \frac{y}{H}\right)^2 + y^2 = H^2. \tag{13}$$

The width of the ditch is, therefore,

$$B = Q - 2H, \tag{14}$$

while its maximum depth is H.   The maximum width of the sheet of liquid seeping down into the porous medium is

$$B_1 = 2|x|_{y=\infty} = Q = B + 2H. \tag{15}$$

The above simple results will, of course, strictly apply only if the shape of the ditch belongs to the family defined by Eq. (13),

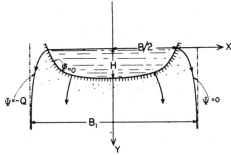

FIG. 114.—The seepage out of a ditch with free surfaces bounded by vertical asymptotes.   (*After Kozeny, Wasserkraft u. Wasserwirtschaft.*)

and if the porous medium is of very great thickness, so that the liquid can maintain indefinitely its vertical downward seepage. Although slight deviations in the shape of the ditch from those given by Eq. (13) will in themselves cause no serious errors in the use of Eq. (14) or (15), the assumption of the effectively infinite thickness of the porous medium definitely limits their applicability to cases where the underground water table lies at a great depth below the base of the ditch.   In many practical situations, however, the water seeping down from the ditch will reach the normal ground-water level at a relatively shallow depth, thus forcing the streamlines to assume a horizontal rather than a vertical trend.

A different approximation to the problem, especially that of the latter type, is given by the complex variable relation

$$z = He^{-\frac{\pi \omega}{Q}} - i\omega + \frac{Q}{2} \tag{16}$$

which is equivalent to

$$\left. \begin{array}{l} x = He^{-\frac{\pi \Phi}{Q}} \cos \frac{\pi \Psi}{Q} + \Psi + \frac{Q}{2} \\[2mm] y = -He^{-\frac{\pi \Phi}{Q}} \sin \frac{\pi \Psi}{Q} - \Phi. \end{array} \right\} \tag{17}$$

Here, the two symmetrical free-surface streamlines, $\Psi = 0$, $-Q$, are given by

$$x = He^{\frac{\pi y}{Q}} + \frac{Q}{2}; \qquad x = -He^{\frac{\pi y}{Q}} - \frac{Q}{2}, \tag{18}$$

and hence increase in depth logarithmically with increasing distances from the ditch (*cf.* Fig. 115), whereas in the case of

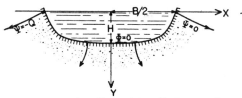

Fig. 115.—The seepage out of a ditch with radially spreading free surfaces.
(*After Kozeny, Wasserkraft u. Wasserwirtschaft.*)

Eq. (12) the free surfaces approach vertical asymptotes as the distance $Q/2$ from the center of the ditch is approached. The shape of the ditch ($\Phi = 0$) corresponding to Eqs. (16) and (17) is defined by the equation

$$\left(x - \frac{\Omega}{\pi} \cos^{-1} \frac{y}{H} \right)^2 + y^2 = H^2, \tag{19}$$

so that its maximum depth is H, while its width is

$$B = Q + 2H. \tag{20}$$

Thus here, too, a knowledge of the width and depth of the ditch will give the seepage flux out of it, provided its shape is described by Eq. (19).[1]

[1] It should be noted that while Eq. (19) is formally identical with Eq. (13), one must take the positive radical $\sqrt{H^2 - y^2}$ in solving Eq. (19) for x, whereas in Eq. (13) the negative values of this radical must be used in solving for x.

The shapes of the ditches and the limiting free surfaces for both Eqs. (10) and (16) with $H = 1$, $B = 8$ are shown in Fig. 116.[1] While the shapes of the ditches do differ somewhat for the two cases, the radical differences between the corresponding free surfaces and characters of the flow must not be attributed to those in the shapes of the ditches. For the fundamental features of the flow, such as the general nature of the free surfaces and the flux through the system, are determined largely by the boundary conditions rather than by the detailed shape of the boundary surfaces.[2] Although the assumption has been made in the case of

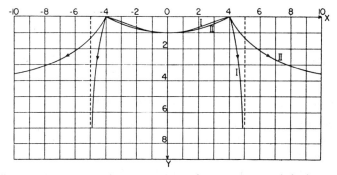

Fig. 116.—Free surfaces resulting from the seepage out of ditches under special conditions. I. Shape of ditch and free surface defined by Eq. 6.7(11). II. Shape of ditch and free surface defined by Eq. 6.7(17).

both Eqs. (10) and (16) that the porous medium is effectively of infinite thickness, these equations nevertheless imply radically different types of potential distribution at great depths from the

[1] Further examples of the system defined by Eq. (11) may be found in the paper by V. V. Wedernikow (*loc. cit.*).

[2] Of course, in the method of analysis used here, the behavior of the free surfaces and the potential distributions at large depths—*cf.* Eqs. (21) and (22)—are definitely related to the detailed shapes of the ditches, for Eq. (13) will give a shape for the ditch which has an inflexion and is convex upward near $y \sim 0$, thus initiating the vertical asymptotic fall for the free surfaces, whereas Eq. (19) will give a ditch profile that is everywhere convex downward with free surfaces tending toward a radial spread (*cf.* also footnote on p. 328, and Fig. 116). However, it is clear that in the general treatment of the problem in which the character of the boundary conditions are introduced *ab initio*, either type of asymptotic behavior, Eq. (21) or (22), could be developed regardless of the shape of the ditch.

surface. Thus from Eq. (11) it follows that the equipotentials at great depths are the horizontal parallel lines

$$\Phi \sim -y. \tag{21}$$

This evidently represents, therefore, an effectively free fall of the seepage liquid as a vertical sheet under the action of gravity which is bounded by vertical asymptotes, as is represented by the free surfaces of Fig. 116.

In the case of Eq. (16), on the other hand, the equipotentials at great depths are given by

$$\Phi \sim -\frac{Q}{2\pi} \log \frac{x^2 + y^2}{H^2}. \tag{22}$$

The flow at great depths must, therefore, be radial, thus explaining the lateral spread of the free surfaces of Fig. 116 for this case. Furthermore, the potential along the y axis, $-\dfrac{Q}{\pi} \log \dfrac{y}{H}$, will clearly be greater than that at the same depth along the free surfaces,[1] $-y$; this difference provides the potential drop for the horizontal motion superposed on the vertical free fall.

Although Eq. (16), with its implication, Eq. (22), gives a *possible* flow system, it hardly can be considered even as a rough approximation to the real seepage problem, in which the free surfaces as well as the intermediate streamlines must become asymptotic to a horizontal or sloping line representing the normal ground-water level, or in which the stream of seeping water flows in an effectively vertical direction to a very deep-lying water table or to a very permeable bed carrying the water table, and through which the seepage water merely drips until it impinges on the ground-water surface (see next section). The shape of the free surface near the ditch can be somewhat modified so as to have an inflexion by taking a linear combination[2] of Eqs. (10) and (16), and a strictly horizontal asymptote ($y = y_0$) for the free surfaces can be obtained by adding to Eqs. (10) or (16) a term as

[1] This higher value of Φ along the y axis for this case also explains why the flux here is less than that for Eq. (10) [*cf.* Eqs. (15) and (20)], even though the average width of the fluid-filled medium is much greater here than in the case of free-fall drainage.

[2] *Cf.* P. Nemenyi, "Wasserbauliche Strömungslehre," p. 201.

$$\frac{A}{\sinh \frac{\pi}{Q}(\Phi + i\Psi + y_0)} \cdot$$ But these will still give potential dis-

tributions at great depths either of the free-fall type of Eq. (21), or the artificial radial type of Eq. (22). Although it has thus far not been possible to construct ab initio conjugate-function relations which give the proper physical variations of the potential distributions for all types of practical seepage problems, the above examples should at least serve to indicate the nature of this indirect method of seeking solutions to flow problems, as well as its possibilities[1] and limitations.

6.8. The Seepage of Water from Canals or Ditches into Sands with Deep-lying Water Tables.—A

somewhat more direct method of treating the problem of the seepage of water from canals into sands with deep-lying water tables (cf. Fig. 117) than that given in Sec. 6.7c is one which is quite similar to the method of Hopf and Trefftz described in Sec. 6.2. Here, too, a consistent set of potential and streamline distributions and the shapes of the free surfaces are

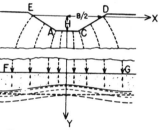

FIG. 117.—The seepage out of a ditch into sands with deep-lying water tables. (*After Wedernikow, Wasserkraft u. Wasserwirtschaft.*)

found rigorously by applying a succession of complex variable transformations, but the exact profile of the ditch or canal to which they correspond is found only at the end of the

---

[1] The difficulty of obtaining potential distributions with free surfaces by the direct solution of Laplace's equation becomes apparent when it is noted that for the simple case of a circular cross section for the ditch and a vertical-sheet seepage, as in Fig. 116, the direct solution of Laplace's equation gives

$$\Phi = -y\left(1 - \frac{a^2}{x^2 + y^2}\right); \qquad \Psi = -x\left(1 + \frac{a^2}{x^2 + y^2}\right)$$

where a is the radius of the ditch (cf. N. K. Bose, *Mem. Punjab Irrigation Res. Lab.*, Lahore, **2**, no. 1, 1929; and V. I. Vaidhianathan, H. R. Luthra, and N. K. Bose, *Proc. Ind. Acad. Sci.* **1**, 325, 1934). Although these solutions give vertical asymptotes for the streamlines $\Psi = $ const., none of the streamlines represent free surfaces. To have found the transcendental functions $\Phi(x, y)$, $\Psi(x, y)$ implied by Eqs. (11) by the solution of $\nabla^2\Phi = 0$, $\nabla^2\Psi = 0$, individually, without the aid of the complex-variable methods, would evidently have been a most formidable task.

solution.[1]   To illustrate the method, we shall first treat the case where the highly permeable gravel bed containing the water table lies at an effectively infinite depth below the surface of the canal.

Thus, in a manner similar to the introduction of the complex variable z_1 by Hopf and Trefftz [*cf.* Eq. 6.2(4)], the variable

$$\tau = \theta_1 + i\theta_2 = -z - i(\Phi + i\Psi - H); \qquad \theta_1 = \Psi - x;$$
$$\theta_2 = -(y + \Phi - H), \qquad (1)$$

is introduced, where H is the fluid level in the canal.   Taking $+y$ as downward and, for simplicity, $\bar{k}$ as unity, it is clear that along the free surfaces of the gravity-flow system, $\theta_2 = H$, while along the base of the canal AC, $\theta_2 = 0$, the atmospheric pressure being

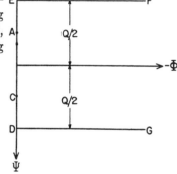

given the value zero.   The sides of the ditch CD and AE are now

FIG. 118.—The τ-plane map of Fig. 117.   FIG. 119.—The ω-plane map of Fig. 117.

---

[1] This analysis and that of the next section are due to V. V. Wedernikow (*loc. cit.*).   The effect of the capillarity on the problem of the seepage out of sands is considered by this author in the paper in the same journal, **30,** 245, 1935.   In a still later paper (*Zeits. angew. Math. u. Mech.*, **17,** 155, 1937) Wedernikow discusses the problem of the seepage of water out of canals by the application of inversion transformations to the hodograph planes, which eliminate the circular segments and lead to triangular recti-linear diagrams.   By subsequent mapping of the latter by means of the ordinary Schwarz-Christoffel transformations onto auxiliary planes he is thus able to treat cases where the canal profile has a preassigned rectilinear trapezoidal shape.   The numerical results, however, differ only inappreci-ably from those presented here for the equivalent curved profiles.   On the other hand, it may be noted that both of the methods of Wedernikow are applicable to the present problem only because of the absence of surfaces of seepage from the flow systems.   (*Cf.* footnote on p. 301).

assumed to be of such a shape that along them θ_1 has the constant values c_1 and c_2 determined by the points D and E. The map of Fig. 117 onto the τ plane will, therefore, be that of Fig. 118. In the $\omega = \Phi + i\Psi$ plane, the boundary $GDCAEF$ will clearly map on the infinite strip shown in Fig. 119.

Mapping both the τ and ω planes onto the real axis of the $\zeta = \xi + i\eta$ plane, by means of the Schwarz-Christoffel theorem, with the correspondences that

$$\begin{array}{ll} A \to -k^*; & C \to +k^* \\ E \to -1; & D \to +1, \end{array} \right\} \tag{2}$$

it is readily found that

$$\tau = \theta_1 + i\theta_2 = C_1 \int_0^\zeta \frac{d\zeta}{\sqrt{(1-\zeta^2)(k^{*2}-\zeta^2)}} \tag{3}$$

$$\omega = \Phi + i\Psi = C_2 \int_0^\zeta \frac{d\zeta}{\sqrt{1-\zeta^2}} = C_2 \sin^{-1}\zeta. \tag{4}$$

To determine the constants in these transformations, it is noted that at D, where $\zeta = +1$:

$$\tau = \frac{Q}{2} - \frac{B}{2} + iH = C_1 \int_0^1 \frac{d\zeta}{\sqrt{(1-\zeta^2)(k^{*2}-\zeta^2)}} = C_1(K + iK'),$$

$$\omega = i\frac{Q}{2} = C_2 \sin^{-1} 1 = \frac{\pi}{2}C_2; \qquad C_2 = \frac{iQ}{\pi},$$

where K, K' are the complete elliptic integrals of the first kind with moduli k^*, $\sqrt{1-k^{*2}}$. Hence,

$$C_1 K = \frac{Q-B}{2}; \qquad C_1 K' = H; \qquad \frac{Q-B}{2H} = \frac{K}{K'}. \tag{5}$$

Further, at C where $\zeta = k^*$,

$$\omega = \Phi + i\Psi = i\Psi = i\left(\frac{Q}{2} - \frac{B}{2} + \frac{b}{2}\right) = \frac{iQ}{\pi} \sin^{-1} k^*,$$

so that

$$k^* = \sin\frac{\pi}{Q}\left(\frac{Q}{2} - \frac{B}{2} + \frac{b}{2}\right) = \cos\frac{\pi m H}{Q} \right\}$$

and

$$Q = \frac{\pi m H}{\cos^{-1} k^*} = B + \frac{2HK}{K'}, \tag{6}$$

where b is the width of the base of the canal and $1/m$ the average slope of its sides, in the sense that: $m = (B - b)/2H$.

Equations (3) and (4), together with Eq. (1), give the potential and streamline distributions within the flow systems, as well as the shapes of the free surfaces bounding them.   The flux is given by Eqs. (6), the calculation being most readily carried out by assuming Q/H, calculating the associated k^* by the first of Eqs.

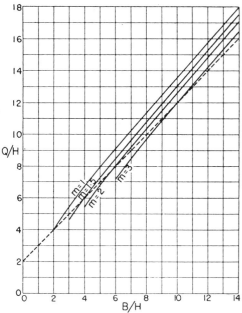

Fig. 120.—The variation of the seepage flux, Q, from canals or ditches draining into deep-lying sands according to Eq. 6.8(6). B = total width of the canal or ditch at the top of the free surface; H = maximum depth of the free water; $m = (B - b)/2H$; b = width of flat part of base.   Dotted curve gives Q according to Eq. 6.7(14). $\bar{k}$ has been taken as 1.

(6), and then the value of B/H to which the Q/H corresponds by the second equation.   Several such curves, for different values of m, are given in Fig. 120.[1]

The maximum or asymptotic width of the downward seeping sheet of water is, as to be expected, readily found to be given by

[1] The curves of Wedernikow (loc. cit.) give only B/H versus K/K', and still require the application of the second of Eqs. (6) to give Q/H.

$$B_1 = 2|x|_{y=\infty} = Q = B + \frac{2HK}{K'}, \qquad (7)$$

in which it must be remembered that in the numerical interpretation of the equivalence between the flux Q and the lengths such as B one must multiply the latter by $\bar{k}$, which has been taken here as unity.

It will be noted that the expression for the flux derived here, Eq. (6), differs from that found by the simple analysis of Sec. 6.7,

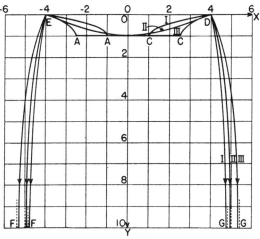

Fig. 121.—Canal or ditch profiles and their free surfaces for the seepage into sands with deep-lying water tables. I. Profile of Sec. 6.8, $b = 2$, $Q/H = 9.68$. II. Profile of Sec. 6.7, Eq. 6.7(13), $Q/H = 10.0$. III. Profile of Sec. 6.8, $b = 5$, $Q/H = 10.8$. B = total width at the top of the free surface; H = maximum depth of the free water; b = width of flat part of base (for I and III). $B/H = 8$ for all cases, and $\bar{k} = 1$.

i.e., Eq. 6.7(14), only by the factor K/K'. As this may be either less or greater than one (as $Q/H <$ or $> 4m$), the canal or ditch profiles implied by the theory of this section will lose either more or less[1] water by seepage than will those of Eq. 6.7(13). That these differences in the values of Q are actually due to those in the canal or ditch profiles is clearly shown by Fig. 121, in which are

[1] Wedernikow seems to exclude all cases for which $K/K' < 1$, and explains the excess (for $K/K' > 1$) over that of Eq. 6.7(14) by the efflux through a crevice (analytically a "cut") in the profiles at the points C and A. The basis for concluding that Eqs. (3) and (4) imply a cut at the points C and A is, however, not clear, as Figs. 118 and 119 map continuously from AC onto CD and AE as $|\zeta|$ in Eqs. (3) and (4) crosses the value k^*.

drawn the profiles for drainage surfaces of the same total width
($B = 8$) and depth ($H = 1$) but with different shapes of the sides,
corresponding in one case (II) to Eq. 6.7(13), and in the other two
to Eqs. (3) and (4), that of curve I being for $m = 3$ and that of
curve III for $m = 1.5$. The corresponding values of Q are 9.68,
10.0, and 10.8 for the cases I, II, and III, respectively. The order
of these fluxes are clearly in the order of the average depth of the
profiles below the top free surface, as is to be expected, since the
greater this average depth the less will be the effective sand

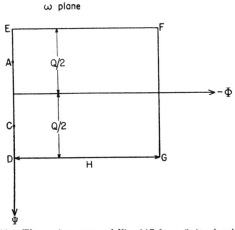

Fig. 122.—The ω-plane map of Fig. 117 for a finite depth to FG.

resistance which the seepage water must overcome in its total
drop in potential of H units.

**6.9. The Seepage of Water from Canals or Ditches into Sands
Underlain by Highly Permeable Gravel Beds at Shallow Depths.**
When the depth of the underlying highly permeable gravel bed
that contains the water table is comparable to the dimensions of
the canal or of the fluid head in it, the analysis of the last section
must be slightly modified.

As the gravel bed is considered to be of a much higher permeability than the sediments above it, and can carry much more fluid
than the latter can supply, the water that does seep into the gravel
bed will simply drip in it as a stream of drops impinging upon the
water table. The pressure at the top of the gravel bed may then
be taken as atmospheric. The value of θ_2 along this surface FG
(cf. Fig. 117) will therefore be H, as it is along the free surfaces,

and Fig. 118 for the τ plane will be unchanged except that the points F, G will be separated and lie along ED at equal distances from the θ_2 axis.   The mapping of the τ onto the ζ plane will then be given again by Eq. 6.8(3).

The ω-plane diagram will, however, now be changed to the rectangle of Fig. 122.   Denoting the points on the real axis of the ζ plane on which the points G, F are mapped by Eq. 6.8(3) by

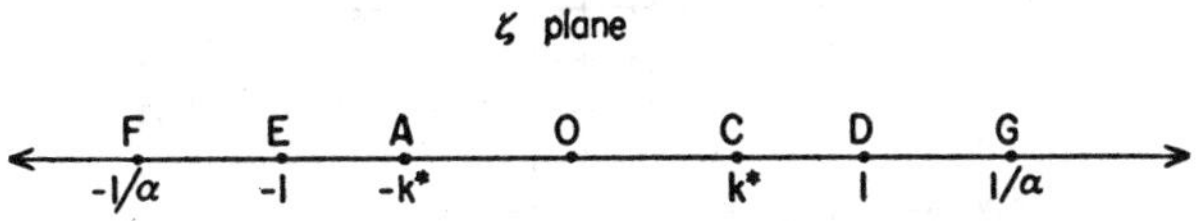

ζ plane

FIG. 123.—The ζ-plane map of Fig. 122.

$\pm 1/\alpha$, the mapping of Fig. 122 onto the ζ plane (Fig. 123) will be effected by the relation

$$\omega = \Phi + i\Psi = C_2 \int_0^\zeta \frac{d\zeta}{\sqrt{(1 - \zeta^2)(1 - \alpha^2\zeta^2)}}. \tag{1}$$

The constants C_1, C_2, α, k^* may be determined as follows: At d, where $\zeta = 1$, Eqs. (1) and 6.8(3) give

$$\frac{iQ}{2} = C_2 K(\alpha); \qquad \frac{Q - B}{2} + iH = C_1[K(k^*) + iK'(k^*)], \tag{2}$$

where α, k^* are the moduli of the elliptic integrals.   At C, where $\zeta = k^*$, it follows that

$$i\Psi = \frac{i(Q - B + b)}{2} = C_2 F(\sin^{-1} k^*, \alpha), \tag{3}$$

and at G

$$-T + i\frac{Q}{2} = C_2[K(\alpha) + iK'(\alpha)],$$

so that

$$\frac{2T}{Q} = \frac{K'(\alpha)}{K(\alpha)}. \tag{4}$$

Solving these equations for Q, it is found that:

$$Q = \frac{2Hm}{1 - \dfrac{F(\sin^{-1} k^*; \alpha)}{K(\alpha)}} = B + 2H\frac{K(k^*)}{K'(k^*)}. \tag{5}$$

To calculate Q as a function of B, H, and T, for given m, from these equations, one may begin by assuming Q/T, obtaining α

from Eq. (4), then k^* from the first of Eqs. (5), after assuming Q/H, and then finally B/H from the second of Eqs. (5).

Several curves of Q/H *versus* B/H, calculated in this way are shown in Fig. 124. It will be seen that, as is to be expected, the fluxes for the seepage through a bed of finite thickness, as given by Eq. (5), exceed those for the seepage into a bed of infinite

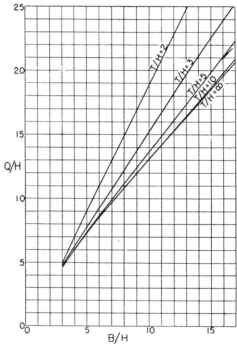

Fig. 124.—The variation of the seepage flux, Q, with the widths of canals or ditches overlying shallow gravel beds. B = total width at the top of the free surface; H = maximum depth of the free water; T = depth of gravel bed below the top of the free water; $m = (B - b)/2H$; b = width of flat part of base; $\bar{k} = 1$.

thickness calculated in the last section, the excess becoming especially large when the depth T is of the order of H.

6.10. An Approximate Theory of the Seepage of Water through Dams with Sloping Faces.—Although, as was seen in Sec. 6.6, the potential distributions in dams with sloping faces can be obtained quite rigorously by means of electrical models, after a trial-and-error cutting of the upper free-surface boundary, each model gives the results only for a single fluid-inflow head, fluid-outflow head, and slopes of the inflow and outflow faces. To obtain the data in tabular or graphical form to cover a wide range of these variables would clearly require a laborious and lengthy empirical program. It is, therefore, of some value to have available a method of calculation which, even though approximate, will give the correct order of magnitude for the flux, and can be carried through for any special case without undue labor.

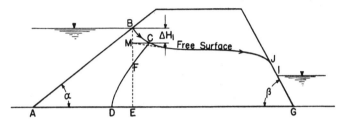

FIG. 125.—The free surface in a dam with sloping faces. (*After Dachler, Die Wasserwirtschaft.*)

Such a method is that developed by Dachler,[1] in which the inflow face, main body, and outflow face of the dam are treated as separate flow systems by different approximations, and are then synthesized by the requirements that the fluxes through

[1] DACHLER, R., *Die Wasserwirtschaft*, p. 37, 1933; *cf.* also p. 41, 1934. In view of the importance of the exact location of the terminal *J* (*cf.* Fig. 125) of the free surface on the outflow face from the point of view of stability, it may be mentioned that for zero outflow-fluid levels an approximate formula for the position of *J* has been derived by L. Casagrande (*Die Bautechnik*, no. 15, 1934) which is supported by experimental tests for $\beta \lessgtr 90$ deg. This formula is: $y_0 = m \sin \beta - \sqrt{m^2 \sin^2 \beta - H^2}$, where y_0 is the height of *J* above the base of the dam, *m* is the length of the free surface plus the distance *JI*—for which an approximation must be made—and *H* is the total fluid head at the inflow face. Furthermore, A. Casagrande (*Jour. New Eng. Water Works Assoc.*, **51**, 131, 1937) has developed curves giving y_0 as a function of β and *H* for all values of β. These were obtained by comparing the values of y_0 as given by a Kozeny parabola [Eq. 6.7(8)] for the free surface with those following from a direct graphical solution for the network of equipotentials and streamlines of the flow system. The differences between these two solutions were then plotted as functions of β and *H*. In this latter paper will also be found a number of very practical remarks concerning the design of earth dams, based upon the hydrodynamics of the seepage problem, as well as many helpful suggestions for the construction of graphical solutions for flow problems involving free surfaces in both isotropic and anisotropic media.

each should be equal, and that the fluid heads should be continuous in passing from one part to the other. The inflow part of the system is taken to be bounded by $ABCD$ in Fig. 125. The flow through this region is computed by means of a conjugate-function transformation appropriate to a wedge of angle α, bounded on one side by the streamline AD and on the other by the equipotential AB. BC is taken as that streamline of the system leaving B, no attempt being made to correct for the fact that it is not a free-surface streamline.

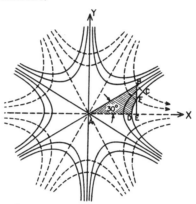

Fig. 126.—Approximate potential and streamline distributions near the inflow face (30 deg. inclination) of a sloping dam. (*After Dachler, Die Wasserwirtschaft.*)

If $\alpha = \pi/2n$, where n is an integer, the appropriate complex-variable transformation is:

$$H + i\Psi = (x + iy)^n, \qquad (1)$$

where for convenience the velocity potential has been replaced by the fluid head H. The typical equipotentials and streamlines for $n = 3$ are represented in Fig. 126, where

$$x^3 - 3xy^2 = H; \qquad 3x^2y - y^3 = \Psi. \qquad (2)$$

Although the equipotential CD bounding this region is not determined a priori by the above procedure, its choice can be made definite by the requirement[1] that the loss of energy by

[1] This condition has, of course, no physical significance, and is chosen only for convenience, as it permits the calculation of the flow to the right of BCD to be approximated by that to the right of BE with the same average fluid head along BE as along BCD.

friction in passing through $ABCD$ should be the same as that in passing through ABE. Now this loss of energy suffered by the fluid in passing between two streamlines separated by $\Delta\Psi$, across the drop ΔH in the fluid head is

$$dW = \gamma g \Delta H \Delta \Psi, \tag{3}$$

where γ is the fluid density and g is the acceleration of gravity. Hence the above condition becomes equivalent to the requirement that BCF and FDE have the same area in the (H, Ψ) plane. Taking the length of EB as one, the value of Ψ at B from Eq. (2) or its equivalent, for other values of n, and extending the line

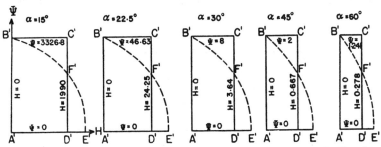

FIG. 127.—The graphical determination of ϵ_1 in Eq. 6.10(4). (*After Dachler, Die Wasserwirtschaft.*)

$B'C'$ in the (H, Ψ) planes (*cf.* Fig. 127) so that the area of $B'C'F'$ equals that of $F'D'E'$, the proper value of H for CD is readily obtained. The resultant conductivity of $ABCD$, which is the ratio $\dfrac{A'B'}{A'D'}$ in Fig. 127, is found in this way to be expressible as a function of the angle α by the equation

$$\epsilon_1 = \frac{Q}{\Delta H_1} = 1.12 + 1.93 \tan \alpha, \tag{4}[1]$$

which is in good agreement with the values found by direct experiments. It is to be noted that ΔH_1 in Eq. (4) refers to the potential drop only over the segment BC of the free surface.

The flow in the dam to the right of BE is now resolved into two parts. The first, from BE to KL, confined in the main body of the dam, is considered to be of an essentially linear character

[1] It is assumed for convenience here and throughout this section that the effective permeability $\bar{k}$ has the value 1.

(*cf.* Fig. 125), the flux being related to the fluid-head drop across it, ΔH_2, by

$$Q = \frac{H_m \Delta H_2}{l}, \tag{5}$$

where H_m is the mean height of the free surface along CK, and l is the distance EL.

For the flow from KL to the outflow face, resort is made to model experiments. The point K is chosen for definiteness as the intersection between the actual free surface and the line GK issuing from the toe of the dam at an inclination equal to two-thirds that of the outflow face, β (*cf.* Fig. 128). Such model

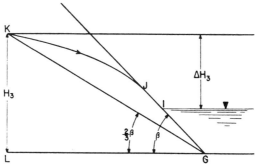

FIG. 128.—(*After Dachler, Die Wasserwirtschaft.*)

experiments carried out with vertical inflow faces KL, and angles β of 30, 45, and 60 deg., for various values of $\Delta H_3/H_3$ (*cf.* Fig. 128) gave for the quantity

$$\epsilon_3 = \frac{Q}{\Delta H_3} \tag{6}$$

the data indicated by the points in Fig. 129. The solid curves are those given by the equation

$$\epsilon_3 = 0.068 \, \beta° \left(0.86 + 0.39\frac{\Delta H_3}{H_3} - \sqrt{\frac{\Delta H_3}{H_3} + 0.36} \right). \tag{7}$$

These results are now to be used as follows (*cf.* Fig. 130 where $\Delta H_3/H_3 = 1$). The position of L is first assumed; H_3 is then given by $GL \tan \frac{2}{3}\beta$, so that with a knowledge of the outflow fluid level, ΔH_3 is determined, and hence ϵ_3 by Eq. (7). Returning to Eq. (6), Q can be calculated, and then from Eq. (4), ΔH_1. As the total fluid-head differential H is given by

$$\Delta H_1 + \Delta H_2 + \Delta H_3 = H, \qquad (8)$$

a knowledge of ΔH_1 and ΔH_3 will give ΔH_2, H being preassigned initially, as well as $H_m = H_3 + \Delta H_2/2$. When these are put into Eq. (5) the same value of Q should be found as was originally given by Eq. (6). A lack of agreement indicates an error in the choice of the position of L, and the process should be repeated until consistency is obtained between the various quantities entering the formulas.

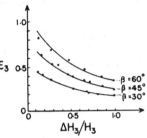

FIG. 129.—The results of model experiments for the determination of ϵ_3. Solid curves are those given by Eq. 6.10(7). (*After Dachler, Die Wasserwirtschaft.*)

When the outflow fluid level is so high that $\Delta H_3/H_3 < 0.1$, the limit at which the formula of Eq. (7) has empirical justification, the approximate calculation of the flux can be carried through by assuming the flow from BE to IN to be linear, where IN is the vertical drawn from the intersection of the outflow fluid body with the outflow face. The added drop due to the wedge to the right of IN can then be approximated by the same expres-

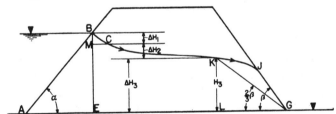

FIG. 130.—The significant quantities used in the calculation of the seepage flux through sloping dams. (*After Dachler, Die Wasserwirtschaft.*)

sion giving that due to the inflow wedge to the left of BE, namely, Eq. (4).

Unfortunately there are available no really suitable exact solutions to the problem of the seepage through dams with sloping faces with which to test the accuracy of the approximate theory presented above. Due to the high inflow-fluid levels corresponding to the models of Figs. 110 and 111, of Sec. 6.6, the regions

$KJGL$ with the form factor ϵ_3, constructed in the manner indicated above, will extend to or overlap the region ABE, so as to lead to an excessive outflow resistance or to values of the flux that are too low. It is nevertheless of interest to show that even in these unfavorable cases the approximate theory gives at least the correct order of magnitude for the flux. Thus noting that $\alpha = \beta = 30$ deg., and $\Delta H_3/H_3 = 1$, for Fig. 110, it follows from Eqs. (4) and (7) that $\epsilon_1 = 2.235$ and $\epsilon_3 = 0.171$. Now since the region $GJKL$ of Fig. 130 overlaps ABE for this case, the section $MKLE$ may be omitted, and the value of Q will be determined by the equations

$$Q = 2.235\Delta H_1 = 0.171\Delta H_3: \qquad \Delta H_3 + \Delta H_1 = 1,$$

where the total fluid head has been taken as 1. The result is: $Q = 0.159$.

Noting similarly that for Fig. 111, $\alpha = \beta = 45$ deg., so that $\epsilon_1 = 3.05$ and $\epsilon_3 = 0.257$, and again omitting the section $MKLE$ of Fig. 130, it is readily found that here $Q = 0.238$.

To get the corresponding values of Q from the potential and streamline distributions given by the electrical models, one need only modify slightly the procedure outlined in Sec. 4.17. Thus recalling that the flux Δq between any two neighboring streamlines is

$$\Delta q = \frac{\Delta \Phi}{\Delta n}\Delta s, \qquad (9)$$

where $\Delta \Phi$ is the potential drop over the length Δn, and Δs is the separation between the streamlines, it is clear that by taking the equipotentials separated by $\Delta \Phi$ such as to form squares in cutting across the streamlines, so that $\Delta s = \Delta n$, the total flux will simply equal the number, m, of squares lying between the two limiting streamlines multiplied by $\Delta \Phi$, i.e.,

$$Q = m\Delta \Phi. \qquad (10)$$

Applying this procedure to Fig. 110 for the strip lying between the equipotentials $\Phi = 0.75$ and $\Phi = 0.80$, it is found that $m = 5.1$. Hence, as for a total unit ΔH, $\Delta \Phi = 0.05$, it follows at once that

$$Q = 5.1 \cdot 0.05 = 0.255.$$

Proceeding similarly in the case of Fig. 111, along the strip between $\Phi = 0.70$ and $\Phi = 0.75$, it is found that $m = 5.6$, so that

$$Q = 5.6 \cdot 0.05 = 0.28.$$

These values are to be compared with 0.159 and 0.238, given by the approximate theory. As was previously anticipated, the latter values are too low, the discrepancy being smaller in the more favorable case of Fig. 111. The orders of magnitude, however, are correctly given by the approximate theory, and it appears reasonable that for lower upstream fluid levels and for more extended free surfaces, even the numerical values will not be greatly in error.

A final point to be noted relating to the problem of the seepage of water through dams with sloping faces is the implication of Eq. (1) with respect to the seepage velocities along the inflow face. Thus, Eq. (1) gives for the fluid-head distribution in the region ABE,

$$H = r^n \cos n\theta, \tag{11}$$

where r is the radial distance from the point A, and θ is the polar angle. The velocity normal to the inflow face is, therefore, proportional to

$$-\frac{1}{r} \frac{\partial H}{\partial \theta}\bigg|_{\theta=\alpha} = nr^{n-1} = \frac{\pi}{2\alpha} r^{\frac{\pi}{2\alpha}-1}. \tag{12}$$

It follows that the velocity will vanish at the heel and increase to a maximum in going up the slope, the rate of rise increasing as the slope of the face decreases.[1] Approximately the same type of variation will occur at the outflow face below the surface of seepage. At the junction of the surface of seepage and the outflow-fluid level the velocity will be infinite, but will rapidly decrease as the outflow terminus of the free surface is approached, the velocity there being tangent to the outflow face and equal to the component along the face of the velocity of free fall[2] $\bar{k}$. As the regions of maximum inflow or outflow velocity will be those of maximum erosion, it is clear that to control the latter, particular

---

[1] This result, of course, breaks down for a vertical-inflow face, where the velocity decreases from a maximum at the heel to zero at the free surface.

[2] *Cf.* also Sec. 6.1.

attention should be given to the intersections of the upstream and downstream fluid levels with the dam faces.[1]

6.11. Seepage Streams from Canals and Ditches Which Merge with Shallow Water Tables.[2]—The theory of the ground-water flow in a sloping sand drained by a ditch was presented in Sec. 6.2, the net drainage by the ditch being given by Eq. 6.2(8). The reverse problem arises in the consideration of the seepage out of a ditch or canal, set above a shallow normal ground-water level, with respect to the tendency of such seepages to water-log the surrounding sand. Although it has been possible to give a treatment of such cases where the water table lies in a bed of much higher permeability than that which is the original carrier of the seepage stream by direct analytical methods (*cf.* Secs. 6.8 and 6.9), those situations in which the seepage stream merges with the water table demand the application of approximate and semi-empirical methods of solution. One such type of solution will be presented here.

Strictly speaking, a steady-state condition of gravity flow out of an infinitely extended homogeneous sand requires that the sand have a nonvanishing slope at points distant from the ditch except when the seepage ultimately develops into a vertical free fall through the sand as assumed in Sec. 6.8. For at large distances from the ditch the free surface must evidently become asymptotically parallel to the normal unperturbed fluid level, and hence to the impermeable base. If the latter is horizontal, however, and the free surface is parallel to it, the velocities away from the ditch will vanish, which is, of course, contradictory to the assumption that a steady-state flux is being introduced into the sand at the ditch or canal.

In addition to the physical requirement that the free surface of a gravity-flow system corresponding to the above type of seepage out of a ditch or canal must become asymptotic to the impermeable base of the sand, the potential distribution must asymptoti-

---

[1] For a practical discussion of methods of keeping the free surface from impinging on the outflow face, when the tailwater head is zero, by the insertion of tile drains or pervious filter blankets, and thus eliminating the surface erosion, see the paper by A. Casagrande (*loc. cit.*).

[2] It must be explicitly understood that the results of this section and those of the others in this chapter are to be applied only after a steady-state condition has been established in the flow system.

cally become that appropriate to a linear sheet flow. Thus the equipotentials at large distances from the ditch or canal must become normal to the impermeable base with a uniform spacing proportional to the slope of the impermeable bed. It is clear, therefore, that the exact solutions of Laplace's equation given in Secs. 6.7 to 6.9, which imply an asymptotic approach to vertical free fall, cannot be taken as physical representations of practical seepage-flow systems in sands of finite thickness[1] in which the seepage stream merges with the normal ground-water table.

If, however, instead of considering the sand to be of infinite extent, with the boundary conditions at infinity mentioned above, it is supposed that the region of interest is of width $2L$, and that the height H_0 of the free surface at $x = \pm L$ is known, no difficulty will arise in taking the impermeable base for $|x| < L$ as horizontal, and the system as a whole as symmetrical about the axis of the ditch or canal. Beyond $|x| = L$, the sand will indeed have to assume a downward slope in order to carry away the seepage fluid, if it should continue indefinitely. But in specifying the height H_0 at $x = L$, the region beyond $|x| = L$ is eliminated from the problem, and the analysis can be chosen so as to be appropriate to a system of finite dimensions. Unfortunately even the

---

[1] Likewise one must reject the potential and streamline distributions:

$$\Phi = \frac{2b}{\pi} \cos \alpha \cosh \frac{\pi x}{2b} \sin \frac{\pi y}{2b} + x \sin \alpha;$$

$$\Psi = -\frac{2b}{\pi} \cos \alpha \sinh \frac{\pi x}{2b} \cos \frac{\pi y}{2b} + y \sin \alpha,$$

where x, y are the coordinates parallel and normal to the impermeable base of the porous bed, of inclination α and depth b, which are given by a direct solution of Laplace's equation (*cf.* N. K. Bose; and V. I. Vaidhianathan, H. R. Luthra, and N. K. Bose, *loc. cit.*). These do, indeed, give a streamline and potential distribution in the neighborhood of the canal possessing some of the characteristics that are to be physically expected. However, as none of the streamlines $\Psi =$ const. represent free surfaces, and the exponential increase of Φ with large values of x is not consistent with the normal unperturbed sheet flow, similar to that given by Eq. 6.7(2), that must take place at large values of x (*cf.* also Fig. 131), it is difficult to properly evaluate the practical significance that one should attribute to these solutions. The approximate treatment presented here, on the other hand, deals more directly with the value of the seepage flux, which is obtained in such a manner that only its exact quantitative value should be affected by the approximations made in the analysis.

problem limited in this manner is difficult to treat exactly, and though the methods of Hopf and Trefftz or Hamel are applicable, in principle, the necessary analysis has not yet been carried through. In the following we shall, therefore, present an approximate treatment, in which both model experiments and an approximate analysis are combined to give a simple procedure for computing the seepage flux without entering in detail into a study of the free surface of the system.

This theory, also due to Dachler,[1] is similar to that given in the last section for the seepage through dams, in that the flow system

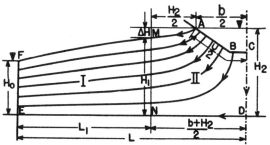

Fig. 131.—Diagrammatic representation of the seepage stream out of a ditch which overlies a shallow water table. (*After Dachler, Die Wasserwirtschaft.*)

is again divided into separate parts, to each of which is applied a different method of approximation.[2] Owing to symmetry—resulting from the assumption that the unperturbed ground-water level is horizontal—one need consider only half of the system, diagrammatically represented in Fig. 131. As already mentioned, it will be supposed that in addition to the width b of the canal and the height H_2 of the water level in it above the impermeable base,

[1] DACHLER, R., *Die Wasserwirtschaft*, p. 110, 1933.

[2] One may also note here the treatment of J. M. Burgers (*De Ingeneuir*, no. 22, 1926) in which a similar separation of the flow system is made. The flow in part II for a flat-bottomed canal ($DC = H_2$) is compared with that in the linear-flow system corresponding to a canal with vertical sides dug to the impermeable base, ED, of the sand. The difference in potential at a distant point for the two cases, with the same flux and same potential at the canal—that of the actual canal being computed from Eq. (4)—is then interpreted as representing an "efflux resistance" arising from the particular geometry of region II. However, as no account is taken in this calculation of the depression of the free surface, and hence decrease of the average thickness of the flow channel, at large distances from the canal, this theory appears to be less satisfactory than that of Dachler.

one also knows the height H_0 of the free surface at a distance L from the center of the canal.

The separation of the regions I and II at the distance $(b + H_2)/2$ from the axis of the canal, as indicated in Fig. 131,

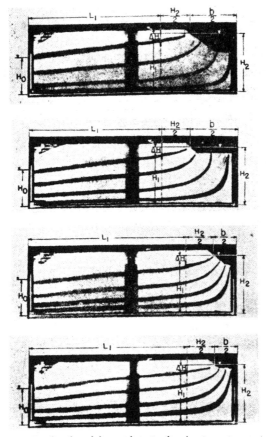

FIG. 132.—Photographs of model experiments showing the seepage streams out of ditches overlying shallow water tables.   (*After Dachler, Die Wasserwirtschaft.*)

though strictly arbitrary, finds justification in the fact that beyond the distance $(b + H_2)/2$ the flow is very approximately linear. This is verified by the traces of the streamlines in sand models of the system of Fig. 131, as shown in Fig. 132. Denoting by H_1 the height of the free surface at MN, the flow through region I can, therefore, be approximated by

$$\frac{Q}{2} = \bar{k}\left(\frac{H_1 - H_0}{L_1}\right)\left(\frac{H_1 + H_0}{2}\right); \qquad L_1 = L - \frac{b + H_2}{2}. \quad (1)^1$$

For region II, in which the streamlines undergo the major change in their directions, one may tentatively set

$$\frac{Q}{2} = \bar{k}(H_2 - H_1)\epsilon, \tag{2}$$

where the "form factor" ϵ is determined by the detailed geometry of the region near and including the canal. Eliminating the unknown quantity H_1, Q is given by

$$Q = 2\bar{k}\epsilon[H_2 + \epsilon L_1 - \sqrt{(H_2 + \epsilon L_1)^2 - H_2{}^2 + H_0{}^2}]. \tag{3}$$

It is readily seen from this expression that the fractional error in Q will in general be less than that in ϵ, so that those made in the computation of the latter will not lead to excessive errors in the resultant value of Q.

To obtain the value of ϵ, the flow in region II will be approximated by that corresponding to the potential and streamline distribution given by

$$\left.\begin{array}{c} H + i\Psi = \log\left(\sinh z + \sqrt{\sinh^2 z - \sinh^2 f}\right) \\ z = x + iy, \end{array}\right\} \tag{4}$$

where the fluid head is again being used in place of the velocity potential Φ. A typical set of the equipotentials and streamlines defined by Eq. (4) is shown in Fig. 133. It will be seen that the boundary CDN is exactly reproduced by the streamline $\Psi = \pi/2$, where DN is given by $y = \pi/2$. For the boundary of the canal one can evidently take any curve $H = $ const. while the free surface AM must be approximated by the ordinary streamline passing through A, the extremity of the canal boundary. The point M in a specific problem will be chosen so that its height H_1 will give the same flux by Eq. (1) as by Eq. (2).

---

[1] Although the actual average potential along FE will be less than the equivalent of the fluid head H_0, approximately the same error will enter at MN, so that $H_1 - H_0$ should give a good representation of the correct average fluid-head difference across region I. In the case of Eq. (2), the corresponding errors in $H_2 - H_1$ are taken into account by the method of determining ϵ.

The values of ϵ obtained from Fig. 133 by choosing various values of $b/2$ and H_2 and reading off the corresponding values of H_1 are given in Fig. 134, in the form of curves of constant ϵ, with

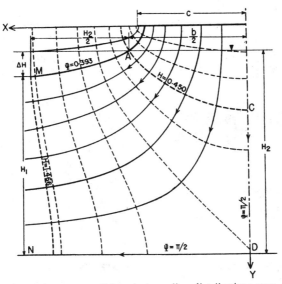

FIG. 133.—Approximate potential and streamline distributions near the ditch shown in Fig. 131. (*After Dachler, Die Wasserwirtschaft.*)

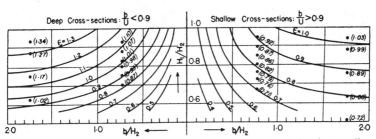

FIG. 134.—Curves of constant ϵ as calculated from the potential and streamline distributions of Fig. 133. Points are values found by direct model experiments. (*After Dachler, Die Wasserwirtschaft.*)

coordinates b/H_2 and H_1/H_2. Thus, for the heavy marked contours of Fig. 133, the corresponding value of ϵ would be

$$\epsilon = \frac{\frac{\pi}{2} - 0.393}{1.466 - 0.450} = 1.16. \qquad (5)$$

In addition to the ratio b/H_2, which is a measure of the extension of the ditch as compared to the thickness of the porous stratum, the shape of the ditch may be characterized by the ratio b/U, where U is the length of the wetted perimeter of the ditch. In Fig. 134, however, the results for various values of b/U are lumped together into a group with $b/U > 0.9$, corresponding to shallow ditch profiles, and one with $b/U < 0.9$, corresponding to deep profiles. The values of ϵ are, therefore, to be considered as

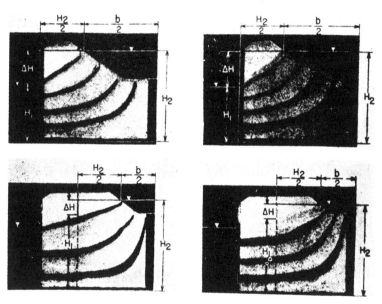

Fig. 135.—Photographs of model experiments showing the streamlines near the base of a drainage ditch overlying a shallow water table. (*After Dachler, Die Wasserwirtschaft.*)

averages for profiles with $b/U > 0.9$ or < 0.9; the use of these averages should suffice for most practical purposes, although more accurate values for any specific choice of b/U can be readily obtained by returning to Fig. 133 and computing ϵ directly, as in the example of Eq. (5). For $b/U = 0.9$ the values of ϵ obtained from the two halves of Fig. 134 may be averaged.

That the above manner of calculation gives ϵ to a close approximation was verified by direct model experiments, which gave the results shown by the points in Fig. 134. Figure 135 shows photographs of typical cases of these model experiments.

Once ϵ has been determined, the flux can be computed from Eq. (3). However, the value of Q will also be incidentally obtained in finding the value of H_1 which will make the right sides of Eqs. (1) and (2) equal. Thus, choosing

$$b = \frac{1}{2}; \qquad H_2 = 1; \qquad H_0 = \frac{1}{2}; \qquad L = 5; \qquad \frac{b}{U} > 0.9,$$

it is found by reference to Fig. 134 that for $H_1 = 0.915$, $\epsilon = 0.8$, Eqs. (1) and (2) give within 1 per cent the same value of Q, namely,

$$Q = 1.37\bar{k}.$$

To get a numerical value of practical significance, consistent units must be used for the various constants. If, for example, with the above ratios of b/H_2, H_2/L, and H_0/H_2, H_2 is taken numerically equal to 10 ft. = 304.8 cm., k is 10 darcys, and $\gamma/\mu = 1$, the value for water, Q, would be

$$Q = 1.37 \cdot 10 \cdot 0.968 \cdot 10^{-3} \cdot 304.8 = 4.042 \text{ cc./sec./cm. length of ditch,}$$

$$= 1.95 \text{ gal./min./ft. length of ditch.}$$

This type of calculation not only gives the drainage capacity of the ditch or canal for a given ground-water level H_0, at the distance L, but also by inversion it will permit the determination of H_0 if the flux Q is initially specified. One can thus predict the limit of the drainage capacity which can be handled by any particular ditch or canal without producing a condition of waterlogging at distant points.

6.12. The Approximate Treatment of Some Irrigation and Drainage Problems.—Some different types of problems of gravity flow requiring direct approximate methods of analysis are those involved in questions of irrigation and drainage. Such problems have been discussed by Gardner, Collier, and Farr,[1] and in the main their treatment will be followed here. It will be noted that the analyses are based upon such direct physical simplifications as make it possible to avoid the explicit solution of the Laplace equation which would be required for the rigorous discussion of the problems. In particular, no account is taken of the free

---

[1] GARDNER, W., T. R. COLLIER, and D. FARR, *Utah Agr. Exp. Sta. Bull.* **252,** 1934.

surfaces that will be present in all the systems discussed and the associated distortions of the geometrical definitions of the original flow systems. Since, however, the appropriate exact analyses are not yet available, approximate treatments must serve the purpose of giving at least the essential features of the solutions.

6.13. Subirrigation.—The particular question involved here is the extent x_1 of the lateral drainage out of a canal into a coarse sand before the leakage from the canal all seeps into the underlying beds. Making the assumptions that the drainage bed is completely saturated, without upward leakage or a free surface, up to the point x_1, and that the seepage into the clay below is simply that of a uniform vertical free flow as fast as the clay can

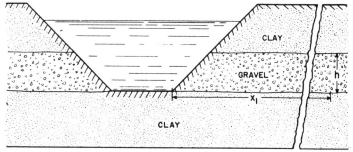

FIG. 136.—Diagrammatic representation of the problem of subirrigation. (*After Gardner, Collier, and Farr, Utah Agr. Expt. Sta. Bull.*).

carry the fluid, the equation of continuity together with Darcy's law gives

$$hk_g\frac{d^2p}{dx^2} = k_c\gamma g, \tag{1}$$

where h is the thickness of the gravel bed, k_g its permeability, k_c that of the clay, γ the density of the water, and g the acceleration of gravity (*cf.* Fig. 136). Integrating Eq. (1), with the conditions that both the pressure and its gradient should vanish at x_1, it is readily found that x_1 is given in terms of the pressure p_0 at the face of the canal at the level of the gravel bed by

$$x_1 = \sqrt{\frac{2p_0hk_g}{k_c\gamma g}} = \sqrt{\frac{2h_0hk_g}{k_c}}, \tag{2}$$

where h_0 represents the fluid-head equivalent of the pressure p_0. Thus, if p_0 represents a fluid head of 2 ft., the gravel bed is

1 ft. thick, and $k_y/k_c = 100{,}000$, x_1 will have the value 632 ft. Although the neglect of the upward seepage into the overlying loam will tend to make the value given by Eq. (2) too high, it should still give results of the right order of magnitude.

6.14. The Problem of Water Logging.—From the definition of the velocity potential, with the vertical coordinate y directed upward,

$$\Phi = \frac{k}{\mu}(p + \gamma gy), \tag{1}$$

it follows that the height above the plane $y = 0$ to which the liquid would rise in an open tube or crevice piercing the porous medium, at the plane y, will be given by

$$H = \frac{p - p_0}{\gamma g} + y = \frac{\mu}{k\gamma g}\Phi - H_0 = \frac{\Phi}{k} - H_0, \tag{2}$$

where p_0 is the atmospheric pressure and H_0 the corresponding fluid head. Choosing then Φ—the absolute value of which is always subject to the addition of an arbitrary constant—so as to include the term H_0 in Eq. (2), it is seen that the curve representing the fluid heights H—the "piezometric surface"—is simply proportional to the velocity potential in the porous medium. Hence, as H gives the lower limit of the height to which the zone of complete liquid saturation—the "zone of water logging"—in gravity-flow systems could extend, a calculation of the potential distribution Φ will give at least a limiting lower bounding surface to that part of a sand that will be water-logged. This definition of the condition of water logging is also equivalent to that depending on the value of the fluid pressure, namely, the sand will be water-logged or dry at any point as the fluid pressure there is greater or less than the atmospheric pressure.

6.15. The Erosion Problem.—If the soil covering a hill or mountain side is of insufficient depth to carry away the rainfall falling on it, the excess water will flood over the surface of the ground and may result in a serious erosion of the top soil. An approximate calculation of the minimum soil depths required for the removal of a rainfall intensity q can be made as follows, assuming that the surface soil lies on the hillside of inclination θ_2 in the form of a wedge, as shown in Fig. 137: Denoting the inclination of the top surface of the soil at the distance x from the crest of the hill by $\theta_1(x)$, the equation of continuity gives

$$q \cos \theta_1(x) = k\frac{d}{dx}\left[x \int_{\theta_1(x)}^{\theta_2} \sin \theta d\theta \right],\qquad (1)$$

or

$$\left(1 - \frac{q}{k}\right) \cos \theta_1(x) = \cos \theta_2 + x \sin \theta_1(x)\frac{d\theta_1(x)}{dx}.\qquad (2)$$

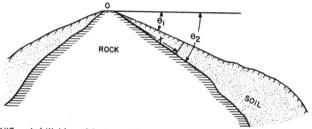

Fig. 137.—A hillside subject to soil erosion. (*After Gardner, Collier, and Farr, Utah Agr. Expt. Sta. Bull.*)

The solution of this equation is readily verified to be given by

$$\cos \theta_1 = \frac{\cos \theta_2}{1 - \frac{q}{k}} + \frac{C}{x^{1-\frac{q}{k}}}.\qquad (3)$$

As $\cos \theta_1$ must remain finite even for $x = 0$, C must be taken as zero. θ_1, therefore, has the constant value given by

$$\cos \theta_1 = \frac{\cos \theta_2}{1 - \frac{q}{k}}.\qquad (4)[1]$$

It is thus seen that if the rainfall intensity q is so great that the right side of Eq. (4) exceeds unity the rainfall will necessarily flood over the soil, even if its top surface is horizontal.

6.16. Tile Drainage.[2]—The essential purpose of systems of tile drains is to protect the top soil of a region from becoming

[1] In the treatment of Gardner, Collier, and Farr it is explicitly assumed that the top surface is plane, so that $\frac{d\theta_1}{dx} = 0$.

[2] The discussion given here differs from that of Gardner, Collier, and Farr, in that the latter synthesize the flow from the gravel bed into the drains as

water-logged. Thus if the ground water in the surface sediments is subjected to an upward potential gradient, the water will rise to the surface and water-log the top soil, unless the drains are capable of gathering in all the upward flow. An approximate calculation of the proper spacing of the drains can be made as follows: Representing the drains by an infinite array of sinks of spacing a, set at the depth d below the surface, they will contribute to the resultant potential distribution of the system a term as

$$\Phi = q \log \left[\cosh \frac{2\pi(y - d)}{a} + \cos \frac{2\pi x}{a} \right] \quad (1)$$

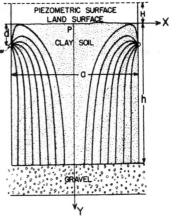

Fig. 138.—A pair of an array of tile drains set over a gravel bed to prevent waterlogging. (*After Gardner, Collier, and Farr, Utah Agr. Expt. Sta. Bull.*)

(*cf.* Secs. 4.9 and 9.8), where the flux into each drain is $4\pi q$ (*cf.* Fig. 138).

The condition that there be no water logging or seepage of water through the top surface will be imposed on the system by supposing that the top surface is impermeable. This is accomplished analytically by placing an image of the array of drains at $y = -d$, or by adding to the Φ of Eq. (1) an exactly similar term but with d replaced by $-d$. The resultant potential can then be written as

$$\Phi = c + q \log \left[\cosh \frac{2\pi(y - d)}{a} + \cos \frac{2\pi x}{a} \right] \left[\cosh \frac{2\pi(y + d)}{a} + \cos \frac{2\pi x}{a} \right]. \quad (2)$$

The fact that this prevention of the seepage through the top surface is really due to the gravity character of the flow and not

---

the sum of simple linear and radial types of flow, whereas here the flow into the system of drains is represented rigorously by Eq. (2), except for the neglect of the free surface that will form the upper boundary of the flow system.

to an impermeability of the top surface will be given cognizance by adding the condition that the pressures at both the central point P (*cf.* Fig. 138) and the drain surfaces, of radii r_w, be atmospheric, p_0. First, however, the constant c in Eq. (2) will be chosen so that the total upward driving potential from the gravel bed at the depth h be equivalent to the fluid head H. To effect a strictly uniform potential at the depth h would require the reflection of the two arrays contributing to Eq. (2) in the line $y = h$ by negative images, then in the x axis, again in $y = h$, etc., forming an infinite set of parallel arrays. If, however, the depth of the gravel bed below the line of drains is of the order of or greater than the drain spacing a, the array at $y = +d$ alone will already give a potential at $y = h$ that is uniform for all practical purposes, as will be shown in detail in Sec. 9.8. Thus the potential at $y = h$, corresponding to Eq. (2), may be expressed as

$$\Phi(y = h) \cong c + q \log \cosh \frac{2\pi(h - d)}{a} \cosh \frac{2\pi(h + d)}{a}. \quad (3)$$

The potential at the surface midway between the drains is

$$\Phi(0, 0) = c + 2q \log 2 \cosh^2 \frac{\pi d}{a}, \quad (4)$$

so that q is given by

$$q = \frac{\bar{k}H}{\log \dfrac{\cosh 2\pi(h - d)/a \cosh 2\pi(h + d)/a}{4 \cosh^4 \pi d/a}}. \quad (5)$$

Applying now the condition that the pressure at P is the same as that at the drain surfaces, it follows that

$$\frac{2q}{\bar{k}} \log 2 \cosh^2 \frac{\pi d}{a} = d + \frac{2q}{\bar{k}} \log 2 \sinh \frac{\pi r_w}{a} \sinh \frac{2\pi d}{a}.$$

Combining this with Eq. (5), it may be written as an equation determining the value of the spacing a, namely,

$$\frac{\log \dfrac{\cosh^2 \pi d/a}{\sinh \pi r_w/a \sinh 2\pi d/a}}{\log \dfrac{\cosh 2\pi(h - d)/a \cosh 2\pi(h + d)/a}{4 \cosh^4 \pi d/a}} = \frac{d}{2H}. \quad (6)$$

The simplest procedure for finding the drain spacings a, as a function of the other parameters, r_w, d, h, and H from this equation, is evidently to assume values of r_w, d, h, and a, and compute the corresponding values of H. For a depth of the gravel bed, h, of 50 ft., and for drain radii, r_w, of $\frac{1}{4}$ and $\frac{1}{2}$ ft., the spacings a computed in this manner are plotted in Fig. 139 against the driving-fluid-head difference H for drain depths, d, of 5 and 10 ft. It will be noted that as the driving head H increases the spacing rapidly decreases. The cost of effective drainage will, therefore, rapidly become prohibitive as H increases. A comparison of the curves, for different d and r_w, shows furthermore that close spacing will become still more essential as the depth of the drains or their radius is decreased.

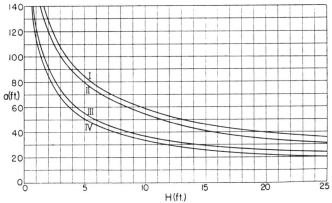

FIG. 139.—Spacing, a, of a line of tile drains required to prevent waterlogging. H = fluid head inducing the upward flow. I. d = 10 ft., r_w = $\frac{1}{2}$ ft. II. d = 10 ft., r_w = $\frac{1}{4}$ ft. III. d = 5 ft., r_w = $\frac{1}{2}$ ft. IV. d = 5 ft, r_w = $\frac{1}{4}$ ft. d = depth of drains; r_w = radius of drains; depth of gravel source bed = 50 ft.

6.17. The Dupuit-Forchheimer Theory of Gravity-flow Systems.

—Although the assumptions of the Dupuit-Forchheimer theory of gravity-flow systems now appear to be so questionable as to make the whole theory quite untrustworthy unless applied with great care, its widespread and indiscriminate use even at present demands at least a brief outline of its fundamental features. These are, of course, all implied in the assumptions, made by Dupuit[1] in 1863, that for small inclinations of the free surface of a gravity-flow system the streamlines can be taken as horizontal, and are to be associated with velocities which are

[1] DUPUIT, J., Études theoriques et pratiques sur le mouvement des eaux, 1863.

proportional to the slope of the free surface, but are independent of the depth ("shell" flow). These two assumptions permitted Dupuit to derive a formula for the gravity flow in a radial system, and in the hands of Forchheimer[1] gave a general equation for the free surface of any gravity-flow system. Forchheimer's result may be conveniently derived by simply applying the equation of continuity to the fluid in any column of liquid of height $\bar{h}$ above the impermeable bed of the system. For if v_x, v_y are the steady-state velocity components in this column, and are the same over its whole length, the equation of continuity requires that

$$\frac{\partial}{\partial x}(\bar{h}v_x) + \frac{\partial}{\partial y}(\bar{h}v_y) = 0. \tag{1}$$

Further, Dupuit's assumptions imply that

$$v_x = -\bar{k}\frac{\partial \bar{h}}{\partial x}; \qquad v_y = -\bar{k}\frac{\partial \bar{h}}{\partial y}. \tag{2}$$

It then follows at once that

$$\frac{\partial^2 \bar{h}^2}{\partial x^2} + \frac{\partial^2 \bar{h}^2}{\partial y^2} = 0, \tag{3}$$

which is Forchheimer's result.

This equation for the free surface in gravity-flow systems has been applied to a variety of gravity-flow problems.[2] The analytical procedures that may be used for its solution are evidently the same as those developed in Chap. IV for treating the two-dimensional Laplace equation for the pressure distribution in a liquid-bearing porous medium, as it is, in fact, formally identical with that equation. Thus, in particular, Eq. (3) implies that for a linear flow system of length L, with boundary values for $\bar{h}$ of h_e, h_w,

$$\bar{h}^2 = \frac{h_e{}^2 - h_w{}^2}{L}x + h_w{}^2, \tag{4}$$

while for the corresponding radial system bounded by concentric circles of radii r_e, r_w,

[1] FORCHHEIMER, PH., Zeits. Arch. Ing. Ver., Hannover, 1886.
[2] FORCHHEIMER, PH., "Hydraulik," 3d ed., Chap. III, 1930.

$$\bar{h}^2 = \frac{h_e{}^2 - h_w{}^2}{\log r_e/r_w} \log \frac{r}{r_w} + h_w{}^2. \tag{5}$$

Returning now to the Dupuit assumptions, it is clear that, since

$$\frac{\partial v_x}{\partial z} = \frac{\partial v_z}{\partial x}; \qquad \frac{\partial v_y}{\partial z} = \frac{\partial v_z}{\partial y}, \tag{6}$$

the assumption of "shell flow" $\left(\dfrac{\partial v_x}{\partial z} = \dfrac{\partial v_y}{\partial z} = 0\right)$ implies that the vertical velocity remains uniform in the horizontal planes. Now at an external boundary maintained at a uniform potential, v_z will evidently vanish. Hence v_z will vanish throughout the system. Without vertical velocities, however, the free surface will be horizontal and the flow will have no gravity characteristics. The absence of the vertical velocities in gravity-flow systems is, therefore, entirely contradictory to the implications of Darcy's law.

The general nature of the actual vertical variation of the resultant velocity in a gravity-flow system may be seen from the relation

$$\frac{\partial}{\partial z}(v_x{}^2 + v_y{}^2 + v_z{}^2) = 2\left[v_x{}^2\frac{\partial}{\partial x}\left(\frac{v_z}{v_x}\right) + v_y{}^2\frac{\partial}{\partial y}\left(\frac{v_z}{v_y}\right)\right]. \tag{7}$$

For from this it follows that at points where the streamlines are convex upward the resultant velocity will decrease with increasing height, and conversely at points where the streamlines are concave upward. Hence near the high fluid-level inflow surfaces the resultant velocities must decrease with the height, while near low fluid-level constant-potential outflow surfaces, where the streamlines will be inflected so as to strike these surfaces at right angles, the resultant velocities will increase with increasing height.[1]

Even more forceful than these general considerations showing the inconsistencies of the Dupuit assumptions with the direct implications of Darcy's law are the results of the specific calculations carried through rigorously in Sec. 6.5 for the seepage of

---

[1] These general deductions, which should apply to all steady-state liquid-flow systems, may be verified for the particular problem of the seepage through dams by reference to the velocity distributions along the faces of the dams, as shown in Figs. 103 to 106.

water through dams with vertical faces. Thus recalling Figs. 103 to 106, it will be seen that even along the inflow face, where the slope of the free surface is a minimum, the velocity distribution is far from uniform.[1] Nor is there any relation between the average of the velocities along the inflow face and the slope of the free surface, which is always zero there. The failure of these assumptions is even more striking at the outflow face where the velocities vary from zero to infinitely high values, although such discrepancies were probably anticipated by Dupuit, as he originally proposed his theory for applications only in regions of small free-surface slope.

As is to be expected, the generalization of Forchheimer, as represented by Equation (3), fares no better than the original Dupuit assumptions. For as already noted in Sec. 6.5, the free surfaces given by Eq. (4), shown by the curves $\bar{h}$ in Figs. 103 to 106, are clearly but poor approximations of the correct free surfaces. The discrepancies are especially large in those cases where the outflow-fluid head is zero. As the Dupuit-Forchheimer theory takes no account of the surface of seepage, the boundary values of $\bar{h}$ to be used with the solutions of Eq. (3) must evidently be those of the fluid head, so that when the latter vanishes, so will the free surface heights derived from Eq. (3). On the other hand, this neglect of the surface of seepage is not the sole cause of the difficulty, as even if the boundary fluid heights are taken to be those given by the rigorous theory or by the model experiments, the parabolic variation of Eq. (4) will still not give a close approximation to the true free surfaces. However, as anticipated in Sec. 6.5, if all the above-mentioned difficulties regarding the assumptions underlying and consequences of Eq. (3) are disregarded, and the Dupuit assumptions are applied to the solutions of Eq. (3) so as to get the flux, the resulting values will nevertheless be surprisingly close to those given empirically or by exact calculation. Thus for linear flow, this procedure gives, in conjunction with Eq. (4),

$$Q = \bar{k}\bar{h}\frac{\partial \bar{h}}{\partial x} = \frac{\bar{k}(h_e{}^2 - h_w{}^2)}{2L},\tag{8}$$

[1] Indeed a logical application of the Dupuit assumptions would lead to the absurd result that since the free-surface slope along the inflow face is zero—as the free surface issues from a constant potential boundary—the inflow velocity would vanish everywhere.

and for the case of radial flow, the same procedure applied to Eq. (5) gives

$$Q = 2\pi \bar{k} r h \frac{\partial \bar{h}}{\partial r} = \frac{\pi \bar{k}(h_e^2 - h_w^2)}{\log r_e/r_w}. \tag{9}$$

As has already been seen, Eq. (8) reproduces the values of Q given by exact calculation with an accuracy sufficient for all practical purposes (cf. last 2 rows in Table 14). And as will be found in Sec. 6.18, the fluxes obtained empirically with radial sand models are predicted quite accurately by Eq. (9). While it is difficult to see why the various errors in the solution for $\bar{h}$, as given by Eq. (3), will cancel so as to give close approximations to the resultant fluxes, we shall find in Sec. 6.20 that both Eqs. (8) and (9) can also be derived rigorously by an entirely different method for systems which, from physical considerations, may well be expected to have fluxes but slightly different from those for the corresponding gravity-flow systems.

Although there is an extensive literature in which the free surfaces of various gravity-flow systems have been derived from solutions of Eq. (3),[1] it hardly appears worthwhile, in view of the above discussion, to reproduce these derivations here. However, when it is known a priori that the slopes of the free surfaces will be small throughout the flow system or through any particular part of it, Eq. (3) or the original Dupuit assumptions may be justifiably used in approximate calculations of the free surface, provided the correct values are known for the fluid heights at the boundaries of the region of interest. The methods for finding the appropriate solutions of Eq. (3) are identical with those described and illustrated in detail in Chap. IV. If the geometry of the system and the boundary conditions are similar to those of a problem treated in Chap. IV, the solution derived there may be formally taken over for the solution of Eq. (3), with the single change that the functions p or Φ of Chap. IV are to be replaced by the $\bar{h}^2$ of Eq. (3).

Before leaving the discussion of the Dupuit-Forchheimer theory, it may be noted that the same assumptions have been

---

[1] Cf. in particular: Ph. Forchheimer, "Hydraulik," 3d ed., Chap. III, and J. Kozeny, *Ing. Zeits.* **1,** 97, 1921.

used[1] to construct theories of the time variations of the ground-water level in the neighborhood of canals or wells. Thus if $\bar{h}(x, y)$ is the height of the ground-water level at (x, y) above the impermeable base—taken as horizontal—the equation of continuity, applied in exactly the same manner as in the derivation of Eq. (3), gives for the variation of $\bar{h}$ both with time and the coordinates

$$f\frac{\partial \bar{h}}{\partial t} = \bar{k}\left\{\frac{\partial}{\partial x}\left(\bar{h}\frac{\partial \bar{h}}{\partial x}\right) + \frac{\partial}{\partial y}\left(\bar{h}\frac{\partial \bar{h}}{\partial y}\right)\right\}, \qquad (10)$$

where f is the porosity of the sand. As this equation is nonlinear and cannot be solved exactly, further approximations have been introduced. First, however, $\bar{h}$ is expressed as $H + z$, where z is the height of the ground-water level (free surface) above the zero plane, and H is the height of that plane above the impermeable base, so that Eq. (10) becomes

$$f\frac{\partial z}{\partial t} = \bar{k}\left\{\frac{\partial}{\partial x}\left[(H + z)\frac{\partial z}{\partial x}\right] + \frac{\partial}{\partial y}\left[(H + z)\frac{\partial z}{\partial y}\right]\right\}. \qquad (11)^2$$

If now H is chosen so as to represent the major part of the thickness of the water-saturated zone, *i.e.*, so that $z \ll H$, Eq. (11) may be approximated by

$$\frac{f}{\bar{k}H}\frac{\partial z}{\partial t} = \frac{\partial^2 z}{\partial x^2} + \frac{\partial^2 z}{\partial y^2}. \qquad (12)$$

This linear equation is identical with that for two-dimensional heat conduction or that for the two-dimensional flow of a compressible liquid through a porous medium [*cf.* Eq. 3.4(6)]. Although solutions to this equation can be derived by the methods to be developed in Chap. X for quite arbitrary boundary and initial conditions for z, a detailed discussion of these solutions hardly seems warranted in view of the questionable character of the Dupuit assumptions underlying the construction of the original Eq. (10). In spite of the practical importance of a

---

[1] *Cf.* Ph. Forchheimer, *op. cit.*, p. 104; V. Felber, *Die Wasserwirtschaft* **25**, 25, 60, 1932; *Wasserkraft u. Wasserwirtschaft*, **26**, 73, 1931; J. Kozeny, *ibid.*, **28**, 102, 1933.

[2] This equation was first proposed by J. Boussinesq, *Jour. math.*, **10**, 14, 1904, and also applies to the case where the impermeable base is not level, so that H varies with (x, y).

knowledge of the variations in ground-water levels in problems of drainage and irrigation, it seems better to await the development of a more satisfactory theory than to present the implications of Eq. (12) which will involve errors of unknown magnitude.

6.18. Sand-model Experiments with Three-dimensional Gravity-flow Systems.—Although it would appear now, in view of the discussion of Sec. 6.17, that Eqs. 6.17(5) and 6.17(9), giving the shape of the free surface and flux in a gravity-radial-flow system, as based upon the Dupuit-Forchheimer theory, could hardly have claimed any degree of validity without a direct empirical or rigorous analytical confirmation, it was not until 1927 that they were seriously questioned. It was then that Kozeny[1] published the first attempt to attack the flow problem by direct potential-theory methods. Thus beginning with Laplace's equation, Eq. 6.1(2), he attempted to synthesize a solution satisfying the boundary conditions of the gravity-flow system by means of elementary solutions of the type used in the problem of the partially penetrating well [*cf.* Eq. 5.3(7)].

Unfortunately, however, the correct boundary conditions were not applied to the problem. Thus the flux through the system was taken as that corresponding to the streamline entering the well at the fluid level in the well. But, as has already been pointed out, there will be a definite discontinuity at the well, so that the free surface of the system will enter the well above the fluid level in the latter, giving rise to a "surface of seepage." Then, too, this solution consisted only of constant terms and a series of Hankel functions, so that the radial velocities at large distances from the well vanished exponentially. Physically, however, it is clear that at points distant from the well surface the radial velocities should asymptotically approach those in a strict two-dimensional radial-flow system. The potential function at such points must, therefore, asymptotically approach a logarithmic variation or contain a logarithmic term explicitly, as does, for example, Eq. 6.20(5) (*vide infra*). Finally, Kozeny's potential function did not possess the characteristic to be required of every rigorous solution of a gravity-flow problem, namely, that the uppermost streamline be a *free-surface* streamline with a potential proportional to its vertical coordinate.

[1] KOZENY, J., *Wasserkraft u. Wasserwirtschaft*, **22**, 120, 1927.

Kozeny's final expression for the total flux Q was

$$Q = 2\pi \bar{k} \zeta h r_w F\left(\zeta, \frac{\pi r_w}{2h}\right),\tag{1}$$

where $\bar{k}$ is the "effective" permeability of the sand, *i.e.*, $k\gamma g/\mu$, h is the sand thickness, r_w the well radius, and $\zeta = 1 - \dfrac{h_w}{h}$, where h_u is the fluid height in the well. $F(\zeta, \pi r_w/2h)$ is a rather complicated function of both arguments, and has to be computed numerically. Although the proportionality of Q to r_w, which Eq. (1) implies when the function F is independent of $\pi r_w/2h$, appears to be one of the important predictions of this theory, it explicitly depends on the assumption that $\pi r_w/2h$ is large, just the opposite of which, however, corresponds to most cases of practical interest. As to the other implications of Eq. (1), one need simply note that, as will be seen presently, the most recent experimental investigation of the problem of gravity flow[1] unambiguously establishes a formula that does not at all correspond to Eq. (1).

While Kozeny's theory, therefore, cannot be considered as giving a satisfactory analytical solution to the problem of radial gravity flow,[2] it did, however, furnish the impetus for the first direct empirical investigations of the problem. Thus to test the prediction of this theory that Q should be a maximum when $\zeta = \frac{1}{2}$, and its variation for values of $\zeta < \frac{1}{2}$ (fluid heights in the well bore higher than half of the sand thickness), and to see what changes would be introduced for $\zeta > \frac{1}{2}$, Kozeny himself made some experimental measurements on an actual radial tank. His experiments do seem to check his theory for values of $\zeta < \frac{1}{2}$. To represent all the data, however, the empirical formula was proposed as

---

[1] WYCKOFF, R. D., H. G. BOTSET, and M. MUSKAT, *Physics* **3**, 90, 1932.

[2] Mention should also be made of Kozeny's later work (*Wasserkraft u. Wasserwirtschaft*, **28**, 88, 1933) in which a formula is obtained for the pressure distribution at the base of a radial-flow system reproducing quite closely that found in the experiments of Wyckoff, Botset, and Muskat (*vide infra*). The analysis, however, which is developed by analogy with the complex variable theory of the parabolic linear gravity flow (*cf.* Sec. 6.7), appears to be difficult to justify, as the resultant potential distributions do not satisfy Laplace's equation.

$$Q = Q_{max}\left[1 - \left(\frac{h_w}{h}\right)^2\right]^{3/4}, \qquad (2)$$

which evidently has little physical significance.

A more complete test of Kozeny's theory was carried through by Ehrenberger.[1] Although they were beautifully performed and exhaustive in many respects, they still do not seem to give a satisfactory answer to the problem. For the essential result is again an empirical correction factor to be applied to Kozeny's formula. Since this correction factor also lacks physical significance, its extension to other dimensions of the flow apparatus or to other geometrical forms of the flow system would have to be determined by new experiments. Furthermore, Ehrenberger's experiments are difficult of interpretation, since in most cases the fluid was pumped from the well with a tube just underneath the fluid level, thus probably disturbing the flow in the neighborhood of the well.

The next attempts to derive a more satisfactory answer to the problem of gravity flow were two investigations by Wenzel,[2] and Wyckoff, Botset, and Muskat,[3] which were carried out almost simultaneously. Wenzel was interested in the problem of inverting Eq. 6.17(9) and using it to determine the permeability of a water-bearing sand, rather than in the general question of gravity flow. Thus Wenzel measured the drawdowns in some 80 observation wells lined up radially about two wells that were being pumped at a known rate. He then computed the permeability from pairs of observations for wells at distances r_1 and r_2 from the pumping well by a formula essentially equivalent to an inversion of Dupuit-Forchheimer's Eq. 6.17(9), namely,

$$\bar{k} = \frac{Q \log r_2/r_1}{\pi(h_2{}^2 - h_1{}^2)}, \qquad (3)$$

where h_1 and h_2 are the fluid heights at the two points r_1, r_2, and may be found by subtracting the drawdowns from the undisturbed water level in the sand.

[1] EHRENBERGER, R., *Zeits. Oster. Ing. Arch. Ver.*, nos. 9/10, 11/12, 13/14, 1928.

[2] WENZEL, L. K., *Trans. Amer. Geophys. Union*, p. 313, 1932, published by the National Research Council.

[3] WYCKOFF, BOTSET, and MUSKAT, *loc. cit.*

As the result of these computations Wenzel found that the permeability $\bar{k}$ apparently increased uniformly as the distance r_2 of the second observation well increased, and concluded that at least under the conditions of his experiments Eq. (3) must need some correction. The failure of Eq. (3) to give a constant value of $\bar{k}$ was, however, attributed to the lack of equilibrium in the flow system rather than to a fundamental error in Eq. (3). Thus, since there will evidently be a time lag in the establishment of equilibrium in the free surface at distant points from the pumping well, after the pumping at the central well is begun, this will, according to Wenzel, give an apparent increase in the value of $\bar{k}$ when computed from observations distant from the well.

Although in field measurements, as those of Wenzel, one must certainly take care that equilibrium has been reached in the system, a strict constancy in $\bar{k}$ could not, of course, have been expected even after the attainment of complete equilibrium unless the basic Dupuit formula, Eq. 6.17(5), were rigorously valid. That this, however, is not the case is clearly shown by the results of Wyckoff, Botset, and Muskat.[1] These investigators, in contrast to Wenzel, returned to the laboratory methods of Kozeny and Ehrenberger. A diagram of their apparatus is given in Fig. 140.

The essentially new features introduced in this work were the manometers connected to the bottom of the sand tank with which to measure the radial pressure distribution along the bed of the system, and the continuous circulation of the fluid in the

[1] In fact, it would seem that the transient effect mentioned by Wenzel would rather cause an apparent decrease in $\bar{k}$ with increasing r_2, assuming the correctness of Eq. (3) for a steady-state condition of flow. For if the distant points r_2 lag in reaching their equilibrium drawdowns, the corresponding values of $h_2{}^2 - h_1{}^2$ would be too large, and hence $\bar{k}$ will be found to be smaller than for small values of r_2 where the lag will be inappreciable. Furthermore, it should be noted that while in the complete report of these investigations—U. S. Geol. Surv. Water-supply Paper, no. 679-A, 1936—Wenzel shows that the averages of the apparent values of the permeability over opposite radii from the pumping well, which eliminate at least partially the effect of the normal ground-water gradient in the water-bearing stratum, show a much more satisfactory constancy, it is nevertheless difficult to evaluate the significance of such a result in view of the fact that the penetration of the pumping well was only about 40 per cent. This circumstance alone should invalidate Eq. (3), even if the Dupuit assumptions were to be taken for granted.

tank so that perfectly steady states could be maintained without difficulty. Careful precautions were also taken to pack the sand so it would be not only of uniform permeability but free throughout of trapped air as well. For convenience a 15-deg. sector of a complete radial-flow system was used. To study the strict gravity-flow problem, the tank, filled with fine sand, was fitted with glass sides. When in later experiments the effect of added pressure drives was studied, an all-metal (6.3-mm. brass) tank was constructed with standpipes at both the input and outlet ends. The fluid flow was measured with a carefully calibrated flow meter connected in the circulation system.

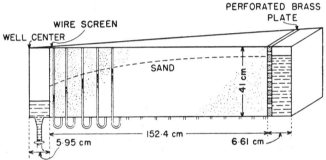

Fig. 140.—Diagram of tank used in the experimental study of radial gravity flow. Distance between each of first ten manometers = 5.08 cm.; distance between each of second nine manometers = 10.16 cm. (*From Physics,* **3,** 93, 1932.)

Although in the experiments with the tank with the glass sides observations were made on the heights of the water in the sand at various distances from the well center, the most quantitative data were the readings in the manometer tubes, which evidently gave the fluid *heads*[1] in the system, as measured at the bottom of the sand. The analysis of these data showed that within experimental errors they satisfied the relation

$$\bar{h}^2 - h_w{}^2 = \frac{h_e{}^2 - h_w{}^2}{\log r_e/r_w} \log \frac{r}{r_w}, \tag{4}$$

where $\bar{h}$ is the manometer reading or fluid *head* at the distance r, and h_e and h_w readings at r_e and r_w, the latter being conveniently chosen as the well radii. The character of the agreement of a

[1] These are frequently called "piezometric heights" in the hydrological literature (*cf.* Sec. 6.14).

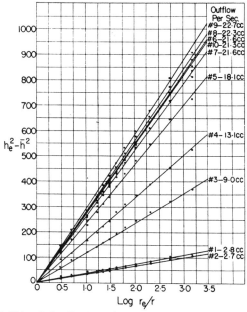

FIG. 141.—Fluid-head distributions, $\bar{h}(r)$, at the base of the radial gravity-flow model of Fig. 140. h_e = fluid head at $r = r_e$. (*From Physics*, **3**, 99, 1932.)

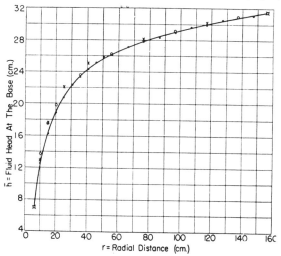

FIG. 142.—The fluid-head distribution at the base of a radial gravity-flow system. Dots: observed data; circles: calculated by Eq. 6.18(4); x's: calculated by Eq. 6.20(5); r_w = 6.4 cm.; r_e = 156 cm.; h_w = 7.0 cm.; h_e = 31.5 cm (*From Amer. Geoph. Union Trans.*, p. 394, 1936.)

typical set of empirical data with Eq. (4) is shown in Fig. 141. It will be seen that in spite of deviations of individual points, the data as a whole may be considered as establishing Eq. (4) beyond any reasonable question.

A Cartesian plot of the manometer readings for one of the experiments is shown in Fig. 142. The dots give the observed data, the circles those computed by Eq. (4), and the x's those calculated from the approximation theory given in Sec. 6.20 [cf. Eq. 6.20(5)].

To determine the relation between the total flow Q through the system and the boundary conditions, the permeability of the sand was measured in place,[1] making use of the fact that when the fluid level in the well is above the sand height the flow assumes a two-dimensional radial character, for which the pressure distribution and flux Q were already known. For by inverting Eq. 4.2(10), it is seen at once that $\bar{k}$ may be computed by the equation

$$\bar{k} = \frac{Q \log r_e/r_w}{2\pi(h_e - h_w)h}, \quad (5)$$

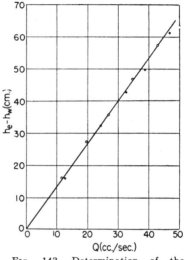

FIG. 143.—Determination of the sand permeability in the gravity-flow experiments by measurements with strictly radial-flow conditions. $h_e - h_w$ = fluid-head differential; Q = flow rate through the sector. (From Physics, **3**, 101, 1932.)

where Q is the flow through the system when the fluid heads at r_e and r_w are h_e and h_w, h being the sand thickness. Hence on plotting Q versus $h_e - h_w$, $\bar{k}$ is given by the slope of the resulting straight line, after taking account of the factor $(\log r_e/r_w)/2h\pi$. Such a plot for some actual experiments is shown in Fig. 143.

[1] In these experiments, as well as in all those in which an attempt was made to relate the flux Q to the physical and geometrical constants of the system the tank of Fig. 140 was replaced by an all-metal pressure-tight 15-deg. sector, as the effective permeability of the sands in the open tank of Fig. 140 were found to differ widely from those computed from packed-tube experiments.

A more difficult matter was the question as to how to take care of the capillary layer that necessarily overlies the main fluid body in the sand. Although one might at first suppose that the capillary layer would take no part in the flow, a little consideration shows that on the contrary it must behave much as a siphon acting in the direction of the main flow and thus augmenting the latter. Diagrammatically, the composite system of the capillary layer and main fluid body may be represented as in Fig. 144. It will be seen that both the top and bottom of the capillary layer are essentially free surfaces at atmospheric pressure. The fluid *heads* at the sand bottom are indicated by $\bar{h}$, and the fluid heights

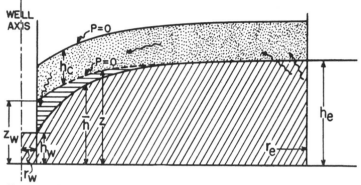

Fig. 144.—Diagrammatic representation of the effect of the capillary layer on a gravity-flow system. $\bar{h}$ = fluid head at base; z = fluid height in the sand below the capillary layer. (*From Physics*, **3**, 106, 1932.)

in the sand by z. h_c represents the thickness of the capillary layer and is approximately equal to the capillary rise of the liquid in the particular sand through which it is flowing.

The result of the analysis, the further details of which may be found in the original paper, is that when the fluid heights in the well and especially in the input reservoir are close to the top of the sand, so that the capillary layer is not completely developed, the flow in the capillary layer may be neglected, and the strict gravity flow may be accurately represented by the equation

$$Q_g = \frac{\pi \bar{k}(h_e{}^2 - h_w{}^2)}{\log r_e/r_w},\tag{6}$$

where the notation is the same as used in Eq. (5). On the other hand, when both the fluid heights in the well and input reservoir are below the top of the sand by at least the capillary rise of the fluid h_c, the flow in the capillary layer can no longer be neglected, and the combined flow may be quite closely represented by the equation

$$Q_{g+c} = \frac{\pi \bar{k}(h_e - h_w)(h_e + h_w + h_c)}{\log r_e/r_w}. \tag{7}$$

The necessity for using Eq. (7) rather than Eq. (6) for relatively small inflow heads is shown in the following table of data taken in runs where the capillary layer was completely developed. h_c for these experiments was 9.5 cm.

TABLE 15.—COMPARISON OF OBSERVED AND CALCULATED FLUXES
THROUGH A RADIAL GRAVITY-FLOW SYSTEM

Run no.	h_e	h_w	h	Calc. Q cc./sec., Eq. (6)	Calc. Q cc./sec., Eq. (7)	Obs. Q
1	23.67	18.07	29.0	3.06	3.76	3.90
2	15.37	0.17	29.0	3.09	4.98	4.95
3	14.37	0.17	29.0	2.70	4.47	4.55
4	9.57	0.17	29.0	1.18	2.37	2.50

When $h_c = 0$, or when it is negligible as compared to $h_e + h_w$, Eq. (7) evidently reduces to Eq. (6), as indeed it should.

Equation (6), which had originally been derived on the basis of the erroneous Dupuit assumptions, and which will be deduced also by means of a physically reasonable approximation method (*cf.* Sec. 6.20), is thus established empirically, except for the perturbation due to the capillary layer. Since in practical flow systems the sand thickness will usually be considerably greater than the height of capillary rise in the sand, the simpler formula of Eq. (6) will suffice in most cases.

With respect to the shape of the free surfaces, however, these empirical investigations have failed to give any simple analytic representation, although numerous observations on the free-surface heights were made. A typical set of streamlines, including those in the capillary zone, traced by ink streamers injected along the inflow face are reproduced in Fig. 145. While the Dupuit-Forchheimer theory explicitly gives an equation for the free surface, its failure to take into account the surface of seepage at the well should alone suffice to invalidate its implications with respect to the shape of the free surface. This indeed follows also from the empirical fact that since, as has been seen above, the fluid-head distribution at the base may be formally represented by identically the same expression, Eq. (4), as the Dupuit-Forchheimer formula for the free surface, the validity of the latter would require a coincidence between the piezometric heights at the base and the free surface heights, which the experiments conclusively show to be not even approximately true.

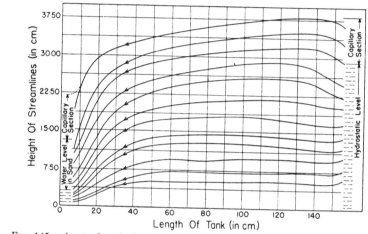

Fig. 145.—A set of typical streamlines observed in the experiments on gravity flow. (*From Physics*, **3**, 98, 1932.)

As to the Dupuit assumption of shell flow in the distant parts of the radial-flow system, however, an indirect confirmation may be drawn from the empirical establishment of Eqs. (4) and (6). For a little inspection will show that the latter is simply equal to the velocity at the base corresponding to Eq. (4) multiplied by the inflow head h_e. This, in turn, implies the uniformity of the velocity along the inflow face as is required by the "shell-flow" hypothesis. Furthermore, it will be seen below that the approximation theory of Sec. 6.20 also leads to a practically uniform inflow velocity if the external boundary radius is appreciably greater than the sand thickness, as is readily deduced from Eq. 6.20(5). On the other hand, it should be noted that the value of this average velocity is not that equivalent to the free-surface hydraulic gradient—which should theoretically be zero at the external boundary—but is rather given by the gradient of the fluid head at the base of the system.

6.19. Composite Pressure-head and Gravity-flow Systems.— Not only were the pressure and flow relations governing a radial gravity-flow system established in the investigation described in the last section, but these results were extended to systems in which radial drives were superposed upon the simple gravity flow. These correspond to systems in which the fluid levels or fluid heads at the input ends of the systems are greater than the sand height. The fluid heads at the wells must, of course, still

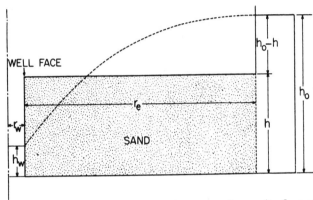

Fig. 146.—Diagrammatic representation of a composite gravity-flow system. (*From Physics*, **3**, 103, 1932.)

be less than the sand height, or the flow will be strictly two-dimensional and radial, with no gravity component.

A little consideration will show that to a first approximation a composite flow system as defined above may be treated analytically as a strict radial flow simply superposed on a gravity flow. Thus, as is diagrammatically indicated in Fig. 146, if h is the sand height, h_0 the fluid head at the inflow surface, and h_w the fluid head in the well, the radial component of the flux will be given by [*cf.* Eq. 4.2(10)]

$$Q_r = \frac{2\pi \bar{k} h(h_0 - h)}{\log r_e/r_w}. \tag{1}$$

The gravity-flow component will be, by Eq. 6.18(6),

$$Q_g = \frac{\pi \bar{k}(h^2 - h_w^2)}{\log r_e/r_w},$$

so that the resultant and sum is

$$Q_{gr} = \frac{\pi \bar{k}(2hh_0 - h^2 - h_w{}^2)}{\log r_e/r_w}. \tag{2}$$

Since the capillary layer cannot be completely developed when the fluid head at the input face exceeds the sand height, its effect will usually be negligible. Table 16 shows a typical set of data testing Eq. (2), with

$$h = 29.0 \text{ cm.}; \quad r_e = 39 \text{ cm.}; \quad r_w = 3 \text{ cm.}; \quad \frac{\pi \bar{k}}{24} = 0.0335.$$

The data were taken with an all-metal pressure-tight radial 15-deg. sector, at the bottom of which were distributed the manometer tubes to measure the values of $\bar{h}$, similar to the scheme of Fig. 140. It will be observed that the agreement here is even considerably better than in the case of the strictly gravity-flow tests.

TABLE 16.—COMPARISON OF OBSERVED AND CALCULATED FLUXES THROUGH A COMPOSITE PRESSURE-HEAD AND GRAVITY-FLOW RADIAL SYSTEM

h_0	h_w	Calculated Q, cc./sec.	Observed Q, cc./sec.
89.47	27.47	46.97	46.1
78.38	27.47	38.56	38.4
59.67	27.47	24.38	23.7
39.37	27.47	8.99	8.6
70.87	22.57	36.07	36.1
47.57	22.57	18.41	18.5
78.37	17.17	44.56	44.2
58.47	17.17	29.48	29.0
44.97	17.17	19.25	19.3
74.37	12.57	43.32	43.5
59.57	12.57	32.09	31.5
48.17	12.57	23.46	23.9
68.17	6.97	40.05	40.8
50.87	6.97	26.94	27.0
35.55	6.97	15.32	16.4
61.47	0.17	35.61	36.1
56.47	0.17	31.82	32.1
48.17	0.17	25.52	26.7

Carrying out a similar analysis of the fluid-head distribution on the composite-flow system, it readily follows that the resultant fluid head at the radius r at the base should be given by

$$\bar{h} = \frac{h_0 - h}{\log r_e/r_w} \log \frac{r}{r_w} + \sqrt{h^2 - \frac{(h^2 - h_w{}^2)}{\log r_e/r_w} \log \frac{r_e}{r}}. \qquad (3)$$

Table 17 shows an experimental test of this equation,

TABLE 17.—COMPARISON OF OBSERVED AND CALCULATED FLUID-HEAD DISTRIBUTION AT THE BASE OF A COMPOSITE PRESSURE-HEAD AND GRAVITY-FLOW RADIAL SYSTEM

r, cm.	$\bar{h}$ (calculated), cm.	$\bar{h}$ (observed), cm.	r, cm.	$\bar{h}$ (calculated), cm.	$\bar{h}$ (observed), cm.
6.4	0	0	55.8	39.9	40.0
10.2	16.4	15.7	66.0	41.8	41.7
15.3	23.5	22.8	76.2	43.0	43.1
20.3	27.8	27.4	86.4	44.4	44.3
25.4	30.6	30.6	96.5	45.6	45.4
30.5	33.0	33.2	106.7	46.7	46.5
35.6	34.9	35.1	116.8	47.6	47.4
40.6	36.5	36.8	127.0	48.3	48.5
45.7	37.6	38.0	137.2	49.1	49.3
50.8	39.0	38.9	147.3	49.7	50.0
			156.0	50.4	50.4

for the boundary conditions

$h_0 = 50.4$ cm.; $\qquad h = 40.5$ cm.; $\qquad h_w = 0$;

$r_w = 6.4$ cm.; $\qquad r_e = 156.0$ cm.; $\qquad Q = 50.0$ cc./sec.

Here again the agreement is even better than in the case of the simple gravity flow. One can therefore consider both equations as satisfactorily established.

6.20. An Approximate Potential Theory of the Flux through Gravity-flow Systems.—It has been pointed out that owing to the questionable character of the assumptions underlying the Dupuit-Forchheimer theory its success in leading to formulas for the flux through linear and radial gravity-flow systems— Eqs. 6.17(8), (9)—which give remarkably close approximations of the values of the flux found by rigorous analytical or direct empirical methods must be considered as largely fortuitous. These assumptions have been shown to be fallacious on the basis of both general considerations and specific calculations such as

those of Sec. 6.5.   To resolve the paradoxical position in which
the flux formulas are thus placed, we shall now briefly outline a
theory which also leads to these formulas, but which is free of
the Dupuit assumptions, involving only approximations which
may be a priori expected to cause but slight errors in the resulting
calculations of the flux.

This theory[1] is based upon a replacement of the actual gravity-
flow system by one with the same boundary conditions, but without the free surface.

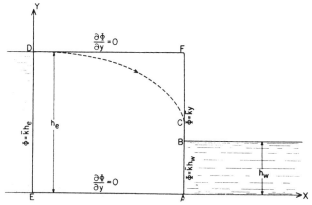

Fig. 147.—Diagrammatic representation of the approximation system for
calculating the flux through a dam with vertical faces. (*From Amer. Geoph.
Union Trans.*, p. 393, 1936.)

Thus, in the case of linear flow, with inflow head h_e and outflow
head h_w, the approximation system replacing the physical gravity-
flow system $ABCDE$ will be $AFDE$, with the same boundary
conditions along AF and ED as in $ABCDE$, as shown in Fig.
147.   The boundaries EA and DF will be taken as impermeable.
Recalling the discussion of the electrical model of gravity-flow
systems given in Sec. 6.6, it will be seen that the approximation
system $AFDE$ is exactly equivalent to the electrical model of
the gravity-flow system before the corner FCD has been cut away
to simulate the free surface CD.

As the potential along CF rises continuously to equal, at F,
the value along ED, it is clear that the total flux along CF must

[1] *Cf.* Muskat, M., *Physics* **6**, 402, 1935, and *Trans. Amer. Geophys.
Union*, p. 391, 1936.

be very small.[1] The region DCF will, therefore, contribute but little additional flux beyond that passing through the physical gravity-flow system $ABCDE$. Hence, although the flux through $AFDE$ will necessarily exceed that through $ABCDE$—since $AFDE$ has a greater permeable cross section than $ABCDE$—it is reasonable to expect that this excess will be of no practical significance, so that the former will give a very close approximation to the latter. That this is indeed so is confirmed by the direct electrical measurements[2] of the conductivity of the model of the approximation system $AFDE$ and of that of $ABCDE$, corresponding to case I of Sec. 6.5, which show that the excess of the former over that of the latter is no greater than $\frac{1}{4}$ per cent. This fact is further verified by a comparison of the flux theoretically implied by this approximation theory and that calculated in Sec. 6.5 by the rigorous method of hodographs.

The theoretical value of the flux through the approximation system $AFDE$ is in turn readily found after determining the internal potential distribution appropriate to the boundary conditions, which are, namely,

$$y = 0, \quad h_e: \quad \frac{\partial \Phi}{\partial y} = 0; \quad x = 0, \quad 0 \leqslant y \leqslant h_e: \quad \Phi = \bar{k}h_r, \left.\right\}$$
$$x = L: \quad \begin{cases} \Phi = \bar{k}h_u: & 0 \leqslant y \leqslant h_w \\ = \bar{k}y: & h_w \leqslant y \leqslant h_e, \end{cases} \left.\right\} \quad (1)$$

where $\bar{k} = k\gamma g/\mu$ is the "effective" permeability of the sand-water system.

Making use of the theory of the Fourier series,[3] it is readily found that the potential function satisfying these boundary conditions may be expressed as

$$\Phi = \bar{k}h_e - \frac{\bar{k}}{2h_eL}(h_e^2 - h_w^2)x +$$
$$\frac{2\bar{k}h_e}{\pi^2}\sum_1^\infty \frac{[(-1)^n - \cos n\pi h_u/h_e]}{n^2 \sinh n\pi L/h_e} \cos \frac{n\pi y}{h_e} \sinh \frac{n\pi x}{h_e}. \quad (2)$$

---

[1] In fact, as will be seen below, the flux density near F is even negative, since the potential along FD, for example, decreases in leaving F.

[2] *Cf.* WYCKOFF, R. D., and D. W. REED, *Physics* **6**, 395, 1935.

[3] *Cf.* Sec. 4.3.

The associated flux is evidently

$$Q = -\int_0^{h_e} \frac{\partial \Phi}{\partial x} dy = \frac{\bar{k}(h_e{}^2 - h_w{}^2)}{2L}, \tag{3}$$

which is identical with Eq. 6.17(8), also implied by the Dupuit-Forchheimer theory, and gives the values $\bar{Q}/\bar{k}^2$ of the last row in Table 14. Recalling again the comparison between these values of $\bar{Q}/\bar{k}^2$ and those, of the adjacent row, given by the rigorous theory, it will be seen that within the errors of the calculations (1 per cent) the two rows agree exactly, thus proving that the excess of the flux in the approximation system over that in the physical flow system cannot exceed the order of magnitude of these errors.

That the success of the prediction of the flux by this approximation theory is not also fortuitous is proved by the fact that it also gives a good reproduction of the pressure or potential distribution along the base of the gravity-flow system EA. Thus setting $y = 0$ in Eq. (2), and using the values of h_e, h_w, and L, corresponding to the case I of Sec. 6.5, the resultant pressure distribution along EA is shown by the circles in Fig. 103. The agreement with the corresponding solid curve calculated by the exact theory of Sec. 6.5 is evidently as close as may be expected from an approximate theory, and indeed shows that the latter really does give a physically satisfactory approximate representation of the linear gravity-flow system in those features not directly related to the free surface itself. In fact, the velocity distribution along the inflow face ED, as implied by Eq. (2), and which is represented for the same case by the corresponding circles of Fig. 103, is also a fair approximation to that in the actual flow system.[1] Finally, even the velocity distribution along the outflow face AF represents an approximate reproduction of the correct distribution except in the region about and above the free surface, as shown again by the circles in Fig. 103. The inversion in the sign of the velocity near the top of the outflow face is, of course, due to the fact that the approximation theory implies that there is an actual fluid source available near the top of the outflow face to maintain the high potential there.

---

[1] The equivalent implication of the Dupuit-Forchheimer theory would be a straight line parallel to ED (in Fig. 103) at a distance $u/\bar{k} = 0.925$.

A similar approximation system can be constructed for the problem of radial gravity flow. Thus taking again for the boundary conditions those appropriate to the system before the introduction of the free-surface boundary, $i.e.$,

$$y = 0, \quad h_e: \frac{\partial \Phi'}{\partial y} = 0; \quad r = r_e, \quad 0 \leqslant y \leqslant h_e: \quad \Phi = \bar{k}h_e,$$

$$r = r_w \begin{cases} \Phi = \bar{k}h_w: & 0 \leqslant y \leqslant h_w \\ = \bar{k}y: & h_w \leqslant y \leqslant h_e, \end{cases} \tag{4}$$

where $r = r_w$, r_e are the bounding radii, the internal potential distribution is found to be

$$\Phi = \bar{k}h_e + \frac{\bar{k}(h_e^2 - h_w^2) \log r/r_e}{2h_e \log r_e/r_w} +$$

$$\frac{2\bar{k}h_e}{\pi^2} \sum_1^\infty \frac{\left\{ (-1)^n - \cos \dfrac{n\pi h_w}{h_e} \right\} U(\alpha_n r) \cos \dfrac{n\pi y}{h_e}}{n^2 U(\alpha_n r_w)}, \tag{5}[1]$$

where

$$U(\alpha_n r) = I_0\left(\frac{n\pi r}{h_e}\right) K_0\left(\frac{n\pi r_e}{h_e}\right) - I_0\left(\frac{n\pi r_e}{h_e}\right) K_0\left(\frac{n\pi r}{h_e}\right), \tag{6}$$

I_0, K_0 being the Bessel functions[2] of the third kind and of zero order.

Here, too, the flux is given by the Dupuit-Forchheimer formula, since

$$Q = 2\pi r \int_0^{h_e} \frac{\partial \Phi}{\partial r} dy = \frac{\pi \bar{k}(h_e^2 - h_w^2)}{\log r_e/r_w}, \tag{7}$$

the series terms again vanishing on integration. And as has been seen in Sec. 6.18, this formula represents the empirically observed values of the flux through radial gravity-flow systems with an accuracy sufficient for all practical purposes. Again the rigorous derivation of this formula for the approximation system defined by Eq. (4) shows that it should give a flux somewhat

[1] For purposes of practical computations, the term $\dfrac{U(\alpha_n r)}{U(\alpha_n r_w)}$ may be approximated by $\dfrac{K_0(n\pi r/h_e)}{K_0(n\pi r_w/h_e)}$, for $r_e \geqslant h_e$.

[2] $Cf.$ Chap. XVII, of "Modern Analysis," by E. T. Whittaker and G. N. Watson.

in excess of that for the physical gravity-flow system, although the discrepancy might well be a priori expected to be no larger than the errors of computation of the flux from the empirical data.

The potential or pressure distribution at the base of a particular radial-flow system corresponding to one of the sand-model experiments described in Sec. 6.18 is indicated by the x's in Fig. 142. While the agreement with the empirical distributions —the solid curve and dots—is not as close here as that found between the approximate and rigorously calculated distributions for the case of the linear flow, the discrepancies are still of limited magnitude. In fact, the better agreement in the case of linear flow is to be expected when it is observed that the decrease of the thickness of the actual flow channel near the outflow surface, owing to the formation of the free surface in the physical gravity-flow system, will more seriously affect the details of the flow in a radial system than in one of linear flow, as a consequence of the higher concentration of the pressure gradients in the former and hence greater sensitivity of the region near the outflow surface to changes in its geometry.[1] On the other hand, the approximation of Eq. (5) to the correct pressure distribution at the base can certainly be considered to be sufficiently close to show that its implication of the still more accurate formula for the flux, Eq. (7), is not simply fortuitous, as it undoubtedly is with respect to the Dupuit-Forchheimer theory.

A still more simple and yet physically satisfactory method of deriving Eqs. (3) and (7) may be based on the observation that the use of the above approximation theory in deriving the Eqs. (3) and (7) for the flux in linear and radial gravity-flow systems is essentially equivalent to a generalization of a theorem derived in Sec. 4.5 for the flux through a plane radial system with arbitrary pressure distributions over the circular boundaries. For there, it will be recalled, it was shown that the flux was given by the same formula as for strict radial flow [Eq. 4.2(10)], provided only that one used for the boundary pressures the averages of the actual boundary distribution [cf. Eq. 4.5(12)]. Equations (3) and (7) can be derived by the same procedure. Thus from

---

[1] This effect is quite similar to the greater sensitivity of the flux in radial systems to the sand permeability immediately about the outflow surface than in the linear systems (cf. Secs. 7.3, 8.3, and 8.4).

either Eqs. (1) or (4) it readily follows that the average value of the potential over the outflow surface, of height h_e, subjected to the constant potential $\bar{k}h_w$ to the height h_w, is given by

$$\bar{\Phi}_w = \frac{\bar{k}(h_e{}^2 + h_w{}^2)}{2h_e}. \tag{8}$$

As the potential over the inflow face is ch_e, the difference of the average potentials is

$$\Delta\bar{\Phi} = \frac{\bar{k}(h_e{}^2 - h_w{}^2)}{2h_e}. \tag{9}$$

Applying this now to the formula for the flux in a strict linear-flow system [*cf.* Eq. 2.5(4)], Eq. (3) is obtained at once. And Eq. (7) follows from an application of Eq. (9) to the expression for the flux in a strictly radial system [*cf.* Eq. 4.2(10)]. Similarly, one may treat the type of composite gravity-flow systems discussed in Sec. 6.19. For on taking the external boundary potential to be $\bar{k}h_0$, where h_0 is greater than the sand thickness $h = h_e$, the average potential drop becomes, in place of Eq. (9),

$$\Delta\bar{\Phi} = \frac{\bar{k}(2hh_0 - h^2 - h_w{}^2)}{2h}. \tag{10}$$

Applying this also to the formula for the flux in a strictly radial-flow system, Eq. 6.19(2), which has been established empirically, follows immediately.[1]

Another useful application of this method may be made to the gravity flow into partially penetrating wells. Here again, regardless of the details of the potential distribution over the well and external boundary surface, the production capacity of a well of fractional penetration $\bar{h}$ in a sand of total thickness h may be expressed as

$$Q = \frac{2\pi h\bar{h}\Delta\bar{\Phi}}{C \log r_e/r_w}, \tag{11}$$

where C is the geometrical factor taking into account the partial penetration of the well, as derived in Sec. 5.4, and expressed by the curves of Fig. 17, and $\Delta\bar{\Phi}$ is the average potential difference

---

[1] This result can, of course, also be obtained by the above more formal approximation theory by simply substituting in Eq. (4), $\Phi = \bar{k}h_0$ for the boundary condition at $r = r_e$ in place of $\Phi = \bar{k}h$ [h_e in the notation of Eq. (4)].

between the well and external boundary surfaces. Now if h_w ($\leqslant h\bar{h}$) is the fluid head in the well above its bottom, and the potential is taken such that it would be zero at the bottom of the well if h_w were zero, the average potential over the well surface will be $\bar{\Phi}_w = \dfrac{\bar{k}(h^2\bar{h}^2 + h_w{}^2)}{2h\bar{h}}$. The uniform and average potential over the external boundary is evidently $\bar{\Phi}_e = \bar{k}h\bar{h}$. $\Delta\bar{\Phi}$ is therefore given by

$$\Delta\bar{\Phi} = \frac{\bar{k}(h^2\bar{h}^2 - h_w{}^2)}{2h\bar{h}}. \tag{12}$$

Hence the production capacity of the well will be

$$Q = \frac{\pi\bar{k}(h^2\bar{h}^2 - h_w{}^2)}{C \log r_e/r_w}. \tag{13}$$

Comparison of this expression with Eq. (7) shows that the effect of partial penetration in a well flowing by a gravity drive can be taken care of by exactly the same correction factor—Fig. 17—as takes care of this effect for normal pressure-drive wells. And in a similar manner it may be shown that Eq. 6.19(2) combined with the same correction factors C will give the production capacities of partially penetrating wells producing under the composite action of gravity and pressure drives.

Finally it may be noted that the approximation methods of this section can be applied to gravity-flow systems of more complex geometry than those specifically discussed above. Thus, for example, the production capacities of the various members of groups of artesian wells producing by simple or composite gravity flow may be obtained directly from those to be derived in Chap. IX for wells producing by pressure drives by simply replacing the applied pressure differentials in those formulas by the values of $\gamma g\Delta\bar{\Phi}/\bar{k}$ given in Eq. (9) or (10). Or, the production capacity of a well producing by gravity flow near an extended line source as a river or canal can be similarly obtained from Eq. 4.7(8) by the same substitution. In fact, this procedure should give close approximations to the correct value of the flux in any gravity-flow system in which the free surface does not drop so sharply in leaving the inflow surface or external boundary that the volume of the corresponding approximation system would greatly exceed that of the physical gravity-flow system.[1]

6.21. Summary.—When the average vertical height at which a liquid leaves a porous medium is smaller than that at which it enters the medium, the force of gravity will play the role of a driving agent inducing the flow into and through the outflow surfaces. This force is really the gradient of the potential energy of the liquid as it moves from the higher inflow to the lower outflow levels. As the liquid moves toward the outflow surface it uses up its potential energy in overcoming the resistance of the porous medium. Furthermore, in its tendency to fall to the lower levels, it will in general break away from the uppermost boundaries of the porous medium and give rise to a "free surface" which will form a natural boundary to the liquid-saturated zone—capillarity effects being neglected—above which the medium will be dry and free of liquid. It is the presence of such free surfaces that characterizes all gravity-flow systems, in contrast to those subjected to external fluid pressures in which the liquid is forced to fill out the whole of the porous medium.

As the free surface represents a boundary of the flow system across which no flow takes place, it must clearly be a limiting streamline of the system. Moreover, as the region above it is free of liquid, the pressure over it is uniform, so that the potential at each point of the free surface will vary linearly with its height above any arbitrary horizontal plane. While in nongravity-flow systems the potential variations over the streamline boundaries are a priori unknown, and must be found from the solution for the potential distribution in the interior, the additional knowledge of the variation of the potential over a free surface may be considered as compensating for the fact that the *shape* of the free surface is not known a priori. From an analytical point of view, however, the problem of determining the shape of the free surface, even though the potential variation over it is known, is much more difficult of solution than that of finding the potential variation when the shapes of all the boundary elements are preassigned. It is indeed this feature of the

---

[1] While the same resultant formulas have already been obtained in a number of cases by using the Dupuit-Forchheimer theory ("Hydraulik," Chap. III), great caution must be used in drawing conclusions from the associated free surfaces given by that theory.

problem which makes the mathematical treatment of gravity-flow systems particularly difficult.

While the determination of the exact shape of the free surface for even the simplest systems is a very formidable task, a number of its properties may be derived from general considerations. Perhaps the most significant of these is the fact that the free surface will always—excluding only certain exceptions that might arise when the outflow surfaces are inclined to the horizontal by less than 90 deg.—terminate on the outflow surface above the outflow-fluid level. The flow through the part of the outflow surface between this terminus and the outflow-fluid level will be a seepage into a region which is free of both porous medium and liquid. It will, therefore, be exposed to an uniform atmospheric pressure, but will not be a streamline surface. This part of the outflow surface is termed a "surface of seepage," and as the shape of the free surface, including its terminus at the outflow surface, is unknown a priori, so will be the length of the surface of seepage. Although in such cases where the outflow-fluid level is zero the whole of the flux through the gravity-flow system must pass through the surface of seepage, it is only recently that its existence has been recognized. And it has thus far been taken into account in the analytical treatment of gravity-flow systems in only relatively few cases.

The manner in which the free surface must terminate at the surface of seepage can also be deduced without any special analysis. Thus it will become tangential to the outflow surface if the slope of the latter is negative or infinite, and will cut it vertically if the slope is positive (*cf*. Sec. 6.1).

The total velocity along a free surface in a porous medium can never exceed the free-fall velocity due to gravity in that medium, and its horizontal component can never be greater than half of the free-fall velocity. The velocity at the lower terminus of the surface of seepage, however, would always be infinitely high, except for the breakdown of Darcy's law when the critical velocities appropriate to the particular liquid and medium constituting the flow system are exceeded (*cf*. Sec. 2.2).

The first practical gravity-flow problem for which a satisfactory analytical treatment was carried through is that of the drainage of an inclined water sand by a ditch dug at the top of the sand. The only essential approximation of the solution of this problem,

given by Hopf and Trefftz, lies in the neglect of the surface of
seepage at the entry of the upstream free surface into the drainage
ditch. The analysis is based upon a preliminary complex-
variable transformation of the original z-plane representation
of the flow system onto an intermediate plane, where the region
of interest assumes the form of a trapezoidal figure, and in which
all the boundary segments including those for the free surfaces
are uniquely determined, except for that corresponding to the
shape of the ditch. This plane and that giving the equipotential
and streamline-distribution representation of the original flow
system are then both mapped onto a quadrant of an auxiliary
plane, thus giving implicitly the desired relation between the
velocity potential or stream function and the coordinates in the z
plane. The mapping of the trapezoidal figure, however, requires
a choice for the shape of the segment corresponding to the
boundary of the ditch, which in turn implies a unique shape of
the ditch itself. In the practical application of the theory it is
not convenient to preassign the shape of the ditch, and the
simplest procedure is to choose a transformation function and
then at the end determine the shape of the ditch implied by the
choice.

The result of most practical interest of this analysis is the
relation between the net fluid caught by the ditch and the drawdown at its upstream edge. This relation is found to be that
the fraction of the original fluid capacity of the sand that is
caught by the ditch is equal to the ratio of the drawdown of the
free surface at the ditch to the thickness of the undisturbed layer
of water-saturated sand [cf. Eq. 6.2(8)]. This result is to be
contrasted to that expressed by previous formulas which have
also involved the width of the ditch and the depth of the water
in it.

A very powerful though rather difficult method of treating
gravity-flow systems is the method of hodographs. A hodograph
is a representation of a dynamical system in which the coordinates
are the velocity components. Its use in the study of gravity-flow
systems is based on the fact that although the shapes of the free
surfaces are unknown a priori, their hodographs are always segments of circles of radius equal to half of the free-fall velocity
of gravity, with center on the negative half of the vertical
velocity axis, and passing through the origin. The hodograph

of a rectilinear impermable boundary is a straight line—in the hodograph plane—parallel to the boundary and passing through the origin. A rectilinear constant-potential surface, as formed by a stationary body of liquid, has for a hodograph a line through the origin and normal to the boundary. Finally, a surface of seepage is represented in the hodograph plane by a line normal to the surface, if it is rectilinear, and passing through the lower intersection with the vertical velocity axis of the circular hodograph corresponding to the free surface.

In this manner a complete mapping can be effected of the original system onto the hodograph plane, the only unknown elements being the numerical values of the coordinates corresponding to some of the corners of the physical flow system. The difficulty in treating the hodograph, even though its shape is known, is that it contains the circular segment corresponding to the free surface, and the mapping of such figures onto a half plane cannot in general be carried out by means of elementary functions or by such as are given by the Schwarz-Christoffel theorem. However, when the physical flow system is a permeable dam with vertical faces, the hodograph assumes a form (cf. Fig. 99) that can be mapped onto a half plane by means of the modular elliptic function. On the real axis of this plane one can then specify the intermediate potential function giving the sum of the inclinations of the velocity and acceleration vectors along the boundary. From these boundary values the potential function and its conjugate can be determined in the whole of the half plane. The relation between these functions and the velocity components in the flow system [cf. Eqs. 6.4(2), (3)] ultimately leads to integral representations for the internal velocity distributions.

Carrying through such a procedure, one can derive analytically all the properties of interest of the flow system representing the seepage of water through a dam with vertical faces, including the exact shape of the free surface, the velocity distributions along the faces of the dam, the pressure distribution along its base, and finally the seepage flux passing through the dam. While the calculations have thus far been carried through numerically for only six specific cases, they suffice to show all the significant features of the general gravity-flow problem. Thus the free surface is found to leave the inflow face horizontally, approaching the outflow face with a continuously increasing slope, finally terminating there tangentially. The surface of seepage is of pronounced magnitude, its length being of the order of half of the inflow-outflow fluid-head differential. The velocities along the inflow face increase uniformly from zero at the top of the inflow-fluid level to a maximum at the base. Along the outflow face the horizontal velocity is zero at the terminus of the free surface, and increases to infinitely high values in going down to the lower terminus of the surface of seepage. If the outflow-fluid level is nonvanishing, the velocity then decreases uniformly in going down still further along the outflow face until the impermeable base is reached (*cf.* Figs. 103 to 106).

An important practical result of these calculations is that the flux per unit length through such dams with vertical faces is given with remarkable accuracy by the simple formula

$$Q = \frac{k\gamma g(h_e^2 - h_w^2)}{2\mu L}, \quad [\textit{cf.} \text{ Eq. } 6.5(16)]$$

where k is the permeability of the dam, γ, μ the density and viscosity of the liquid, g the acceleration of gravity, h_e, h_w are the fluid heads at the inflow and outflow faces of the dam, and L is the width of the base of the dam. This result is at first glance very surprising, since the above equation had originally been predicted by a theory—the Dupuit-Forchheimer theory—all other features of which are in marked contradiction to those given by the calculations with the rigorous hydrodynamic theory. Thus this theory fails in its prediction of the shape of the free surface, the pressure distribution at the base of the dam, and in the velocity distribution along the inflow face. In fact, one is forced to the conclusion that the implication of the above equation by the Dupuit-Forchheimer theory is entirely fortuitous. On the other hand, the equation itself does possess a physical significance for, as explained below, it can be also derived by a different approximation theory which is free from the Dupuit-Forchheimer assumptions, and involves only such as would be a priori expected to lead to but slight errors in its predictions of the flux.

While the application of the method of hodographs to the treatment of the problems of gravity flow through porous media represents the most important theoretical development yet made in the study of gravity-flow systems, it suffers from the very formidable difficulty of carrying through numerically the various processes involved. And when the geometry of the flow system becomes even slightly more complex than that corresponding to a dam with vertical faces as, for example, in the case of a dam with sloping faces, the analysis becomes still more difficult and necessitates the use of graphical or numerical methods for the derivation of specific results. Fortunately, however, the method of electrical models (Sec. 6.6) can be adapted also to problems of gravity flow—even though there is no direct electrical analogue of the effect of gravity in a flow system—and rigorously equivalent electrical models can, in fact, be constructed for any specific gravity-flow system of almost arbitrary complexity.

The immediate difficulty in the construction of such models is, of course, the lack of an a priori knowledge of the complete boundary of the gravity-flow system to be simulated by the electrical model. In particular, neither the shape of the free surface nor the upper terminus of the surface of seepage—the outflow terminus of the free surface—are given in the formulation of the gravity-flow problem, and must, in fact, be determined in the course of the solution to the problem. However, it is a priori known from the definition of the free surface that the potential along it must vary linearly with the vertical coordinate. If, then, the other boundary conditions are simulated electrically as in the nongravity-flow models (cf. Sec. 4.17), and the upper portion of the model is cut away by trial and error so that the potential along the boundary so formed does vary linearly with the vertical coordinate, this boundary will be the rigorous equivalent of the free surface in the actual flow system. Of course, one must also adjust the potential along part of the outflow face so that it, too, increases linearly with the vertical coordinate, for the pressure along the outflow face above the outflow-fluid level must also be uniform, just as it is over the free surface. However, this is easily accomplished by attaching to the outflow face, above the outflow-fluid level, a highly conducting strip and imposing at its lower and upper terminae the outflow and inflow potentials, respectively.

Once the free surface has been found, the internal potential and streamline distribution can be mapped just as in the case of electrical models of nongravity-flow systems. Likewise the cur-

rent flux through the model can be measured to give the equivalent fluid flux in the corresponding flow system.

The particular advantage of such models, readily made of
cardboard sheets sprayed with colloidal graphite, is their flexibility. Thus, while the analytical theory has so far been carried
through only for the dam with vertical faces, it is just as easy
to study electrically dams with sloping faces (Figs. 110 and 111)
as those with vertical faces (Figs. 108 and 109). Moreover, dam
structures fitted with central impermeable cores or aprons and
piling can be simulated electrically without any difficulty whatever. And even permeability differences between the various
parts of the structure can be taken into account by simply varying
the number of graphite coatings sprayed on the corresponding
parts of the model.

In addition to the direct analytical procedures such as those of
Hopf and Trefftz and the method of hodographs, in which potential functions are constructed and derived so as to give the solution for a preassigned gravity-flow problem, one may use the more
primitive inverse procedure of constructing potential functions
and then fitting them to their corresponding physical-flow system.
The essential point of this procedure is the construction of complex-variable relations between the vector $z = x + iy$ and the
complex potential $\omega = \Phi + i\Psi$, as, $\omega = f(z)$, or, $z = F(\omega)$, in such
a manner that along one of the streamlines $\Psi = $ const. the poten-
tial varies linearly with y, the vertical coordinate. This streamline will then represent the free surface of the corresponding
flow system, and if the latter is of physical significance, the
complex potential ω will also be of physical significance.

Such relations have been constructed for the following cases:
(a) linear sheet flow down a sloping terrain; f is a linear function of
z; (b) systems with parabolic free surfaces; F is a quadratic function of ω; (c) special types of seepage out of ditches; F is the sum of
exponential and linear functions of ω. While the solutions with
the parabolic free surfaces do approximate those for thick dams
with vertical faces, the solutions corresponding to the seepage out
of ditches constructed thus far in this manner give either an
entirely artificial radial type of potential distributions at large
distances from the ditches, or one that asymptotically approaches
a free-fall type of seepage at great depths.

Problems of the seepage of water out of ditches or canals in which the seepage stream is of an essentially free-fall type can also be treated by a more direct procedure (Secs. 6.8, 9). Such systems correspond physically to situations in which the ground-water table either lies at a very great depth or is carried by a gravel or other bed of a permeability so much higher than that of the sediments immediately underlying the base of the ditch or canal that the seepage out of the latter does not suffice to maintain a continuous stream in the former. In such cases the seepage stream itself will not merge with the ground-water table, but will simply be dispersed in the gravel bed as a stream of droplets impinging on the water table. One may treat these systems by a method that is very similar to that of Hopf and Trefftz, the solution being rigorous insofar as the condition that the stream function along the bank of the ditch or canal varies linearly with its horizontal coordinate is strictly satisfied. Assuming that this is so, the analysis can be carried through without difficulty, and ultimately one can compute the shape of the banks which is implied by the assumption. To the extent that the shape so found does correspond with the practical case of interest—and for given depths and top widths of the canal or ditch there is still an arbitrary parameter determining the shape of the banks [cf. Eqs. 6.8(6), 6.9(5)]—the solution giving the seepage flux and shape of the free surfaces may be considered as accurate. In this manner it is found that the flux increases in an approximately linear manner with the width of the ditch at the top of the water surface and, for fixed ratio of width to depth of water in the ditch, it is directly proportional to this depth. Furthermore, it increases as the average slope of the banks increases, for given top width and water depth. Finally, the seepage flux increases with decreasing depth of the highly permeable bed carrying the water table, the rate of increase becoming quite large when this depth is of the order of the depth of the free water within the ditch or canal profile (cf. Fig. 124).

Although the method of the electrical models is capable of giving a complete description of the solution to any specific gravity-flow problem, a new model is required for each case with different geometrical dimensions. For the problem of the seepage of water through dams with sloping faces a complete study would require the construction of different models for series of values of the base width, inflow- and outflow-fluid heads, and inflow- and outflow-face angles. As such a series of studies has not been carried out as yet, it is necessary to resort to approximation methods.

An approximate method of calculating the flux through a dam with sloping faces can be constructed as follows. The complete flow system—the part of the cross section of the dam occupied by the flowing water—is supposed to be separated into three parts, to each of which a different approximate treatment is applied. The first is taken as the triangular wedge-shaped region bounded by the inflow face, base of the dam, and the normal to the base from the uppermost point of the inflow face reached by the liquid. When the inclination of the inflow face is an integral fraction of 90 deg. $(\pi/2n)$, the flow in this part can be rigorously described by complex potentials, if treated as a nongravity-flow system. The free surface near the inflow face is taken as following the upper-most streamline of this first region. The contiguous and main body of the seepage stream is then approximated by a linear seepage with a linearly decreasing section—the free surface dropping at a constant rate. Finally, the region including the outflow face is given a still different treatment, its characteristics being determined by sand-model experiments. Although the systematic procedure of carrying through such an analysis requires that certain details with regard to the junctions of the various regions be taken into account (cf. Sec. 6.10), the exact position of these junctions is ultimately determined by the condition that the fluxes calculated as flowing through each of the three regions must be equal. However, the trial-and-error adjustment thus required involves only very simple algebraic processes.

Some interesting information given by this method, in addition to the value of the seepage flux, concerns the character of the velocity distribution along the inflow face. Thus it is found that except when the inflow face is vertical, the velocity will increase uniformly from zero in rising from the base of the dam, the rate of rise increasing as the slope of the face decreases. Below the surface of seepage at the outflow face the variation will be of a similar character, though the velocity just at the top of the outflow-fluid level will theoretically always be infinite. These regions of high velocity are, of course, those where erosion will take place most rapidly and hence those where it is most important that preventive measures be applied.

The fluxes through dams with sloping faces computed by the above-described method will, of course, not be quantitatively exact owing to the various approximations involved. They should, however, be of the correct order of magnitude, the accuracy being better for lower inflow-fluid heights. This is verified by carrying through the computations for the two cases studied by means of the electrical models.

This method of dividing the whole region of flow into separate parts to which may be applied different types of approximate treatment may also be used in computing the seepage flux out of ditches or canals (*cf.* Sec. 6.11) in which the seepage stream merges directly with the ground-water table. Here the flow system may be divided into two parts: (1) A region immediately surrounding the ditch, in which the streamlines show large curvatures, and extending from its center to a distance equal to half of the sum of the width of the ditch and the fluid level above the impermeable base, and (2) a region of approximately linear flow extending to the extremity of the flow system of interest, where it is supposed that one knows the fluid height in the sand. The flow in the former is approximated analytically by a complex potential function, and the associated fluxes are confirmed by sand-model experiments. That in the latter is taken as approximately linear, its inflow head being so chosen, by trial and error, that the flux through it be equal to that through the region surrounding the ditch. Again, this procedure, though involving several approximations, should give at least the correct orders of magnitude for the seepage fluxes.

Even more approximate methods of analysis must be resorted to in order to determine the essential features of some of the more practical problems of irrigation, drainage, erosion, etc. Thus to find the extent of the lateral drainage into a coarse gravel bed bisected by a canal before it seeps into the underlying beds, it is necessary to neglect completely the free surface formed in the gravel bed. By supposing, on the contrary, that the flow in the latter consists of a linear seepage outward from the canal on which is superposed a downward lateral leakage, it is found that the square of the distance to which this seepage will extend is directly proportional to the pressure at the face of the gravel bed, to its thickness, and to its permeability, and inversely proportional to the permeability of the underlying clay [*cf.* Eq. 6.13(2)].

If the problem of soil erosion be expressed as that of finding the thickness of a soil layer which will carry away all the rainfall impinging on it without any surface flooding, an approximate treatment shows that such a soil layer on a hillside must be wedge-shaped if it is to be free of surface flooding. As should be expected, the angular thickness of this wedge increases with the rainfall intensity but decreases as the soil permeability increases [*cf.* Eq. 6.15(4)].

The essential features of the problem of tile drainage for the prevention of the water logging of the top soil by upward water seepage from underlying gravel can also be deduced if the presence of free surfaces is neglected (*cf.* Sec. 6.16). Except for capillary effects the zone of water logging may be defined as the region which would be completely saturated with liquid if connected to the main channel of flow by a crevice or open hole. The upper surface bounding this zone of water logging is equivalent to what is usually termed the "piezometric surface," and its height is proportional to the velocity potential in the main flow channel. In the region below this piezometric surface, or within the water-logged zone, the fluid pressures will evidently exceed the atmospheric pressure, so that the requirement that an array of tile drains prevent the water logging of the top surface is equivalent to the condition that the fluid pressures at the surface be no greater than atmospheric. Considering the depth of the drains and their radii to be known, as well as the depth of the underlying flow channel supplying the upward seepage, and the fluid-head differential inducing the seepage, it is possible to give an approximate calculation of the drain spacing necessary for the complete removal of the upward flow by the drains without any seeping through and water logging the top soil [*cf.* Eq. 6.16(6)]. As is to be expected, the necessary spacing decreases with increasing driving head, decreasing drain depth or gravel-bed depth, and decreasing drain radius (*cf.* Fig. 139), and if these latter parameters are less than certain limits, the cost of an effective tile-drain array will become prohibitive.

An approximate method of treating gravity-flow systems used for many years and even at present by those concerned with such flow systems is that which may be termed the Dupuit-Forchheimer theory. This theory, founded in 1863 by Dupuit and later extended by Forchheimer, is essentially based upon the assumptions that (1) the liquid in a gravity-flow system moves in cylindrical shells, the horizontal velocity being independent of the depth; and (2) the value of the horizontal velocity is proportional to the slope of the free surface. With these assumptions one may deduce both the shape of the free surface of a gravity-flow system and the flux through it [*cf.* Fig. 6.17 (3)].

Although these assumptions appear to have some plausibility for the purpose of deriving an approximation theory, both general considerations and rigorous calculations nevertheless prove them and their consequences to be in rather violent contradiction to the actual characteristics of gravity-flow systems. Thus it may be readily shown that the uniformity of the horizontal velocity with depth in a system exposed at a vertical boundary to a uniform potential implies the vanishing of all vertical velocities, which, of course, removes the effect of gravity in determining the characteristics of the system. Furthermore, the rigorous calculations for the case of the seepage through dams with vertical faces (*cf.* Sec. 6.5) shows very definitely that the horizontal velocities are far from uniform, and that their average moreover bears no relation to the slope of the free surface, which is always zero at the inflow face. Finally, the shapes of the free surfaces predicted by this Dupuit-Forchheimer theory give but poor approximations to the correct free surfaces (*cf.* Fig. 103), the discrepancies being accentuated by the total neglect in this theory of the surfaces of seepage at the outflow surfaces.

In view of these difficulties, it is clear that the success of the Dupuit-Forchheimer theory in predicting formulas for the flux which for practical purposes reproduce exactly the correct values of the flux through linear and radial gravity-flow systems must be considered as wholly fortuitous. And indeed, the Dupuit-Forchheimer assumptions are not a unique set leading to these remarkably accurate formulas for the flux. For, as will be seen below, a radically different set of assumptions, and such as appear physically to be particularly appropriate for the purpose of computing the flux through a gravity-flow system, also lead to identically the same flux formulas.

In spite of the fundamental importance of the problem of radial gravity flow into a well nothing further than the application of the above-mentioned Dupuit-Forchheimer theory was attempted until 1927. It was then that this theory was first questioned and the problem was attacked by direct potential-theory methods. Although the theoretical investigations were not successful from the point of view of providing a rigorously satisfactory treatment, they were the impetus for experimental studies of the problem. The most recent of these (*cf.* Sec. 6.18), performed with sand models of the actual flow system, has given the following results: The free surfaces do not follow those predicted by the Dupuit-Forchheimer theory. In particular, they terminate not at the outflow-fluid level as assumed in the latter theory but leave the outflow surface at a height above the outflow-fluid level of the order of half of the total fluid-head differential. However, the pressure or fluid-head distributions at the base can be expressed by a formula identical in form with that given by the Dupuit-Forchheimer theory for the shape of the free surface, namely,

$$\bar{h}^2 = \frac{h_e^2 - h_w^2}{\log r_e/r_w} \log \frac{r}{r_w} + h_w^2, \qquad [cf.\ Eq.\ 6.18(4)]$$

where $\bar{h}$ is the fluid head at the radius r along the base of a system bounded by the radii r_e, r_w, where the fluid heads are h_e, h_w. In the Dupuit-Forchheimer theory $\bar{h}$ represents the height of the free surface at the radius r. On the other hand, the formula of this theory for the flux

$$Q = \frac{\pi k \gamma g (h_e^2 - h_w^2)}{\mu \log r_e/r_w}, \qquad [cf.\ Eq.\ 6.18(6)]$$

where k is the permeability of the sand, γ, μ, the density and viscosity of the liquid, and g the acceleration of gravity, was found to reproduce accurately—after correcting for the flow in the capillary zone—the measured fluxes.

These experiments were also extended to the cases where pressure drives were superposed on the simple gravity flow, modifications of considerable practical importance that had been previously treated neither theoretically nor experimentally. The results of these extensions were equally unambiguous and remarkably simple. Thus the fluid-head distribution at the base of the composite system was found to be simply the sum of the contributions of the strictly gravity-flow fluid head and that due to a nongravity flow with outflow head equal to the sand thickness. Likewise, the resultant flux could be represented by the sum of those due to these individual systems, *i.e.*, by

$$Q = \frac{\pi k \gamma g (2hh_e - h^2 - h_w{}^2)}{\mu \log r_e/r_w}, \qquad [cf. \text{ Eq. } 6.19(2)]$$

where h_e, here, is the external fluid head which exceeds the sand thickness h.

Although, as has already been noted, the above formula for the flux in the simple radial gravity-flow system had already been predicted by the Dupuit-Forchheimer theory, the case of composite flow does not seem to fall within its scope unless one assumes the superposition of the nongravity and gravity types of flow indicated above. Furthermore, it has been seen that the success of this theory even in the simple case of strict gravity flow is hardly more than fortuitous. Fortunately, however, both the simple and the composite cases can be treated by a different approximation method which not only leads to the empirically established flux formulas but which appears to be physically reasonable as well.

This theory is simply based on the observation that owing to the relatively high potentials along the outflow face of a gravity-flow system, such as a dam with vertical faces, above the point where the free surface terminates, very little fluid would pass through that upper part of the outflow face even if the free surface were not to drop below the inflow-fluid level. Thus one may physically expect that if the potential along the outflow face be continued with its linear variation to the level of the inflow head, and if the free surface were not allowed to drop—as is actually the case in the electrical model of the flow system before the upper boundary is cut away to simulate the free surface—the resultant flux would only be but slightly greater than that in the physical gravity-flow system. Now this hypothetical approximation system can be treated rigorously, and the value of the flux is found to be exactly that given by the Dupuit-Forchheimer theory. Equivalent results are also obtained on applying this method to the problem of radial gravity flow (*cf.* Sec. 6.20).

Of course, this method explicitly neglects the existence of the free surface, and in this respect could be given no preference to the Dupuit-Forchheimer theory. However, aside from the fact that its assumptions appear to be intrinsically reasonable from the point of view of calculating the flux, it is physically satisfactory because it also gives close approximations to the correct pressure distributions along the bases of both the linear and radial gravity-flow systems and the velocity distribution along the inflow faces of vertical-faced dams, which have been calculated rigorously. And even along the outflow faces of the latter the velocity distribution is well reproduced by the approximation theory below the top of the outflow-fluid level, and breaks down only immediately below and above the terminus of the free surface. From the point of view of this theory, therefore, it is not at all surprising that its predictions of the flux should be so accurate.

Finally, it may be noted that even this theory can be applied still more simply by summarily replacing the actual potential variation along the outflow surface of total height h_e, which is exposed up to the height h_w to a uniform potential equivalent to the fluid head h_w, and a uniform pressure from there on, by its *average* value, which is equivalent to the fluid head $(h_e^2 + h_w^2)/2h_e$. If, then, the flow systems are treated as nongravity flow systems, with the total fluid-head differential of

$$h_e - \frac{(h_e^2 + h_w^2)}{2h_e} = \frac{(h_e^2 - h_w^2)}{2h_e},$$ the above-mentioned flux for-

mulas will be obtained at once. Likewise, in the case of the composite pressure-drive and gravity-flow systems one need apply only the average fluid-head difference

$$h_e - \frac{(h^2 + h_w^2)}{2h} = \frac{(2h_c h - h^2 - h_w^2)}{2h}$$

to the corresponding nongravity-flow systems to obtain the empirically established flux formulas. As this method can be shown to be rigorously equivalent to the more detailed procedure of the formal approximation theory outlined above, one may thus derive the value of the flux in any gravity-flow system of the general type of that of the seepage through a dam or radial flow into a well of partial or complete penetration, in which the inflow surface is a vertical cylinder (or plane) cutting through the whole of the sand body, without formally taking into account the peculiar and unknown upper boundary formed by the free surface.

CHAPTER VII

SYSTEMS OF NONUNIFORM PERMEABILITY

7.1. Introduction. Surfaces of Discontinuity.—While most flow systems occurring in practice will be sufficiently "ideal" as to justify the assumption that the porous medium is homogeneous, there are certain typical deviations from homogeneity which are not only of interest as physical variations from the idealized systems, but are moreover known to occur with sufficient frequency as to warrant some detailed study of problems involving such deviations from homogeneity. Of course, it is clear that all fluid-bearing sands are in detail far from homogeneous and uniform, and that their associated permeabilities may vary between wide limits within a relatively small extent of sand. But these local inhomogeneities with their random distribution will, on a large scale, give an average effect as if the sand were perfectly uniform throughout. Of practical interest, therefore, are only such large-scale variations in which the permeability may suffer abrupt changes as certain geometrical boundaries are crossed or may vary with the coordinates in a regular manner. While at the same time the permeability may also vary with the direction of flow, the discussion of the present chapter will presuppose that the sand is isotropic. The effect of the anisotropy, in a homogeneous sand, has been treated in Sec. 4.15.

When the permeability varies continuously within the medium, the pressure distribution in the system may be found and discussed just as in the case of homogeneous media, except that the fundamental Laplace's equation for the pressure must be replaced by a somewhat more general equation, as will be seen in the next section. However, when the sand is composed of two or more distinct regions with different but uniform permeabilities, certain conditions must be imposed at the boundaries separating the regions. Although the details of the solution will obviously depend on the particular shapes of the individual regions, the method of attack is the following. Solutions of Laplace's equation are assumed for each region separately and then these are adjusted at the boundaries separating the regions—the "surfaces of discontinuity"—so that the combined set of solutions corresponds to the combined fluid system. This "adjustment" may be formulated more precisely by the following two conditions that must be satisfied at any surface of discontinuity separating two regions (1) and (2):

$$p(1) = p(2), \tag{1}$$

$$k_1 \frac{\partial p(1)}{\partial n} = k_2 \frac{\partial p(2)}{\partial n}, \tag{2}$$

at all points of the boundary, the normal to which is indicated by n.

Equation (1) simply expresses the requirement that the pressure be continuous as one passes across the surface of discontinuity since, in general, such a geometrical boundary can of itself stand no stress. In view of Darcy's law, Eq. (2) states that the normal velocity must be continuous at the surface of discontinuity. This requirement follows at once when it is observed that each fluid element impinging from either region on the surface of discontinuity must of necessity pass into the other region at the point where it enters the interface.[1] When these two conditions are satisfied at each point of all the surfaces of discontinuity, the resultant set of pressure distributions for the several regions will correspond to a composite flow system in which the various parts of different permeabilities are connected physically as well as geometrically.

7.2. Continuous Variations in the Permeability.—Although it may not occur frequently in practical situations that enough

---

[1] The streamlines, however, will be refracted according to the law $k_2 \tan \theta_1 = k_1 \tan \theta_2$, where θ_1, θ_2 are the angles made at the interface by the streamlines in regions (1) and (2) with the normal to the interface. This law, which readily follows from Eqs. (1) and (2), also governs the refraction of the lines of force in electrostatic systems, where k/μ corresponds to the dielectric constant, and of the streamlines in electrical conduction systems, where k/μ represents the conductivity (*cf.* Sec. 3.6). If the streamline is at the same time a free surface, the additional condition must be imposed that: $\sin \theta_1 \cos (\alpha + \theta_1) = \sin \theta_2 \cos (\alpha + \theta_2)$, where α is the inclination of the surface of discontinuity. The angles θ_1, θ_2 for a free surface are therefore definitely determined by k_1/k_2 and α. (*Cf.* A. Casagrande, *Jour. New Eng. Water Works Assoc.*, **51**, 131, 1937.)

data about an underground sand will be available to determine
for it a functional variation of its permeability with the distance
from a particular well, it is of interest to show that when the
permeability variation is known, the flow problems for such a
sand may be still. solved, in principle, by the application of the
same methods used for homogeneous media.[1] Thus, recalling the
general formulation of Darcy's law developed in Chap. III, it
may be expressed by the set of equations

$$\left. \begin{array}{l} v_x = -\dfrac{k}{\mu}\dfrac{\partial p}{\partial x} \\[2mm] v_y = -\dfrac{k}{\mu}\dfrac{\partial p}{\partial y} \\[2mm] v_z = -\dfrac{k}{\mu}\dfrac{\partial p}{\partial z} + \dfrac{k}{\mu}\gamma g. \end{array} \right\} \tag{1}$$

These equations are, of course, valid even if k depends on the
coordinates, since they in effect define the local permeability
$k(x, y, z)$. Hence applying again the equation of continuity,
Eq. 3.1(2), without assuming that k is constant, it is found that

$$\frac{\partial}{\partial x}\left(k\frac{\partial p}{\partial x}\right) + \frac{\partial}{\partial y}\left(k\frac{\partial p}{\partial y}\right) + \frac{\partial}{\partial z}\left(k\frac{\partial p}{\partial z}\right) - \gamma g\frac{\partial k}{\partial z} = 0. \tag{2}$$

Assuming that $k(x, y, z)$ is known, this is again an equation in p
which, when solved, will give the details of the pressure
distribution.

Although no general solution of Eq. (2) can be given,[2] the simple
case of radial two-dimensional flow does allow an explicit solution. Thus in the two dimensions (x, y) Eq. (2) becomes

$$\frac{\partial}{\partial x}\left(k\frac{\partial p}{\partial x}\right) + \frac{\partial}{\partial y}\left(k\frac{\partial p}{\partial y}\right) = 0, \tag{3}$$

[1] It may be mentioned that while such analyses may not be of great
practical value when dealing with homogeneous fluids, recent work shows
that certain features of flow systems involving gas-liquid mixtures may be
studied by a translation of the fluid mixture characteristics into equivalent
permeability variations in the porous medium. (*Cf.* R. D. Wyckoff and
H. G. Botset, and M. Muskat and M. W. Meres, *Physics*, **7**, 325, 346, 1936.)

[2] When k can be expressed as a product of functions of (x, y, z) and gravity
effects (the last term) are neglected, Eq. (2) will be separable and hence
permit general solutions. A particular case of such forms for k is that when
it is variable only in one direction and hence depends on only one coordinate.

or more particularly, if k depends only on the radius and the flow is radial, it follows from Eq. 4.2(2) and the equation of continuity that

$$\frac{1}{r} \frac{\partial}{\partial r}\left[rk(r)\frac{\partial p}{\partial r} \right] = 0, \tag{4}$$

so that

$$rk(r)\frac{\partial p}{\partial r} = \text{const.} = \frac{Q\mu}{2\pi},$$

where Q is the rate of flow per unit sand thickness.

Finally,

$$p(r) = \frac{Q\mu}{2\pi} \int_{r_w}^{r} \frac{dr}{rk(r)} + p_w, \tag{5}$$

where p_w is the pressure at the well, of radius r_w. Thus to get the explicit form for p, one need perform only a single integration. When it is carried through, one can make the usual analysis and interpretation as made in the case when k is taken as uniform.

7.3. Discontinuous Radial Variations in the Permeability.— Of considerably more practical interest than the problem of continuous permeability variations discussed in the last section are those in which the fluid-bearing sand is composed of two or more distinct homogeneous parts of different permeabilities. As a first example of such systems we shall treat one possessing two-dimensional radial symmetry,[1] so that the producing formation

---

[1] For the almost trivial case of a linear system, as a filter bed, composed of a section of porous medium of length L_1 and permeability k_1, contiguous to and in "series" with another of length L_2 and permeability k_2, with inflow and outflow pressures P_1, P_2, it is readily verified that the pressure distributions in the two sections are given by $p_1 = P_1 - \dfrac{k_2(P_1 - P_2)x}{k_2L_1 + k_1L_2}$, and $p_2 = P_2 + \dfrac{k_1(P_1 - P_2)(L - x)}{k_2L_1 + k_1L_2}$, so that the flux is $Q = \dfrac{k_1k_2(P_1 - P_2)}{\mu(k_2L_1 + k_1L_2)}$, and the resultant effective permeability is $k_s = \dfrac{k_1k_2(L_1 + L_2)}{k_2L_1 + k_1L_2}$. When the two sections, of areas A_1, A_2, are contiguous and in "parallel," the pressure distributions are the same in both parts and are given by $p = P_1 - (P_1 - P_2)x/L$, L being their common length. The total flux is then

$$Q = \frac{(k_1A_1 + k_2A_2)(P_1 - P_2)}{L\mu},$$

with an effective permeability $k_p = (k_1A_1 + k_2A_2)/(A_1 + A_2)$. If there are n sections of permeabilities k_i and thicknesses L_i in "series," it is easily may be considered as separated into two adjacent concentric annular regions of permeabilities k_1 and k_2 (*cf.* Fig. 148). Such a system will correspond to a well which has been drilled into a region having a permeability higher or lower than that of the reservoir as a whole.   Or, it may represent the conditions of flow

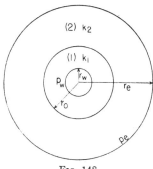

about a well which was initially drilled into a homogeneous sand, the inhomogeneity having been caused by a partial plugging or mudding off of the region immediately surrounding the sand face during the course of production or in the process of drilling, respectively, or conversely by an increase in the permeability near the well bore consequent upon the introduction of acid, as in the method of artificial stimulation of

Fig. 148.

the production from limestone reservoirs known as "acid treatment."[1]

As the pressure distributions in each of the annular regions will evidently possess radial symmetry, the most general solution of Laplace's equation which can be used in constructing the resultant distribution is that given by Eq. 4.2(6).   Denoting the solutions for the two regions of different permeability by subscripts 1 and 2, that in region (1) may be expressed as

$$p_1 = a_1 \log r + b_1, \tag{1}$$

and that in region (2) as

$$p_2 = a_2 \log r + b_2. \tag{2}$$

If the boundary conditions are

$$p = p_w: \quad r = r_w; \quad p = p_e: \quad r = r_e, \tag{3}$$

it is clear that since r_w lies in region (1) and r_e in region (2), the first condition must be applied to p_1 and the second to p_2.   Thus

---

shown by analogy with the analogous problem of electrical conductors that the resultant permeability is given by $1/k_s = (\Sigma L_i/k_i)/\Sigma L_i$, while if they are all of the same length and of areas A_i and are placed in "parallel," their resultant permeability will be given by $k_p = (\Sigma A_i k_i)/\Sigma A_i$.

[1] *Cf.* Sec. 7.7.

at $r = r_w$, Eq. (3) implies that

$$p_1 = a_1 \log r_u + b_1 = p_w, \tag{4}$$

and at $r = r_e$, that

$$p_2 = a_2 \log r_e + b_2 = p_e. \tag{5}$$

On choosing a_1, b_1, a_2, b_2 so that they satisfy these equations we shall have taken care of the boundary conditions. But in addition, the conditions of Eqs. 7.1(1) and 7.1(2) must be applied at the surface of discontinuity, given by $r = r_0$. Equation 7.1(1) evidently gives

$$p_1 = a_1 \log r_0 + b_1 = p_2 = a_2 \log r_0 + b_2, \tag{6}$$

and from Eq. 7.1(2) it follows that

$$\frac{k_1 a_1}{r_0} = \frac{k_2 a_2}{r_0}. \tag{7}$$

There have thus been derived four equations for the four constants a_1, b_1, a_2, b_2. When they are solved and the constants are put back into Eqs. (1) and (2), it is found that

$$p_1 = \frac{p_e - p_w}{\log \dfrac{r_0}{r_w} + \dfrac{k_1}{k_2} \log \dfrac{r_e}{r_0}} \log \frac{r}{r_w} + p_w: \qquad r_w \leqslant r \leqslant r_0, \tag{8}$$

$$p_2 = \frac{\dfrac{k_1}{k_2}(p_e - p_w) \log \dfrac{r}{r_e}}{\log \dfrac{r_0}{r_w} + \dfrac{k_1}{k_2} \log \dfrac{r_e}{r_0}} + p_e: \qquad r_0 \leqslant r \leqslant r_e \tag{9}$$

which give the resultant pressure distribution in the composite system of Fig. 148.

The rate of fluid flow through the sand into the well per unit sand thickness is evidently given by

$$Q = \frac{2\pi k r}{\mu} \frac{\partial p}{\partial r} = \frac{2\pi k_1 (p_e - p_w)/\mu}{\log \dfrac{r_0}{r_w} + \dfrac{k_1}{k_2} \log \dfrac{r_e}{r_0}}. \tag{10}$$

To illustrate these relations the pressure distribution given by Eqs. (8) and (9) has been plotted in Fig. 149 for the system defined by

$$r_w = \tfrac{1}{4} \text{ ft.;} \qquad r_0 = 50 \text{ ft.;} \qquad r_e = 500 \text{ ft.;}$$

$$p_e = 70 \text{ atm.;} \qquad p_u = 0; \qquad \frac{k_i}{k_2} = 3.0.$$

The dotted curve represents the pressure distribution that would exist in the sand if it were homogeneous. It will be seen that the high permeability near the well results in a slower rise of the pressure near the well, or smaller pressure gradients. This is, of course, to be expected, when it is observed that since the same

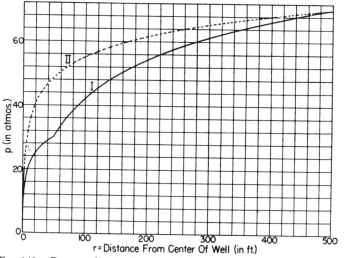

Fig. 149.—Pressure distributions in radial-flow systems. I: distribution in a sand in which the annulus of 50-ft. radius surrounding the well has a permeability three times that of the rest of the sand. II: distribution in a homogeneous sand; well radius and pressure = $\tfrac{1}{4}$-ft., 0; external boundary radius and pressure = 500-ft., 70 atm.

flux must pass through the high-permeability zone near the well as through the more distant low-permeability regions, the corresponding pressure gradients, except for that necessarily associated with the geometry of the system, will naturally be lower while still sufficing to carry the fluid into the well.

The effect of the inhomogeneity of the sand under the conditions represented by the above equations may be seen still more clearly by comparing the flux Q through the inhomogeneous sand with that, Q_0, from a sand of the same overall dimensions and under the same pressure differential, but with the uniform permeability k_2. Denoting the ratio k_1/k_2 by α, it follows from

Eqs. (10) and 4.2(10) that

$$\frac{Q}{Q_0} = \frac{\alpha \log r_e/r_w}{\log \dfrac{r_0}{r_w} + \alpha \log \dfrac{r_e}{r_0}}; \qquad \alpha = \frac{k_1}{k_2}, \tag{11}$$

the difference from the value unity giving the relative change in the production capacity caused by the presence of a cylinder of

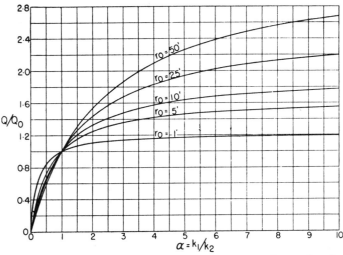

Fig. 150.—Variation of the production capacity of a well as a function of $k_1/k_2 = $ (the permeability within the annulus of radius r_0)/(permeability from r_0 to 500 ft.). $Q/Q = $ (production capacity of well in sand where $k_1/k_2 \neq 1$) (production capacity of well in sand with permeability $= k_2$ everywhere); well radius $= \frac{1}{4}$ ft.

permeability k_1 in the immediate neighborhood of the well, and extending to the distance r_0 in a sand which has the permeability k_2. Thus for the above numerical example, Eq. (11) gives $Q/Q_0 = 1.87$, showing that the 50-ft. annulus surrounding the well with a permeability three times that of the main sand body will increase the flow by 87 per cent.

Equation (11) is plotted in Fig. 150 for several values of r_0. The marked effect on the production capacities due to very localized permeability anomalies will be evident. Thus if the zone about the well has a permeability of 2.5 times that of the remainder of the sand, it need be only 5 ft. in radius—only 0.01 per cent of the total sand volume for $r_e = 500$ ft.—to increase the production capacity by 30 per cent. Or, if the inner zone has a one-fourth normal permeability, the production capacity will be cut to less than two-thirds of its normal value even though the zone is only 1 ft. in radius—0.0004 per cent of the sand volume.

These large effects, caused by small zones about the well bore, are evidently due to the highly localized character of the pressure drop in a radial-flow system about the well center. For it will be recalled from Fig. 149 that 60 per cent of the total pressure drop in a radial flow system of 500 ft. radius about a $\frac{1}{4}$-ft. well is concentrated in the first 25 ft. of the sand, the remaining 40 per cent being distributed about the external zone of 475 ft. width. This fact is also shown by Fig. 150 where it is seen that the heights above the line $Q/Q_0 = 1$ of the several curves rise much more slowly than the values of r_0. Thus, whereas the first 25 ft. with a permeability anomaly of 2.0 raises the production capacity by 43 per cent, the addition of the next 25 ft. gives an increase of only 10 per cent, and likewise the cut in production capacity of 64.5 per cent due to a 25-ft. zone of permeability anomaly of one-fourth is increased to only 67.6 per cent if the next 25 ft. are added to the low-permeability zone.

As to the significance of the actual value of the permeability anomaly, k_1/k_2, it is to be noted that for the zones of small radius, it is not at all necessary to have very large values of k_1/k_2 in order appreciably to increase Q/Q_0. Thus the curves of Fig. 150 have a constantly decreasing slope as α is increased, and approach the limiting values $Q/Q_0 = (\log r_e/r_w)/\log (r_e/r_0)$ for $\alpha \rightarrow \infty$, corresponding to a well of radius r_0 compared to that of radius r_w. On the other hand, the limit for Q/Q_0 as α decreases from 1 is simply 0. For if the inner zone is made strictly impermeable, the flow through the whole system must necessarily vanish, however narrow the low-permeability zone may be.

Aside from the application of these results to the problem of acid treatment of wells, which will be discussed in a later section, the above results are of practical significance in that they afford explanations for large variations in the production capacities of apparently identical wells in the same oil or water sand. Thus a very localized high-permeability zone may give rise to a well of abnormally high production capacity, whereas the sand as a whole may be quite uniform and yield average wells of considerably lower production capacities. Similarly, a well of abnormally low production capacity may be due to a localized low-permeability zone confined to the immediate vicinity of that particular well.

7.4. Adjacent Beds of Different Permeability. Fluid Flow in Fractured Limestones.—As has already been pointed out in Chap. I, limestone reservoirs, in contrast to those composed of ordinary consolidated sands, are usually permeated by numerous fractures and crevices, some of which may extend for considerable distances through the producing formation. Furthermore, as the permeability of the limestone proper is frequently very low, the principal reason for the appreciable production capacities of such formations is the presence of these crevices and fractures which, in spite of their very small width, have effective permeabilities very much greater than that of the limestone itself. The general nature of the mechanism of the flow of fluids in limestones may thus be described as being one in which the main body of the reservoir feeds its fluid into the highly permeable fractures, these latter bringing the fluid directly or by complex interconnection into the outlet wells. The details of this mechanism will clearly be determined by the detailed structure of the limestone. Thus a uniformly disseminated network of crevices would give a resultant effect equivalent to a radially homogeneous sand, whereas if there are only a few but extended fractures, the flow system may assume at least some of the characteristics of a linear channel. It is this latter situation which will be discussed in the present section, under the assumptions that there is but one fracture in the system and that the only direct outlet for the fluid is the part of the well surface pierced by the fracture.

In the idealized representation of a limestone with extended fractures, it is convenient to consider the limestone proper and fracture as two adjacent porous media of different permeability, that of the fracture being an "effective" permeability corresponding to that of an equivalent free linear channel carrying a fluid under viscous conditions. Although in the practical realization of such a system the fracture will be of very small width, this fact will be anticipated only to the extent that the boundary condition to be imposed at the outlet terminal of the medium representing the fracture will correspond to the more convenient requirement of a uniform flux density which, in view of the small fracture width, is quite equivalent to that of a uniform pressure which would be required physically. It will also be convenient to take the fracture width as the unit of length and both the media representing the fracture and limestone to be of infinite extent, the latter also being of infinite width. Because of the symmetry of the system—containing but a single fracture—only a single quadrant of the (x, y) plane, the horizontal section of the limestone-fracture system (Fig. 151) need be considered. The analytical problem may then be formulated thus: Find pressure distributions p_1 and p_2 such that

$$
\left.
\begin{aligned}
\nabla^2 p_1 &= 0; \quad \nabla^2 p_2 = 0 \\
\frac{\partial p_1}{\partial y} = 0: \quad y &= 0; \quad \frac{\partial p_1}{\partial x} = 1: \quad x = 0, \quad y \leqslant \tfrac{1}{2} \\
p_1 = p_2, \quad k_1 \frac{\partial p_1}{\partial y} &= k_2 \frac{\partial p_2}{\partial y}: \quad y = \tfrac{1}{2} \\
\frac{\partial p_2}{\partial x} &= 0, \quad x = 0, \quad y \geqslant \tfrac{1}{2}.
\end{aligned}
\right\} \quad (1)
$$

It will suffice for the present purposes to suppose that medium

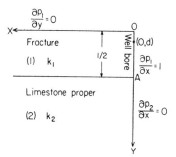

(1)—the fracture—extends to the same depth as medium (2), and that none of the constants of the system or physical conditions vary with the depth, so that the problem becomes two-dimensional. Furthermore, we shall synthesize the resultant solution from the preliminary and simpler ones corresponding to a single flux element at $(0, d)$ in medium (1). The starting point for the development of these preliminary solutions is the representation in integral form

Fig. 151.—Diagrammatic representation of a limestone bisected by a fracture, showing a single quadrant of the cross-section plane. (*From Physics*, **7**, 11, 1936.)

of the pressure distribution due to the primary flux element at $(0, d)$.[1] Such a representation is given by

[1] One may also develop the solution by an extension of the method of images illustrated in Secs. 4.7 and 4.9 for homogeneous media, although it would be somewhat more cumbersome than that given here. This method is based on the observation that a primary unit pole at $(0, d)$ in medium (1) requires an image at $(0, 1 - d)$ of strength $(k_1 - k_2)/(k_1 + k_2)$ contributing to p_1 and an image at $(0, d)$ of strength $2k_1/(k_1 + k_2)$ contributing to p_2 in order to satisfy the boundary conditions at the surface of discontinuity at $y = \tfrac{1}{2}$. These images are then successively reflected in the other boundaries so as to give a resultant satisfying the other boundary conditions of the problem. The analysis presented here, however, follows that given by M. Muskat, *Physics*, **6**, 14, 1935, for the analogous electrical problem.

$$p_0 = \log\left[x^2 + (y - h)^2\right] = 2\int_0^\infty \frac{dz}{z}\left\{e^{-z} - \cos xz\left[\begin{matrix}e^{-(d-y)z:\ y\,\ell\,d}\\ e^{-(y-d)z:\ y\,\iota\,d}\end{matrix}\right]\right\}. \quad (2)$$

This would be the pressure distribution in the system if it were strictly uniform throughout. However, the requirement of vanishing flux at $y = 0$ (because of the symmetry of the system about $y = 0$) and the different permeability for $y > \frac{1}{2}$ induce perturbations in the pressure distributions for both $y < \frac{1}{2}$ and $y > \frac{1}{2}$. Denoting these perturbations by $\bar{p}_1$ and p_2, they may be expressed as

$$\left.\begin{aligned}\bar{p}_1 &= 2\int_0^\infty \frac{dz}{z}[\psi_1(z) + (A_1 e^{-yz} + B_1 e^{yz})\cos xz]\\ p_2 &= 2\int_0^\infty \frac{dz}{z}[\psi_2(z) + A_2 e^{-(y-\frac{1}{2})z}\cos xz],\end{aligned}\right\} \quad (3)$$

where ψ_1 and ψ_2 are functions independent of (x, y) to correspond to the term e^{-z} in p_0, and the coefficients A_1, B_1, A_2 are functions of the variable of integration, z, to be chosen in such a way that the resulting pressure-distribution functions satisfy the boundary conditions of Eq. (1).[1] As p_2 can be taken to give the whole of the pressure distribution for $y > \frac{1}{2}$, the boundary conditions for the preliminary problem involving only a single flux element can now be written, with $p_0 + \bar{p}_1$ denoted by p_1, as

$$\frac{\partial p_1}{\partial y} = 0: \quad y = 0; \quad p_1 = p_2, \quad k_1\frac{\partial p_1}{\partial y} = k_2\frac{\partial p_2}{\partial y}: \quad y = \frac{1}{2}. \quad (4)$$

Applying these conditions to the integrands of Eqs. (2) and (3), it is readily found that they will be satisfied if the arbitrary functions ψ, A, B, are solutions of the equations

$$\left.\begin{aligned}-e^{-dz} - A_1 + B_1 &= 0\\ -e^{-(\frac{1}{2}-d)z} + A_1 e^{-\frac{z}{2}} + B_1 e^{\frac{z}{2}} &= A_2\\ e^{-z} + \psi_1 &= \psi_2\\ k_1\{e^{-(\frac{1}{2}-d)z} - A_1 e^{-\frac{z}{2}} + B_1 e^{\frac{z}{2}}\} &= -k_2 A_2.\end{aligned}\right\} \quad (5)$$

[1] That the integrals in Eq. (3) are solutions of Laplace's equation may be readily verified by substitution. In fact, any convergent linear discrete (series form) or continuous (integral form) superposition of the elementary solutions $e^{\pm yz}\genfrac{}{}{0pt}{}{\cos}{\sin} xz$ will satisfy Laplace's equation.

Furthermore, in order to insure the convergence of $\bar{p}_1$ and p_2, ψ_1 and ψ_2 must satisfy the relations

$$\psi_1(0) + A_1(0) + B_1(0) = 0; \qquad \psi_2(0) + A_2(0) = 0. \tag{6}$$

The solution of these equations gives

$$
\left.
\begin{aligned}
A_1\Delta &= 2\delta \sinh \left(\tfrac{1}{2} - d\right)z + 2 \cosh \left(\tfrac{1}{2} - d\right)z \\
B_1\Delta &= 2(1 - \delta)e^{-\frac{z}{2}} \cosh dz; \qquad A_2\Delta = 4 \cosh dz \\
\Delta &= -2\left(\delta \cosh \frac{z}{2} + \sinh \frac{z}{2}\right); \qquad \delta = \frac{k_2}{k_1} \\
\psi_1 &= \left(\frac{2}{\delta} - 1\right)e^{-z}; \qquad \psi_2 = \frac{2}{\delta}e^{-z}.
\end{aligned}
\right\} \tag{7}
$$

With these values for A_1, B_1, A_2, and ψ_1, ψ_2, the resultant pressure distributions may be written in the form

$$
\left.
\begin{aligned}
p_1 &= 4\int^{\infty}\frac{dz}{z}\left[\frac{e^{-z}}{\delta} - \frac{\cos xz}{\delta \cosh \dfrac{z}{2} + \sinh \dfrac{z}{2}}\right. \\
&\left\{
\begin{array}{ll}
\left[\delta \sinh\left(\dfrac{1}{2} - d\right)z + \cosh\left(\dfrac{1}{2} - d\right)z\right] \cosh yz: & y \leqslant d \\
\left[\delta \sinh\left(\dfrac{1}{2} - y\right)z + \cosh\left(\dfrac{1}{2} - y\right)z\right] \cosh dz: & y \geqslant d
\end{array}
\right\} \right] \\
p_2 &= 4\int^{\infty}\frac{dz}{z}\left[\frac{e^{-z}}{\delta} - \frac{e^{-(y-\frac{1}{2})z}\cosh dz \cos xz}{\delta \cosh \dfrac{z}{2} + \sinh \dfrac{z}{2}}\right]
\end{aligned}
\right\} \tag{8}
$$

Setting now $\delta = \tanh \epsilon$, Eqs. (8) take the form

$$
\left.
\begin{aligned}
p_1 &= 4\int^{\infty}\frac{dz}{z}\left\{e^{-z} \operatorname{ctnh} \epsilon - \frac{\cos xz}{\sinh \left(\dfrac{z}{2} + \epsilon\right)}\right. \\
&\left[
\begin{array}{ll}
\cosh \left(\dfrac{z}{2} + \epsilon - dz\right) \cosh yz: & y \leqslant d \\
\cosh \left(\dfrac{z}{2} + \epsilon - yz\right) \cosh dz: & y \geqslant d
\end{array}
\right] \\
p_2 &= 4\int_0^{\infty}\frac{dz}{z}\left[e^{-z} \operatorname{ctnh} \epsilon - e^{-(y-\frac{1}{2})z}\frac{\cosh \epsilon \cosh dz \cos xz}{\sinh \left(\dfrac{z}{2} + \epsilon\right)}\right].
\end{aligned}
\right\} \tag{9}
$$

As already suggested, the actual width of the strip (1) will be taken as very small in the practical application of the present problem, so that the flux along OA will be taken as uniform, for a uniform pressure. Assuming then a uniform unit flux distribution along OA, the resultant pressure distributions will be given by the integrals of Eq. (9) with respect to d from $d = 0$ to $d = \frac{1}{2}$. In the case of p_1, however, it is convenient to subtract p_0 and leave it in the logarithmic form of Eq. (2). Denoting also the integrated values by p_1 and p_2, it is found that

$$
\left.
\begin{aligned}
p_1 &= \left(\frac{1}{2} - y\right) \log\left[x^2 + \left(\frac{1}{2} - y\right)^2\right] \\
&\quad + y \log\,(x^2 + y^2) + 2x \tan^{-1}\frac{x}{2x^2 + 2y^2 - y} - 1 \\
&\quad + 2\int^{\infty}\frac{dz}{z}\left\{\left(\frac{1}{\delta} - \frac{1}{2}\right)e^{-z} + \frac{2\cos xz}{z\Delta}\right. \\
&\qquad\qquad \left[e^{-yz}\left(\delta\cosh\frac{z}{2} + \sinh\frac{z}{2} - \delta\right)\right. \\
&\qquad\qquad\qquad \left.\left. + (1 - \delta)e^{-\frac{z}{2}+yz}\sinh\frac{z}{2}\right]\right\} \\
p_2 &= 2\int_0^{\infty}\frac{dz}{z}\left[e^{-z}\operatorname{ctnh}\epsilon - 2e^{-(y-\frac{1}{2})z}\frac{\cosh\epsilon\sinh z/2\cos xz}{z\sinh(z/2 + \epsilon)}\right],
\end{aligned}
\right\} \quad (10)
$$

where Δ is given by Eq. (7).

For purposes of computation it is convenient to remove the constants by subtracting the values of p at $x = 0$. Thus for $y = \frac{1}{2}$

$$
\begin{aligned}
p_1(x, \tfrac{1}{2}) - p_1(0, \tfrac{1}{2}) &= \tfrac{1}{2}\log\,(1 + 4x^2) + 2x\tan^{-1}1/2x \\
&\quad + 4\int_0^{\infty}\frac{dz\,\sin^2 xz/2}{z^2\sinh(z/2 + \epsilon)}\left[e^{-\frac{z}{2}}\sinh\left(\frac{z}{2} + \epsilon\right) + \sinh\left(\frac{z}{2} - \epsilon\right)\right] \\
&= \pi x + 2\int_0^{\infty}\frac{\sin^2 xz\,\sinh(z - \epsilon)}{z^2\sinh(z + \epsilon)}dz, \quad (11)
\end{aligned}
$$

$$
p_2\left(x, \frac{1}{2}\right) - p_2\left(0, \frac{1}{2}\right) = 8\cosh\epsilon\int_0^{\infty}\frac{\sin^2 xz/2\,\sinh z/2}{z^2\sinh(z/2 + \epsilon)}dz. \quad (12)
$$

Several pressure distributions (along $y = \frac{1}{2}$), as given by the above formulas,[1] are plotted in Fig. 152. The pressure variation

[1] For $\epsilon \leqslant 10^{-3}$ the integrals of Eqs. (11) or (12) may be closely approximated by combinations of simpler ones which may be either evaluated

is linear at first and then assumes a logarithmic form, the pressure for very large x and $\epsilon \leqslant 10^{-3}$ approaching the values given by

$$p\left(x, \frac{1}{2}\right) - p\left(0, \frac{1}{2}\right) = \frac{2 \cosh \epsilon}{\epsilon}\,(1.2704 + \log \epsilon x). \qquad (13)$$

However, the linear variation of p persists for larger and larger values of x as ϵ is made smaller, and in fact becomes strictly

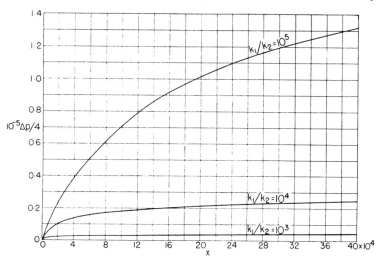

Fig. 152.—Pressure distributions along a liquid-producing fracture bisecting a limestone reservoir. Δp = pressure drop from the well terminal of fracture to the distance x (measured in units of fracture width), for a production rate of $4\pi k_1/\mu$ units per unit thickness of limestone; k_1/k_2 = (effective permeability of fracture)/(permeability of limestone proper). (*From Physics*, **7**, 111, 1936.)

linear throughout for $\epsilon = 0$, when the medium (2) has a vanishing permeability. For then it follows from Eq. (11) that

$$p(x, \tfrac{1}{2}) - p(0, \tfrac{1}{2}) = 2\pi x. \qquad (14)$$

On the other hand, in the limiting case when the medium (2) has the same permeability as medium (1), $\epsilon = \infty$, and ·

$$p\left(x, \frac{1}{2}\right) - p\left(0, \frac{1}{2}\right) = \log(1 + x^2) + 2x\tan^{-1}\frac{1}{x}. \qquad (15)$$

Observing that Eq. (13) gives the pressure differential between the well and the points along the fracture at the distance x, for a

exactly or derived from tables of the "sine" and "cosine integrals." (*cf.* 'Funktionen Tafeln," by E. Jahnke and F. Emde, 1928.)

flux of $2\pi k_1/\mu$ into the well from either side of the y axis, the pressure differential for unit production rate or the effective resistance of the system can be easily calculated. For the range of k_2/k_1 of practical interest—$\leqslant 10^3$—and for $x = 4 \cdot 10^5$, this resistance is plotted as curve I in Fig. 153, for $k_1/\mu = 1$. The decreasing slope for decreasing k_2/k_1 in this figure is clearly due to the fact that, as the permeability of the second medium continues to

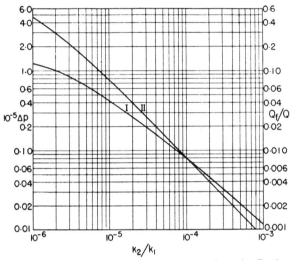

FIG. 153.—Resistivity of limestone-fracture systems for unit effective permeability of the fracture (k_1). I. Pressure drop Δp between the well and a distance $4 \cdot 10^5$ (in units of fracture width) from the well, for a unit flux per unit limestone thickness. Δp is in atmospheres when $k_1/\mu = 1$. II. Fraction Q_f/Q of total flow entering the well which has traveled through the fracture for the whole distance of $4 \cdot 10^5$ units. $k_2 =$ permeability of limestone proper; $\mu = 1$ c.p. (*From Physics*, **7**, 112, 1936.)

decrease, more and more of the fluid is carried directly by medium (1)—a fracture in the practical realization of the system under consideration—so that the effective conductivity of the combined system will not be reduced by as large a factor as will that of medium (2). In fact, it follows from Eq. (12) that

$$\frac{\partial p}{\partial x} = 4 \cosh \epsilon \int_0^\infty \frac{\sin xz \sinh z/2}{z \sinh \left(\dfrac{z}{2} + \epsilon\right)} dz$$

$$= \frac{\operatorname{ctnh} \epsilon}{x} \left[2 - \frac{1}{x^2}\left(\operatorname{ctnh}^2 \epsilon - \frac{1}{3}\right) + 0\left(\frac{1}{x^4}\right) \right]. \tag{16}$$

Assuming that $\frac{\partial p}{\partial x}$ is uniform for large x over the width of the fracture, it gives the flux entering the fracture directly for large values of x. Dividing by 2π, one obtains the fraction of the total

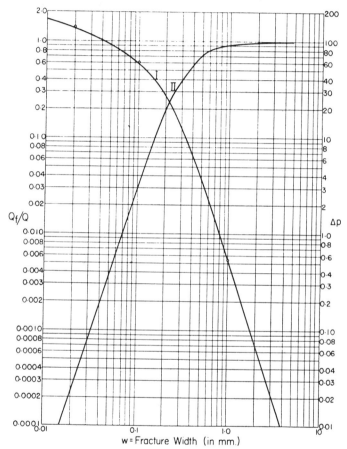

FIG. 154.—Resistivity of limestone-fracture systems for fixed limestone permeability (0.01 darcys). I. Pressure drop (atmospheres) over first 300 ft. from well for a unit flux (cubic centimeter per second) per unit limestone thickness. II. Fraction Q_f/Q of total flow entering the well which has traveled through the fracture for the whole distance of 300 ft. (*From Physics*, **7**, 113, 1936.)

flow into the well which has come directly through the fracture, at least beyond the distance x from the well. The curve showing how this fraction varies with k_1/k_2 for $x = 4 \cdot 10^5$ is given as curve II in Fig. 153. Thus it is seen that, whereas for $k_1/k_2 = 10^4$

99.2 per cent of the total flux is fed into the fracture from the limestone within the distance x, only 54 per cent of the total flux is so derived if $k_1/k_2 = 10^6$.

As curve I of Fig. 153 is based on a fixed (unit) permeability of the medium (1), it gives the variation of the composite resistance of the limestone-fracture system in which the fracture is kept fixed, and only the limestone permeability is allowed to vary. If, however, the limestone is considered as of fixed permeability and the size of the fracture is varied, the effective resistance of the system follows a curve as I in Fig. 154.[1] The limestone permeability has been taken here as 0.01 darcys and the "reservoir radius" as 300 ft. It will be noted that for relatively large fractures the resistance of the composite system varies inversely as the cube of the fracture width. This is to be expected since, as will be seen in Sec. 7.8, the effective permeability of the fracture varies as the square of its width, and the total flux through it, for a given pressure differential, is also proportional to its absolute width.

Curve II of Fig. 154 gives the fraction Q_f/Q of the total flow which enters the well directly through the fracture. $1 - Q_f/Q$ gives the fraction which has been fed into the fracture by the surrounding limestone within 300 ft. of the well. It will be seen that for fracture widths greater than 0.75 mm., less than 14 per cent of the total flow is fed into the fracture within 300 ft. of the well.

7.5. Bounded Limestone-fracture Systems.—The analysis of the last section was based on the assumption that both the limestone and its fractures extended for infinite distances from the producing well. Although the results showed the pressure to vary logarithmically at large distances from the well, so that the composite conductivity of the system will be insensitive to the exact dimensions, if finite, it will be instructive to derive the corresponding solutions for the case where it is assumed at the outset that the dimensions of the system are finite. The method of Fourier series may then be applied in a manner quite similar to that presented in the last section for what was in effect a Fourier integral[2] analysis.

---

[1] The effective permeability k_1 of the fracture has here been translated into equivalent widths w, by means of Eq. 7.8(2), *infra*.

[2] *Cf.* W. E. Byerly, "Fourier Series and Spherical Harmonics," Chap. IV.

Thus it will be specifically assumed that over the distant boundaries the pressure is kept at exactly the value p_e (*cf.* Fig. 155). Appropriate solutions for p_1 and p_2 may then be expressed as

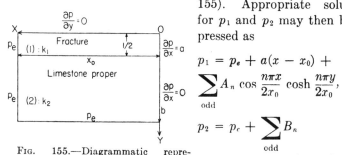

Fig. 155.—Diagrammatic representation of a bounded limestone-fracture system.

$$p_1 = p_e + a(x - x_0) + \sum_{\text{odd}} A_n \cos \frac{n\pi x}{2x_0} \cosh \frac{n\pi y}{2x_0}, \quad (1)$$

$$p_2 = p_e + \sum_{\text{odd}} B_n \cos \frac{n\pi x}{2x_0} \sinh \frac{n\pi(b - y)}{2x_0}. \quad (2)$$

As these expressions already satisfy the conditions that p_1, $p_2 = p_e$ at $x = x_0$, $p_2 = p_e$ at $y = b$, $\frac{\partial p_1}{\partial y} = 0$ at $y = 0$, $\frac{\partial p_2}{\partial x} = 0$ for $x = 0$, $\frac{1}{2} \leqslant y \leqslant b$, and $\frac{\partial p_1}{\partial x} = a$ for $x = 0$, $0 \leqslant y \leqslant \frac{1}{2}$, the remaining conditions to be satisfied are those of the continuity of p and $k\frac{\partial p}{\partial y}$ at $y = \frac{1}{2}$. These give

$$a(x - x_0) + \sum A_n \cosh \frac{n\pi}{4x_0} \cos \frac{n\pi x}{2x_0} =$$
$$\sum B_n \sinh \frac{n\pi(b - \frac{1}{2})}{2x_0} \cos \frac{n\pi x}{2x_0},$$

$$k_1 A_n \sinh \frac{n\pi}{4x_0} = -k_2 B_n \cosh \frac{n\pi(b - \frac{1}{2})}{2x_0}.$$

Observing that

$$a(x - x_0) = \frac{-8ax_0}{\pi^2} \sum_{\text{odd}} \frac{1}{n^2} \cos \frac{n\pi x}{2x_0},$$

solving for A_n, B_n, and then computing the corresponding pressure differential at $y = \frac{1}{2}$, it is found that

$$p_e - p_w = \frac{8ax_0}{\pi^2} \sum_{\text{odd}} \frac{1}{n^2\left[1 + \frac{k_2}{k_1} \operatorname{ctnh} \frac{n\pi}{4x_0} \operatorname{ctnh} \frac{n\pi(b - \frac{1}{2})}{2x_0}\right]}, \quad (3)$$

where

$$p_w = p_1(0, \tfrac{1}{2}) = p_2(0, \tfrac{1}{2}).$$

As x_0 is measured in units of the width of medium (1)—the fracture—and will, therefore, be a very large number, Eq. (3) may be very closely approximated by the expression

$$\Delta p = \frac{8ax_0}{\pi^2} \sum_{\text{odd}} \frac{1}{n^2 \left(1 + \frac{4k_2x_0}{k_1 n\pi} \right)}, \tag{4}$$

where b has been taken to be of the order of $2x_0$.

Now the total flux in the system is evidently

$$Q = \frac{2ak_1}{\mu} = \frac{\pi^2 k_1 \Delta p}{4\mu x_0 \sum_{\text{odd}} \frac{1}{n \left(n + \frac{4k_2x_0}{k_1\pi} \right)}}. \tag{5}$$

Hence, noting that

$$\sum_{\text{odd}} \frac{1}{n(n + \alpha)} = \frac{1}{\alpha} \left[0.2886 + \psi(\alpha) - \frac{1}{2}\psi\left(\frac{\alpha}{2}\right) \right], \tag{6}$$

where ψ is the logarithmic derivative of the Gamma function,[1] and setting $s = 2x_0 k_2 / k_1\pi$, Eq. (5) may be rewritten as

$$\frac{\Delta p}{Q} = \frac{\mu}{\pi k_2} \left[0.2886 + \psi(2s) - \frac{1}{2}\psi(s) \right]. \tag{7}$$

The value of $\Delta p/Q$ for fracture widths of 1.0, 0.1 and 0.02 mm. as given by Eq. (7) are plotted in Fig. 154 as the circles on curve I. It will be noted that for practical purposes the values agree well with those computed by the method of Sec. 7.4. The fact that the resistances given by Eq. (7) are somewhat higher is clearly due to the assumption made here that the lower boundary of medium (2), where the pressure is supposed to be p_e, is at a distance $2x_0$ or greater from the well, whereas the analysis of Sec. 7.4 essentially implies that the pressure would be p_e on a circular boundary of radius x_0 (provided x_0 is very large). Although this latter assumption appears to be somewhat less arbitrary than that used in deriving Eq. (7), the form of Eq. (7)

[1] *Cf.* Sec. 5.3.

is more convenient for numerical computation than Eqs. 7.4(11) or 7.4(12).

7.6. The Theory of the Acid Treatment of Limestone Wells.[1]— An interesting practical application of the analysis developed in the last several sections is that of physically explaining the reaction of oil wells producing from limestone formations to "acid treatment." Acid treatment is a means of artificial stimulation of the production from wells producing at small or unprofitable rates through the agency of acid—usually hydrochloric acid— introduced into the well bore. The method has thus far only been appreciably successful in the "treatment" of wells from limestone formations, which are readily attacked by the acid.[2] The elementary explanation usually given for the effectiveness of the method is that the acid "cleans out" the well and hence increases its production capacity. However, the wide divergence of results obtained with apparently identical wells evidently requires a more detailed examination of the whole phenomenon. Thus in some cases the treatment appears to be quite ineffective, while in others, under apparently identical conditions, it is remarkably productive of higher recovery rates. Then, too, in some cases the treatment has been found to increase only the *economic* ultimate recovery, while in others the *physical* ultimate recovery seems to have been definitely increased as well. In the present section an attempt will be made to correlate some of these widely different results of field experience on the basis of differences in the initial mechanisms of production of the treated wells and in their initial conditions before treatment.

First, however, it should be emphasized that the discussion to be given here presupposes that no new fluid sources are opened by the acid treatment, so that the economic rather than the physical ultimate recovery from the reservoir is increased by the treatment. If the acid penetrates fractures in the producing formation and unseals new and untapped sources of fluid, not only will the physical ultimate recovery of the system be increased, but the whole character of the response will be so modified as to mask the features characteristic of the various mechanisms to be discussed below. The wells will show appreciably increased production

[1] MUSKAT, M., and R. D. WYCKOFF, *Physics*, **7**, 106, 1936.
[2] Hydrofluoric acid has been tried in wells producing from sandstone reservoirs, but with little success up to the present.

capacities regardless of the detailed effect of the acid upon the part of the reservoir being drained initially.   Furthermore, the decline rate after treatment may become considerably less than that before the treatment, whereas if new sources of fluid are not opened, the decline rate should always be increased after an effective acid treatment.   In all cases, of course, the wells will show an initially high production rate immediately after treatment, owing to the rebuilding of the pressure about the well bore in the course of the treatment.   However, these transients will be quite short-lived, and one should use only the "settled" production rates for comparison with the initial rates in order to test the effectiveness of the treatment.

Considering then only such cases where only the *economic* ultimate recovery (increased production capacity) is increased by the acid, it is clear that the effect of such treatments must be essentially nothing more than that of increasing the effective permeability of the producing horizon surrounding the well. For, since the acid treatment itself can have no effect upon the reservoir pressure, and hence no effect upon the pressure differential available to force fluid into the well, the entire gain in production capacity must be assigned to an effective decrease in the resistance of the flow system.   From this point of view the theory of acid treatment—for reservoirs which do not show increased *physical* ultimate recoveries—becomes one of analyzing the effect on the resultant permeability of a porous medium owing to the introduction into the well of acid capable of reacting with and dissolving those parts of the medium with which it comes in contact.

A theoretical deduction of these effects must evidently be based upon a definite mechanism of production and, in particular, a definite geometric configuration of the flow system, presupposed for the initial condition of the producing formation before the introduction of the acid.   For practical purposes such mechanisms may be divided into two types, although in reality intermediate cases may be of even more likely occurrence.   These types are: (1) That in which the crevices or major pores of the limestone are widely and uniformly disseminated throughout the pay, so that the flow is of an *essentially radial character;* and (2) that in which the major part of the flow into the well is carried by a limited number of extended fractures which are fed laterally by the limestone, the resultant flow system being a mean between the strictly linear and radial types. In the first case the acid will simply increase the permeability of a relatively small annular ring concentric with the well bore—of volume of the order of one-fifth of the volume of acid used. In the second case the acid will penetrate considerable distances along the fractures[1] and will thus widen them and increase their effective permeability over extended distances along their length.

Thus assuming that 500 gal. of 18°Bé. hydrochloric acid is injected into a well and that it penetrates uniformly into the surrounding formation of 20 per cent porosity and 10-ft. thickness, it will occupy an annular ring of radius 3.26 ft., and will, upon complete reaction, dissolve 13 cu. ft. of bulk limestone. One thousand gallons of acid would penetrate an annulus of approximately 4.6-ft. radius with the solution of 26 cu. ft. of limestone. However, 500 gal. of the same acid could widen a fracture in the same formation by 0.156 in. for a distance of 100 ft.

The numerical magnitude of the effect of the treatment will be quite sensitive to the condition of the formation with respect to its actual permeability before treatment. Thus a tight limestone will respond differently from one of greater permeability, and one with narrow fractures differently from one with fractures of appreciable width. Furthermore, a formation which has been plugged or mudded off in the immediate vicinity of the well bore will show different results from a homogeneous pay. The analytical theory for these various effects which has been developed in the last several sections will now be applied to these separate cases.

7.7. The Effect of Acid Treatment in Radial Systems.—The diagrammatic representation of a radial-flow system which has been treated by acid is given in Fig. 156. The annulus between the well bore, of radius $r_w = \frac{1}{4}$ ft., and the radius r_0 is supposed to have been affected by the acid, while the limestone beyond r_0 remains with its original permeability k_2. The resultant effect of the acid may be conveniently represented by the ratio of the production capacity of the well after treatment, Q, to that before treatment, Q_0. These production capacities, in turn, depend upon

---

[1] Field evidence exists in which offset wells as far as 1,200 ft. distant from the treated well have been affected, thus definitely proving the existence of extended fractures.

the permeabilities within and beyond the annular acid-affected region. It is convenient, however, to consider separately the cases in which the initial permeability of the annular region is normal and equal to that of the main body of limestone, so that the acid simply increases this permeability to a higher value, and that in which the annular region has an initial low permeability owing to a natural inhomogeneity in the reservoir or mudding and plugging in the course of drilling, the effect of the acid here being essentially that of raising the permeability approximately to its normal value.   In the first case the effect of the acid treatment is given graphically in Fig.

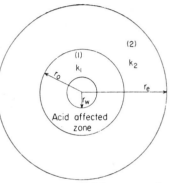

157,[1] where the abscissas k_1/k_i give the ratio of the final annular permeability to its initial value, k_i being also equal to k_2, that of the main body of limestone. It will be seen that the effects as a whole are not very large unless the radius of the affected annulus is quite large, which would require very great amounts of acid. Furthermore, Q/Q_0 does not depend markedly upon the absolute value of the permeability increase as long as k_1/k_i is of the order of

FIG. 156.—A radial-flow system (in limestone) that has been subjected to acid treatment.

5 or greater. In fact, even if the inner annulus is completely dissolved away, so that $k_1/k_i = \infty$, the increase in Q_0 would equal only that corresponding to an increase in well-bore radius from $\frac{1}{4}$ ft. to r_0. Thus the production capacity will not be doubled unless the affect of the acid extends to the radius 11.2 ft., even though the acid *completely removes* the limestone within that radius.

The corresponding curves for the second case are given in Fig. 158. Here the abscissas are the ratios of the initial permeability of the inner ring to the value after acid treatment, it being assumed that this latter value is equal to that of the main body of limestone. Large values of Q/Q_0 are here more readily obtained, even for small r_0, if the inner ring is initially of very low permeabil-

[1] Figures 157 and 158 are essentially equivalent to Fig. 150, having been replotted for convenience in application to the problem of acid treatment.

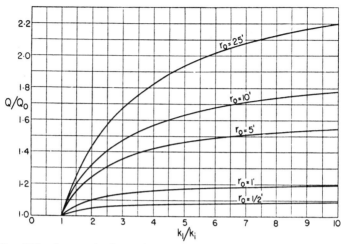

Fig. 157.—Increase in the production capacity of a radial-flow system due to acid treatment if the initial permeability is everywhere uniform. Q/Q_0 = (production capacity after treatment)/(production capacity before treatment); k_1/k_i = (permeability of affected zone of radius r_0 after treatment)/(permeability of affected zone before treatment); well radius = ¼ ft.; external-boundary radius = 500 ft. (*From Physics*, **7**, 108, 1936.)

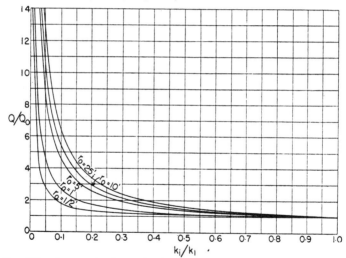

Fig. 158.—Increase in the production capacity of a radial-flow system due to acid treatment, if the acid-affected zone, of radius r_0, is initially of lower permeability than rest of limestone and is raised to latter by the acid. Q/Q_0 = (production capacity after treatment)/(production capacity before treatment). k_i/k_1 = (permeability of affected zone before treatment)/(permeability after treatment). Well radius = ¼ ft.; external-boundary radius = 500 ft. (*From Physics*, **7**, 109, 1936.)

ity and is not limited by the value of r_0.   For evidently in the limiting case when the well face has been completely plugged or mudded off, $k_i = 0$, and Q/Q_0 would become infinitely large.

7.8. The Effect of Acid Treatment in Highly Fractured Limestones.—That extended fractures in a limestone reservoir may play a significant role in the production from such reservoirs becomes fairly obvious when it is observed that a fracture of even a small width may have an effective permeability hundreds of times as great as that of the limestone proper.   For a real fracture of width w may evidently be considered as equivalent to an open linear channel of equal width.   Now for viscous-flow conditions the carrying capacity of such a linear channel per unit pressure gradient may be shown by the classical hydrodynamics to be given by

$$Q = w^3/12\mu, \tag{1}[1]$$

where μ is the viscosity of the liquid.   The equivalent permeability of the channel is, therefore,

$$k = \frac{w^2}{12} = \frac{10^8 w^2}{12} \text{ darcys}, \tag{2}$$

if w is expressed in centimeters.   Hence a fracture of only 0.1 mm. width will have a permeability of 833 darcys, whereas the permeability of the limestone proper will usually be of the order of 0.01 darcy.   In fact, the total fluid-carrying capacity of a complete radial system of radius 45 ft. consisting of a limestone of permeability 0.01 darcy can be carried by a *single* linear fracture 45 ft. long, of depth equal to that of the radial system, and of width 0.126 mm.

If acid is introduced into a well drilled into a limestone formation, it will evidently tend to flow rapidly into, and widen, any fractures leading into the well bore as well as reacting with the limestone immediately surrounding the bore.   Hence, in view of the fact that the radial permeability increase about the well bore will have only a relatively small effect upon the production·capacities, unless the affected region be initially abnormally tight, it becomes of interest to examine the effects to be expected owing to the widening of the fractures.   While in practical cases these fractures will be of limited extent and the acid will not, in general,

[1] *Cf.* H. Lamb, "Hydrodynamics," 6th ed., p. 582, 1932.

penetrate the fractures over their whole length, it is to be recalled
that in the theoretical analysis it was assumed that not only do the
fractures extend initially to the effective reservoir boundary—
300 to 500 ft.—but that furthermore the effect of the acid is to
uniformly widen the fractures over their whole lengths. This
latter assumption will lead to larger increases in the production
capacities than would be observed if the acid did not completely
penetrate the fractures. However, these errors should be small
owing to the high concentration about the well of the pressure
distributions along the fractures, especially in fractures of small
width. The exact value of the effective or real permeabilities at
points distant from the well bore will, therefore, affect but little
the overall resistance of the system.

The final results based on the above assumptions[1] are shown
graphically in Fig. 159. The ordinates Q/Q_0 give the ratios of
the production capacities after the acid treatment to those before
treatment, and the abscissas give the added width of the fractures
after treatment, which will be roughly proportional to the amount
of acid used. The separate curves refer to different initial fracture widths, and for all cases, except the dotted curve, it was
assumed that the limestone permeability is 0.01 darcy, that for
the case of the dotted curve being 0.083 darcy.

Perhaps the most striking feature of these curves is the enormous effect possible, especially for the smaller fracture sizes, for
even moderate increases in their widths. Although, as already
mentioned, the values of Q/Q_0 of Fig. 159 must be somewhat
high owing to the assumptions made in the analysis, it seems
certain, however, that even if strictly corrected for these assumptions, it would still be found that increases in the production
capacity as high even as 100-fold can be reasonably explained by
the fracture mechanism without invoking any *ad hoc* hypotheses.

It will also be clear from Fig. 159 that the effects are larger for
the systems whose fractures are initially narrower and hence were
initially producing at lower rates. The fact that the dotted
curve, which was computed for a limestone permeability of
0.083 darcy, falls lower than the corresponding curve
($w_i = 0.1$ mm.) for a limestone permeability of 0.01 darcy,

[1] The calculation of Q/Q_0 simply consists in computing the ratio of the
reciprocals of the ordinates for curve I, Fig. 154, for the final and initial
values of the fracture width.

shows that if the initially low production rate is due to a low limestone permeability rather than small fracture widths, the effect will again be relatively large as compared to that for a well with initially higher production rate (higher limestone permeability).

In the case of the radial-flow mechanism it was seen that the effects will be greatest for wells in which the inner-zone permeability is a small fraction of that of the rest of the limestone, *i.e.*, for

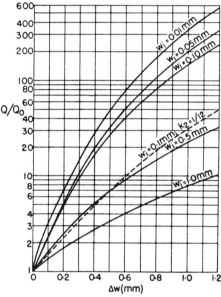

FIG. 159.—Increase in the production capacity of a fractured limestone due to acid treatment. Q/Q_0 = (production capacity after treatment)/(production capacity before treatment). Δw = increase in fracture width caused by the acid injection. w_i = initial fracture width; ———— limestone permeability = 0.01 darcy; .... limestone permeability = $\frac{1}{12}$ darcy. . *(From Physics, 7, 114, 1936.)*

those which show a relatively serious plugging. However, for a fixed inner-zone permeability the wells with the highest permeability beyond this zone, and hence highest initial production capacities, will show the greatest responses to the acid treatment, while if the permeability of the outer zone is considered as fixed, those in which the inner ring is of lowest permeability—and hence of lowest initial production capacity—will show the greatest response. As field experience has shown that the smaller producers usually react best to acid treatment, it must be concluded that insofar as the mechanism may correspond to the radial-flow case, the small producers are such because of serious plugging about the well bore rather than because of a low permeability in the main body of the limestone.

These results may be summarized as folllows:

1. Small increases in the production capacity—up to about 50 per cent—due to acid treatment *may* be explained on the assumption that the permeability of a small radial zone about the well bore has been increased from normal to higher values, as well as by removal of radial plugging or widening of extended fractures fed laterally by the limestone proper. Unless the limestone does have extended fractures or is appreciably plugged near the well bore, acid treatment should be relatively ineffective in stimulating the production.

2. Moderate increases in the production capacity—50 to 500 per cent—can be explained on a radial-flow basis only on the assumption that the wells were initially plugged, the extent of the plugging being the principal factor in determining the initial production capacity, so that small producers will show larger responses. They can also be explained equally well by the assumption of the extended-fracture flow.

3. Increases in the production capacity appreciably larger than 500 per cent, for wells of initially moderate capacity, can be explained only on the assumption that there are extended fractures in the limestone which are penetrated and widened by the acid. Here the smaller producers should show the greater responses, whether their initially small production capacities are due to low limestone permeabilities or small widths of the fractures. For very small producers, increases higher than 500 per cent could also be explained on the radial-flow mechanism of production, although it would have to be assumed that there was initially a condition of *almost complete plugging* near the well bore.

It should be pointed out that from preliminary "flow tests" one can make no a priori prediction as to the mechanism of production and hence foretell the probable effect of the acid. All mechanisms will show an approximately linear relation between the production rate and pressure differential, from which only the resultant resistance can be derived; this resultant resistance can be synthesized from either the radial flow or fracture mechanism by appropriate assignment of the many physical and geometrical constants available for defining the details of the system.

7.9. Partially Penetrating Wells in Stratified Horizons.— Another type of problem involving regions of different permeability and of practical importance is that in which a partially penetrating well is drilled into a sand composed of strata of different permeability. In contrast to the two-dimensional problems thus far treated in this chapter, that of the present section is three-dimensional in character. However, from a formal point of view, the analysis will be quite similar to that which was developed in Sec. 7.4, the only essential difference being that the fundamental elementary solutions from which the final potential distributions will be synthesized will here be the product of an exponential and a Bessel function, whereas in Sec. 7.4 they were products of an exponential and trigonometric function.

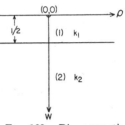

Fig. 160.—Diagrammatic representation of a "nonpenetrating" well tapping a stratified sand.

Although in any actual case the penetration of the well will be of nonvanishing magnitude, and the sand will be of finite thickness and may be of variable permeability, it will be assumed here for simplicity that the well is nonpenetrating, and that the sand is composed simply of a layer (1) of permeability k_1 resting on an infinite homogeneous sand (2) of permeability k_2, as indicated in Fig. 160.

This problem can be treated by a slight generalization of the method of images as used in Sec. 5.3.[1] However, it will be both simpler and more instructive to illustrate the use of Bessel functions[2] in the discussion of potential problems with axial symmetry. At the same time we shall take the occasion to

[1] HUMMEL, J. N., *Zeits. Geophysik*, **5**, 89, 228, 1929, *cf.* also footnote on page 410.

[2] MUSKAT, M., *Physics*, **4**, 129, 1933, where other references are cited. *Cf.* also L. V. King, *Roy. Soc. Proc.* A-**139**, 237, 1933. In these references the problem is treated from the point of view of the electrical analogy of the flow problem.

point out explicitly the essential principles of what is one of the most powerful methods of solving the classical partial differential equations of mathematical physics, namely, the method of the separation of variables. Thus, whereas the elementary solutions of Laplace's equation as $\log r$ (Sec. 4.2), $r^n \frac{\cos}{\sin} n\theta$ (Sec. 4.5), $f(x + iy)$ (Sec. 4.8), $1/r$ (Sec. 5.2), and $e^{\pm yz} \frac{\cos}{\sin} xz$ (Sec. 7.4) thus far used in the construction of pressure or potential distributions appropriate to specific physical problems, have been either derived from degenerate forms of the partial differential equation, as Eqs. 4.2(3) and 5.2(2), or have been simply proposed as solutions to be verified by substitution, the method of separation of variables provides a systematic procedure for deriving such elementary solutions. This method is applied as follows:

Taking Laplace's equation in cylindrical coordinates, Eq. 3.7(3), and introducing the assumption of axial symmetry and the dimensionless variables of Eq. 5.3(3), namely,

$$w = \frac{z}{2h}; \qquad \rho = \frac{r}{2h}, \tag{1}$$

where h is the thickness of the upper layer which is tapped by the well, one has

$$\frac{\partial^2 \Phi}{\partial \rho^2} + \frac{1}{\rho} \frac{\partial \Phi}{\partial \rho} + \frac{\partial^2 \Phi}{\partial w^2} = 0. \tag{2}$$

It is now assumed that the potential function Φ can be expressed as a function in which the variables are "separated," namely, as the product of a function R of the single variable ρ, and one, W, of the single variable w, *i.e.*,

$$\Phi = R(\rho)W(w); \tag{3}$$

this is substituted in Eq. (2) and the result is divided throughout by Φ. One thus finds

$$\frac{1}{R} \frac{d^2R}{d\rho^2} + \frac{1}{R\rho} \frac{dR}{d\rho} + \frac{1}{W} \frac{d^2W}{dw^2} = 0. \tag{4}$$

Next, transposing the last term of Eq. (4) to the right side, and observing that the left side is then a function of ρ alone and the right side is a function of w alone, it is clear that the two sides can be equal only if they are both equal to a constant, *i.e.*, if

$$\frac{1}{R}\frac{d^2R}{d\rho^2} + \frac{1}{R\rho}\frac{dR}{d\rho} = -\frac{1}{W}\frac{d^2W}{dw^2} = -\alpha^2,$$

where α is an arbitrary constant.

It now follows at once that

$$\frac{d^2W}{dw^2} - \alpha^2 W = 0, \tag{5}$$

$$\frac{d^2R}{d\rho^2} + \frac{1}{\rho}\frac{dR}{d\rho} + \alpha^2 R = 0. \tag{6}$$

Thus the problem of solving the *partial* differential Eq. (2) has been reduced, by "separating the variables" in Eq. (2), to that of solving the *ordinary* differential Eqs. (5) and (6) for the component functions R and W, a reduction which, in general, results in very significant analytical simplifications.

The solution of Eq. (5) is evidently

$$W = \text{const. } e^{\pm w\alpha}. \tag{7}$$

Equation (6), on the other hand, does not have a solution that can be represented by a finite number of elementary functions. Rather, its solution is an infinite series in powers of $\rho\alpha$ with uniquely defined coefficients,[1] whose sum is known as the Bessel function of zero order, and is denoted by $J_0(\rho\alpha)$, *i.e.*,

$$R = \text{const. } \sum_0^\infty \frac{(-1)^n (\rho\alpha)^{2n}}{2^{2n}(n!)^2} = \text{const. } J_0(\rho\alpha). \tag{8}[2]$$

---

[1] PIAGGIO, H. T. H., "Differential Equations," 1929, Chap. IX.

[2] The second fundamental solution of Eq. (6), often called the Neumann's function $N_0(\alpha\rho)$, has a logarithmic singularity at $\rho = 0$, and hence need not be considered here. Further, if α is not real, neither of the solutions of Eq. (6) can be used here as one becomes infinite for large ρ and the other becomes infinite for small ρ. If α is imaginary, the solution becoming infinite for large ρ is the function I_0 to be used in the next section, while that becoming infinite for small ρ is the Hankel function K_0 also used there and in Sec. 5.3. Further details concerning the theory of these various functions will be found in Chap. XVII, "Modern Analysis," by E. T. Whittaker and G. N. Watson, and in the treatise "The Theory of Bessel Functions," 1922, by G. N. Watson, which is entirely devoted to functions of this type.

The elementary solutions of Eq. (2) can, therefore, be expressed as

$$\Phi = e^{\pm w\alpha}J_0(\rho\alpha).$$ (9)

Returning now to the original physical problem stated at the beginning of this section, one may represent the nonpenetrating well by a point sink situated at the well center. Such a sink will contribute a potential distribution of the form

$$\Phi_0 = \frac{-1}{\sqrt{\rho^2 + w^2}}.$$ (10)

With this solution one could build up a system of images, as was done in Sec. 5.3, so as to give a resultant potential distribution satisfying the boundary conditions. But since the elementary solution of Eq. (2) derived above—Eq. (9)—is expressed in terms of the Bessel function, it can be combined with Eq. (10) only if the latter is also expressed in terms of the Bessel function. This may be effected by means of the integral relation[1]

$$\Phi_0 = \frac{-1}{\sqrt{\rho^2 + w^2}} = -\int_0^{\infty} e^{-w\alpha}J_0(\rho\alpha)d\alpha.$$ (11)

Combining then Φ_0 with a similar continuous superposition of the elementary solutions as Eq. (9), the resultant potential distribution in the upper layer of Fig. 160 may be expressed as

$$\Phi_1 = \int_0^{\infty} J_0(\rho\alpha)[-e^{-w\alpha} + A(\alpha)e^{-w\alpha} + B(\alpha)e^{(w-\frac{1}{2})\alpha}]d\alpha,$$ (12)

the last two terms of the integrand representing the perturbation in Φ_0 due to the lower layer in Fig. 160. In fact if the bottom layer be of the same permeability as the upper zone the potential due to the well will simply be Φ_0, and indeed, as will be seen later, if $k_2 = k_1$, A and B will vanish.

Similarly one may develop the solution Φ_2 that gives the potential distribution in the lower zone. Since, however, it extends to infinity, the term involving $e^{w\alpha}$ must be excluded, since this would become infinite at infinity. Φ_2 must, therefore, have the form

$$\Phi_2 = \int_0^{\infty} C(\alpha)e^{-(w-\frac{1}{2})\alpha}J_0(\rho\alpha)d\alpha.$$ (13)

[1] WATSON, G. N., "Theory of Bessel Functions," p. 384.

The form of the functions $A(\alpha)$, $B(\alpha)$, and $C(\alpha)$ are to be chosen in such a way that Φ_1 and Φ_2 satisfy the boundary conditions of the problem. These are, physically, that (1) the plane $w = 0$ be impermeable to the flow of the liquid; (2) the pressures, and hence Φ/k, be continuous as one passes across the interface at $w = \frac{1}{2}$; and (3) that the normal velocity $-\dfrac{\partial \Phi}{\partial n}$ be continuous at the interface $w = \frac{1}{2}$. Analytically, these may be stated as

$$\left.\begin{array}{ll} \dfrac{\partial \Phi_1}{\partial w} = 0: & w = 0 \\[2mm] \dfrac{\Phi_1}{k_1} = \dfrac{\Phi_2}{k_2}: & w = \dfrac{1}{2} \\[2mm] \dfrac{\partial \Phi_1}{\partial w} = \dfrac{\partial \Phi_2}{\partial w}: & w = \dfrac{1}{2} \end{array}\right\} \tag{14}$$

Since the integrals of both Φ_1 and Φ_2 have the limits 0 and ∞, and since their integrands have the same term $J_0(\rho\alpha)$, it is only necessary to consider the remainder of the integrands in applying the boundary conditions of Eq. (14).[1]

Applying these conditions to the functions of Eqs. (12) and (13), and setting $k_2/k_1 = \delta$, it is found that the coefficients A, B, C must satisfy the equations

$$\left.\begin{array}{l} A - Be^{-\frac{\alpha}{2}} = 0 \\[2mm] Ae^{-\frac{\alpha}{2}} + B - \dfrac{C}{\beta} = e^{-\frac{\alpha}{2}} \\[2mm] Ae^{-\frac{\alpha}{2}} - B - C = e^{-\frac{\alpha}{2}} \end{array}\right\} \tag{15}[2]$$

These have the solutions:

---

[1] While such procedures lead directly only to *sufficient* conditions upon A, B, C, the theorem of the uniqueness of the solutions to Laplace's equation with given boundary conditions (*cf.* Sec. 3.5) insures that the solution so obtained will be analytically equivalent to the only physical solution of the problem.

[2] The first of these equations has been obtained by applying the first of Eqs. (14) only to the terms A and B in Φ_1, as the first term due to Φ_0 as given by Eq. (10) automatically satisfies the first of Eqs. (14), although the derivative of its integral representation is discontinuous at $w = 0$.

$$A = \frac{-e^{-\alpha}(\delta - 1)}{\Delta\delta}; \qquad B = \frac{-e^{-\frac{\alpha}{2}}(\delta - 1)}{\Delta\delta} \left.\right\}$$

$$C = \frac{2e^{-\frac{\alpha}{2}}}{\Delta}; \qquad \Delta = -\frac{2e^{-\frac{\alpha}{2}}}{\delta}\left\{\sinh\frac{\alpha}{2} + \delta\cosh\frac{\alpha}{2}\right\} \left.\right\} \tag{16}$$

so that A and B vanish for $\delta = 1$ ($k_2 = k_1$), as anticipated above.

After a little reduction, it is found that with these values of A, B, and C, and with the notation

$$\begin{aligned} \delta &= \text{ctnh } \epsilon: & \frac{k_2}{k_1} &= \delta > 1 \\ &= \tanh \epsilon: & \frac{k_2}{k_1} &= \delta < 1, \end{aligned} \left.\right\} \tag{17}$$

Φ_1 and Φ_2 may be expressed as follows:
For $\delta > 1$,

$$\begin{aligned} \Phi_1 &= -\int_0^\infty \frac{J_0(\rho\alpha)\sinh\left(\epsilon + \frac{\alpha}{2} - w\alpha\right)}{\cosh\left(\epsilon + \frac{\alpha}{2}\right)}d\alpha \\ \Phi_2 &= -\cosh\epsilon\int_0^\infty \frac{J_0(\rho\alpha)e^{-(w-\frac{1}{2})\alpha}}{\cosh\left(\epsilon + \frac{\alpha}{2}\right)}d\alpha. \end{aligned} \left.\right\} \tag{18}$$

For $\delta < 1$,

$$\begin{aligned} \Phi_1 &= -\int_0^\infty \frac{J_0(\rho\alpha)\cosh\left(\epsilon + \frac{\alpha}{2} - w\alpha\right)}{\sinh\left(\epsilon + \frac{\alpha}{2}\right)}d\alpha \\ \Phi_2 &= -\sinh\epsilon\int_0^\infty \frac{J_0(\rho\alpha)e^{-(w-\frac{1}{2})\alpha}}{\sinh\left(\epsilon + \frac{\alpha}{2}\right)}d\alpha. \end{aligned} \left.\right\} \tag{19}$$

These equations contain the whole description of the potential distribution in the system. Although we shall not enter into the details of the evaluation of the integrals, the essential results of the analysis will be briefly outlined. Thus for $w = 0$, *i.e.*, at the top of the sand, it is found that

$$\Phi_1(w = 0) = -\left[\frac{1}{\rho} + 2\sum_1^{\infty}\frac{\nu^n}{\sqrt{\rho^2 + n^2}}\right]$$

where

$$\nu = \frac{1 - \delta}{1 + \delta},$$

$$(20)$$

showing that the potential distribution may be considered to be due to a series of images at $w = \pm n$ with strengths ν^n. On expanding Eq. (19) in powers of ρ, or substituting directly in Eqs. (18) and (19) the power-series expansion for J_0 as given by Eq. (8), a series convenient for numerical computation when ρ is small is obtained. It is

$$\Phi_1(w = 0) = -\left[\frac{1}{\rho} - 2\log(1 - \nu) - 2\nu + \frac{2\nu}{\sqrt{1 + \rho^2}} + \right.$$

$$\left. 2\sum_1^{\infty}(-1)^m c_m f_m(\nu)\rho^{2m}\right], \quad (21)$$

where

$$c_m = \frac{1 \cdot 3 \cdots 2m - 1}{2 \cdot 4 \cdots 2m}; \quad f_m(\nu) = \sum_2^{\infty}\frac{\nu^n}{n^{2m+1}}, \quad (22)$$

and the term $2\nu/\sqrt{1 + \rho^2}$ has been separated out so as to increase the radius of convergence of the series from $|\rho| = 1$ to $|\rho| = 2$.

For large values of ρ, an asymptotic expansion of Φ_1 in inverse powers of ρ may be derived. To terms in $1/\rho^9$, it is given by

$$\Phi_1(w = 0) = \frac{-1}{\delta\rho}\left[1 + \frac{1}{4\rho^2\delta^2}(\delta^2 - 1) - \frac{3}{16\rho^4\delta^4}(\delta^2 - 1)(3 - 2\delta^2)\right.$$

$$\left. + \frac{5}{64\rho^6\delta^6}(\delta^2 - 1)(17\delta^4 - 60\delta^2 + 45) - \cdots\right]. \quad (23)$$

If $\delta = 0$ (the lower zone is completely impermeable) Eq. (23) breaks down and one must return to Eq. 5.3(7), which was constructed explicitly for this case. On the other hand, if the lower medium is of infinite permeability—$\delta = \infty$—Eq. (23) again must be replaced. The appropriate expression for this case is found to be

$$\Phi_1(w = 0; \delta = \infty) = -4\sum_{0}^{\infty} K_0((2n + 1)\pi\rho). \quad (24)[1]$$

In order to get a clearer idea of the effect of the lower medium upon the potential distribution, $-\Phi_1$ has been plotted in Fig. 161 as a function of ρ for several representative values of the ratio

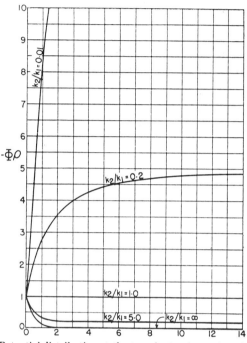

Fig. 161.—Potential distribution at the top of a two-layered sand, tapped by a "nonpenetrating" well. $k_2/k_1 =$ [permeability of lower (infinitely thick) stratum]/(permeability of upper layer); $\rho =$ radial distance measured in units of twice the upper-layer thickness.

k_2/k_1. To avoid difficulties in plotting and overlapping of the curves when ρ is small, $-\rho\Phi_1$ has been plotted rather than $-\Phi_1$ itself.

It will be seen that when the permeability in the upper layer is higher than that below ($k_2/k_1 < 1$), the potential does not fall off as rapidly as in a homogeneous system, whereas in the case

[1] The detailed derivation of this result and of Eq. 5.3(7), as well as a treatment of the electrical analogy of the case where the permeable stratum is composed of three layers will be found in the paper by Muskat (*loc. cit.*).

of lower layers more permeable than those above, the potential drop is more rapid.   This may be understood in a qualitative way when it is observed that a bottom layer of low permeability will tend to concentrate the flow into the upper layer, giving it a radial character and hence small potential gradients, as contrasted with the perfectly spherical distribution when the lower zone has the same permeability as the upper.   Just the opposite effect would be obtained by drawing the flow downward into a lower zone of high permeability.

Of more practical interest, however, is the relation between the effective resistance of the system and the permeability ratio k_2/k_1.   This may be expressed in a practical form by computing the production capacity of a well for a unit pressure or potential differential over the system as a function of k_2/k_1.   Because of the different forms that Φ takes for large values of ρ, depending upon the value of k_2/k_1, the relation may best be represented graphically.   In making these computations it is to be noted first that all the potential expressions given above correspond to a well-production rate of 2π and well potentials given by

$$\Phi_w = \frac{-1}{\rho_w} + 2 \log \frac{2k_2}{k_1 + k_2}. \qquad (25)[1]$$

It has also been assumed for definiteness that

$$\rho_w = 0.005; \qquad \rho_e = 10,$$

where ρ_w is the well radius divided by twice the upper-layer thickness, and ρ_e is the external radius at which the external potential Φ_e is applied, also divided by twice the upper-layer thickness. Thus, for example, for a well radius of $\frac{1}{4}$ ft., $\rho_w = 0.005$ corresponds to an upper layer 25 ft. thick, and $\rho_e = 10$ corresponds to an external radius of 500 ft.

The quantity[2] $Q/\Delta\Phi$ or the flux into the well per unit potential drop across the sand (or unit pressure drop for a value of $k/\mu = 1$ in the upper layer) is plotted against k_2/k_1 in Fig. 162.   It is evident from this curve that the effect of the lower zone may be considered merely as a small correction to the main flow in the

---

[1] Terms in Eq. (21) of the order of $\rho_w{}^2$ or smaller are here neglected.

[2] $\Delta\Phi$ is simply the difference between $\Phi_1(10)$, as given by Eq. (23), and $\Phi_w(0.005)$, as given by Eq. (25).

upper sand.   Thus over the infinite range of permeability, k_2, for the lower zone the total change in Q is only 3 per cent.

Although this last result was derived on the explicit assumption that the well just taps the upper sand, it is not difficult to see in a qualitative way that the increase in the effect of the lower zone when the well penetration is nonvanishing will not be large. For assuming for the moment that the lower sand is of the same permeability as the upper one, we may apply the results of Secs. 5.3 and 5.4 and Fig. 85 which show that for a single sand with a partially penetrating well the lower layers of the sand below the bottom of the well give rapidly decreasing contributions to the

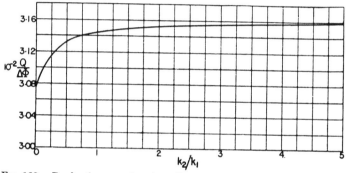

Fig. 162.—Production capacity of a well just tapping a 25 ft. sand of permeability k_1 overlying an infinitely thick sand of permeability k_2. $Q/\Delta\Phi$ = production capacity per unit potential drop in upper stratum (cubic centimeter per second per atmosphere for $k_1/\mu = 1$); well radius = $\frac{1}{4}$ ft.; external boundary radius = 500 ft.

total flow as these layers lie at increasing depths in the sand. In fact, unless the penetration of the well in the upper sand exceeds 50 per cent and is in numerical value less than 25 ft., the addition below it of a very thick sand of the same permeability will probably not increase the total production capacity of the well by more than 15 per cent.   Now returning to Fig. 162 of this section it is seen that, at least for a nonpenetrating well, more than three-fourths of the increase in production capacity due to a lower sand of infinite conductivity is already attained if the lower sand has a permeability no greater than that of the upper layer.   Hence it may be concluded that the effect of a lower zone of high permeability on a well partially penetrating a sand overlying this lower layer will probably not exceed a 20 per cent increase in the total production capacity for penetrations up to 50 per cent, whereas in cases of small well penetrations and lower permeability deeper layers the increases will probably not exceed 10 per cent.[1]

Finally it may be observed that the large differences between the curves of Fig. 161 for the various values of k_2/k_1 do not contradict the small effect just noted in the variation of the flux Q with k_2/k_1. The reason is evidently that the apparently large differences in the values of $-\rho\Phi_1$ for large ρ, among the curves of Fig. 161, involve only small absolute variations in the value of Φ_e. And since Φ_e itself is negligibly small compared to the well potential Φ_w, even relatively large percentage variations in Φ_e will affect only slightly the differential $\Delta\Phi = \Phi_e - \Phi_w$, and the ordinates $Q/\Delta\Phi$ which are plotted in Fig. 162.

7.10. The Effect of a Sanded Liner on the Production Capacity of a Well.—As a final problem involving porous media of different permeabilities within the same flow system we shall consider the effect of a sanded liner on the production capacity of a well. Although at first thought it would appear that the presence of a column of clean sand at the bottom of a well bore could have but little effect upon the production capacity of the well, since the permeability of such a column of clean sand will be much higher than that of the usual pay sand, it is nevertheless frequently observed that a flowing or pumping well will suffer a considerable decrease in production rate owing to the entry of sand into the well bore. That the loss in production rate, or even actual failure of the well, is in many cases due to the presence of the sand within the liner and not to the "mudding off" or clogging of the screen itself is proved by the fact that complete recovery often results from the removal of the accumulated debris. It is, therefore, of considerable practical interest to analyze in detail the physical explanation of this phenomenon.

The physical system may be represented diagrammatically as in Fig. 163. It will be assumed that the sand fills the well

---

[1] A close approximation to a quantitative solution when the well penetration is nonvanishing should be given, as in Sec. 5.3 for the case of a single sand, by assuming a uniform flux along the well surface and then taking the well potential as that at three-fourths of the depth of penetration from the top of the sand. However, for most practical purposes the qualitative results given above should suffice.

bore to the top of the pay and is of uniform permeability k_1, while the permeability of the pay sand is k_2. Although the sand in the well bore is maintained there by gravity alone, so that near the top of the column the upward flowing liquid will tend to loosen the packing, and at high rates of flow may even remove the sand from the well, the essential features of the problem will evidently be displayed by a system in which the sand column has a uniform permeability from top to bottom. It will also be convenient to suppose the pay sand to extend indefinitely outward from the well bore and then take the reservoir pressure

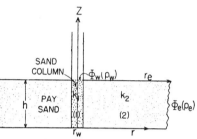

Fig. 163.—Diagrammatic representation of a well with a sanded liner.

to be that at a large distance r_e from the well bore. The problem may then be analytically formulated as follows: Find potential functions Φ_1 and Φ_2 such that

$$\frac{\partial \Phi_1}{\partial z} = \begin{cases} 2c: & z = 1; \\ 0: & z = 0; \end{cases} \quad \frac{\partial \Phi_2}{\partial z} = 0; \quad z = 0, 1 \\ \frac{\Phi_1}{k_1} = \frac{\Phi_2}{k_2}; \quad \frac{\partial \Phi_1}{\partial r} = \frac{\partial \Phi_2}{\partial r}: \quad r = r_w; \quad \Phi_2 \to \Phi_e, \text{ for } r \to r_e \gg r_w \quad \Bigg\} \quad (1)$$

where the unit of length has been taken as the sand thickness h. The condition that $\dfrac{\partial \Phi_1}{\partial z} = 2c$ at $z = 1$ has been chosen instead of the requirement that $\Phi_1 = \Phi_w$ (the well potential) in order to simplify the analysis. However, in view of the small value of the radius r_w as compared to the other dimensions of the system, the two conditions are physically quite equivalent.

The first set of conditions of Eq. (1) will clearly be satisfied by the potential distributions

$$\left.\begin{aligned}
\Phi_1 &= B_0 + c\left(z^2 - \frac{r^2}{2}\right) + \sum_1^\infty B_n I_0(n\pi r)\cos n\pi z \\
\Phi_2 &= \bar{A} + A_0 \log \frac{r}{r_w} + \sum_1^\infty A_n K_0(n\pi r)\cos n\pi z,
\end{aligned}\right\} \quad (2)$$

where I_0 and K_0 are Bessel functions of the third kind, the former becoming exponentially infinite for large arguments and the latter vanishing exponentially for large arguments. K_0 is also known as the Hankel function, which was introduced in the analysis of partially penetrating wells (*cf.* Sec. 5.3).[1] B_n, A_n, and $\bar{A}$ are constants to be determined so as to satisfy the other conditions of Eq. (1). These give

$$\left.\begin{aligned}
\sum_1^\infty [\delta B_n I_0(n\pi r_w) - A_n K_0(n\pi r_w)]\cos n\pi z &= \bar{A} - \delta B_0 - \\
&\qquad c\delta\left(z^2 - \frac{r_w^2}{2}\right) \\
\pi \sum_1^\infty n[B_n I_1(n\pi r_w) + A_n K_1(n\pi r_w)]\cos n\pi z &= \frac{A_0}{r_w} + c r_w; \\
\delta = \frac{k_2}{k_1}; \qquad \Phi_e &= \bar{A} + A_0 \log \frac{r_e}{r_w}.
\end{aligned}\right\} \quad (3)$$

The first two equations being Fourier series, the coefficients of the $\cos n\pi z$ may be determined by the method of Sec. 4.3. Thus it is found that

$$\left.\begin{aligned}
\bar{A} &= \Phi_e + c r_w^2 \log \frac{r_e}{r_w}; \qquad A_0 = -c r_w^2; \qquad B_0 = \frac{\Phi_e}{\delta} - \\
&\qquad c\left[\frac{1}{3} - \frac{r_w^2}{2} - \frac{r_w^2}{\delta}\log \frac{r_e}{r_w}\right] \\
A_n &= -\frac{I_1(n\pi r_w) B_n}{K_1(n\pi r_w)}; \quad B_n = \frac{-4c(-1)^n/n^2\pi^2}{\left[I_0(n\pi r_w) + \dfrac{I_1(n\pi r_w) K_0(n\pi r_w)}{\delta K_1(n\pi r_w)}\right]}.
\end{aligned}\right\} \quad (4)$$

[1] For further properties of the Bessel functions of general order n, *i.e.*, I_n, K_n, see "Modern Analysis," Sec. 17.70 and 17.71. The functions K_n used here, however, are taken to be those defined in "The Theory of Bessel Functions," and hence are $(-1)^n$ times those defined in "Modern Analysis" (*cf.* also footnote on page 431).

Taking the well potential Φ_w as that at the center of the sand column, Eq. (2) gives

$$\Phi_w = \frac{\Phi_e}{\delta} + c\left[\frac{2}{3} + \frac{r_w^2}{2} + \frac{r_w^2}{\delta}\log\frac{r_e}{r_w}\right] + \sum_1^\infty (-1)^n B_n \quad (5)$$

as $I_0(0) = 1$.

Introducing the notation

$$\bar{B}_n = \frac{1}{n^2\left[I_0(n\pi r_w) + \dfrac{I_1(n\pi r_w)K_0(n\pi r_w)}{\delta K_1(n\pi r_w)}\right]},$$

and observing that the flux rate from the well is related to the constant c by the relation

$$Q = -\pi r_w^2 h\frac{\partial \Phi_1}{\partial z}\bigg|_{z=1} = -2\pi chr_w^2, \quad (6)$$

it is found that the conductivity of the system may be written in the form

$$\frac{Q}{\Delta p} = \frac{2\pi k_1 hr_w^2/\mu}{\dfrac{2}{3} + \dfrac{r_w^2}{2} + \dfrac{r_w^2}{\delta}\log\dfrac{r_e}{r_w} - \dfrac{4}{\pi^2}\displaystyle\sum_1^\infty \bar{B}_n}, \quad (7)$$

where $\Delta p = p_e - p_w$. Denoting by Q_0 the flux rate for a sand-free well bore, the effect of the sand in the well bore may be expressed by the ratio

$$\frac{Q}{Q_0} = \frac{r_w^2 \log r_e/r_w}{\delta\left\{\dfrac{2}{3} + \dfrac{r_w^2}{2} + \dfrac{r_w^2}{\delta}\log\dfrac{r_e}{r_w} - \dfrac{4}{\pi^2}\displaystyle\sum_1^\infty \bar{B}_n\right\}}; \quad \delta = \frac{k_2}{k_1}. \quad (8)$$

Before giving a discussion of Eq. (8) it is of interest to note its behavior as limiting cases are approached. Thus the limit when there is no sand at all in the well bore evidently corresponds to a vanishing value of δ. As the $\bar{B}_n$ vanish for $\delta = 0$, Eq. (8) reduces to

$$\frac{Q}{Q_0}(\delta = 0) = 1, \quad (9)$$

showing that the system degenerates into one of strict radial flow, as indeed it should.

The case $\delta = 1$ clearly corresponds to an ordinary non-penetrating well, as then the sand column is of the same permeability as that of the main sand body. Here

$$\bar{B}_n = \frac{1}{n^2\left\{I_0(n\pi r_w) + I_1(n\pi r_w)\dfrac{K_0(n\pi r_u)}{K_1(n\pi r_w)}\right\}} = \frac{\pi r_w K_1(n\pi r_w)}{n}. \quad (10)[1]$$

Summing $\Sigma \bar{B}_n$ by the Euler[2] summation formula, it is found that

$$z\sum_{1}^{\infty}\frac{K_1(nz)}{n} = \frac{\pi^2}{6} - \frac{\pi z}{2} + O(z^2). \quad (11)$$

Hence, as $r_w \ll 1$, Eq. (8) becomes

$$\frac{Q}{Q_0}(\delta = 1) = \frac{r_w \log r_e/r_w}{2}; \qquad Q = \frac{\pi k_2 r_w \Delta p}{\mu}, \quad (12)$$

where in the second equation r_w is the real value of the well radius. This latter value for Q is identical with that following from an analysis similar to that of Sec. 5.3, applied directly to the problem of a nonpenetrating well, except for a factor of 2.[3] This factor evidently arises first from the fact that the "nonpenetrating well" here is a disk of area πr_w^2, whereas in the treatment just mentioned it is taken as a hemisphere of area $2\pi r_w^2$, and secondly because the "well potential" Φ_w of Eq. (5) has been taken as that at the center of the disk, so that the Δp of Eq. (12) is the maximal Δp in the system, not the average of Δp over the boundary surfaces.

The values of Q/Q_0 for general values of δ, with $r_w/h = 0.01$ and $r_e/r_w = 2,000$, are plotted in Fig. 164. The tremendous effect of even very permeable sand columns is shown in this figure by the almost vertical descent of the curve for small values of k_2/k_1. Thus a sand column of permeability as high as 200 times that of the sand pay will reduce the production

---

[1] *Cf.* "Theory of Bessel Functions," p. 80.

[2] *Cf.* E. T. Whittaker and C. Robinson, "The Calculus of Observations," 2d ed., 1926, Chap. VII, or "Modern Analysis," Sec. 7.21.

[3] For the details of this analysis *cf.* M. Muskat, *Physics* **2**, 329, 1932, pt. I. Eq. 5.2(11) also gives the same result when divided by 2 to give the flux into each hemispherical surface.

capacity[1] of the well to only 34 per cent of its original value.
Furthermore, if the permeability of the sand column is of the

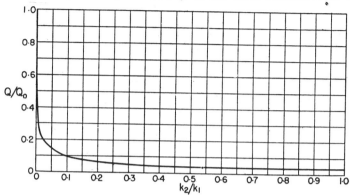

Fig. 164.—Effect of a sanded liner on the production capacity of a well.
Q/Q_0 = (production capacity of well with sanded liner)/(production capacity of
sand-free well). k_2/k_1 = (permeability of pay sand)/(permeability of sand
column in liner); sand thickness = 25 ft.; well radius = $\frac{1}{4}$ ft.; external-boundary
radius = 500 ft.

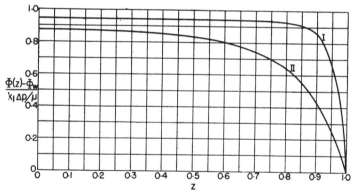

Fig. 165.—The potential distribution along the axis of a well bore filled with
sand of permeability k_1 to the top of the producing pay. $\dfrac{\Phi(z) - \Phi_w}{k_1 \Delta p/\mu}$ = (potential
drop between top of pay at well axis and depth z)/(total potential drop in system).
z = depth (in units of sand thickness) along well axis from bottom of sand column.
(Permeability of sand column)/(permeability of pay sand) = 10 for curve I, and
100 for curve II.

same order as that of the pay, the production capacity is effec-

[1] Although the problem of a sanded liner has been considered here from
the point of view of a well draining the sand, all the results, including
Figs. 164 and 165, will obviously apply also to the case when liquid is
being injected *into* the sand through the well, as in salt-water disposal or
flooding systems.

tively that of a nonpenetrating well regardless of the exact value of the ratio of the permeabilities.

The physical reason for the marked effects of sand columns of high permeability becomes clear when one considers the potential distributions in the well bores when filled with sand. Figure 165 gives curves showing the vertical potential distribution along the axes of well bores for which $k_2/k_1 = 0.01$ and 0.1, the latter value corresponding to a sand column containing appreciable amounts of fines or silt. The distributions are given by

$$\frac{\Phi_1(z) - \Phi_w}{k_1(p_e - p_w)/\mu} = \frac{1 - z^2 - \dfrac{4}{\pi^2}\sum \bar{B}_n + \dfrac{4}{\pi^2}\sum (-1)^n \bar{B}_n \cos n\pi z}{\dfrac{2}{3} + \dfrac{r_w{}^2}{2} + \dfrac{r_w{}^2}{\delta}\log\dfrac{r_e}{r_w} - \dfrac{4}{\pi^2}\sum \bar{B}_n},$$

$$(13)$$

which readily follows from Eqs. (2) and (5), and gives the fraction of the total potential drop in the sand which exists between the top of the sand column and the depth z. It will be seen that the potential rises very rapidly as z decreases, so that the lower parts of the sand are effectively producing against back pressures which are very high as compared to the well pressure at the top of the sand column, and hence cannot contribute appreciably to the production from the well. The penetration of the well is thus reduced from one of physically complete penetration to an effective value in a homogeneous sand of less than 20 per cent of the thickness of the pay horizon, if the sand column permeability is less than 200 times that of the pay sand.

While it has been possible to solve rigorously the problem of the sanded liner—under the single simplifying assumption that the flux is uniform over the top of the sand column—it is instructive to note that the essential features of the problem can be derived by a much more elementary analysis based on some reasonable physical approximations. Thus, assuming that the flow within the well bore is strictly linear while that in the sand pay is strictly radial, the pressure distribution along the axis of the well bore must satisfy the equation

$$\frac{d^2 p(z)}{dz^2} = \frac{-2\delta}{r_w{}^2 \log r_e/r_w}[p_e + \gamma g(h - z) - p(z)], \qquad (14)[1]$$

where the quantities z, r_w, and r_e are now absolute lengths, with the solution

$$p_e + \gamma g(h - z) - p(z) = \frac{(p_e - p_w)\cosh bz}{\cosh bh}, \qquad (15)$$

[1] It is supposed that p_e is measured at the top of the sand, $z = h$.

where

$$b^2 = \frac{2\delta}{r_w^2 \log r_e/r_w},\qquad(16)$$

so that

$$\frac{Q}{Q_0} = \frac{\tanh bh}{bh}.\qquad(17)$$

As should be expected, this approximate theory gives lower values for Q/Q_0 than the exact analysis, since it does not take into account the vertical flow in the sand pay toward the upper part of the sand column where the pressures are lowest. Thus while Eq. (8) gives $Q/Q_0 = 0.258$ for $\delta = 0.01$, Eq. (17) gives a value of 0.195. The order of magnitude of the effect, however, is the same as that given by the exact theory.

FIG. 166.—Diagrammatic representation of a well bore filled with sand to a depth h_e above the top of the pay.

When the sand column extends above the top of the pay, it is no longer possible to give a rigorous treatment of the system, and one must resort to approximations of the type just indicated. Thus it will be supposed that up to the top of the pay the flow is that given by the theory developed for the problem when the sand column extends only to the top of the pay, while the flow in the well bore above the pay will be taken as strictly linear.

Hence if the sand column extends to a height h_e above the top of the pay (*cf.* Fig. 166), it will require an additional pressure differential of

$$\delta p = \frac{h_e Q \mu}{\pi r_w^2 k_1} + \gamma g h_e,\qquad(18)$$

to carry the flux Q through the system beyond that required if $h_e = 0$. It follows, therefore, that now

$$\frac{Q}{Q_0} = \frac{r_w^2 \log r_e/r_w}{\delta\left(\frac{2}{3} + \frac{r_w^2}{2} + \frac{r_w^2}{\delta} \log \frac{r_e}{r_w} - \frac{4}{\pi^2} \sum_1^\infty \bar{B}_n + 2h_e\right)},\qquad(19)$$

where here h_e, as well as r_w, and r_e, is to be measured in units of the sand thickness. That the additional sand column above the top of the pay acts as a very effective choke upon the well is readily seen when it is noted that, if the column extends only 5 ft. above the top of a 25-ft. sand, the production capacity will be cut to 42 per cent of what it would be if the sand column extended only to the top of the pay and had a permeability 100 times that of the pay, and to 16 per cent if its permeability is 10 times that of the pay. The reason for this marked effect is clearly the very high effective resistance of the well bore, resulting from its small cross section.

The practical value of keeping the well bore free of sand follows at once from the above results. During the flush-production stage the velocities of flow will probably be sufficient to flush out any sand entering the well bore, but in the later stages of production the accumulation of sand may be initiated and will rapidly tend toward an aggravated case of "sanding." Of course, any plugging of the screen will act as an added choke beyond that caused by the sand within the liner.

It should also be observed that although a sand column does have a very pronounced plugging action, it does not eliminate completely the flow at points even considerably below the top of the sand column. Hence, while a large part of edge waters might be eliminated by such a simple plug, sufficient water could still enter the well to produce serious emulsion troubles. The above analysis, therefore, is not to be interpreted as indicating the use of uncemented sand plugs for the elimination of bottom water entering a well by edge-water encroachment. However, such unconsolidated sand columns will tend to suppress bottom water that may be "coning" into the well, although they will not be quite as effective as cemented plugs.[1] In fact, a preliminary test of the success to be expected for a particular plugback job may be made by pouring unconsolidated sand into the well bore and observing its effect on the water production. If its effect is negligible it is unlikely that the cemented plug will be successful, and drastic "pinching" of the well will be required.

7.11. Summary.—When there are large-scale variations in the permeability of a porous medium, the pressure distributions must be obtained somewhat differently than in the case of systems of

---

[1] *Cf.* Secs. 8.10 and 8.11 for the detailed discussion of this problem.

uniform permeability. For continuous variations of the permeability a generalization of Laplace's equation [cf. Eq. 7.2(2)] must be used to find the pressure distribution. This equation may be solved analytically when the permeability depends on only one of the Cartesian coordinates or has axial symmetry, depending only on the radial distance from a point [cf. Eq. 7.2(5)]. If the permeability varies discontinuously, having uniform but different values in the various parts comprising the whole system, the pressure distribution may still be found by means of Laplace's equation. Separate solutions are set up for each region of uniform permeability and are then adjusted at the internal boundaries between the regions—the "surfaces of discontinuity"—in such a way that the pressures and normal velocities are the same on each side at each boundary point [cf. Eqs. 7.1(1), 7.1(2)]. When this adjustment is made the resultant set of pressure distributions for the several regions will correspond to a composite flow system in which the various parts of different permeabilities are connected physically as well as geometrically.

A case of discontinuous variations in permeability of some practical interest is that in which the liquid-bearing sand may be considered as separated into two adjacent concentric annular regions of different permeabilities. Such a system may represent a well which has been drilled into a region having a permeability higher or lower than that of the producing sand as a whole. It will also correspond to a well which was initially drilled into a homogeneous sand, the inhomogeneity having been caused by a partial plugging or mudding off of the region immediately adjacent to the well bore during the course of production or drilling, or conversely by an increase in the permeability about the well bore as the result of "acid treatment." The analysis for this problem shows that, as is to be expected in view of the highly localized character of the pressure drop in a radial-flow system about the well center, the production capacity of a well is very sensitive to the value of the permeability of the zone immediately surrounding the well bore. Thus, if the annular zone adjacent to the well bore has a permeability 2.5 times that of the remainder of the sand and is only 5 ft. in radius—occupying only 0.01 per cent of the total sand volume—the production rates for given total pressure differentials will be 30 per cent higher than for a homogeneous sand. Likewise if this zone has a permeability of one-fourth that of the main sand body the production capacity of the well will be reduced to only 46 per cent of its normal value. Furthermore, these effects do not increase in proportion to the radius of the zone of abnormal permeability; rather, the major effect is caused by the first few feet about the sand face and additions to the zone give successively smaller changes in the production capacity. Similarly for zones with relatively high permeabilities, the increase in production capacity—decrease in total resistance of the system —as the permeability of the inner zone is increased rapidly approaches the limiting value corresponding to a well in which the inner zone has been removed entirely and hence has a radius equal to that of the inner zone (*cf.* Fig. 150).

These results provide the explanation for the large variations in the production capacities which are frequently observed for neighboring and apparently identical wells. For if the sand is not homogeneous in detail it is not unlikely that neighboring wells may penetrate zones of appreciably different local permeabilities and hence show markedly different production capacities, even though the sand as a whole may be considered as homogeneous.

Another problem involving systems composed of regions of different permeability arises in the study of the flow of fluids in limestone reservoirs. Limestone rocks themselves are usually of very low permeability, and one must attribute the production capacities of wells penetrating limestone reservoirs to the fractures and crevices which permeate and are disseminated through the limestone proper. When such fractures are of limited extent and uniformly distributed through the pay, they will give a resultant effect equivalent to that of a homogeneous porous medium. However, when they are of extended length and limited in number they may be considered separately as linear channels which are fed laterally with the fluid issuing from the limestone proper. The fractures themselves may then be represented as distinct zones of the porous medium with permeabilities equal to the effective permeability of a linear free channel carrying a liquid under viscous-flow conditions. From the classical hydrodynamics it follows that a linear free channel of width w (centimeters) has an effective permeability of $10^8 w^2/12$ darcys.

By extending the methods of Fourier integrals[1] or Fourier series, already developed in Chap. IV for the treatment of homogeneous systems, the pressure distributions may be derived within a system—of infinite or finite extent—composed of a homogeneous two-dimensional porous medium (the limestone proper) bisected by a linear strip of a different permeability (the fracture) which is pierced by the producing well. The pressure variation along the strip is linear near the well and changes into a logarithmic type at larger distances from the well. However, the linear variation persists for longer distances from the well as the permeability of the strip (fracture) is increased relatively to that of the rest of the system (cf. Fig. 152).

For a fixed permeability of the linear strip—fixed width and permeability of the fracture—the resistance of the composite system increases as the main limestone permeability decreases. If, however, the limestone permeability is kept fixed, the resultant resistance decreases with increasing fracture width, the variation for widths exceeding 0.5 mm. following the inverse cube of the fracture width. The significance of the fracture as a carrier of the fluid into the well is further shown by a computation of the total flow through the system which enters the well after traveling directly through the fracture at least beyond a certain distance from the well. As is to be expected, this fraction increases with the fracture width. In fact, for fracture widths greater than 0.75 mm., less than 14 per cent of the total flow is fed into the fracture within 300 ft. of the well; however, for widths of 0.1 mm. or less, more than 97 per cent of the total flow into the well is fed into the fracture within the 300 ft. nearest the well (cf. Fig. 154).

An especially interesting application of the analysis of systems of nonuniform permeability may be made to the theory of the acid treatment of oil wells producing from limestone reservoirs. The effect of hydrochloric acid in increasing the production capacities of wells producing from limestone reservoirs may be given a ready explanation on the basis of the results derived analytically for systems of nonuniform permeability. Thus, if the flow into the well is essentially radial—owing to an approximately uniform distribution of fractures of limited extent—and the pay immediately surrounding the well bore is of normal permeability, as

[1] Cf. footnote on page 227.

compared to the main body of limestone, an increase of the permeability in the zone adjacent to the well bore due to the introduction of acid will give a relatively small increase in the production capacity of the well. In fact, even if the acid should completely dissolve away the limestone up to a radius of 5 ft., the production capacity of the well will be increased by only 65 per cent. If, however, the well is plugged or if the limestone surrounding the well bore is of an abnormally low permeability, the introduction of acid should be considerably more effective in increasing the production capacity of the well. If, for example, the zone about the well bore has a permeability one-tenth that of the main body of limestone and is raised to that of the latter by the acid, the production capacity of the well will be increased by 70 per cent even if the affected zone is only 3 in. thick, and by as much as 350 per cent if the effect of the acid penetrates to a radius of 5 ft.

Still larger effects of acid treatment will result if the limestone is permeated by extended fractures which are penetrated and widened by the acid. Thus, if a limestone of permeability 0.01 darcy is cleft by only a single extended fracture of 0.5 mm. width, the production capacity of a well cutting the fracture will be increased by 570 per cent if the width of the fracture is doubled. If the fracture is initially 0.1 mm., a 0.5-mm. addition in width will result in a 2,500 per cent increase in production capacity. Furthermore, smaller wells will in general show larger responses than good producers, a result well verified by field experience. Thus the theoretical analysis of the flow in porous media of nonuniform permeability is fully capable of explaining the wide range of field observations as to the effects of acid treatment in increasing the production capacities of wells producing from limestone reservoirs simply on the basis of the differences in detail of the mechanism of production among individual wells. In fact, the theory indicates that insofar as fractured limestones will show greater responses to acid treatment than such as will give an essentially radial flow, "shooting" wells penetrating reservoirs of the latter type should be an effective preliminary to acid treatment.

Another practical problem involving a system composed of parts of different permeability is that of the effect of a sanded liner on the production capacity of a well. For, although one would hardly expect that a column of clean sand at the bottom of a well bore would appreciably reduce the production capacity of the well, it is a common field observation that both flowing and pumping wells suffer very considerable decreases in production rate upon the entry of sand into the well bore. A closer analysis of this question, however, fully explains the observed effects. For as all the production from the well must pass through the narrow well bore, the pressure gradients in the well bore, considered as a porous medium, must be very high and will be very sensitive to its effective permeability. In fact, even if the sand column permeability is 100 times as great as that of the main sand body, 87.5 per cent of the total pressure drop in the system will take place in simply passing along the well axis from the top to the bottom of the pay. This situation is evidently equivalent to one in which a high back pressure is imposed on the lower parts of the sand, thus cutting down the flow coming from these parts. The effective well penetration is thus reduced and with it the associated production capacity of the system. In particular, the analysis shows that if the sand column extends to the top of a pay of 25 ft. thickness and has a permeability 100 times as great as that of the pay, the production capacity will be cut to 26 per cent of that of a system with a sand-free well bore; if the permeability ratio is 10, the production capacity of the sanded well will be only one-tenth of a similar one which is sand-free. Furthermore, these effects are greatly accentuated if the sand column extends beyond the top of the pay. For an additional sand-column height of only 5 ft. above the top of the pay will cut the production capacities of the above system to 42 and 16 per cent in the first and second cases, respectively, of what they would be if the sand column extended no higher than the top of the pay.

CHAPTER VIII

TWO-FLUID SYSTEMS

8.1. Introduction.—The almost universal occurrence of ground waters in the immediate neighborhood of oil-bearing sands lends considerable practical interest, from the point of view of oil-production technology, to the question of the mutual interaction between the movements of ground waters and the production of oil from sands contiguous to water-bearing sands. Although it is physically somewhat arbitrary, it is convenient to distinguish among the various problems in which underground waters influence the production of oil those in which the water is essentially confined to the edges of the field, and is "encroaching" into the oil sand, from those in which the underground waters underlie the oil sand in such a way that both oil-saturated and water-saturated sands may be penetrated by the same vertically directed well. Of course, this latter situation may be the direct result of an edge-water encroachment and is to some extent present in all edge-water-encroachment problems insofar as the water-oil interfaces are never strictly vertical. Nevertheless, these two types of problems lead to practical consequences which differ so materially that we shall consider them separately as those of water encroachment and those of water "coning," the latter getting its name from the fact that when the water underlies the oil zone in the same sand, the production of the oil tends to raise the water into the general form of a cone in the region immediately below the well bore.

8.2. Edge-water Encroachment. General Nature and Formulation of the Problem.—That the interaction between the movements of the edge waters and the oil in the sands contiguous to the water-bearing sands is really mutual and reciprocal becomes evident upon a little consideration. For, on the one hand, until an oil pool is tapped and the oil in it begins to move toward the wells there is evidently no physical reason for the neighboring waters to leave their previous states of equilibrium and encroach upon the oil pool. On the other hand, the influence of the water movements upon those of the oil becomes apparent when it is observed that water-bearing sands usually extend to outcrops which supply large and mobile hydrostatic heads to the deep-lying waters contiguous to the oil sands. The underground waters, therefore, possess almost limitless stores of mobile energy which may be applied to help in the production of the oil, both in such cases when the oil no longer has sufficient amounts of its more active driving agent—the dissolved gas—to maintain the production, and in others when the production is taken at high back pressure and the oil is produced much as a dead fluid driven by the encroaching water, rather than by the expansion of its dissolved gases. In the former cases, we have the situation corresponding to that of "natural water flooding" such as was encountered in the later stages of the production from the Powell and Mexia fields in Texas and in many other similar and well-recognized examples. In the latter cases we have the type of production usually described as a "water-drive production." Such are, for example, the production from the limestone fields of Mexico and the artificially controlled production of the Bradford field in Pennsylvania, derived from a direct water-flooding program.

A detailed analysis of the general problem of water encroachment, therefore, necessitates that a clear distinction be drawn between the various possible situations which may be involved, as these must be treated separately. Thus we shall discuss first the specific idealized problem in which the encroaching water plays the role of a "drive" upon the oil which is either gas-free, as in a practically depleted field, or is being produced at such a high back pressure that there is but little free gas evolving from the oil in the sand body. The oil is to be considered as "dead" and its flow to be due entirely to the hydrostatic head of the encroaching water, which, of course, must be greater than the pressure maintained at the outflow wells. As a consequence, it follows that the water and oil are always in contact along a moveable interface until the interface reaches an outflow well.

It should again be emphasized that in practically all cases the oil in actual underground reservoirs is saturated with gas at the prevailing reservoir pressure, and in addition frequently has associated with it, in the structurally higher portions of the reservoir, so-called "free gas caps," also under the prevailing reservoir pressure. Hence in addition to the available driving energy of the adjacent water horizon there is present the expansive energy of the gas content of the reservoir. It is clear, therefore, that in the present discussion where it is explicitly assumed that we are dealing with "dead" liquids, this assumption predicates the existence of a complete water drive. That is, we assume the condition in which the water influx to the reservoir is equal in volume to that of the fluid produced from the wells (or substantially so) and that, therefore, the reservoir pressure is maintained constant, to the extent that the liquids involved are incompressible. Under this condition a "gas drive" is evidently either nonexistent or at best provides only a latent source of energy for the production.

.One can, therefore, again take Darcy's law as the basis of the analysis, as in the discussion of flow problems in the last several chapters, and the analytical problem again becomes one in potential theory. However, it is a new type of problem, for it involves two regions—that occupied by the oil and that by the water—in each of which there is a fluid flowing which is governed by a velocity potential Φ, the two being always in contact at a moveable interface. At this interface there is continuity of the pressure and normal velocities, but in the interior of the regions separated by the interface, the potential functions are different because of the difference in the viscosities of the oil and water, although the sand or porous medium is considered to be of the same permeability in both the water and oil zones. In the last chapter a treatment was given of some systems composed of two regions of different permeability, such as the problem of a local permeability variation in the sand in the immediate neighborhood of a well (Sec. 7.3), or that in which the sand drained by the well is composed of two layers of different permeabilities (Sec. 7.9). The new feature here is that the difference between the regions is due to a difference between the *fluids* rather than between the permeabilities of the sand, so that as the one encroaches upon and displaces the other the boundary separating the regions (*i.e.*, the two fluids) continuously changes. The analogues of an electrostatic system with two dielectrics or of heat- or current-conduction systems composed of parts of different conductivities correspond hydrodynamically to a porous medium having *different permeabilities* on the two sides of a geometrical boundary, but flowing a *single* liquid. The inverse case, which is important here, seems to have no analogue of physical significance in other fields of potential theory; it is probably for this reason that no treatment of this type of potential problem is to be found in texts on potential theory.[1]

The general problem may now be formulated as follows: Given two closed surfaces, S_e and S_w, bounding a region in which a potential function Φ is to be determined under the following conditions: (1) The values of Φ over S_e and S_w are specified as a function of time;[2] (2) at a given instant, a surface S_i enclosing S_w is supposed to divide the region between S_e and S_w into two parts with "constants" c_1 and c_2 and potential functions Φ_1 and Φ_2 such that Φ and $c\,\dfrac{\partial \Phi}{\partial n}$ are continuous across S_i, S_i itself possessing a velocity at each point given by $\bar{v} = -c\nabla\Phi$ (*cf.* Fig. 167).[3]

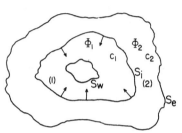

FIG. 167.—Diagrammatic representation of a general two-fluid encroachment system.

A graphical and numerical procedure for solving this problem is not difficult to devise, although to carry it out would be extremely laborious. Thus by graphical or numerical means the potential distributions Φ_1 and Φ_2 could be computed for the

[1] The discussion presented here follows that given by M. Muskat, *Physics*, **5**, 250, 1934.

[2] In a general case the boundary values on S_e and S_w may be "mixed" and over part of them the normal gradient of Φ rather than Φ itself may be specified. For the present purposes, however, it will suffice to suppose that Φ itself is given over all of S_e and S_w.

[3] The velocity potential is here considered to be the function $k(p - \gamma gz)$, so that the constant c is the reciprocal of the viscosity μ. It should also be carefully noted that this velocity, $\bar{v} = -c\nabla\Phi$, is equivalent to the volume rate of flow in cubic centimeters per second as if the fluid were flowing through a medium of 100 per cent porosity—whereas the actual velocity is given by $\bar{v}/f$, where f is the porosity of the medium. This particular definition of the velocity is used here and throughout the remainder of the discussion for analytical convenience, except where specifically noted otherwise. The exceptions are those cases where the actual velocity is of importance, and the porosity f is then introduced explicitly in the equations.

initial position of S_i. Constructing the velocity vectors $c\nabla\Phi$ over S_i, the position of S_i at a short interval of time after the initial instant could be determined. Then with the new regions (1) and (2), new functions Φ_1 and Φ_2 could be calculated and the whole process repeated. The difficult step would be the computation of Φ_1 and Φ_2 even at the initial instant when S_i is given. For unless S_i and the regions (1) and (2) are of the simplest geometrical form the "stationary" potential problem of determining Φ_1 and Φ_2 is itself still an unsolved problem, unless one resorts to graphical and numerical methods.

Although it is quite equivalent to that already given, a somewhat more analytical formulation suggests a method for solving at least a few simple systems. Here it is convenient to focus the attention directly upon the interface S_i. Thus we shall suppose the surface S_i at the time t to be represented by

$$F(x, y, z, t) = 0. \tag{1}$$

At a later time, $t + \delta t$, each point (x, y, z) will evidently be displaced to $(x + \delta x, y + \delta y, z + \delta z) = (x + v_x\delta t, y + v_y\delta t, z + v_z\delta t)$ where v_x, v_y, v_z are the velocity components at (x, y, z) on the interface. The new interface will, therefore, be given by

$$F(x + v_x\delta t, y + v_y\delta t, z + v_z\delta t, t + \delta t) = 0 = F(x, y, z, t), \tag{2}$$
whence

$$\frac{\partial F}{\partial t} + \bar{v} \cdot \nabla F = 0, \tag{3}$$

which is the well-known Kelvin relation governing the motion in a liquid of a surface containing a given set of particles. Adding now the condition that

$$\bar{v} = -c\nabla\Phi, \tag{4}$$

Eq. (3) becomes

$$\frac{\partial F}{\partial t} - c\nabla\Phi \cdot \nabla F = 0. \tag{5}$$

The problem may now be restated as:

Find the potential function Φ_1 between a surface S_w and $F(x, y, z, t) = 0$ and a potential function Φ_2 between $F(x, y, z, t) = 0$ and a surface S_e, such that

$$\begin{aligned}
\Phi_1 &= \theta_w(x,\,y,\,z,\,t) \text{ on } S_w \\
\Phi_2 &= \theta_e(x,\,y,\,z,\,t) \text{ on } S_c \\
\Phi_1 &= \Phi_2, \\
c_1\frac{\partial\Phi_1}{\partial n} &= c_2\frac{\partial\Phi_2}{\partial n}
\end{aligned}\Bigg\} \text{ on } F(x,\,y,\,z,\,t) = 0,\qquad (6)^1$$

where

$$\frac{\partial F}{\partial t} - c_{1,2}\nabla\Phi_{1,2}\cdot\nabla F = 0.$$

Before proceeding now with the solution of three simple cases, the following two points may be noted. The first is, that the function F cannot be permanently represented by a set of equipotential surfaces unless a special condition is satisfied. This is easily seen when it is observed that if $\Phi_0(x,\,y,\,z) = V$ represents that one of the set of equipotential surfaces which at time t coincides with F, the change in V experienced by a particle on F in time dt will be given by

$$dV = \frac{\partial V}{\partial t}dt + \frac{\partial V}{\partial n}\frac{dn}{dt}dt \doteq \left[\frac{\partial V}{\partial t} - c\left(\frac{\partial V}{\partial n}\right)^2\right]dt, \qquad (7)$$

where n is the element of length along the normal traversed by the particle. Evidently, then, unless the quantity in brackets is constant along the interface F, the values of V for the various points on F will change unequally after the time dt. At that time, F will, therefore, no longer be an equipotential if it were one at the initial instant.

The following point, although applicable only to a limited group of problems defined as above, is of considerable practical importance. Thus, supposing that the surfaces S_c and S_w are maintained at uniform potentials Φ_e and Φ_w, respectively, then, since the potential gradient in the region between S_c and S_w is, under these conditions, directly proportional[2] to the difference

[1] Equation (6) really implies that gravity is being neglected so that $\Phi = kp$. If gravity is to be taken into account the continuity of the pressures, $p = \dfrac{\Phi}{k} + \gamma gz$, must be required instead of the continuity of Φ itself. However, if the difference in density between the fluids in regions (1) and (2) may be neglected, one can still use the condition $\Phi_1 = \Phi_2$.

[2] It is readily verified that if for a stationary F, Φ_1 and Φ_2 satisfy Eqs. (6) with $\Phi_1 = \Phi_w$, $\Phi_2 = \Phi_e$ on the boundaries, the solutions $\Phi_1{}'$, $\Phi_2{}'$ for the conditions that $\Phi_1{}' = \Phi_w{}'$ on S_w and $\Phi_2{}' = \Phi_e{}'$ on S_e, are

$\delta\Phi = \Phi_e - \Phi_w$, one may set

$$\nabla\Phi = \delta\Phi\nabla_0\Phi, \qquad (8)$$

where $\nabla_0\Phi$ refers to a unit total potential difference. The equation for F may then be written as

$$\frac{\partial F}{\partial(t\delta\Phi)} - c\nabla_0\Phi \cdot \nabla F = 0, \qquad (9)$$

and hence, since t enters explicitly only in the first term, it follows that

$$F = F(x, y, z, t\delta\Phi). \qquad (10)$$

Now the *rate* of encroachment of the fluid in region (2) into the region (1) is evidently proportional to $\delta\Phi$. Nevertheless, it is seen from Eq. (10) that the *shape* of the interface is independent of the rate of encroachment but depends only on the product $t\delta\Phi$. Hence a system with a small $\delta\Phi$ will possess the same family of interfaces F as one where $\delta\Phi$ is large except that they will be crossed at proportionately later instants.

In presenting now the solutions to specific examples, we shall return, for physical clarity, to the hydrodynamic phraseology and suppose that the region between S_e and $F = 0$ is occupied by a liquid of fluidity[1] c_2 "encroaching" on and displacing a liquid of fluidity c_1 between $F = 0$ and S_w, the latter being termed the outflow surface or the "well."

8.3. Linear Encroachment.—The simplest case of the type of problem which has just been formulated is evidently to be found in a one-dimensional system. Physically, such a system, as indicated in Fig. 168, may correspond either to a case where the two fluids are moving in a channel of areal extent that is small as compared to its length, or in just the opposite case where the lateral extent of the interface, over which the pressure is uniform, is so large compared to its irregularities that it may be taken as a

$$\Phi_1' = \frac{\Phi_e' - \Phi_w'}{\Phi_e - \Phi_w}\Phi_1 + b$$

$$\Phi_2' = \frac{\Phi_e' - \Phi_w'}{\Phi_e - \Phi_w}\Phi_2 + b, \qquad b = \frac{\Phi_e\Phi_w' - \Phi_w\Phi_e'}{\Phi_e - \Phi_w}.$$

[1] The constants c—the reciprocals of the viscosity—are here termed "fluidities" in conformity with the conventional terminology in the rheological literature.

perfectly plane infinitely extended interface. The former situation might actually be at least approximated in a narrow channel of high permeability in which the oil is being driven out by the encroaching water ahead of the general advancing interface. The latter situation, on the other hand, may be taken as a representation of a general edge-water encroachment over the lateral extent of a field, as, for example, in the case of the "East Texas" field.[1]

To treat this one-dimensional problem analytically, it is convenient to turn around the formal statement of Eq. 8.2(6) and

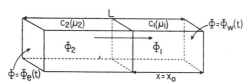

Fig. 168.—A linear encroachment system.

consider first the determination of the surface F, defined for this case by

$$\frac{\partial F}{\partial t} - c\frac{\partial \Phi}{\partial x}\frac{\partial F}{\partial x} = 0. \tag{1}$$

Because of the simplicity of the problem the potential gradient $\partial\Phi/\partial x$ can be given directly in terms of the position of F. Thus it may be easily verified that at the interface given by $x = x_0$

$$c_2\frac{\partial \Phi_2}{\partial x} = c_1\frac{\partial \Phi_1}{\partial x} = \frac{c_2 c_1(\Phi_e - \Phi_w)}{(c_2 - c_1)x_0 + c_1 L}, \tag{2}$$

where the distance L is simply a convenient reference point from which to measure the advance of the interface. Further, in the present notation

$$F = x - x_0(t) = 0: \qquad \frac{\partial F}{\partial t} = -\frac{dx_0}{dt}; \qquad \frac{\partial F}{\partial x} = 1.$$

Hence Eq. (1) becomes

$$\frac{dx_0}{dt} + \frac{c_2 c_1(\Phi_e - \Phi_w)}{(c_2 - c_1)x_0 + c_1 L} = 0, \tag{3}$$

[1] A detailed analysis of the production of the "East Texas" field, based, however, on a physical representation of the encroaching water as an expansible liquid, will be given in Sec. 10.8.

with the solution

$$\frac{c_2 - c_1}{2}(x_0{}^2 - L^2) + c_1 L(x_0 - L) = -c_1 c_2 \int_0^t (\Phi_e - \Phi_w)dt, \quad (4)$$

where the initial instant has been taken as that when the interface is at the external boundary, $x_0 = L$.

For the case when the difference in the boundary values, $\Phi_e - \Phi_w$, is kept constant, Eq. (4) takes the form

$$\left(\frac{x_0}{L} - 1\right)\left[\frac{x_0}{L}(1 - \epsilon) + \epsilon + 1\right] + \frac{2\Delta\Phi c_1 t}{L^2} = 0; \qquad \epsilon = \frac{c_1}{c_2} = \frac{\mu_2}{\mu_1}. \quad (5)$$

The fact that t enters in the combined term $2\Delta\Phi c_1 t/L^2$ shows that for corresponding states of the encroachment the time

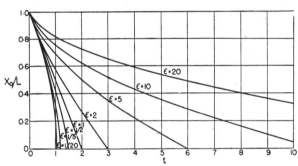

FIG. 169.—The progress of the interface in a linear encroachment system as a function of the time. x_0/L = fraction of total length of system occupied at the time t (arbitrary units) by the original liquid. ϵ = (viscosity of encroaching liquid)/(viscosity of liquid displaced). (*From Physics*, **5**, 253, 1934.)

required to reach these states varies inversely as the average potential gradient $\Delta\Phi/L$, directly as the total length of the system L, and directly as the viscosity of the displaced liquid. With this in mind, the details of the encroachment may be studied by simply plotting x_0/L against t, setting for convenience $2\Delta\Phi c_1/L^2$ equal to unity. The curves so computed are given in Fig. 169 for various values of ϵ. The individual curves are parabolic, as Eq. (5) indicates. And as is to be expected, the encroachment of the interface is accelerated as it advances (x_0 decreases) if the encroaching liquid has the higher fluidity ($\epsilon < 1$), whereas it is retarded if the encroaching liquid has the lower fluidity ($\epsilon > 1$).

The actual flux through the system as the interface moves toward $x_0 = 0$ is given essentially by the slopes of the curves of Fig. 169, and are presented more explicitly in Fig. 170. They have been computed from the equation

$$-\frac{Q}{L} = \frac{dy}{dt} = \frac{-1}{2(\epsilon + y - \epsilon y)}; \qquad y = \frac{x_0}{L}. \qquad (6)$$

The marked acceleration of the interface and corresponding increase in the flux through the system if $\mu_2 < \mu_1 (\epsilon < 1)$ is clearly shown in Fig. 170 in the rapid rise of $-\dfrac{dy}{dt}$ as y approaches 0.

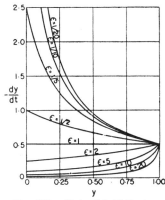

FIG. 170.—Rate of fluid displacement $(-dy/dt)$ in a linear encroachment system. y = fraction of total length of system occupied by the original liquid; ϵ = (viscosity of encroaching liquid)/(viscosity of liquid displaced). (*From Physics*, **5**, 253, 1934.)

The total time required for the interface to move across the system is given by the value of t at $x = 0$, or

$$t_{\max.} = \frac{\mu_1\left(1 + \dfrac{\mu_2}{\mu_1}\right)fL^2}{2\Delta\Phi}, \qquad (7)$$

f being the porosity of the medium. When $\mu_2 < \mu_1$, this is evidently less than the single-fluid "encroachment time"

$$t_{\max.}{}^0 = \frac{\mu_1 fL^2}{\Delta\Phi}. \qquad (8)$$

It may be noted also from [Eq. (7) that the maximum decrease in $t_{\max.}$ due to the encroachment is attained when $\mu_2 \ll \mu_1$, so that

$$t_{\max.}/t_{\max.}{}^0 \sim \tfrac{1}{2}. \qquad (9)$$

Hence in this limiting case the drive by an encroaching fluid of very low viscosity would cut by half the time required for the column of original liquid to pass through the system. When $\mu_2 > \mu_1$, the $t_{\max.}$ will clearly exceed $t_{\max.}{}^0$, and will ultimately become infinite as μ_2 becomes infinitely large.

8.4. Two-dimensional Radial Encroachment.—In artificial flooding processes or in later stages of the natural encroachment process the situation may arise where a particular well is surrounded completely by a ring of water, encroaching radially and driving the oil in front of it into the well. For simplicity of treatment this problem will be idealized so as to possess complete radial symmetry. Thus the functions Φ_2 and Φ_1 will be taken as uniform over the external and internal boundaries, defined by the cylinders $r = r_e$ and $r = r_w$, respectively (cf. Fig. 171). Again it is to be noted that the external boundary $r = r_e$ simply represents a convenient initial position from which to measure the progress of the encroachment. $r = r_w$, however, represents the actual surface of the well.

In view of the radial symmetry of the problem one may set

$$F = r - r_0(t) = 0;$$

$$\nabla\Phi \cdot \nabla F = \frac{\partial\Phi}{\partial r}\frac{\partial F}{\partial r}, \qquad (1)$$

so that $r = r_0(t)$ gives the position of the interface at the time t. Equation 8.2(5) then takes the form

$$\frac{\partial r_0}{\partial t} + c\left(\frac{\partial\Phi}{\partial r}\right)_{r_0} = 0. \qquad (2)$$

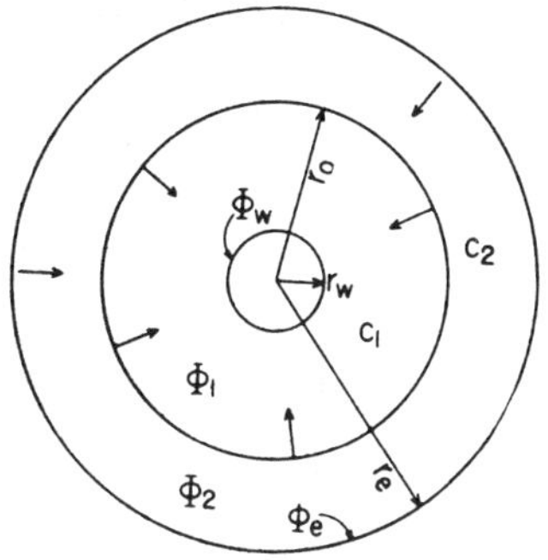

Fig. 171.—Diagrammatic representation of a radial-encroachment system.

Now the potential distribution in such a system as the above has already been found in Sec. 7.3 for any value of r_0. Recalling these results, it is easily seen that

$$c_1\frac{\partial\Phi_1}{\partial r} = \frac{c_1(\Phi_e - \Phi_w)/r_0}{\log\dfrac{r_0}{r_w} + \dfrac{c_1}{c_2}\log\dfrac{r_e}{r_0}} = c_2\frac{\partial\Phi_2}{\partial r}. \qquad (3)$$

Putting this in Eq. (2) and integrating, it follows that

$$r_0{}^2(b - a + a\log r_0{}^2) = -4c_1\int_0^t\Delta\Phi dt + \text{const.,}$$

where

$$a = 1 - \epsilon; \qquad b = \epsilon\log r_e{}^2 - \log r_w{}^2; \qquad \epsilon = \frac{c_1}{c_2}. \left.\vphantom{\int}\right\} \qquad (4)$$

Taking now $\Delta\Phi$ as constant—no change in the well or external pressure with the time—and assuming that at $t = 0$ the interface r_0 is at the external boundary r_e, the integration of Eq. (4) may be carried out and the result written as

$$\frac{r_0{}^2}{r_e{}^2}\left(\log \frac{r_e{}^2}{r_w{}^2} - a + a \log \frac{r_0{}^2}{r_e{}^2}\right) = \frac{-4c_1 t \Delta\Phi}{r_e{}^2} + \log \frac{r_e{}^2}{r_w{}^2} - a. \quad (5)$$

Setting for convenience $4c_1\Delta\Phi/r_c{}^2 = 1$, and adding the numerical assumption that $r_e/r_w = 2{,}000$, which is of the order of magnitude involved in cases of practical interest, curves representing Eq. (5) for various values of $\epsilon = \mu_2/\mu_1$ are plotted in Fig. 172. It will be observed that as the encroachment proceeds, i.e., as r_0/r_e decreases, the rate of encroachment is accelerated. It should be noted, however, that here the actual acceleration is the resultant of the natural acceleration due to the converging character of the flow and the effect of replacing the original

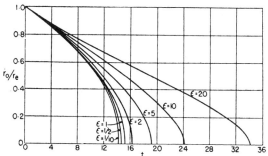

Fig. 172.—The progress of the interface in a radial-encroachment system as a function of the time. r_0 = radius of interface at time t (arbitrary units). r_e = external-boundary radius. ϵ = (viscosity of encroaching liquid)/(viscosity of liquid displaced). r_w = well radius = $r_e/2{,}000$. (*From Physics*, **5**, 255, 1934.)

fluid of viscosity μ_1 by that of viscosity μ_2, which will aid or oppose the former as $\epsilon < 1$ or > 1.

The rate of fluid flow during the encroachment, which is proportional to $\dfrac{\partial r_0{}^2}{\partial t}$, may be computed by the equation

$$\frac{-Q}{\pi r_e{}^2} = \frac{\partial (r_0/r_e)^2}{\partial t} = \frac{-2c_1\Delta\Phi/r_e{}^2}{(1 - \epsilon)\log \dfrac{r_0}{r_e} + \log \dfrac{r_e}{r_w}}; \qquad \epsilon = \frac{c_1}{c_2} = \frac{\mu_2}{\mu_1}. \quad (6)$$

Setting again

$$\frac{4c_1\Delta\Phi}{r_e{}^2} = 1, \qquad \text{and} \qquad \frac{r_e}{r_w} = 2{,}000,$$

Eq. (6) is plotted in Fig. **173** for various values of ϵ, with r_0/r_e replaced by y. The interpretation is similar to that of

Fig. 170, the principal point to be noted being the fact that here the increase in the flow for the case $\mu_2 < \mu_1$ does not become appreciable until $y \sim 0$.

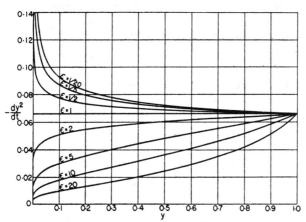

Fig. 173.—Rate of fluid displacement $(-dy^2/dt)$ in a radial-encroachment system. y = (radius of interface)/(radius of external boundary, r_e). ϵ = (viscosity of encroaching liquid)/(viscosity of liquid displaced). Well radius = $r_e/2,000$. (*From Physics*, **5**, 255, 1934.)

The total time required for the interface to move from $r_0 = r_e$ to $r_0 = r_w$ in an actual medium of porosity f is, by Eq. (5),

$$t_{max.} = \frac{r_e^2 f}{4c_1\Delta\Phi}\left[\left(1 - \epsilon\frac{r_w^2}{r_e^2}\right)\log\frac{r_e^2}{r_w^2} + (\epsilon - 1)\left(1 - \frac{r_w^2}{r_e^2}\right)\right]$$

$$\cong \frac{r_e^2 f}{4c_1\Delta\Phi}\left[\log\frac{r_e^2}{r_w^2} + \epsilon - 1\right]: \quad \frac{r_w}{r_e} \ll 1. \quad (7)$$

The ratio of this value to that for the single liquid of fluidity c_1 is evidently

$$\frac{t_{max.}}{t_{max.}(\epsilon = 1)} = 1 + \frac{\epsilon - 1}{\log r_e^2/r_w^2}. \quad (8)$$

It appears, therefore, that for the case where the encroaching fluid has a viscosity less than that of the fluid displaced ($\epsilon < 1$), the decrease in $t_{max.}$ due to the entry of the liquid of higher fluidity cannot exceed the value $\dfrac{1}{2\log r_e/r_w} \sim 6.6$ per cent, for $r_e/r_w = 2,000$, whereas in the corresponding linear-flow problem the encroachment by the liquid of higher fluidity could cut $t_{max.}$

to one half of that in a homogeneous system. The physical basis for this difference may be understood on observing that since the acceleration in the encroachment does not become appreciable until the liquid of high fluidity enters the region of high-pressure gradients near the well, the value of t_{max}. will differ from that for a homogeneous liquid ($\epsilon = 1$) only because of these last stages of the encroachment; but since these stages consume only a negligible fraction of the total time required for the whole process to be completed, the net effect on t_{max} is small.

For $\epsilon \geqslant 1$, on the other hand, it follows again as in the case of linear encroachment that t_{max}. may be made arbitrarily large as ϵ is increased indefinitely.

8.5. The History of a Line of Fluid Particles in a Homogeneous System.—Although the problem of encroachment may also be rigorously solved when the system has spherical symmetry,[1] more general systems in which the shape of the interface is not immediately evident from the geometry of the problem involve great difficulties. For in these the shape of the interface must be found *simultaneously* with its instantaneous position and with the potential distribution on the two sides of it. While it is possible to give a method of successive approximations[2] in which it is necessary only to find in each step solutions of the potential equation and the shape and position of the interface separately, the difficulties still remain, for practical purposes, almost unsurmountable.

It is, therefore, of some interest to develop an approximation to the real encroachment problem which will indicate, at least qualitatively, the shape of the oil-water interface and the nature of its advance. Such an approximation may be developed by neglecting the difference in viscosities between the two fluids. The equivalent of the water-oil interface will then be a geometrical surface or curve which at a given initial instant does coincide with the real water-oil interface, and then moves along with all the fluid particles on this original surface as they would advance in a single-fluid system toward the outlet well. The analytical process is then one of tracing the history of this line of fluid particles—originally lying along the real oil-water interface—as it

---

[1] *Cf.* M. Muskat, *Physics, loc. cit.*

[2] *Cf.* M. Muskat, *Jour. of Applied Physics,* **8**, 434, 1937.

advances toward the well.   There will then be no real physical interface and no separation of the whole region of interest into two parts in which separate potential functions must be found. The system will rather be of a uniform fluidity and permeability throughout, with a single potential distribution applicable to the whole of the region of interest—from the external boundaries to the well outlets.

Since this approximation still involves the motion of a curve containing a given set of fluid particles—as does the water-oil interface in a real encroachment problem—one may again represent this curve by a function F which will satisfy the Kelvin relation [Eq. 8.2(5)].   Taking then c as unity and denoting by Φ the potential function valid for the normal homogeneous system, the equation for F becomes

$$\frac{\partial F}{\partial t} - \nabla\Phi \cdot \nabla F = 0. \tag{1}$$

If the boundary pressures are kept constant there will evidently be in the system a permanent set of equipotential and streamline surfaces.   If, then, an orthogonal set of curvilinear coordinates are introduced as

$$u = u(x, y, z); \qquad v = v(x, y, z); \qquad w = w(x, y, z) \atop \nabla u \cdot \nabla v = \nabla u \cdot \nabla w = \nabla v \cdot \nabla w = 0, \Bigg\} \tag{2}$$

$\nabla\Phi \cdot \nabla F$ will take the form

$$\nabla\Phi \cdot \nabla F = \frac{\partial \Phi}{\partial u}\frac{\partial F}{\partial u}|\nabla u|^2 + \frac{\partial \Phi}{\partial v} \cdot \frac{\partial F}{\partial v}|\nabla v|^2 + \frac{\partial \Phi}{\partial w} \cdot \frac{\partial F}{\partial w}|\nabla w|^2. \tag{3}$$

Hence, if one of the set of surfaces, u, for example, is taken as the set of equipotential surfaces, $\Phi = \text{const.}$, $\nabla\Phi \cdot \nabla F$ will reduce to the form

$$\nabla\Phi \cdot \nabla F = \frac{\partial F}{\partial u}|\nabla u|^2.$$

With this system of coordinates the equation for F may be expressed as

$$\frac{\partial F}{\partial t} - |\nabla u|^2\frac{\partial F}{\partial u} = 0, \tag{4}$$

where $|\nabla u|^2$ is to be expressed in terms of u, v, w.

Equation (4) is easily integrated with the result

$$F = \alpha t + \alpha \int \frac{du}{|\nabla u|^2} + g(v, w) = \text{const.,} \qquad (5)^1$$

where g is an arbitrary function to be adjusted so that F assumes its initial form at $t = 0$ (α may be always taken as unity). All the later history of the particles lying initially on this line may then be found by simply plotting Eq. (5) for successive values of t.

This theory will now be illustrated by two examples of practical interest. For simplicity the discussion will be confined to two-dimensional problems. In such cases the curvilinear coordinates may be taken as the equipotential curves $\Phi = \text{const.}$ and streamlines $\Psi = \text{const.}$ Equation (5) then may be written as

$$F(\Phi, \Psi) = t + \int \frac{d\Phi}{|\nabla\Phi|^2} + g(\Psi) = \text{const.} \qquad (6)$$

8.6. The Line Drive into a Single Well.—As the first example we shall consider the case of the encroachment into a single well, from an infinite line source, of a group of particles lying originally on the line source (*cf.* Fig. 35). This corresponds to an idealized edge-water flood advancing into an oil reservoir containing a single well located near the edge water, and which is produced by the water drive alone.

From the principles of the theory of conjugate functions (*cf.* Sec. 4.8) it will be clear that the equipotential (Φ) and streamline (Ψ) distributions for the present problem may be derived from the complex-variable function

$$\Phi + i\Psi = q \log \frac{z - id}{z + id} + \Phi_0, \qquad (1)$$

which represents the complex potential due to a sink at $(0, d)$ and its negative image at $(0, -d)$. The constant q gives the strength of the source or sink, and Φ_0 represents the value of the potential at the position of the line drive, $y = 0$.

Resolving Eq. (1) into its real and imaginary parts, it is readily found that

---

[1] Of course, any function of the right side of Eq. (5) will also satisfy Eq. (4), but since the resulting function must equal a constant, there is no loss in generality in simply setting its argument equal to a constant.

$$\left.\begin{aligned}
\Phi &= \frac{q}{2} \log \frac{x^2 + (y-d)^2}{x^2 + (y+d)^2} + \Phi_0 \\
\Psi &= q \tan^{-1} \frac{-2dx}{x^2 + y^2 - d^2}.
\end{aligned}\right\} \tag{2}$$

It further follows from the theory of conjugate functions that [cf. Eq. 4.8(3)]

$$\begin{aligned}
|\nabla\Phi|^2 &= \left(\frac{\partial\Phi}{\partial x} + \frac{i\partial\Psi}{\partial x}\right)\left(\frac{\partial\Phi}{\partial x} - \frac{i\partial\Psi}{\partial x}\right) \\
&= \left|\frac{2qid}{z^2 + d^2}\right|^2 = \frac{4d^2q^2}{(x^2 + d^2 - y^2)^2 + 4x^2y^2}.
\end{aligned} \tag{3}$$

Introducing now the notation

$$\frac{\Phi_0 - \Phi}{q} = \eta; \qquad \frac{\Psi}{q} = \xi; \qquad \frac{x}{d} = x; \qquad \frac{y}{d} = y, \tag{4}$$

Eqs. (2) can be rewritten as

$$\left.\begin{aligned}
x^2 + y^2 + 1 &= 2y \operatorname{ctnh} \eta \\
x^2 + y^2 - 1 &= -2x \operatorname{ctn} \xi,
\end{aligned}\right\} \tag{5}$$

so that

$$x = \frac{\sin \xi}{\cosh \eta \pm \cos \xi}; \qquad y = \frac{\sinh \eta}{\cosh \eta \pm \cos \xi}. \tag{6}$$

As we shall take the streamline $\Psi = 0$ to lie along the y axis between the well and the line drive, the upper sign must be used. Putting then Eq. (6) into Eq. (3), it is found that

$$|\nabla\Phi|^2 = \frac{q^2}{d^2}(\cosh \eta + \cos \xi)^2. \tag{7}$$

Applying this now to Eq. 8.5(6) and integrating, the result is

$$F(\eta, \xi, t) = t - \frac{d^2}{q \sin^2 \xi}$$

$$\left[\frac{\sinh \eta}{\cos \xi + \cosh \eta} - 2 \operatorname{ctn} \xi \tan^{-1} \tan \frac{\xi}{2} \tanh \frac{\eta}{2}\right] + g(\xi) = 0. \tag{8}$$

If the initial position of the line of particles of interest is the x axis, i.e., the curve $y = 0$, we must have $g(\xi) = 0$. The final result is then

$$t = \frac{d^2}{q \sin^2 \xi}\left[\frac{\sinh \eta}{\cos \xi + \cosh \eta} - 2 \operatorname{ctn} \xi \tan^{-1} \tan \frac{\xi}{2} \tanh \frac{\eta}{2}\right]. \tag{9}$$

Along the y axis, i.e., $\xi = 0$, Eq. (9) is indeterminate. The usual methods give

$$t(x = 0) = \frac{d^2}{q} \frac{\tanh \eta/2}{6}\left(3 - \tanh^2 \frac{\eta}{2}\right)$$

$$= \frac{yd^2}{6q}(3 - y^2). \qquad (10)$$

Setting $d = q = 1$, the curves $t = $ const. are plotted in Fig. 174 in a Cartesian-coordinate system. At the initial instant the line of particles, whose history has been traced, lies on the the x axis

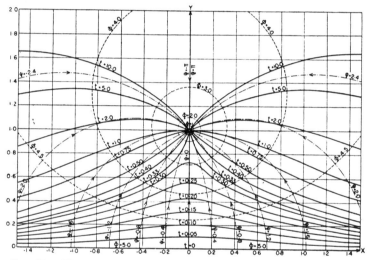

FIG. 174.—History of a line of fluid particles as they emerge from an infinite line source and travel in a homogeneous medium toward an isolated sink (well). $\Phi = $ const: equipotentials; $\Psi = $ const: streamlines. (From Physics, **5**, 260, 1934.)

(the line drive). For convenience the potential of the drive has been taken as $\Phi = 5.0$, which corresponds to setting $\Phi_0 = 5.0$. The other equipotentials, $\Phi = $ const., and the streamlines $\Psi = $ const., defined by Eqs. (5), have been drawn as the dotted curves. As may be readily verified, these are a system of mutually orthogonal circles, their centers lying upon the y and x axes, respectively (cf. Fig. 36).

As might have been anticipated, the curves $t = $ const. $\leqslant \frac{1}{3}$ have maximal coordinates along the y axis. This simply means that the particles near the y axis move faster than those distant from it; and this, clearly, is a consequence of the higher gradients of the potential along the y axis. Furthermore, as the particles approach the well at $y = 1$, they move into regions of increasing gradients and their velocities are continually accelerated. The maximum for increasing values of t becomes more and more pronounced, until in the limiting case when $t = \frac{1}{3}$ the particles along the y axis experience such large gradients that the curve is drawn in as a cusp as the first particle, of those initially on the line $y = 0$, enters the well.

Once the curve $t =$ const. goes through the point of convergence, $y = 1$, all the succeeding curves will also pass through this point, since all the streamlines must end there. On the other hand, since this point remains fixed and the other points along the curves $t =$ const. must follow the streamlines, the curves for $t > \frac{1}{3}$ will flatten out again near the y axis and those for high values of t will even follow the streamlines above the point $(0, 1)$ and produce a cusp in the direction opposite to that for $t = \frac{1}{3}$. These details may be verified by reference to Fig. 174.

Finally it is of interest to note the order of magnitudes involved when the above analysis is applied to a system with dimensions of practical interest. Thus from Eq. (9) it is clear that if the line drive is located at a distance d from the well and the total pressure differential is Δp, the times involved will be

$$\bar{t} = \frac{fd^2\mu t}{k\Delta p} \log \frac{2d}{r_w}, \tag{11}$$

where t is the time used in Fig. 174, r_w is the well radius, k is the actual permeability of the sand, f its porosity, and μ is the viscosity of the fluid. Hence if $f = 0.20$, $\mu = 1$ centipoise, $k = 1$ darcy, $d = 500$ ft., $r_w = \frac{1}{4}$ ft., and $\Delta p = 700$ lb., the time $t = \frac{1}{3}$ for the line of particles to first reach the well will be equivalent to

$$\bar{t} = 31.2 \text{ days.}$$

With regard to the area swept out by the line of particles by the time they first reach the output well, it need simply be noted that this will be equal to the total flux leaving the driving source or entering the outlet well during that time, divided by the porosity of the medium. Thus by Eq. (2) the flux entering the well per unit sand thickness is given by

$$Q = 2\pi r_w \left(\frac{\partial \Phi}{\partial r}\right)_{r_w} = 2\pi q = \frac{2\pi k \Delta p}{\mu \log \dfrac{2d}{r_w}}. \tag{12}$$

which, when combined with Eq. (11), gives for the total area swept out during the time $t = \frac{1}{3}$, the value

$$A = \frac{Q\bar{t}}{f} = \frac{2\pi d^2}{3}. \tag{13}$$

Thus in reaching a well which is twice as distant from the line drive as another well, the area swept out will be four times as large. Evidently in the corresponding problem of a real water encroachment Eq. (13) gives the vol-

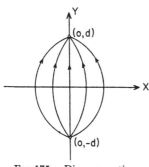

ume of oil displaced or produced by the well before the water reaches it, on the assumption that there is no retention of the oil and that the flooding is complete.

8.7. The Direct Drive between Two Wells.—As a second example we shall treat the problem of the direct flooding between two wells alone in an infinite two-dimensional reservoir. That is,

FIG. 175.—Diagrammatic representation of a direct fluid drive between two wells.

we shall trace the history of a line of fluid particles just emerging from a "driving" or "flooding" well at $(0, -d)$ as they travel toward the output well at $(0, d)$ (cf. Fig. 175). As in the previous case, the instantaneous traces assumed by these particles will correspond in a real flooding problem to the water-oil interfaces at the corresponding instants, except, of course, for the correction due to the viscosity difference between the oil and water, which shall be neglected here.

Upon a little inspection it will be observed that the potential distribution for the present problem will be identical with that for the case of the line drive into a well. For as Eq. 8.6(1) indicates, the potential distribution for the latter case is really nothing more than that between a source and sink symmetrically placed about the line "drive." One can, therefore, carry over the analysis of the previous case up to Eq. 8.6(8). The peculiar feature characterizing the present problem will then enter only in the statement of the initial conditions, *i.e.*, that instead of requiring that at $t = 0$ the line of particles lies on the x axis, it will be required that at $t = 0$ they have just emerged from the "flooding" well at $(0, -d)$. Specifically, it is to be required that

$$\eta = \eta_0 \qquad \text{at} \qquad t = 0,$$

where η_0 is an equipotential encircling the input well or even coincident with it. With this requirement, $g(\xi)$ in Eq. 8.6(8) may be determined with the result that

$$t = \frac{d^2}{q \sin^2 \xi}\left(\frac{\sinh \eta}{\cosh \eta + \cos \xi} - 2 \operatorname{ctn} \xi \tan^{-1} \tan \frac{\xi}{2} \tanh \frac{\eta}{2}\right)$$
$$- \frac{d^2}{q \sin^2 \xi}\left(\frac{\sinh \eta_0}{\cosh \eta_0 + \cos \xi} - 2 \operatorname{ctn} \xi \tan^{-1} \tan \frac{\xi}{2} \tanh \frac{\eta_0}{2}\right). \quad (1)$$

This equation is diagrammatically represented in Fig. 176 for the constants $d = q = 1$, and $\eta_0 = - \sinh^{-1} 100$, which implies that the initial position of the line of particles is the equipotential circle of radius $\frac{1}{200}$ of the distance between the wells.

Perhaps the most striking feature of the curves of this figure is their small distortion from complete radial symmetry. Although the curves are actually oval-shaped, the sharpness of the advancing front of the line of particles does not become appreciable until the particle on the y axis—the line of centers between the wells—has traveled to the mid-point between the wells. And even at $t = \frac{2}{3}$, when the trace first reaches the outflow well and the particle on the line of centers has advanced the total distance of separation between the wells, the receding front of the curve has traveled as much as half of that distance, although it has been advancing continuously in regions of progressively decreasing gradients.

The reason for this relatively small distortion of the curves $t = \text{const.}$ from radial symmetry becomes evident when one notes the nature of the potential distribution in the system, represented in Fig. 176 by the dotted curves $\Phi = \text{const.}$ Thus for $t \leqslant \frac{1}{3}$ when all the particles are still below the mid-line $y = 0$, *all* the particles find progressively decreasing gradients as they recede from the input well. Furthermore, the equipotentials remain approximately concentric with the input well for an appreciable distance out from it, so that during the first period of its history the line of particles is moving in a potential field of almost complete radial symmetry. The distortion that does occur

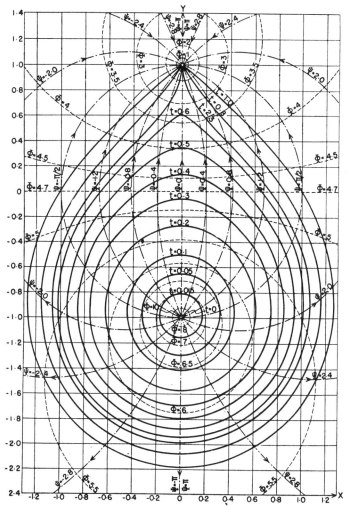

FIG. 176.—The history of a ring of fluid particles as they emerge from a source and travel in a homogeneous medium toward a sink. Φ = const.: equipotentials; Ψ = const.: streamlines. (*From Physics*, **5**, 261, 1934.)

is due simply to the *differential* variation between the gradients along the streamlines $\Psi \sim 0$, near the y axis, and those distant from it.

On the other hand, when $t > \frac{1}{3}$ the particles near the y axis cross the mid-line and enter a field of increasing gradients, whereas those distant from the y axis are still traveling in regions of decreasing gradients. These particles above the x axis become accelerated while those below the x axis are still being retarded, and a real distortion begins to develop. However, the lateral "spread" of the curves has by this time already become so large that the maximal linear distortion that can be attained by the time the output well is reached is given by the ratio of 2 to 1.

The order of magnitude of the actual times represented by the values of t in Fig. 176 is given for the present problem by

$$\bar{t} = \frac{2fd^2\mu t}{k\Delta p} \log \frac{2d}{r_w}, \tag{2}$$

where the notation is that used in Eq. 8.6(11).

Hence, if

$k = 1$ darcy; $f = 0.2$; $\mu = 1$ centipoise; $2d = 500$ ft.;

$r_w = \frac{1}{4}$ ft.; $\Delta p = 700$ lb.; $\bar{t} = 42.9t$ days,

so that $t = \frac{2}{3}$ gives $\bar{t} = 28.6$ days for the line of particles to first reach the output well. Comparing this result with that of Sec. 8.6, it is seen that it would take less time for fluid to pass between two wells than between a line drive and a well separated by the same distance as in the two-well system and flowing under the same pressure differential.

Finally, it may be observed that the area swept out by the line of particles by the time they first reach the output well can be shown, in a manner similar to that used in the last section, to be given by

$$A = \frac{Q\bar{t}}{f} = \frac{4\pi d^2}{3}. \tag{3}$$

Although more complex systems can also be treated analytically, they may be much more readily studied by the aid of an electrolytic model. Since, however, these more complex systems of practical interest involve groups and networks of wells, their discussion will be deferred to the next chapter. On the other hand, we shall return now to the original problem of water encroachment and see what modifications may be expected to be introduced in the above idealized analysis when such factors as gravity and the viscosity difference between the two fluids are taken into account.

8.8. The Effect of Gravity on the Shape of the Encroaching Interface.—The particular examples treated in detail in the last several sections have been taken as two-dimensional. This implies that even though the sand is of appreciable thickness the dynamical conditions in the sand are exactly the same in all planes parallel to the bedding plane or that chosen as the plane of reference. As an immediate consequence it follows that in the encroachment problems that have been considered the water-oil interfaces must all be perfectly vertical and perpendicular to the plane of reference. But such an interface, however, can be maintained only when the density of the oil and water are assumed to be the same, a condition which obviously does not obtain in practice. Although a quantitative analysis of this problem is extremely difficult, the general effect of gravity on the encroachment picture may be seen qualitatively from the following considerations.

The encroachment fronts in a general two-fluid system are subject to two types of distortion as the result of the action of gravity: (1) A tendency for the assumed vertical wall of water to flatten out into a level surface; and (2) in an inclined producing horizon there will be a tendency toward the suppression of the "fingering," so that the entire front will tend to advance along lines parallel to the structural contours.

Considering case (1) in a little more detail, it is clear that if, for example, a difference in density of 0.2 exists between the oil and water, the assumed vertical front will behave as if it bounds a fluid of density 0.2 advancing through the porous medium. Thus, considering the system shown in Fig. 177, which represents a linear porous channel with pressures P_1 and P_2 applied at the terminals, it may be assumed that in the initial stage of the flood the water-oil interface a is a vertical plane which is caused to advance along the channel by the pressure differential $(P_1 - P_2)$. Clearly the difference in density of the fluids will result in an increasing pressure differential with depth, and the water front will progress with greater rapidity at the bottom than at the top of the channel. Thus, if the front was originally vertical, later stages will show an aspect of the water-oil interface somewhat as shown at b and c. In a homogeneous stratum, therefore, edge water should always be expected to make its first appearance at the bottom of the wells, but considering any particular plane within the stratum, the shape of the encroaching flood will be quite similar to that obtained analytically above.

The type of distortion involved in case (2) is also due to the difference in density between the oil and water and is really only a broader phase of the distortion just discussed. In Fig. 178 is

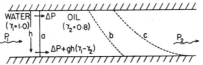

Fig. 177.—Diagrammatic representation of the effect of gravity in a horizontal linear encroachment system.

represented a section of an inclined stratum with a line drive advancing into a single well. The front a indicates what the general shape of the water-oil interface would be at a late stage if the stratum were horizontal. Because of the inclination of the stratum and the difference in density between the oil and water, a component of gravity equal to $g \sin \theta$ will act along the plane of the stratum on particles of water in the elevated cusp. This will give rise to an effective pressure head equal to the weight of a

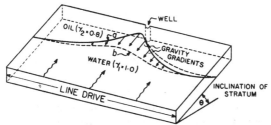

Fig. 178.—Diagrammatic representation of the effect of gravity in an inclined-encroachment system.

fluid column of height equal to the vertical height of the particle in the cusp above the general undistorted level, and of density $\Delta\gamma$, which will oppose the dynamical pressure gradients due to the line drive and hence tend to suppress the cusp. The result will be less pronounced fingering such as indicated by the front b. The steeper the inclination of the stratum, the greater will be this suppressing tendency and the nearer will the water front approach a level surface. It will also be apparent upon further consideration, that the greater the ratio of gravity gradient to dynamical pressure gradient, the less will be the cusping or fingering tendency. This, of course, is the main reason for regulating the output of wells to reasonably low values (low dynamical pressure gradient) in order to obtain a more uniform advance of the water.

8.9. Effect of the Differences in Viscosity between the Fluids on the Two Sides of the Interface.—It is also of interest to consider the nature of the modifications which would be introduced in the above idealized systems, such as the line drive into a single well or direct drive between two wells, if the differences in viscosities between the fluids on the two sides of the interface were not neglected. These modifications will evidently involve both the shape of the interface and the rate of its advance.

With regard to the shape of the interface the general observation may be made that even in a two-fluid system a line of particles will not become distorted unless its various parts are in regions of different pressure gradients. Thus in the case of radial flow (Sec. 8.4), where all particles on the initial circle are subject to the same pressure gradient, the interface contracts uniformly without any distortion, whether or not there is a difference in viscosity on the two sides of the interface. The viscosity difference cannot of itself produce a distortion. Rather it can only modify distortions that are caused by variations in the pressure gradient which are due essentially to the geometry of the system, although the modifications will increase with the magnitude of the distortion that would be present even without a viscosity difference.

In particular, the effect on the encroachment of the water into the oil of higher viscosity will be an accentuation of the distortion and the sharpening of the "fingering" or "cusping" appearing in Figs. 174 and 176 for the single-fluid system. For the replacement of the higher viscosity oil by the encroaching water will evidently decrease the total resistance of the system and hence increase the average fluid velocities. The increase of the velocities, however, will be maximal along the main axis of the fingering where more of the oil has already been displaced by the water. This differential increase in the fluid velocities will, therefore, accentuate the distortion in such a way as to make the fingering more pronounced, although the effect will not become large until the interface approaches the vicinity of the output wells where the natural asymmetry in the pressure gradients, and hence natural distortion, is of appreciable magnitude.

As to the effect of the viscosity difference upon the rate of encroachment, the exact solutions of Secs. 8.3 and 8.4 indicate the nature of the modification. For although the entry of the lower viscosity encroaching fluid will immediately begin to lower the overall resistance of the system—and hence increase the overall rate of encroachment—the magnitudes of the increases will not be large except when the oil is replaced in regions of geometrical convergence. Thus in the case of the line drive into a single well (*cf.* Sec. 8.6) the increase in production rates and rates of encroachment will not be appreciable until the interface begins to cusp into and enter the immediate vicinity of the output well. On the other hand, in the case of the direct drive between two wells, treated in Sec. 8.7, the flooding out of the oil immediately surrounding the input well will at once give an appreciably larger production rate than if the water had as high a viscosity as the oil. However, this rate will increase but slowly during most of the following course of the drive and will rise again noticeably only when the interface will have come into close proximity to the output well.

Although these considerations give a qualitative description of the nature of the effect of the viscosity difference between the encroaching and displaced liquids, it must be remembered that in considering the quantitative effects on the simplified encroachment picture, one must distinguish carefully between the idealized representation in which it is supposed that the encroaching water completely displaces the oil originally in the sand, and the actual situation in which the oil is only partially replaced by the incoming water. In the former case, which is really based on the assumption of the complete miscibility of the two liquids, the porous medium will be the same after it is flooded by the water as before, the effect of the encroachment being simply the replacement of the oil by the water which will usually be of a lower viscosity. In truth, however, water and oil are by no means perfectly miscible and hence display a differential surface behavior which results in a retention by the sand of part of its original oil in spite of the flooding action of the encroaching water. This remnant oil will clearly lower the effective permeability of the flooded zone for the water which tends to pass through it, so that in the real encroachment problem the encroaching liquid is not only of a different viscosity than that of the liquid displaced, but, moreover, the flooded zone has a lower permeability than that still containing all of the original liquid. And although for analytical purposes this lowering of the original permeability due to the remnant oil may be immediately translated into an equivalent increase in viscosity of the encroaching water, it is important to note that this increase may well be so large as to more than counterbalance the fact that the water is strictly of lower viscosity than the oil. In fact, it is quite possible that in actual encroachment systems the effective k/μ for the flooded zone is *less* than that in the oil zone in spite of the lower real viscosity of the water. Under these conditions, then, the idealized picture in which the viscosity difference was neglected will have to be modified to correspond to a flooding of the oil zone by a liquid of *higher* viscosity. The modifications outlined above for the case where the encroaching fluid is of lower viscosity will then be just reversed, and fingering or cusping as well as the rate of encroachment will be retarded as compared to the results predicted for a system with no viscosity difference between the two fluids.

The application of these same general principles to the practical problem of water flooding will be given in the next chapter where the general theory of multiple-well systems will be presented.

8.10. Water Coning; Physical Basis of the Theory.—As suggested at the beginning of this chapter, the phenomenon called water coning is that observed in many oil wells, usually when producing at high rates, in which water gradually, and frequently suddenly, displaces a part or all of the oil production and comes to the surface in place of the oil.[1] Insofar as the water, being of greater density than the oil, would under static conditions remain at the bottom of the sand, its rise into the oil zone and thence into the well implies a dynamical effect due to the motion of the oil above it.

---

[1] The same phenomenon may be demonstrated with viscous liquids in sand-free vessels, but in developing the theory of such cases one would have to use the classical hydrodynamics rather than Darcy's law. However, the physical picture and Eq. (1) below would be the same whether the oil flowed through a sand or not.

To follow in detail the process of the formation of a water cone as it breaks through the oil zone and enters the well bore is a problem so complex as to make a theoretical analysis practically impossible.   However, an analytical treatment of the flow system before the water has broken through and during the time that it lies statically beneath the oil zone with an elevated or "coned" surface can be carried through under certain approximations.[1]

For the purpose of the analysis the physical problem will be idealized as is indicated in Fig. 179, which represents a homogeneous sand formation in which the upper portion is saturated with oil and the lower portion with water.   We shall be interested particularly in the condition under which oil may flow

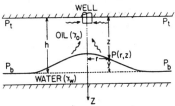

FIG. 179.—Diagrammatic representation of a water-coning system in a homogeneous sand.

into the well without the production of water.   This condition obviously implies that the water assumes a condition of static equilibrium, and hence that

$$p(r, z) + \gamma_w g(h - z) = P_b$$

or

$$p(r, z) + \gamma_w g y = P_b, \qquad (1)$$

where $p(r, z)$ is the pressure at the water-oil interface at the point (r, z), γ_w is the density of the water, γ_0 is that of the oil, g is the acceleration of gravity, h is the thickness of the horizon, and P_b is the formation or reservoir pressure as measured at the bottom of the oil horizon at a point remote from the well.   Because of the radial symmetry of the problem the discussion will be referred to a vertical section passing through the axis of the well.

Equation (1) represents a necessary equilibrium condition if the water cone is to remain in a static condition below the oil zone while flow occurs in the latter zone.   Physically it means that if the drop in pressure at any point, as P, below the reservoir pressure at the same level, equals the differential hydrostatic head $g y(\gamma_w - \gamma_0)$, a water column rising to that point will be in *static* equilibrium.   Furthermore, from this same physical picture it is

[1] MUSKAT, M., and R. D. WYCKOFF, *A.I.M.E.*, **114**, Pet. Dev. Techn. 144, 1935.

clear that in order to maintain *dynamical* equilibrium at the water-oil interface, the latter must be a limiting streamline of the oil zone. However, the stability of this interface and of the elevated cone below it is determined by the nature of the pressure gradients in its immediate vicinity, and even an elementary consideration of these suffices to show that a water cone will not be stable under all conditions of flow in the oil zone. Thus each particle of water at the oil-water interface is acted upon by the pressure gradient present in the immediately adjacent oil zone. Because the flow is convergent, this pressure gradient increases rapidly in the immediate vicinity of the well in a manner

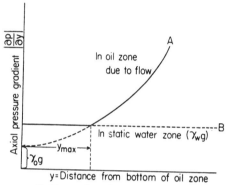

Fig. 180.— Diagrammatic illustration of the equilibrium conditions along the axis of a water-coning system. (*From A. I. M. E.*, **114**, 145, 1935.)

qualitatively indicated by curve *A*, Fig. 180. In the water zone, on the other hand, there is a constant downward vertical pressure gradient due to the acceleration of gravity, its absolute value being shown as curve *B*, Fig. 180. It is evident, from this point of view, that beyond the height indicated as $y_{max.}$, where the pressure gradient in the oil zone is just equal to the opposing differential gravitational force acting upon the water, no water cone will be stable, and that any slight increase in the height of the water cone above this point will result in flow of water into the well. In fact, it is the quantitative definition of this range of cone stability and the determination of the oil-production rate corresponding to the upper limit—the maximum water-free production rate—which is the essential purpose of the quantitative development of these physical considerations.

Assuming $p(r, z)$ to be known, Eq. (1) clearly defines a curve as $z = z(r)$. And although this curve cannot represent in detail the surface of the cone unless it corresponds to a streamline in the oil zone, the complexity of the problem, if solved exactly, makes it necessary to accept Eq. (1) as a sufficient definition of the cone surface and to neglect the discrepancy due to its deviation from an exact streamline. It seems reasonable, however, to suppose that this discrepancy will involve only a small correction to the requirement of Eq. (1), since the latter does give cone surfaces that correspond in all respects to physically reasonable streamline surfaces.

But even Eq. (1) itself cannot be treated without approximations. Thus it cannot be solved for the cone surface unless the pressure function $p(r, z)$ is known at the surface of the water cone and just within the oil zone. On the other hand, the pressure distribution in the oil zone is directly connected, at least in detail, with the shape of the static cone surface which acts like an impermeable boundary to the flow of the oil. Since this combined problem of the exact simultaneous determination of the surface of the water cone and the pressure distribution in the oil zone appears to be too difficult to permit an explicit solution, an approximation will be made.

The simplest approximation, and the only one that seems amenable to treatment[1] without undue labor, is that the pressure-distribution function $p(r, z)$ in the oil zone, *i.e.*, in the region of flow, is effectively the same in the presence of the water cone as the $p(r, z)$ for the case where there is no coning and the water surface is perfectly horizontal. Although the magnitude of the change in $p(r, z)$ due to the presence of the cone is not large, the nature of

---

[1] Although the top of the water cone is an interface between two liquids, it may be noted that the condition of hydrostatic equilibrium, Eq. (1), also makes it equivalent to a free surface such as characterize general gravity-flow systems. In principle, therefore, it should be subject to the methods of treatment developed in Chap. VI. Unfortunately, however, the three-dimensional character of the coning problem makes inapplicable the rigorous analytical procedures based on the theory of conjugate-function transformations. One must therefore resort either to direct empirical methods as described in Sec. 6.18 for radial gravity-flow systems—of which the coning problem is an exact inversion except for the surface of seepage in the ordinary gravity-flow system—or to an approximate analytical procedure such as presented here.

the change is of considerable physical significance and must be clearly understood if the results of the above-mentioned approximation are not to be misinterpreted.

Thus, focusing attention on the top of the cone, where $r = 0$, the actual pressure at the top of the cone may be taken as $p(y)$. It is further convenient to suppose the formation pressure P_b as held constant while the cone height is varied by changing the well pressure p_w. Equation (1) then takes the form

$$p(y) + \gamma_w g y = P_b = \text{const.}$$

or

$$\Phi(y) + (\gamma_w - \gamma_0) g y = P_b = \text{const.} \tag{2}$$

on introducing the more convenient variable $\Phi(y) = p(y) + \gamma_0 g y$, which is the velocity potential in the oil zone, for $k/\mu = 1$, plus the constant $\gamma_0 g h$.

Assuming then that a stable cone exists with its top at the y satisfying Eq. (2), a lowering in the well potential Φ_w, by $\Delta\Phi_w$, will clearly lower the value of Φ at y, and hence y itself must increase in order that Eq. (2) be still satisfied. An increase in the potential drop across the sand will, therefore, raise the cone height y. But more than this can be said. For Eq. (2) implies that the magnitude of the change in y is given by

$$-g\Delta\gamma\Delta y = \Delta\Phi = \left(\frac{\partial\Phi}{\partial y}\right)_{\Phi_w}\Delta y + \left(\frac{\partial\Phi}{\partial y}\right)\Delta y + \left(\frac{\partial\Phi}{\partial\Phi_w}\right)_y\Delta\Phi_w, \tag{3}$$

where the first term gives the contribution to $\Delta\Phi$ due merely to the elevation of the point of measurement of Φ by Δy, the cone height and Φ_w remaining fixed, the second term gives the additional change in Φ due to the fact that the cone has actually risen to $y + \Delta y$ and thus perturbed the potential distribution for the cone height y, and the last term represents the direct effect of the change in well potential Φ_w which initiated the rise from the position at y; finally, $\gamma_w - \gamma_0$ has been replaced by $\Delta\gamma$.

As the top of the cone is stable when at y, the oil at y can have no upward velocity. Hence,

$$\left(\frac{\partial\Phi}{\partial y}\right)_{\Phi_w} = 0. \tag{4}$$

It follows then from Eq. (3) that

$$\left(\frac{\partial \Phi}{\partial \Phi_w}\right)_y |\Delta \Phi_w| = \left[\Delta\gamma + \frac{1}{g}\left(\frac{\partial \Phi}{\partial y}\right)\right] g\Delta y. \tag{5}$$

Now $\left(\dfrac{\partial \Phi}{\partial \Phi_w}\right)_y$ will clearly always be greater than zero. The sign of $\dfrac{\partial \Phi}{\partial y}$, however, is not so obvious, although qualitative considerations indicate that it should be negative. Fortunately this question can be given a definite answer by an appeal to an electrical model of the coning problem.[1] This model, as actually constructed, consisted of a radial sector of pressed carbon, at the broad end of which was soldered a metal plate to give the effect of a uniform reservoir potential, and at the vertex of which was soldered a thin copper rod representing a partially penetrating well at the uniform well potential, taken for convenience as 0. The equivalent of the water cone was obtained simply by cutting away the lower part of the sector, adjusting the exact shape by trial and error so that the potential distribution at the cut surface satisfied the electrical equivalent of Eq. (2). The potential distributions for several cone heights are shown in Fig. 181. It will be seen at once from these potential distributions that $\partial \Phi/\partial y$ is actually negative. Furthermore, at the well axis it increases in magnitude with increasing cone height.

Returning to Eq. (5), it can, therefore, be concluded that the bracket $\left[\Delta\gamma + \dfrac{1}{g}\left(\dfrac{\partial \Phi}{\partial y}\right)\right]$ will continually decrease with increasing cone height until it ultimately will vanish. When the latter condition is reached, Δy will become infinitely large for a finite $\Delta \Phi_w$. This evidently means that the cone becomes unstable, and instead of continuing to rise at a finite rate as Φ_w is decreased, it rises at once to the bottom of the well, so that the water will be produced together with the oil. There will, therefore, be a critical height for the cone, beyond which the cone cannot rise and remain static and stable without entering the well. This critical cone will be attained when the total pressure differential in the sand reaches a certain critical value, and when this is exceeded, the cone will break into the well. The production rate corre-

---

[1] Cf. Sec. 4.17 for a general discussion of electrical-model experiments.

sponding to this critical-pressure differential will, therefore, be the maximum production rate of oil which can be obtained from the system without the simultaneous production of water.

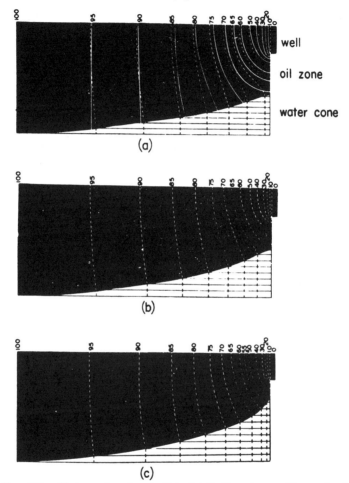

Fig. 181.—Photographs of potential distributions in graphite-conducting models representing radial sectors of oil-sand horizons underlain by water which has coned up beneath the wells. Equipotentials are numbered in per cent of total potential drop over the sand. Solid curves in (a) represent potential distribution when cone is absent. (*From A. I. M. E.*, **114**, 150, 1935.)

We have entered into such a detailed discussion of the proof that the water cone will become critical and unstable after a certain maximum height in order to show that the critical character which will be derived in the approximate theory to be given below will not be simply the consequence of the approximations. For the real approximation of the theory, as already indicated, will be that the term $\frac{\partial \Phi}{\partial y}$ will be neglected entirely. At the same time, however, we shall retain the term $\left(\frac{\partial \Phi}{\partial y}\right)_{\Phi_w}$ corresponding to the unperturbed potential distribution, which is not only negative but in fact decreases rapidly as the bottom of the well is approached. The resultant effect will be at least qualitatively the same as before and will in the same manner lead to a maximum and critical height for the cone. The approximations of the theory will, therefore, only affect the quantitative details of the solution and will not invalidate the general features relating to the critical character of the cone and the relation of the maximal production rates and pressure differentials without coning to such parameters as the well penetration and sand thickness.

8.11. Analytical Development.—It has been seen that to solve Eq. 8.10(1) it is necessary to know the form of the pressure distribution $p(r, z)$, and that to make the analysis possible, the effect of the cone on this distribution must be neglected. Thus it will be supposed that $p(r, z)$ is that given by an analysis of the flow between two perfectly horizontal impermeable boundaries and into a well partially penetrating the sand from the top.

However, before applying the details of the solution to this phase of the problem, which has already been given in Sec. 5.3, we may proceed further with a development of Eq. 8.10(1). Thus introducing again the potential function Φ to replace the pressure p, and the further notation

$$\left.\begin{aligned}\Delta\Phi &= \Phi - \Phi_w = \Phi - \frac{k}{\mu}p_w(z = 0) \\ \Delta P &= P_t - p_w = P_b - \gamma_0 gh - p_w,\end{aligned}\right\} \quad (1)$$

so that ΔP is the pressure drop between the well and the outer sand boundary as measured at the top of the sand, Eq. 8.10(1) can be rewritten as

$$\Delta\Phi = \frac{k}{\mu}\Delta P - \frac{k}{\mu}g\Delta\gamma(h - z). \quad (2)$$

Observing that for large r, $\Delta\Phi = \dfrac{k}{\mu}\Delta P \equiv (\Delta\Phi)_e$, Eq. (2) may be given the final form

$$\frac{\Delta\Phi(r,\, z)}{(\Delta\Phi)_e} = 1 - \frac{g\Delta\gamma}{\Delta P}h\left(1 - \frac{z}{h}\right), \tag{3}$$

which is readily interpreted as the equivalent of Eq. 8.10(1).

As the left-hand side may be found and plotted for fixed values of r as a function of z by means of the theory of Sec. 5.3 (for $r = 0$ it can be read off such curves as are given in Fig. 80),

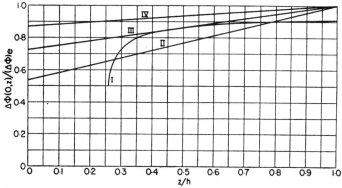

Fig. 182.—Illustration of graphical solution of Eq. 8.11(3) to obtain the equilibrium heights of the cone. (*From A. I. M. E.*, **114**, 148, 1935.)

the solution of Eq. (3) for z may be found as the intersection of the straight lines representing the right side with the curves for the left side. As an example may be taken the case of a well penetrating 31 ft. of a 125-ft. sand (25 per cent penetration). Taking $(\Delta\Phi)_e$ to refer to the potential drop between the well and a point 500 ft. distant, and setting $r = 0$, so as to get the value of z for the top of the cone, the left side of Eq. (3) is plotted as the curve of Fig. 182, while the right side, for three arbitrary values of ΔP, is plotted as the group of straight lines, their slopes being inversely proportional to ΔP.

The values of z at the intersections of the straight lines with curve I are evidently solutions of Eq. (3). Hence they represent the cone heights for the corresponding physical constants of the system and the chosen ΔP which satisfy the hydrostatic equilibrium condition, Eq. 8.10(1).

Physically, however, the two intersections are of quite different significance. For, at the higher rise of cone $z/h = 0.28$, the slope of curve I exceeds that of curve II, which means that at $z = 0.28$ the gradients in the oil zone will exceed the hydrostatic gradient in the water zone. Hence, even though there could be hydrostatic equilibrium at $z = 0.28$, the equilibrium would be dynamically unstable against infinitesimal perturbations in the cone surface. The root at $z/h = 0.28$ can therefore play no role in the physical problem. Conversely, however, the root at $z/h = 0.78$ gives a physically *stable* cone, as the gradient there, in the oil zone, is less

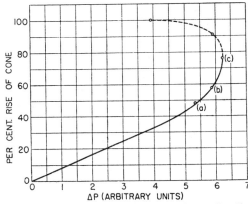

Fig. 183.—Variation of the water-cone height below an oil well of 25 per cent penetration with the pressure drop across the sand, as obtained by means of model experiments. (a), (b), (c) refer to the cases shown in Fig. 181. Total distance between bottom of well and bottom of oil zone is taken as 100 per cent rise of cone. (*From A. I. M. E.*, 114, 151, 1935.)

than that in the water zone. If ΔP is increased, the slope of the lines for the right side of Eq. (3) will decrease until, as in the case of curve III, they will be tangent to curve I, and then for still greater ΔP there will be no intersections with curve I, or no stable heights of cone. The point of tangency of curves I and III ($z/h = 0.48$) then gives the critical and maximum cone height possible for the given r_e, r_w, h, and $\Delta\gamma$.

It is evident that in the vicinity of the critical cone defined above the position of the water surface is very sensitive to small changes in pressure within the adjacent oil zone. And although the analysis here is based upon a pressure distribution undisturbed by the presence of the cone, whereas there will actually be a perturbation in the pressure distribution because of the existence of the cone, it will be recalled from the last section that the unperturbed pressure distribution will give at least qualitatively the same effect in leading to a critical cone as the correct distribution which takes into account the effect of the cone. In fact, by plotting the heights of the cones of the experimental model against their equivalent pressure differentials (in arbitrary units) the solid curve of Fig. 183 is obtained. The point at a cone height of 91 per cent was obtained by extrapolating from Fig. 181 the probable effect on the pressure distribution if the cone should be raised to 86 per cent. The last point is given by the observa-

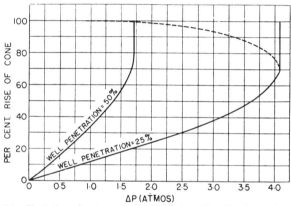

FIG. 184.—Variation of water-cone heights below oil wells of 25 and 50 per cent penetration with the pressure drop ΔP across the sand.   Total distance between bottom of well and bottom of oil zone is taken as 100 per cent rise of cone.   Vertical segments represent unstable cone heights.   Dotted segments correspond to the lower intersections of the curves of Fig. 182.   Sand thickness = 125 ft.; well radius = $\frac{1}{4}$ ft.; formation-boundary radius = 500 ft.; water-oil density contrast = 0.3 g./cc.   (*From A. I. M. E.*, **114**, 153, 1935.)

tion that for a cone height of 100 per cent Eq. 8.10(1) can be satisfied only if the total pressure differential equals the total available differential hydrostatic head in the system.   Comparing the resultant curve with that given in Fig. 184, as derived analytically, it will be seen that all the significant features of the latter are actually confirmed by the model experiments, the latter in fact showing that the actual cones will be somewhat more critical than those deduced theoretically.

Returning now to the further development of the analytical treatment, the behavior of the cones in systems of practical interest will be considered.   The procedure consists in observing the slopes of the lines as of curve II in Fig. 182 giving a certain intersection with potential curves as curve I, these intersections representing cone heights z/h. The slope m of these lines is clearly given by $m = \dfrac{gh\Delta\gamma}{\Delta P}$. Hence, knowing m and the other constants of the system, ΔP can be computed. These may then

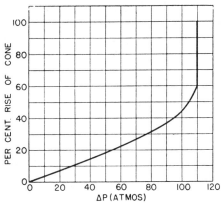

Fig. 185.—Variation of the water cone height below a "nonpenetrating" oil well with the pressure drop ΔP across the sand. Vertical segment represents the unstable cone heights. Sand thickness = 125 ft.; well radius = ¼ ft.; formation-boundary radius = 500 ft.; water-oil density contrast = 0.3 g./cc. (*From A.I.M.E.*, **114**, 153, 1935.)

be plotted against the corresponding z/h. Such curves are given in Figs. 184 and 185 and correspond to the following constants:[1]

$\Delta\gamma = \gamma_w - \gamma_0 = 0.3$ gm./cc.

$g = 980$, and is equivalent to 0.0097 atm. for unit density.

$h = 125$ ft.; r_w = well radius = ¼ ft.

r_e = reservoir-boundary radius = 500 ft.

The fact that the pressure drops required to raise the cones are so small, as appears from Fig. 184, is not surprising when it is noted that a water column of height equal to one half of the sand thickness· of 125 ft. has a head of less than 2 atm. Furthermore, since the ΔP in Figs. 184 and 185 are pressure drops as measured at the top of the sand, the pressure drops in the oil

---

[1] For the details of these calculations, see the original paper by Muskat and Wyckoff, *loc. cit.*

zone available to lift the water are these ΔP plus the hydrostatic head of oil between the bottom of the well and the bottom of the oil horizon.   On the other hand, the relatively large values of ΔP required to raise the cone in the case of the nonpenetrating well are due entirely to the great concentration of the pressure gradients near the well when the latter only taps the sand.   In fact, the pressure gradients near the well vary inversely as the square of the distance from the well.   Hence it may well be that more than 100 atm. of the total pressure drop of 107 atm. (for a cone rise of 54 per cent) is lost within 5 ft. of the well and only a small fraction of it is effective in raising the water in the cone.

The vertical sections of the curves simply indicate the fact that beyond the critical heights corresponding to the points of tangency of curves I and III of Fig. 182, physically stable cones cannot exist.   The dotted portion of the curve for 25 per cent penetration (Fig. 184) represents the upper intersections, as of curves I and II of Fig. 182.   As already anticipated above, this curve agrees in all essential respects with that of Fig. 183 derived from the model experiments, except for the discontinuity in the tangent to the analytical curve at the critical height, which is caused by the neglect of the perturbation in the pressure distribution due to the presence of the cone.

All of the above discussion with regard to the height of the cone has referred to its value along the central axis of the sand, *i.e.*, in the region immediately below the well.   However, it should be clear that the computations can be carried through in exactly the same manner for any distance out from the well, provided only that the curve I in Fig. 182 be appropriately changed and made to refer to the vertical-potential distribution at the chosen distance from the well.   If this is done for a number of such radial positions for the well, one will obtain by plotting the roots of Eq. (3), derived as indicated above, a graphical representation of the shape of the cone.   The results of such computations for the case of a 125-ft. sand when the well penetration is 50 per cent and for various pressure differentials over the sand are presented in Fig. 186.   The actual shapes of the cones will, of course, be obtained by rotating the curves of Fig. 186 around the vertical axis.   Although from the nature of the calculation these curves cannot be accurate in detail, they do have the general shape that would be expected of the streamline curves repre-

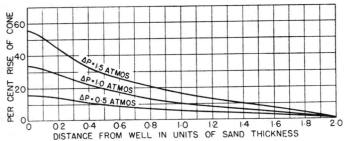

FIG. 186.—Theoretical cross-sectional shapes of water cones below an oil well of 50 per cent penetration for various pressure drops across the sand; sand thickness = 125 ft.; well radius = ¼ ft.; formation-boundary radius = 500 ft.; oil-water density contrast = 0.3 g./cc. (*From A. I. M. E.,* **114,** 154, 1935.)

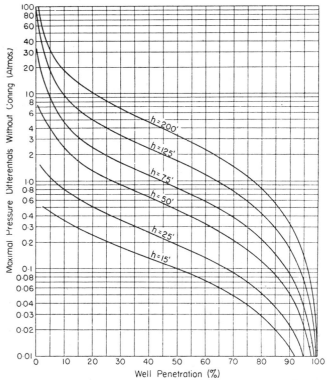

FIG. 187.—Maximal pressure differentials (reservoir pressure − bottom-hole pressure) that can be maintained over an oil sand without water coning, as a function of the well penetration for various oil-zone thicknesses h. Well radius = ¼ ft.; formation-boundary radius = 500 ft.; water-oil density contrast = 0.3 g./cc. (*From A. I. M. E.,* **114,** 155, 1935.)

senting the cone surfaces, and are in fact quite similar to those found by the experimental models (cf. Fig. 181).

The values of the critical-pressure differentials required to bring the cones into the well and obtained by solving Eq. (3), as outlined above, are given in Fig. 187. It will be noted that the critical pressure differential increases very rapidly, especially for the thicker sands, as the well penetration becomes small. Furthermore, it falls steeply to infinitesimal values as the well penetration approaches 100 per cent and the bottom of the well is set near the water level. Finally, the critical-pressure differential not only decreases with decreasing sand thickness, as should be expected, but the variation with the sand thickness becomes increasingly pronounced as the thickness decreases.

Although, from a physical point of view, the pressure differentials and gradients are the controlling factors determining the stability of the water cone, the relation of the rate of oil production to the entry of the water into the well is perhaps of most interest from a practical point of view. To find the production rates corresponding to the ΔP of Fig. 187, it is only necessary to apply the results of Sec. 5.4 as given by Fig. 83 and 84 in which the production capacities per *unit* pressure differential for partially penetrating wells are plotted against the penetrations and sand thicknesses.

The critical or maximal production rates obtained in the above manner are plotted in Fig. 188. Perhaps the most interesting feature of these curves is the fact that the non-penetrating wells will permit the maximal production rates without the cone entering the well. Of course, the fact that the bottom of a nonpenetrating well is at a greater distance from the water horizon than the bottom of a partially penetrating well would naturally tend to make it more favorable for the suppression of the cone, and indeed Fig. 187 does show that the critical-pressure differentials are maximal for the nonpenetrating wells. On the other hand, the very high resistance of the nonpenetrating well system might be expected to more than counterbalance the effect of the large separation between the bottom of the well and the water level, so as to give rise to an optimum nonvanishing penetration at which the critical production rate is a maximum with respect to both smaller and larger penetrations. The curves of Fig. 188 show, however, that the distance of the well bottom from the water level is the most significant factor, although it is to be noted that the curves are quite flat for the small penetrations, and for practical purposes wells with penetrations as high as 15 to 20 per cent will give practically as high water-free production rates as those of extremely small penetrations.   This is indeed

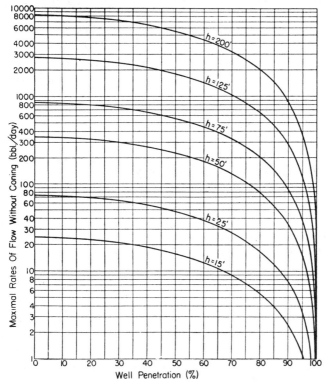

FIG. 188.—Maximal rates of oil flow possible without water coning as functions of the well penetration for various oil-zone thicknesses h.   Well radius = $\frac{1}{4}$ ft.; formation-boundary radius = 500 ft.; water-oil density contrast = 0.3 g./cc.; (*From A. I. M. E.* **114,** 156, 1935.)

a fortunate circumstance, for thus it is possible to obtain in practice the optimum operating conditions without the necessity of attempting to satisfy the almost impractical condition of "nonpenetration" in an actual well.[1]

[1] Noting the parallelism of the curves of Fig. 188 on the logarithmic scale it may be shown that the maximal production rates Q may be expressed very approximately by: $Q = f(\bar{h})h^{2.3}$, where $f(\bar{h})$ depends only on the frac-

The practical significance of the above analysis is that it shows that considerable penetrations into an oil sand should be avoided if it is known that the lower portion contains water. It further shows that if at a certain finite penetration and a given production rate water coning is already taking place, one may reasonably attempt to correct the situation by plugging back the hole to effectively reduce the penetration. This implies the fact that coning, once started in an oil sand, is not necessarily a permanent accompaniment of the production, since the essential reasons for the persistence of the coning are that the pressure drop between the reservoir boundary and points below the bottom of the well exceed the hydrostatic head of the corresponding water column, and that the dynamical pressure gradients in the system exceed the static gradient due to the density difference between the oil and water. Hence, if the well is plugged back or the production rate is "pinched down" so that the total pressure drop and dynamical gradients become insufficient to overcome the hydrostatic head and gravity gradient, the water cone can only fall back toward the bottom of the formation. The time for these transitions to take place is a rather difficult quantity to estimate but it certainly seems that it should be measurable in terms of hours even for relatively tight sands. Thus it appears that at least by the process of controlling the back pressure on a well one should be able either to control with a fair degree of arbitrariness the rate of water entry into the well or to stabilize the oil production with the water lying statically below the oil horizon.

8.12. The Suppression of Water Coning by Shale Lenses.— While field experience as a whole confirms the conclusions developed here concerning the suppression of water coning by decreasing the penetration of a well, it is often observed that the magnitudes of the effects are much larger than would be expected from Figs. 187 and 188. Thus, cases have been observed where plugging back only 2 or 3 ft. has sufficed to eliminate almost completely the water from wells which had previously been producing with rather high water-oil ratios. As an example may be cited the case of a well[1] in "East Texas," in

---

tional penetration $\bar{h}$. This formula may be used to calculate Q for sand thicknesses h, other than those used in Fig. 188.

[1] This well was No. 2 in the Walter Shaw A lease of the "East Texas" field,

which a 20 per cent water content was changed into clean oil flow by simply plugging back 2 ft. of its original penetration of 7 ft. It is clear that if the sand were strictly homogeneous, as has been assumed in the theory given above, the change in total well penetration from approximately 16 to 11 per cent would have but a very small effect in suppressing the water cone, so that the observed effect must be attributed to an inhomogeneity in the sand conditions. In fact, the log of that well shows a tight zone 2 ft. in thickness beginning 4 ft. below the top of the pay, and the well was plugged back to this zone.

That an extended shale streak lying at the bottom of a well would inhibit bottom water from coning into the well is, of course, obvious. In most sands, however, many shale streaks occur as small broken lenses embedded in the main sand body, and the question arises as to the effectiveness of plugging back to such lenses of limited extent. Although it would be very difficult to attempt to derive an analytical solution to the problem involved except by graphical means, the question proposed may be answered in a very simple manner with the aid of the electrical model of the flow system already described in Sec. 8.10. The equipotential contours for a 50 per cent well penetration in a homogeneous sand as found by this model are indicated in Figs. 189a and b as the solid lines where the external potential has been taken for convenience equal to 100. The permeable shale lenses were then introduced by simply cutting narrow slots below the well bottom, in one case with a radius equal to 11.3 per cent of the sand thickness and in another case with a radius 20.6 per cent of the sand thickness. The new potential distributions are shown as the dotted curves.

Although a quantitative application of the changes in the potential distributions caused by the slots or shale lenses will not be attempted, the qualitative implications are quite definite and unambiguous. For it is clear from Fig. 189 that in the presence of the impermeable lens the potential gradients are concentrated and increased in the upper part of the sand and near the edge of the lens, while they are diminished at distant points from the well and in the region below the lens. The particular point of interest, however, is the fact that the presence of the lens very markedly increases the potential in the region below the lens, where the water would tend to rise. Thus in the case of the smaller lens of Fig. 189*a* the point at the bottom of the lens terminating the 55 per cent potential contour originally lay on the 40 per cent contour in the homogeneous sand; in the case of the lens of Fig. 189*b* the same point on the original 40 per cent contour would terminate the 65 per cent contour in the presence of the lens. As the "rising tendency" of the cone at a given point is essentially proportional to the potential difference between that point and the effective sand reservoir boundary,

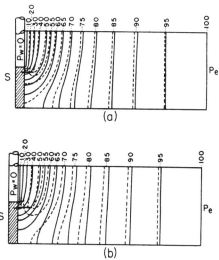

Fig. 189.—Equipotentials, obtained by electrical-model experiments, for 50 per cent well penetration in a homogeneous sand before (solid contours) and after (dotted contours) introduction of an impermeable-shale lens *S*. (*a*) Shale lens radius = 11.3 per cent of sand thickness; (*b*) shale lens radius = 20.6 per cent of sand thickness. (*From A. I. M. E.* **114**, 159, 1935.)

it is clear that the two lenses corresponding to Fig. 189 would lower the coning tendency to values of the order of 75 and 58 per cent, respectively, of what they would be in a homogeneous sand. Such effects might well correspond to a change from a 20 per cent water-oil ratio to the production of pipe-line oil.

Thus it appears that the apparently anomalous success in the elimination of water coning by plugging back a well may be explained on the basis of the same physical principles as the more commonly observed manifestations of the phenomenon. Furthermore, it is seen that the most effective means of eliminating bottom water in a well penetrating a homogeneous sand is to plug the well back to one of the tighter lenses even though such lenses may be only of very limited lateral dimensions.

8.13. Summary.—The occurrence of waters in the neighborhood of oil-bearing sands usually gives rise to either or both of two general types of problem. In the first type the water enters the oil sands as an "edge-water encroachment" and results in a flooding of the oil sand. In the second type, which in a broad sense is but a variation of that of edge-water encroachment, the water has already entered the oil sand and, owing to its greater density, tends to lie at the bottom of the sand, except as the dynamical gradients due to the flow of the oil to the outlet well overcome the density contrast between the oil and water and bring the latter into the well as an elevated cone rising from the water horizon.

In treating the case of edge-water encroachment a new type of potential problem is encountered. For, as the viscosity of the water is different from that of the oil, the water-oil system must be represented as composed of two distinct potential regions separated by the water-oil interface. As the water encroaches into the oil sand the potential region representing the oil zone is constantly swept out by that representing the water zone, and the interface between the two continuously assumes new forms. This motion of the boundary between the two potential regions results in a type of problem quite different from that occurring in other physical problems.

Although a general formulation of this type of potential problem can be readily made, it is difficult to give a procedure for its exact solution except in very special cases. Nevertheless, it is possible to derive the general deduction that the shape of the interface is independent of the absolute value of the pressure differential acting to bring the oil and water into the outlet surfaces. It is true, of course, that the *rate* of the encroachment and the advance of the water-oil interface is directly proportional to this pressure differential, but the shape of the interface is unaffected by it. Hence a system with a small pressure differential will possess exactly the same family of interfaces as one with a large differential except that they will be crossed at proportionately later instants.[1]

[1] The discussion in this chapter deals with liquids which are perfectly miscible (such as two oils) and show no differential surface behavior,

Of the rigorous solutions for the problem of encroachment only those can be developed in closed form in which the symmetry of the system predetermines and makes evident the shape of the interface. These include the problems of linear, strict radial, and spherical encroachment. For other systems, the necessity of finding simultaneously the shape of the interface and the potential distribution in the two regions which it separates leads to great analytical difficulties.

For the linear-encroachment system the analysis (*cf.* Sec. 8.3) shows that as a liquid encroaches into a system containing a liquid of higher viscosity, the *rate* of encroachment and the rate of production increase. This is due to the fact that the initial higher viscosity liquid is being continually replaced by the encroaching fluid of lower viscosity, thus decreasing the resistance of the system and hence increasing the production rate. This effect increases as the encroaching fluid is assigned decreasing values for its viscosity. However, it attains a limit in the sense that the time required for the encroaching fluid to sweep out a given length of sand, even if it has a vanishingly small viscosity, can never fall to less than half of that required if the encroaching fluid be of the same viscosity as that displaced [*cf.* Eq. 8.3(9)].

The problem of radial encroachment may also be solved exactly (*cf.* Sec. 8.4). Here, if the encroaching liquid has a relatively low viscosity, the acceleration in the encroachment, as the encroaching fluid displaces the higher viscosity liquid surrounding the well, is augmented by a natural acceleration due to the converging character of the flow. The increase in production rate, however, does not become appreciable until the interface between the two liquids comes into the immediate vicinity of the well. If the encroaching fluid has a viscosity 0.1 that of the fluid

---

whereas in the case of actual immiscible fluids the flooding liquid will not completely replace that originally contained in the sand. As a result the flooded area will contain a mixture of the two fluids. The retention of a portion of the original fluid within the flooded zone results in a clogging of the pores which, strictly speaking, should be considered as decreasing the permeability of the porous medium. However, for analytical purposes the decrease in permeability may be conveniently represented by an appropriate increase in the viscosity of the flooding fluid. Further details of the dynamical behavior of such systems are beyond the scope of the present discussion.

displaced, the production rate will not be doubled over that at the beginning of the encroachment until 99.96 per cent of the original liquid has been displaced. Because of this, the effect of the lower viscosity of the encroaching fluid in reducing the time required for the interface to reach the well is much less than in the case of linear encroachment. Here, even if the encroaching fluid has zero viscosity the time for it to reach the well is cut down by only 7 per cent [*cf.* Eq. 8.4(8)].

The case of spherical flow leads to results very similar to those for radial flow. The effect of the convergence of the flow channel is still more pronounced in retarding the acceleration in the production rates as the encroachment proceeds with a low-viscosity encroaching liquid, and in diminishing the effect of the encroachment in cutting down the time for the encroaching fluid to reach the outflow surface.

When the symmetry of the system does not suffice to make obvious the form of the water-oil interface it must be determined simultaneously with the potential distributions on either side of it. The analytical difficulties are, however, too great to permit a solution without the aid of numerical or graphical means. Although a perturbation theory can be developed so as to require only separate solutions for the shape of the interface and the potential distributions, it still involves great practical difficulties. Nevertheless, a good qualitative picture of the phenomena may be derived if the difference in viscosity between the encroaching fluid and that displaced is neglected. The interface in the real encroachment system is then replaced by a geometrical line or surface of fluid particles which initially are supposed to lie along the real boundary between the two fluids, and whose subsequent motion is taken to correspond at least qualitatively to that of the real interface if the viscosity difference had not been neglected.

When the boundary conditions are kept fixed, a general solution for this simplified problem—the "zero approximation" to the real encroachment problem—may be found quite generally by simple quadrature provided the equipotential and streamline surfaces of the system can be determined.

As examples of this theory solutions have been found for problems of the line drive into a single well (Sec. 8.6), and of the direct drive between two wells (Sec. 8.7). The former problem corresponds to an idealized edge-water drive into an oil field which contains a single well drilled close to the edge water. The latter represents a flooding system consisting of a single input well and a single output well. The progress of the flood may be followed by tracing the advance of a line of fluid particles, originally encircling the input well, traveling toward the output well (*cf.* Fig. 176).

The interpretation of the solutions to these and similar problems becomes clear when it is observed that a line of fluid particles in a homogeneous system (no viscosity difference between encroaching and displaced fluids), or a water-oil interface in a real two-fluid system, will suffer distortion only when traveling in regions where the pressure distribution is not uniform. For a straight line of fluid particles will remain straight if it happens to be in a system with rectilinear parallel equipotentials and if it lies parallel to them; and similarly, particles lying on a circle concentric with a system of circular equipotentials will continue to lie on the family of concentric circles. On the other hand, if the particles lie in nonuniform fields, those subject to higher pressure gradients will move ahead of the rest, causing a continual distortion of the line of particles. Thus in the case of the line drive into a single well (*cf.* Fig. 35) the pressure gradients are evidently greatest along the normal from the well to the initial line. Hence the particles along and near this line will move ahead of those distant from it, giving rise to a "fingering" of the hypothetical interface. Furthermore, since the gradients increase as the well is approached, this distortion will be continuously accelerated and the interface will finally be drawn into the well as a cusp. Similar considerations may be developed in the interpretation of the solution for the direct flood between two wells (Fig. 176), and other more complex problems.

Although these solutions and considerations have been based upon idealized models in which the effects of gravity and the viscosity difference have been neglected, the types of modification which would be caused by these features are not difficult to see. Thus if the encroachment system is really three-dimensional, the effect of gravity will give an added vertical component to the particle velocities. This, in a horizontal sand, will induce a flattening of the water-oil interface so as to assume an inclined rather than a vertical position. If the sand is not horizontal and the water is encroaching upward along the flanks of a structure it may be shown by a simple analysis that the essential effect of gravity will be a diminution in the "fingering" of the interface.

Similarly the difference in the viscosity between the encroaching and displaced fluids will cause two types of changes in the picture derived from the idealized theory. As already mentioned, the rate of the encroachment will be accelerated or retarded as the encroaching fluid is of lower or higher effective viscosity than that displaced. This effect, however, will not become appreciable until the interface between the two fluids enters regions of highly concentrated pressure gradients.

The other modification to be noted is the effect upon the shape of the interface. It follows from elementary considerations that an encroaching fluid of high viscosity will advance with suppressed fingering, whereas the fingering effect will be accentuated as the viscosity of the encroaching fluid is decreased. This accentuation, however, will take place only in the regions where there is already a natural distortion caused by an asymmetry in the pressure gradients, since if there is no distortion in the shape of a line of particles in a homogeneous system, there will be none in the interface of a real encroachment system, with the same geometry and boundary-pressure distributions.

In addition to the problems of water encroachment which involve the simultaneous treatment of two fluids within the same porous medium, there is the problem of water coning. In the simplest cases it will arise when the oil zone is underlain by a water sand, so that in place of the impermeable base for the oil-producing sand there is another sand or a continuation of the oil sand containing the heavier but equally mobile fluid, water, as the lower boundary of the oil horizon.

Insofar as the oil flows into a well penetrating the oil zone by virtue of a pressure differential existing between the well and points at a large distance from it, the same pressure differential, or at least part of it, must also act upon the water zone and tend to induce the water also to move into the well. Indeed, this tendency is quite real, and is the physical basis of the problem of water coning. However, there are other forces which, in the water zone, oppose this pressure differential, and it is the relative magnitude of the two which determines whether the water will actually enter the well along with the oil.

The opposing force arises from the density contrast between the oil and water. Since the well does not penetrate the water zone, the water must rise against the force of gravity corresponding to the differential density between the water and oil before it can enter the well, whereas in the oil zone the force of gravity ultimately cancels out and does not influence the flow. For example, it may be supposed that a well just penetrates the top of the oil zone, of 125 ft. thickness, and is maintained at 300 lb. pressure, while the reservoir pressure at the top of the sand driving the oil into the well is 741 lb.

Now it may be shown by means of Eqs. 5.3(6) and 5.3(7), that just below and along the axis of a nonpenetrating well, the pressure at a depth z has dropped below that at a point at the bottom of the oil zone, and at a distance of 500 ft. from the well, by the value

$$p_e - p = \gamma_0 g(h - z) + \Delta P\left[\frac{2r_w}{z} - \frac{r_w}{h}\psi\left(1 + \frac{z}{2h}\right) - \frac{\pi r_w}{2h}\operatorname{ctn}\frac{\pi z}{2h}\right],$$

where γ_0 is the density of the oil ($\sim$0.8 gm./cc.), g is the acceleration of gravity, h is the total oil-zone thickness (125 ft.), ΔP is the total pressure differential ($\sim$30 atm.), r_w is the well radius ($\frac{1}{4}$ ft.), and ψ is the logarithmic derivative of the Gamma function. This same pressure differential, neglecting the distortion of the pressure distribution caused by the presence of the water cone, will also act upon the water and tend to sustain it above the general water level to a height whose head will equal $p_e - p$.

Thus at a height of 12 ft. above the general water level ($z = 113$ ft.), the above formula shows that $p_e - p = 0.4023$ atm. whereas 12 ft. of water requires a head of only 0.3894 atm.[1] to sustain it. Hence a cone cannot remain stable after a rise of 12 ft., but will continue to rise. On the other hand, if the cone is brought up to 16 ft. from the water horizon ($z = 109$ ft.) $p_e - p = 0.4975$ atm., whereas the water head of the cone will be equivalent to 0.5192 atm. Hence in this case the difference between the reservoir pressure and that at the assumed height of the cone would not be able to sustain it at this height, and the cone would have to fall. Proceeding in this way by trial and

---

[1] The density of the water is taken here as 1.1 gm./cc., although in the final determination of the stable cone height, only the density contrast is of significance.

error—or more systematically by a graphical solution of Eq. 8.11(3)—it is found that if the cone is elevated to a height of 13.5 ft. above the water level ($z = 111.5$ ft.) the dynamical pressure differential in the oil zone will just balance and sustain the hydrostatic head of the water cone. Similarly the height of the cone may be found at points other than along the axis of the well, and from these a composite surface of the cone may be drawn (*cf.* Fig. 186), separating the flowing oil above and the statically elevated water cone below.

If now the pressure differential of 30 atm. is increased—by lowering the pressure at the well—the height of the cone will evidently be raised to a value given by Eq. 8.11(3) with the new value of ΔP. As ΔP is still further increased the cone will ultimately be raised to such a height that any further increase in ΔP will bring the cone into the well, there being no stable positions of the cone (for the given sand and well penetration) above this critical height. The pressure differential ΔP raising the cone to this critical value will, therefore, be the maximum that can be maintained in the system without the production of water, and the corresponding production rate the maximum that can be attained which will be water-free. That the upper limits to the stable heights of rise of the cones is not introduced by the approximations of the analytical theory (neglect of the perturbation in the pressure distribution due to the presence of the cone) is confirmed by experiments with electrical models of the coning system (*cf.* Figs. 181 and 183).

The same calculations may be carried through for the other oil-zone thicknesses and for wells which actually penetrate the oil zones. Such computations show that not only is the pressure differential required to bring the water into the well a maximum for the nonpenetrating well, decreasing continuously as the well penetration increases—which would be naturally expected (Fig. 187)—but moreover they reveal the rather surprising result that the actual rate of oil production which may be maintained without the simultaneous production of water is also a maximum for the nonpenetrating well and decreases as the well penetration increases (Fig. 188). That is, if an oil sand is underlain by one containing water and if it is penetrated by a well to any given depth, the rate of water-free oil production that can be obtained from the well can be continually increased by plugging back from the bottom of the well until the well just pierces the top of the oil zone. This shows that the effect of the actual recession of the bottom of the well from the water horizon more than counterbalances the increasing resistance of the sand-well system as the exposed sand-face area is reduced to a minimum.

From a practical point of view, however, it is found that the curve of maximal production rate *versus* well penetration is quite flat for small penetrations, so that even penetrations of 20 per cent will permit almost the same production rates as wells of vanishing penetration. Hence if a well has already been plugged back to penetrations of the order of 20 per cent and the water has not yet been eliminated, it will be necessary to cut down the production rate by increasing the back pressure. For if the pressure differentials are diminished to such an extent that they can no longer sustain a fluid head equal to the height of the well bottom above the water horizon, the water must necessarily drop until it reaches a level in equilibrium with the dynamical gradients due to the oil flow.

Although the above practical conclusions are in general confirmed by field observations, apparently anomalous cases frequently arise in which only a few feet of plugging back of a well producing with an appreciable water-oil ratio may suffice to eliminate practically all of the water. An explanation of such cases is readily found if it is assumed that the wells are plugged back to an impermeable lens. For model experiments show that such impermeable lenses, even though of quite limited radial dimensions, will raise the pressures in the region below the lens, thus leaving smaller fractions of the total pressure drop in the system which will be available for lifting the water into the cone. Hence, while it will be in general futile to plug back a coning well which already has a small penetration, a pronounced effect in suppressing the cone can nevertheless be attained if the well can be plugged back to an impermeable lens even though the lens be of small areal extent.

CHAPTER IX

MULTIPLE-WELL SYSTEMS

9.1. Introduction.—Thus far we have treated various systems and types of flow involving single outlet or inlet surfaces by which the liquid could leave or enter the system. In particular, most of the problems discussed in detail have been concerned with a single well as the outlet surface for the liquid supplied at the boundary of the surrounding sand or porous medium. The solutions developed for these problems have served to give descriptions of the behavior of individual wells under various physical and geometrical conditions of the medium as a whole. The features considered in detail have been those which were essentially localized to the immediate vicinities of the individual wells, such as the pressure distributions about partially penetrating wells, or the phenomena involved in the problem of water coning.

However, in the numerical work of the analysis it has always been necessary to define or explicitly assume a definite type and form of exterior boundary at which the fluid leaving the well was being supplied. Although, as frequently noted in the previous chapters, the dimensions of the boundary entered in most expressions of physical interest in a logarithmic manner, and hence could be given only approximate values and yet involve but little error in the final results, even this approximate estimate usually implied certain assumptions which were not otherwise explicitly taken care of in the analysis. Thus, for example, the radius of the external boundary has usually been assumed to be of the order of 500 ft., with the explanation that if it really were anywhere between 250 and 1,000 ft., the computed production rates would differ only inappreciably from those on the assumption of 500 ft. Yet if the wells under consideration were really the only ones in an actual oil reservoir or in an extended water sand, which may reach to or beyond distances as 10 miles, the assumption of an external boundary of 500 ft. would certainly be unwarranted.

It has, therefore, been tacitly assumed that if the whole reservoir or sand were actually much larger in dimensions than 1,000 ft., there were other wells drilled in the neighborhood of the one of interest, and were so located that *they* formed an *effective* external boundary at a distance of the order of 500 ft. Analytically this assumption has been given a formal justification on the basis of the remark that the "external boundary" need possess neither physical nor geometrical significance, but rather its sole function is to define a surface at which the pressure or flux density could be considered as known with reasonable accuracy. In fact, detailed proofs were given to show (*cf*. Sec. 4.5) that even the pressure distribution on the assumed effective boundary does not have to be known in detail in order to compute the production capacity of the well, but that only a knowledge of the *average* pressures over the boundary is required.

Nevertheless it must be admitted that the introduction and use of an "effective" boundary implies at least a qualitative knowledge of the pressure distribution in the complete system in which the mutual interactions of all the wells are taken into account. Thus, as will be seen presently, if there are four wells forming a square with 200-ft. sides near the center of a large area bounded by a circle of 10,000 ft. radius, each well will have a production capacity corresponding to an effective boundary radius of about 10^9 ft., which is very much larger than the actual boundary radius (*cf*. Sec. 9.3). On the other hand, if the four wells constitute a single element of an infinite square network in which the input surfaces are wells at the centers of the squares, the effective boundary radius for each well is 23,200 ft. (*cf*. Sec. 9.24). That these results could have been anticipated without the detailed analysis in which the problems are treated from the beginning as involving multiple-well systems seems quite improbable. Since, however, these types of problems are in some respects of even more practical interest than those dealing with single wells, the theory and solutions for several classes of such problems will be developed in the present chapter.

Both from a practical and analytical point of view, it is convenient to distinguish between, and treat separately, multiple-well systems in which the wells form groups distributed over areas small compared to the total area of the liquid-bearing sand and those in which the well system is distributed over all or a large part of the reservoir or producing sand. The former types of problem will arise during the early stages in the drilling of a large section of a producing horizon, and the analysis will give the interference effects among groups of wells which are confined to areas that are a small part of the whole producing sand. The latter types of problem occur in considerations of the mutual interaction among the wells in a reservoir which is largely or completely drilled, or the field operations involved in flooding and offsetting programs. Because of their simplicity the problem of small groups of wells will be treated first. In this treatment and throughout the whole of this chapter, it will be assumed that the sands are homogeneous, of uniform thickness, and completely penetrated by the wells, so that the flow systems may be taken as two-dimensional.

9.2. Small Groups of Wells. General Theory.—It will be recalled from the previous chapters that a well in a two-dimensional system may be represented by a term in the pressure distribution of the form

$$\frac{Q\mu}{2\pi k} \log r,$$

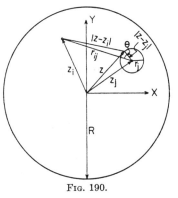

FIG. 190.

where Q represents the flux associated with the well (as source or sink),[1] μ the viscosity of the fluid, k the permeability of the sand, and r is the radial distance as measured from the center of the well. Furthermore, the linearty of Laplace's equation, which the pressure distribution must satisfy, implies that if we add together a number of logarithmic terms of the above type, each of which satisfies Laplace's equation, the sum will still satisfy it, and the sum will be a solution representing the pressure distribution due to a number of wells. Thus, if $z_1, z_2, \ldots$ represent the complex vectors $x_1 + iy_1, x_2 + iy_2, \cdots$ from an arbitrary origin to the center of a series of wells of fluxes $Q_1, Q_2, \ldots$, and if z is the vector to a variable point $x + iy$ (cf. Fig. 190), the resultant

---

[1] Unless otherwise specified, the flux Q in this chapter will refer to a unit sand thickness.

pressure distribution will be given by the expression

$$p(x, y) = c + \frac{\mu}{2\pi k} \sum_j Q_j \log |z - z_j|, \tag{1}$$

where c is a constant which may be chosen so that the average pressure on the external circular boundary takes a preassigned value.

If c and the Q_j are known, Eq. (1) permits a determination of the pressure distributions at all points of the system. From a practical point of view, however, it is of more interest to predict the values of the fluxes Q_j if the well pressures are known. We must, therefore, first determine the well pressures which are implied by Eq. (1) with the Q_j considered as known. Since the well radii r_j will be taken as small compared to all the other dimensions of physical interest, the well pressures, p_j, will be defined as the averages of p taken over the circles of radius r_j, with centers at z_j; *i.e.*,

$$p_j = \frac{1}{2\pi r_j} \oint_{z_j} p\,ds, \tag{2}$$

where the symbol $\oint_{z_j}$ denotes a contour of radius r_j taken about the point $z = z_j$.

If now the distance $|z_i - z_j|$ between the centers of the ith and jth wells be denoted by r_{ij}, it is readily verified that

$$\oint_{z_j} \log |z - z_j|ds = 2\pi r_j \log r_j, \tag{3}$$

while

$$\oint_{z_j} \log |z - z_i|ds = 2\pi r_j \log r_{ij}. \tag{4}$$

Applying these and Eq. (1) to Eq. (2), it follows that

$$p_j = c + \frac{\mu Q_j}{2\pi k} \log r_j + \frac{\mu}{2\pi k} \sum' Q_i \log r_{ij}, \tag{5}$$

where the prime denotes the omission of the term $i = j$. Similarly, taking a contour over the circle $z = R$, which may be considered to represent the external boundary, the average pressure[1] over this boundary is found to be

$$p_e = c + \frac{\mu}{2\pi k} \sum Q_j \log R. \tag{6}$$

This and Eq. (5) will be the fundamental equations for the treatment of small groups of wells, since they permit a determination of the Q_j if the pressures p_j and p_e are specified. They will now be applied to several illustrative examples.

9.3. Examples. *a. A Single Well.*—Although the case of a single well presents no interference features characteristic of the general multiple-well systems, it is not without interest to apply the above equations to this simplest case. Thus setting $j = 1$, Eq. 9.2(5) gives

$$p_1 = c + \frac{\mu Q_1}{2\pi k} \log r_1,$$

and from Equation 9.2(6) it follows that

$$p_e = c + \frac{\mu Q_1}{2\pi k} \log R,$$

so that

$$Q_1 = \frac{2\pi k (p_e - p_1)}{\mu \log R/r_1}. \tag{1}$$

This, it will be recognized, is identical in form with the simple radial-flow formula derived in Chap. IV. It is of interest, however, to compare it with that formula in some detail. Thus it will be noted that no assumption has been made here that the well or external pressures are uniform; p_1 and p_e represent *averages* of the actual pressures over the well and external boundary. Again, Eq. (1) has been derived as a rigorous result without assuming that the external boundary is concentric with the well. On the other hand, it is to be observed that Eq. (1) does not contradict the results of Sec. 4.6 in which was derived the effect on Q due to displacing a well from the center of the external

---

[1] By the use of the Green's function developed in Sec. 4.6, the pressure distribution for a multiple-well system may be derived so as to correspond to a strictly uniform pressure over the external boundary. From a practical point of view, however, this would be rather artificial as compared to that in which only the *average* pressure is preassigned.

circular boundary; for there it was assumed that the well and boundary pressures were uniform and kept fixed, whereas here no such requirement is imposed, so that the independence of Q of the well position is taken care of by the variation of the distribution of the external boundary pressure p_e as the well is moved about within the circle of radius R. Furthermore, Eq. (1) is based upon the pressure distribution of Eq. 9.2(1) with $Q_j = 0$, $j \neq 1$, which means that the pressure is radially symmetrical about the single well in the system, whereas the results of Sec. 4.5 were based on a perfectly general source and sink-free pressure distribution superposed upon the logarithmic contribution due to the well. Equation (1) is, therefore, a reformulation of the simple radial flow formula so as to apply under somewhat

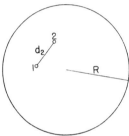

FIG. 191.—A well pair in a sand of large effective radius R.

different conditions, rather than a generalization of the results of Chap. IV.

b. Two Wells.—The real phenomena of interference will evidently occur first in the problem of only two wells within the external boundary (*cf.* Fig. 191). For this case, as well as in those to follow, it will be supposed that the various well pressures p_j and well radii r_j are the same, so that the effects on the fluxes Q due to varying the number of wells, or the spacing between them, may be attributed entirely to these factors rather than to the particular choices of the pressures and dimensions of the individual wells. Furthermore, as a matter of notation, we shall denote the common well pressure by p_w, the common well radius by r_w, the average pressure at the external boundary by p_e, and its radius by R.

Returning then to Eqs. 9.2(5) and 9.2(6), there are obtained for the present case the following equations:

$$\left. \begin{aligned} p_w &= c + \frac{\mu Q_1}{2\pi k} \log r_w + \frac{\mu Q_2}{2\pi k} \log d_2 \\ p_w &= c + \frac{\mu Q_1}{2\pi k} \log d_2 + \frac{\mu Q_2}{2\pi k} \log r_w \\ p_e &= c + \frac{\mu Q_1}{2\pi k} \log R + \frac{\mu Q_2}{2\pi k} \log R. \end{aligned} \right\} \qquad (2)$$

The solution for Q_1 and Q_2 is easily found to be ·

$$Q_1 = Q_2 = \frac{2\pi k(p_e - p_w)}{\mu \log R^2/r_w d_2}. \tag{3}$$

As discussed in detail in the case of a single well, it is found here, as in all cases following from Eqs. 9.2(5) and 9.2(6), that the fluxes or production capacities of the various wells are independent of their absolute positions, and depend only on their mutual separations. As pointed out there, this feature of Eqs. 9.2(5) and 9.2(6) is a direct consequence of the fact that the original pressure distribution of Eq. 9.2(1) consisted only of terms due to wells within the circle chosen as the external boundary. Although this may seem to be an unnecessary restriction of the generality of the analysis, it is actually the analytical expression of the fact that we are considering a "small group of wells." For such systems one may, by definition, consider all the sinks of the system as contained within the external boundary, and hence that the pressure distribution is composed only of the logarithmic terms corresponding to these sinks.

Returning to Eq. (3), it is to be noted that the fluxes Q_1 and Q_2 decrease as d_2 decreases. This effect is, of course, the direct expression of the mutual interference between the two wells. In fact, this interference is such that the flux of each well is that which it would have, if producing alone with the same pressure differential, in a region bounded by a circle of radius R^2/d_2, which, under the assumption of a "small group of wells" $(d_2 \ll R)$, is much larger than the actual boundary radius R. That is, the effective resistance of the system, per well, is that of a circular region of area $\pi R^4/d_2{}^2$ rather than πR^2, which would correspond to no interference between the two wells.

It is of interest to observe that if a single well is to give the same production rate as the total flux from the two-well system, its radius must be $\sqrt{r_w d_2}$. If, for example, the radius of the wells in the two-well system is $\frac{1}{4}$ ft. and their separation is 200 ft., the equivalent single well must have a radius of 7.07 ft. Hence it would be quite impractical to attempt to replace the two-well system by a single well.

c. Three Wells in a Triangular Pattern.—For the case of three wells (*cf.* Fig. 192), it is only necessary to set up four equations similar to the set of Eq. (2). It is readily found that

$$Q_1 = Q_2 = Q_3 = \frac{2\pi k(p_e - p_w)}{\mu \log R^3/r_w d_3{}^2}. \tag{4}$$

Here again the interference increases (Q decreases) as the separations d_3 decrease. Each well has a flux corresponding to

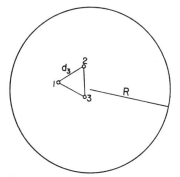

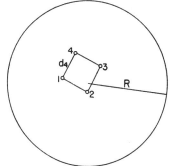

FIG. 192.—A group of three wells in a triangular pattern.

FIG. 193.—A group of four wells in a square pattern.

an effective external boundary, for a single well within the boundary, with a radius $R^3/d_3{}^2$. As this is still larger than R^2/d_2 (for the case $d_2 = d_3$), the interference in the three-well system is greater than that in the two-well system, as would also be expected from general considerations. If the total flux ($3Q_1$)

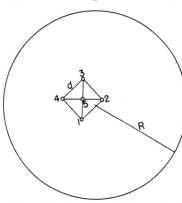

FIG. 194.—A group of four wells in a square pattern with one in the center.

be lumped together, it will correspond to a single well of radius $(r_w d_3{}^2)^{\frac{1}{3}}$.

d. Four Wells in a Square Pattern.—In this case all of the mutual separations are no longer equal (*cf.* Fig. 193); the symmetry, however, is sufficient to give equal fluxes to each of the wells. These are

$$Q_1 = Q_2 = Q_3 = Q_4 = \frac{2\pi k(p_e - p_w)}{\mu \log R^4/\sqrt{2}r_w d_4{}^3}. \tag{5}$$

Thus each well has an effective external radius of $R^4/\sqrt{2}d_4{}^3$, and the wells together are equivalent to a single well with the radius $(\sqrt{2}r_w d_4{}^3)^{\frac{1}{4}}$.

e. Four Wells in a Square Pattern and One in the Center.—
From the symmetry of the system (*cf.* Fig. 194) it is clear that
the four outer wells will have equal fluxes, but different from that
for the central well.   The 6 Eqs. 9.2(5) and 9.2(6) may, therefore,
be at once reduced to the set

$$\left.\begin{aligned}
p_w &= c + \frac{\mu}{2\pi k}\left(Q_1 \log \sqrt{2}r_w d^3 + Q_5 \log \frac{d}{\sqrt{2}}\right)\\
p_w &= c + \frac{\mu}{2\pi k}\left(4Q_1 \log \frac{d}{\sqrt{2}} + Q_5 \log r_w\right)\\
p_e &= c + \frac{\mu}{2\pi k}(4Q_1 \log R + Q_5 \log R),
\end{aligned}\right\} \quad (6)$$

from which it readily follows that

$$\left.\begin{aligned}
Q_1 &= \frac{2\pi k}{\mu\Delta}(p_e - p_w) \log \frac{d}{\sqrt{2}r_w}\\
Q_5 &= \frac{2\pi k}{\mu\Delta}(p_e - p_w) \log \frac{d}{4\sqrt{2}r_w}\\
\Delta &= 4 \log \frac{\sqrt{2}R}{d} \log \frac{d}{\sqrt{2}r_u} + \log \frac{R}{r_w} \log \frac{d}{4\sqrt{2}r_w}\\
\frac{Q_5}{Q_1} &= \frac{\log (d/4\sqrt{2}r_w)}{\log (d/\sqrt{2}r_w)} = 1 - \frac{\log 4}{\log (d/\sqrt{2}r_w)}.
\end{aligned}\right\} \quad (7)$$

In this case it is seen that
there is not only the previous
type of interference in which
the mere presence of a group
of wells decreased the individual flux rates, but here there is
the additional effect of the four
external wells on the central
well.   The latter, owing to its
position, suffers a loss in flux
represented by the term $\log 4/$
$\log (d/\sqrt{2}r_w)$, which amounts
to 22 per cent for a separation
of $d = 200$ ft.   The group as
a whole has an effective single-well radius $\bar{r}_w$ given by

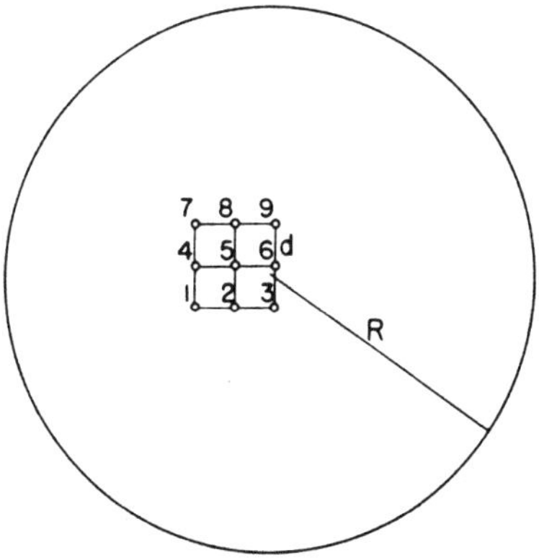

FIG. 195.—A group of nine wells in a square array.

$$\log \frac{\bar{r}_w}{r_w} = \frac{\{\log (d^2/2r_w^2)\}^2}{\log (d^5/2^{5/2}r_w^5)}. \quad (8)$$

f. Nine Wells.—From the symmetry of this system (*cf.* Fig. 195) it is clear that

$$Q_1 = Q_3 = Q_7 = Q_9 \atop Q_2 = Q_4 = Q_6 = Q_8.\left.\right\} \tag{9}$$

To find the interference effects of the wells it is sufficient to know only the ratios $Q_1:Q_2:Q_5$. For this purpose it is necessary to consider only three of the nine equations 9.2(5) as follows:

$$p_w - c = \frac{\mu}{2\pi k}(Q_1 \log 8\sqrt{2}r_w d^2 + Q_2 \log 5d^4 + Q_5 \log d\sqrt{2})$$

$$p_w - c = \frac{\mu}{2\pi k}(Q_1 \log 5d^4 + Q_2 \log 4r_w d^3 + Q_5 \log d) \tag{10}$$

$$p_w - c = \frac{\mu}{2\pi k}(Q_1 \log 4d^4 + Q_2 \log d^4 + Q_5 \log r_w)$$

with the solution

$$Q_1:Q_2:Q_5 = \left[y^2 + y \log \frac{5}{2^{3/2}} - (\log 2)^2 \right]$$

$$: \left[y^2 + y \log \frac{5}{2^{1/2}} + \log \sqrt{2} \log \frac{5}{4} \right]$$

$$: \left[y^2 - y \log 2^{1/2} + \log 2 \log 200 - (\log 5)^2 \right], \tag{11}$$

where

$$y = \log \frac{d}{r_w}.$$

Hence, if $d = 200$ ft., $r_w = \frac{1}{4}$ ft.;

$$Q_1:Q_2:Q_5 = 1.00:0.819:0.615. \tag{12}$$

The effect on the mutual interference of the absolute value of the well spacing is indicated by the fact that if $d = 50$ ft.:

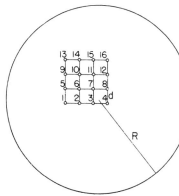

FIG. 196.—A group of sixteen wells in a square array.

$$Q_1:Q_2:Q_5 = 1.00:0.778:0.533.$$

g. Sixteen Wells.—For this case (*cf.* Fig. 196) it may be noted similarly that

$$Q_1 = Q_4 = Q_{13} = Q_{16}$$
$$Q_2 = Q_3 = Q_5 = Q_8 = Q_9 = Q_{12} = Q_{14} = Q_{15} \left.\right\} \quad (13)$$
$$Q_6 = Q_7 = Q_{10} = Q_{11},$$

with the equations

$$p_w - c = \frac{\mu}{2\pi k}(Q_1 \log 27\sqrt{2}r_w d^3 + Q_2 \log 520 d^8 + Q_6 \log 20 d^4)$$

$$p_w - c = \frac{\mu}{2\pi k}(Q_1 \log 2\sqrt{130}d^4 + Q_2 \log 60\sqrt{10}r_w d^7 + \left.\right. \quad (14)$$
$$Q_6 \log 2\sqrt{10}d^4)$$

$$p_w - c = \frac{\mu}{2\pi k}(Q_1 \log 20 d^4 + Q_2 \log 40 d^8 + Q_6 \log \sqrt{2}r_w d^3).$$

and the solution

$$Q_1:Q_2:Q_6 = \left(y^2 + y \log \frac{26\sqrt{20}}{3} + \log 13 \log \sqrt{20} - \right.$$
$$\log \frac{3\sqrt{10}}{2} \log 10\sqrt{2}\right)$$
$$:\left(y^2 + y \log \frac{20\sqrt{13}}{27} + \log \sqrt{20} \log \frac{20}{27\sqrt{2}} - \right.$$
$$\log \frac{\sqrt{2}}{20} \log \frac{\sqrt{130}}{10}\right) \quad (15)$$
$$:\left(y^2 + y \log \frac{2\sqrt{20}}{81} - \log 13 \log \frac{\sqrt{130}}{10} - \right.$$
$$\log \frac{20}{27\sqrt{2}} \log \frac{3\sqrt{10}}{2}\right),$$

so that

$$Q_1:Q_2:Q_6 = 1.00:0.735:0.445, \quad \text{for} \quad \frac{d}{r_w} = 800 \left.\right\}$$
$$Q_1:Q_2:Q_6 = 1.00:0.692:0.362, \quad \text{for} \quad \frac{d}{r_w} = 200. \quad (16)$$

h. Battery of Wells Set on a Circle.[1]—As a generalization of the completely symmetrical arrangements of two, three, and four wells, the system in which n wells are distributed over a circle of radius r ($\ll R$), (*cf.* Fig. 197) will now be analyzed. As, by symmetry, all the Q_j are equal, Eqs. 9.2(5) and 9.2(6) give at once

[1] This problem has also been discussed by Gardner, Collier, and Farr, *loc. cit.*

$$p_w = c + \frac{\mu Q_j}{2\pi k}\left(\log r_w + {\sum}' \log r_{ij}\right)$$
$$p_e = c + \frac{\mu n Q_j}{2\pi k}\log R, \tag{17}$$

so that

$$Q_j = \cfrac{2\pi k \Delta p/\mu}{\left(\log \cfrac{R^n}{r^{n-1}r_w} - \sum_1^{n-1} \log 2 \sin \cfrac{\pi m}{n}\right)}, \tag{18}$$

with a total production capacity for the system of

$$Q^{(n)} = nQ_j = \cfrac{2\pi k \Delta p/\mu}{\left(\log \cfrac{R}{r} + \cfrac{1}{n}\log \cfrac{r}{r_w} - \cfrac{1}{n}\sum_1^{n-1} \log 2 \sin \cfrac{\pi m}{n}\right)}, \tag{19}$$

the logarithm of the equivalent radius being $\log R$ minus the quantity in parentheses. The ratio of the total resultant production capacity of the battery $Q^{(n)}$ to that of a single well is plotted in Fig. 198 as a function of n, for $r/r_w = 80$ and $r/r_w = 200$, with R taken as 5,000 ft. and r_w as $\frac{1}{4}$ ft. Although the limiting values of 2.15 and 1.79 for $Q^{(n)}/Q^{(1)}$, as n becomes infinite, are rather rapidly approached with increasing n, it should be noted that these asymptotic values, corresponding to a single well of diameter r, increase but slowly with r. Hence, while the cost of the battery may be less than that of the equivalent single well, its ultimate capacity will be limited in essentially the same manner as is that of a single well.

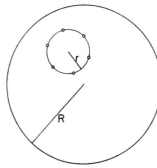

Fig. 197.—A circular battery of wells draining a sand of radius R.

9.4. The Dependence of the Production Capacity on the Number of Wells in the Group.—From the special cases of the last section it is clear that as the number of wells in the group increases the mutual interference increases, so that the production capacities per well decrease. The cumulative effect of the interference is shown by a comparison of the total production capacities of the various groups treated in the last section.   For this purpose it will be assumed that for each case $d/r_w = 800$ and $R = 5,000$ ft. Denoting by $Q^{(n)}$ the total production capacity of a group containing n wells and by $Q_i^{(n)}$ the individual production capacity of the ith type of well in the group of n, we have, first

$$\frac{Q^{(1)}}{Q_1^{(1)}} = \frac{Q^{(2)}}{2Q_1^{(2)}} = \frac{Q^{(3)}}{3Q_1^{(3)}} = \frac{Q^{(4)}}{4Q_1^{(4)}} = \frac{Q^{(5)}}{4Q_1^{(5)} + Q_5^{(5)}}$$
$$= \frac{Q^{(9)}}{4Q_1^{(9)} + 4Q_2^{(9)} + Q_5^{(9)}} = \frac{Q^{(16)}}{4Q_1^{(16)} + 8Q_2^{(16)} + 4Q_6^{(16)}} = 1,$$

and then putting in the individual values of the $Q_i^{(n)}$,

$$Q^{(1)} : Q^{(2)} : Q^{(3)} : Q^{(4)} : Q^{(5)} : Q^{(9)} : Q^{(16)}$$
$$= 1 : 1.509 : 1.818 : 2.061 : 2.152 : 2.778 : 3.333. \quad (1)$$

This equation shows the cumulative effect of the interference among the wells and the decreasing contributions due to the

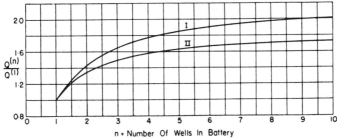

n = Number Of Wells In Battery

FIG. 198.—The variation of the production capacity of a circular well battery with the number of wells in the battery. $Q^{(n)}/Q^{(1)}$ = (production capacity of battery of n wells)/(production capacity of a single well); (radius of battery)/(individual well radius r_w) = 200 for curve I, = 80 for curve II.   External-boundary radius = 5,000 ft.; $r_w = \frac{1}{4}$ ft.

additional wells as the total number is increased.   Or still more clearly, this may be seen from the average fluxes *per well* in the various groups corresponding to Eq. (1).   Denoting these by $\bar{Q}^{(n)} = Q^{(n)}/n$, it follows that

$$\bar{Q}^{(1)} : \bar{Q}^{(2)} : \bar{Q}^{(3)} : \bar{Q}^{(4)} : \bar{Q}^{(5)} : \bar{Q}^{(9)} : \bar{Q}^{(16)}$$
$$= 1 : 0.755 : 0.606 : 0.515 : 0.430 : 0.309 : 0.208. \quad (2)$$

In view of this marked decrease in the average fluxes per well, the significance of the well interferences and interactions will be evident.   For it is clear that after a certain stage in the drilling development the added production obtained from additional wells will not suffice to pay for the additional cost.[1]

9.5. The Pressure Distribution over the External Boundary.— Before leaving the problem of small groups of wells it is of interest to examine in some detail the significance and meaning of the "average pressures" which have been used throughout the above analysis. Thus, for the case of the well surfaces, the fact that the well radii are very small compared to all other dimensions of the system of physical significance implies at once that the actual pressure distribution as given by Eq. 9.2(1) will vary but little over the well surfaces, so that the average pressures p_j are practically equivalent to strictly uniform pressures. In the case of p_e, the average pressure over the external boundary, however, further examination is required.

Although the assumption that the external boundary is circular is entirely unessential and is not involved at all in the Eqs. 9.2(5), it will still be retained in the following discussion, as it gives particularly simple forms for the coefficients of the Q_j in Eq. 9.2(6). Returning then to Eq. 9.2(1), and supposing that z is a point on the external boundary, we have

$$p(z) = c + \frac{\mu}{4\pi k} \sum Q_j \log (R^2 + |z_j|^2 - 2R|z_j| \cos \theta_j), \quad (1)$$

where θ_j is the angle between z_j and z terminating on the boundary. As $|z_j| < R$, Eq. (1) may be expanded as

$$p(z) = c + \frac{\mu}{2\pi k} \sum Q_j \log R - \frac{\mu}{2\pi Rk} \sum Q_j |z_j| \cos \theta_j -$$
$$\frac{\mu}{4\pi R^2 k} \sum Q_j |z_j|^2 \cos 2\theta_j - \cdots. \quad (2)$$

Hence, subtracting the average pressure p_e of Eq. 9.2(6), the residual is found to be

[1] It is to be understood, of course, that the above analysis and results apply only insofar as the fluid taken from the wells is supplied from an external source so that there is no depletion of the fluid in the sand immediately about the wells, as would be the case if the wells were producing by virtue of the evolution and expansion of their dissolved gases or of the expansion of the liquid due to its own compressibility.

$$p(z) - p_e = \frac{-\mu}{2\pi Rk}\left[\sum Q_j|z_j| \cos \theta_j + \right.$$

$$\left. \frac{1}{2R}\sum Q_j|z_j|^2 \cos 2\theta_j + \cdots \right] \quad (3)$$

Now a little inspection will show that the first summation may be resolved into two terms proportional to the linear moments $\Sigma Q_j x_j$, $\Sigma Q_j y_j$ of the wells about the origin, weighted with respect to their production capacities, while the second summation is proportional to the quadratic moments. To a first approximation, therefore, the actual pressure distribution will be uniform over the external boundary if the center of gravity of the wells—weighted with respect to their fluxes—lies at the center of the external boundary. And conversely, it is seen that in any case the pressure will be uniform to this first approximation, over a large circle drawn about the center of gravity of the wells.

These considerations show why it has been tacitly assumed in the previous discussions that the wells were localized near the center of the region of interest and that none lay close to the external boundary. For if the wells lie near the external boundary, their moments about the center may be quite large and the deviation of the boundary pressure from its average value will be so great as to make the latter term lose its physical significance. As the hypothesis of a "small group" of wells also implies a localization in the distribution of the wells, the requirement of an approximate coincidence between the center of the region and the center of gravity of the well system together with a limitation on the magnitude of the quadratic moments may be formulated into a quantitative definition of this type of well system.

9.6. Small Groups of Wells Supplied by an Infinite Line Drive. The treatments given thus far of the production characteristics of small groups of wells have been based on the assumption that the distant external boundary, at which the liquid was supplied, was circular in character. Although this assumption was introduced mainly for analytical convenience and had no effect upon the general features of the interference relations among the wells of the various groups, there is a class of problems in which the noncircular character of the external boundary must be explicitly taken into account. Thus in a flow system composed of several artesian wells drilled into a permeable sand supplied with water by a neighboring river or canal, the effective external boundary is definitely noncircular but must rather be represented by an infinite line source coincident with the effective outcrop of the sand in the canal or river bed.

Analytically, a group of n wells spaced by a distance a and lying parallel to, and at, a distance d from the line source at pressure p_e (*cf.* Fig. 199), gives a pressure distribution of the form

$$p(x, y) = \frac{\mu}{4\pi k} \sum_0^{n-1} Q_m \log \left[\frac{(x - ma)^2 + (y - d)^2}{(x - ma)^2 + (y + d)^2} \right] + p_e, \quad (1)$$

where the Q_m are the fluxes into the individual wells, k the permeability of the medium, and μ the viscosity of the liquid. If

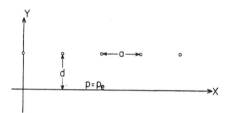

Fig. 199.—A line array of wells parallel to a line drive.

each of the wells is maintained at the pressure p_w, the fluxes Q_m will be determined by the equations

$$p_e - p_w = \frac{\mu}{4\pi k} \left\{ 2Q_j \log \frac{2d}{r_w} + \Sigma' Q_m \log \left[1 + \frac{4d^2}{a^2(m - j)^2} \right] \right\} :$$
$$j = 0, 1, 2 \cdots n - 1, \quad (2)$$

where the prime denotes the omission of the term $m = j$.

Thus if $n = 1$, Eq. (2) gives at once

$$Q_0 = \frac{2\pi k \Delta p}{\mu \log 2d/r_w}, \quad (3)$$

which is the same value found before for this special case [*cf.* Eq. 4.7(8)].

For $n = 2$, Eqs. (2) give

$$Q_0 = Q_1 = \frac{2\pi k \Delta p}{\mu \left[\log \frac{2d}{r_w} + \frac{1}{2} \log \left(1 + \frac{4d^2}{a^2} \right) \right]}. \quad (4)$$

The extra term in the bracket of the denominator evidently represents the mutual interference between the two wells. Thus if $d/a = 1$, $d/r_w = 400$, each well in the two-well system will have a production capacity of only 89.26 per cent of that it would have if it alone were draining the sand. This decrease of 10.74 per cent will clearly become larger as the wells are brought closer together (a decreased).

For $n = 3$, it is readily found that

$$\left.\begin{array}{l} Q_0 = Q_2 = \dfrac{4\pi k \Delta p \, \log\left[\beta^2 \Big/ \left(1+\dfrac{4d^2}{a^2}\right)\right]}{\mu\left\{(\log \beta^2)\left[\log \beta^2\left(1+\dfrac{d^2}{a^2}\right)\right] - 2\left[\log\left(1+\dfrac{4d^2}{a^2}\right)\right]^2\right\}}, \\[2em] \dfrac{Q_1}{Q_0} = 1 - \dfrac{\log\left[\left(1+\dfrac{4d^2}{a^2}\right)\Big/\left(1+\dfrac{d^2}{a^2}\right)\right]}{\log\left[\beta^2\Big/\left(1+\dfrac{4d^2}{a^2}\right)\right]}, \end{array}\right\} \quad (5)$$

where

$$\beta = \frac{2d}{r_w}.$$

Again if $d/a = 1$, $d/r_w = 400$, it is readily found that Q_0 and Q_2 are 86.0 per cent while Q_1 is 79.3 per cent of that which would correspond to a single well draining the sand [cf. Eq. (3)]. That $Q_1 < Q_0 = Q_2$ is evidently due to the fact that the central well is interfered by wells on both sides of it while the end wells suffer interference from wells on only one side.

In a similar manner cases for other values of n can be readily worked out; in Sec. 9.8 the analysis will be given for the limiting case where n becomes infinitely large.

9.7. Infinite Sets of Wells in Linear Arrays.—When the area overlying a fluid-bearing sand is completely traversed at least in one direction by a single or several parallel arrays of wells, extending for distances large compared to the spacing between the wells, the system may be replaced by equivalent sets of infinite well arrays. For if the sand boundary cuts off the array along a line of symmetry as CD (cf. Fig. 200), there will be no reaction of the boundary upon the flow in the interior of the boundary, and it will be exactly the same as if the array were

extended indefinitely without the interruption by the limitation
of the sand. And if the sand boundary interrupts the array
unsymmetrically as AB, the equivalence with the infinite array
will still not be seriously affected, as the edge effects due to the
boundary will be confined essentially to a distortion of the pressure distribution about the particular well closest to the boundary.
It will, therefore, be supposed in the following analysis that we
have real infinite arrays of equally spaced wells, all having the
same radii and maintained at the same pressure.

**9.8. The Pressure Distribution about an Infinite Array of
Wells. The Line Drive.**—To get the fundamental representa-

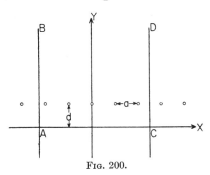

FIG. 200.

tion for the pressure distribution due to an infinite array of wells
each well will again be replaced by a mathematical two-dimensional sink, each of strength proportional to the flux or production
capacity to be associated with the well, all wells being taken as
identical and uniformly spaced. Referring then to the array of
Fig. 200 and supposing that it is at a distance d above the x
axis, the resultant pressure distribution due to the individual
wells of the array is evidently the algebraic sum of the separate
contributions, and is, therefore, given by

$$p(x, y) = q \sum_{-\infty}^{+\infty} \log [(y - d)^2 + (na - x^2)]$$

$$= q \log \prod_{-\infty}^{+\infty} [(y - d)^2 + (na - x)^2]. \quad (1)$$

This sum or product has already been evaluated in Sec. 4.9 with
the result that

$$p(x, y) = q \log \left[\cosh \frac{2\pi(y - d)}{a} - \cos \frac{2\pi x}{a} \right]. \tag{2}$$

This equation will be the fundamental element in the following solutions of problems involving infinite well arrays, just as the term $q \log r$ has been used until now to represent the pressure contribution of a single well. It will, therefore, be of interest to examine it in some detail.

As must be required of the pressure distribution in a porous medium, carrying a liquid under steady-state conditions, and as can be tested by direct substitution, Eq. (2) satisfies Laplace's equation. That it actually corresponds to an array of equally spaced sinks follows from the observation that the argument of the logarithm vanishes at the set of points (na, d), the coordinates of the sinks—the centers of the wells. Furthermore, $p(x, y)$ is symmetrical about the axis of the array, $y = d$, and is periodic in x with the period a, the spacing of the sinks.

An instructive picture of Eq. (2) may be obtained from the equipressure curves which it defines. These may be derived by solving the equation

$$\cosh \frac{2\pi(y - d)}{a} - \cos \frac{2\pi x}{a} = \text{const} = c = e^{\frac{p}{q}}. \tag{3}[1]$$

For the case, $d = 0$, and in the coordinates $x = x/a$, $y = y/a$, Eq. (3) is plotted in Fig. 201. From this figure it is seen at once that at distances from the wells equal to the mutual spacing a, *i.e.*, at $y = \pm 1$, the equipressure curves are for all practical purposes lines parallel to the array ($\cosh 2\pi \gg 1$). This feature may also be shown by plotting the pressure distribution p along lines normal to the array. The two extreme curves of this type giving the pressure along a line normal to the array and through one of the wells and that along a line normal to the array and passing midway between the wells are given in Fig. 202, with $d = 0$, $q = \frac{1}{2}$. These curves show again that the pressure is practically independent of x as soon as one recedes from the line array by a distance of the order of the mutual spacing. Figure

[1] Although the deviations of the equipressure curves of Eq. (3) from the small circular boundaries of the wells are of no physical significance whatever, it may be of academic interest to note that strictly circular equipressure curves could be obtained, if necessary, by the use of conjugate-function methods (*cf.* R. C. J. Howland, *Proc. Cambridge Phil. Soc.*, **30**, 315, 1934.)

202 shows further that at distances of the order of the mutual spacing the pressure gradients become constant as if the well array had been replaced by a continuous line sink.

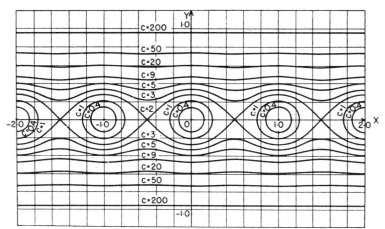

FIG. 201.—The equipressure contours about a single-line array of wells with unit spacing; pressure $p = \log c$.

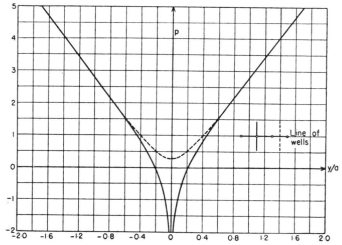

FIG. 202.—The pressure distribution about an infinite array of wells, of spacing a, lying on the x axis. ———— pressure distribution along a normal to the array that passes through a well. - - - - - pressure distribution along a normal to the array that passes midway between two wells.

Another point that may be established here is the exact relation between the flux coefficient q and the actual flux Q into a well.

To derive this relation one simply notes that very near any particular well, such as the one lying on the y axis, the pressure has very approximately the value

$$p = q \log 2 + 2q \log \frac{\pi r}{a}, \tag{4}$$

where r represents the distance from the well. The actual flux into the well will, therefore, be given by

$$Q = -\frac{2\pi k r d p}{\mu \, dr} = \frac{4\pi k q}{\mu}, \tag{5}$$

where the sand thickness is again taken as unity.

Although, as has just been noted, the pressure distribution represented by Eq. (2) gives a system of equipressure curves which are practically parallel to the line of wells at distances from the line of the order of the mutual spacing, it is more suitable and convenient for practical purposes to use a distribution which gives a strictly uniform pressure over a line parallel to the array at a specified distance from it, this line being given the physical interpretation of a "drive" forcing the fluid into the array. To derive such a distribution it is only necessary to reflect the given array into the line which is to be at uniform pressure, and then add the pressure functions due to the actual and image arrays. Thus, if it is supposed that in the coordinate system of Fig. 200 the x axis is the line of uniform pressure (the line drive), the pressures due to the given line at a distance d above the axis and that of its negative image at a distance d below the axis are added. It follows then at once that the pressure due to such a system may be expressed as

$$p = C + q \log \frac{\cosh 2\pi(y - d)/a - \cos 2\pi x/a}{\cosh 2\pi(y + d)/a - \cos 2\pi x/a}, \tag{6}$$

where C is the constant pressure maintained at the "line drive" $(y = 0)$. The equipressure curves of this system are given by

$$\frac{\cosh 2\pi(y - d)/a - \cos 2\pi x/a}{\cosh 2\pi(y + d)/a - \cos 2\pi x/a} = \text{const.}, \tag{7}$$

a special case of which is evidently the line drive at $y = 0$. The pressure distribution along the lines normal to the array, in one case passing through a well, and in another passing midway between the wells, is given in Fig. 203, with the constants C and q having the values 10 and $\frac{1}{2}$, respectively, and the distance h equal to the mutual spacing a.

It will be observed from this figure that for the present case where the well array is supplied by the line drive at $y = 0$, the pressure is no longer symmetrical about the array. Rather, the line drive together with the well array induces a regional gradient in the direction normal to the line drive. However, there is no general migration connected with this gradient beyond the line of

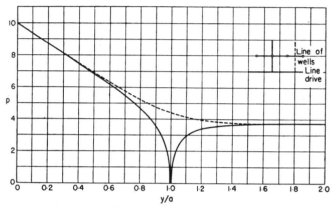

Fig. 203.—The pressure distributions normal to an infinite array of wells, of spacing a, at $y = a$, supplied by a line drive at $y = 0$. ———— pressure distribution along a normal that passes through a well. - - - - pressure distribution along a normal that passes midway between two wells.

wells, since the flow from the line drive for each unit of length equal to the well spacing in the array has the value

$$Q = -\frac{k}{\mu} \int_{-\frac{a}{2}}^{+\frac{a}{2}} \left(\frac{\partial p}{\partial y}\right)_{y=0} dx = \frac{4\pi k q}{\mu},$$

which is the flux per well into the array. (This will be apparent upon observing that while some fluid flows past the line of wells along streamlines in the area between the wells, these streamlines curve backward and approach the wells from the side opposite the drive.)

Finally, it is of interest to compute the effective resistance of the system. By Eq. (6) it is seen that the pressure at the wells, defined by the small circles of radius r_w about the points (na, d), is given by

$$p_w = C + 2q \log \frac{\sinh \pi r_w/a}{\sinh 2\pi d/a}. \tag{8}$$

Since C is the pressure at the line drive, the difference in pressure acting on the system is $C - p_w = \Delta p$. Taking into account Eq. (5), it follows that

$$Q = \frac{2\pi k \Delta p}{\mu \log \dfrac{\sinh 2\pi d/a}{\sinh \pi r_w/a}}. \tag{9}$$

As $r_w/a \ll 1$, and $2d/a$ will usually be of the order of or greater than 1, Eq. (9) can be written as

$$Q = \frac{2\pi k \Delta p/\mu}{\log \left(ae^{\frac{2\pi d}{a}}/2\pi r_w\right)}. \tag{10}$$

It is of interest to observe that Eq. (10) corresponds exactly to the flux for the case of strict radial flow between a well of radius r_w and an effective external concentric circular boundary of radius

$$r_e = \frac{a}{2\pi} e^{\frac{2\pi d}{a}}. \tag{11}$$

Unless $d/a \angle 0.3$, $r_e > a$, as should be expected, since the high pressure is applied to only one side of the well and not symmetrically as in a radial-flow system. The effective resistance of the system is now given by $\Delta p/Q$, or

$$R = \frac{\mu}{2\pi k} \log \frac{ae^{\frac{2\pi d}{a}}}{2\pi r_w}. \tag{12}$$

Finally, the mutual interference effects of the wells in the infinite network follows at once from Eq. (10) on comparing it with the flux Q_0 from a single well given by Eq. 9.6(3). Thus, for $d/a = 1$, $d/r_w = 400$, it is found that

$$\frac{Q}{Q_0} = 0.641.$$

Comparing this with the results of Sec. 9.6 it is seen that the interference due to the wells distant from any given well must be quite small, as the closest wells on either side are alone sufficient to cut the production capacity to 79 per cent of Q_0.

Although the discussion of a single line of wells suffices to give a general picture of the features characteristic of infinite linear arrays, the problems of practical interest usually involve the mutual interactions and interferences among two or more such arrays. However, before giving the analysis of these more complex systems, we must indicate more clearly the physical meaning of the term "line drive," which we shall find convenient to use frequently in the following discussion and which has already been introduced above. For, if interpreted literally, a "line drive" might be taken to represent a concentrated linear fluid source artificially placed parallel to the well arrays in question, and as such would have but limited practical significance.[1] To get a physical interpretation, therefore, it is to be noted that the line drive is really nothing more than a convenient analytical representation of the whole fluid system external to, and acting upon, the well arrays. Thus, in studying the interaction of a large lease or fluid reservoir with a group of neighboring wells or well arrays, the use of the "line drive" representation permits one to replace the complicated description and detailed treatment of the interior of the lease by its analytical equivalent— the pressure distribution along its boundary. Hence, to find the flow from a large reservoir into a system of well arrays, one simply represents the reservoir by the average of its pressure along its boundary parallel to the external well system, and then supposes that there is a "line drive" maintained at this average pressure which is supplying the fluid entering the well system.

Similarly, as will be seen later in treating the problem of offsetting, one may replace several mutually interacting leases by line drives near the boundaries of the leases, these drives being given pressures appropriate to the average pressures observed near the boundaries.

9.9. Two Line Array. Shielding Effects.—To treat the problem of two lines of wells, which may correspond in a practical case to the offset lines across a lease boundary, one need only add to Eq. 9.8(6) another logarithmic term with a different d. Thus for definiteness it will be supposed that the first line is

---

[1] *Cf.* Sec. 9.6 where the line drive really represents an extended linear source as a river or canal into which the permeable sands outcrop.

at a distance d_1 from the line drive, at pressure C, and that the second is at a distance d_2 from the line drive at $y = 0$. As Fig. 204 indicates, we shall take the well spacing within the two lines to have the same value, a, and shall suppose the two lines to be placed so that the wells of the second line are exactly behind those of the first line.

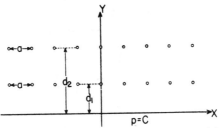

FIG. 204.—Diagrammatic representation of a line drive supplying liquid to two line arrays of wells.

Taking then Eq. 9.8(6) as the fundamental element representing a single array, the two-line system will give a pressure distribution as

$$p = C + q_1 \log \frac{\cosh 2\pi(y - d_1)/a - \cos 2\pi x/a}{\cosh 2\pi(y + d_1)/a - \cos 2\pi x/a}$$
$$+ q_2 \log \frac{\cosh 2\pi(y - d_2)/a - \cos 2\pi x/a}{\cosh 2\pi(y + d_2)/a - \cos 2\pi x/a}, \quad (1)$$

where $4\pi k q_1/\mu$ and $4\pi k q_2/\mu$ are the fluxes per well into the first and second lines, respectively. The fact that q_1 and q_2 are not taken equal arises in the asymmetry of the system, introduced in the assumption that the two arrays are producing under a line drive at $y = 0$, which is evidently closer to the first line than the second. Indeed, the deviation of the ratio q_1/q_2 from unity is a measure of the shielding effect of the first line upon the second, and which is of particular interest from a practical point of view.

To find the actual value of the ratio q_1/q_2 one must evidently predetermine first the pressures under which the wells in the two arrays are being produced. Now unless the two lines of wells are offsets across a lease boundary it may be supposed that they are being produced at about the same bottom-hole pressures, under the uniform sand conditions which are assumed in all the present analysis. Furthermore, since the shielding and interference interactions between the two lines of wells are really caused entirely by the geometrical asymmetry in their positions, with respect to the line drive, the assumption of different well pressures for the two lines would involve additional interactions which would tend to mask those characteristic only of the geometry of the system.

It will, therefore, be required explicitly that the pressures at the wells of the two arrays are all the same. To determine then the ratios q_1/q_2 it is sufficient to simply equate the value of p as given by Eq. (1) at the two wells on the y axis, or more specifically at the two points $(0, d_1 - r_w)$ and $(0, d_2 - r_w)$, where the well radius r_w is to be taken as very small as compared to d_1, d_2, or the well spacing a. In this manner it is found that

$$\frac{q_1}{q_2} = \frac{\log \dfrac{\sinh \pi r_w/a \, \sinh \pi(d_2 + d_1)/a}{\sinh 2\pi d_2/a \, \sinh \pi(d_2 - d_1)/a}}{\log \dfrac{\sinh \pi r_w/a \, \sinh \pi(d_2 + d_1)/a}{\sinh 2\pi d_1/a \, \sinh \pi(d_2 - d_1)/a}}. \tag{2}$$

This equation will now be analyzed to see how q_1/q_2 varies with some of the principal variables involved. The first case to be considered is the variation of q_1/q_2 with d_1, the distance of the line drive from the first array of wells. For this purpose it will be supposed that the other variables remain constant and that the two lines are separated by a distance equal to the mutual spacing. The ratio of the well radius to the spacing will be taken as $1/2640$, which corresponds to a well radius of $1/4$ ft. and a spacing of 660 ft. Equation (2) may then be rewritten as

$$\frac{Q_1}{Q_2} = \frac{q_1}{q_2} = \frac{\log \dfrac{\sinh \dfrac{\pi}{2640} \sinh \pi\left(1 + \dfrac{2d_1}{a}\right)}{\sinh 2\pi\left(1 + \dfrac{d_1}{a}\right) \sinh \pi}}{\log \dfrac{\sinh \dfrac{\pi}{2640} \sinh \pi\left(1 + \dfrac{2d_1}{a}\right)}{\sinh 2\pi d_1/a \, \sinh \pi}}. \tag{3}$$

For the purpose of discussion it will be perhaps clearer to measure the shielding effect as the ratio of the flow into the first line to the total flow. Denoting this ratio by S, it will be given by

$$S = \frac{Q_1}{Q_1 + Q_2} = \frac{1}{1 + Q_2/Q_1}, \tag{4}$$

which can be readily computed since Q_2/Q_1 is known from Eq. (3). It may also be noted that the leakage is given by $1 - S$. Thus, for example, for a shielding of 100 per cent, $S = 1$ and the leakage is zero, as should be expected.

Figure 205 gives the plot of Eq. (4) as a function of d_1/a. The curve brings out two points of interest. The first is that, even when the line drive is very distant from the producing wells and the distances of the two lines from the drive become effectively equal, the shielding falls no lower than about 67 per cent, which gives a leakage of 33 per cent. This may at first thought seem surprising in view of the fact that when the line drive is far from the wells, the latter even though placed behind each other would

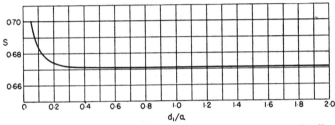

FIG. 205.—The variation of the shielding effect between two lines of wells with the distance d_1 of the line drive from the first well array. S = (production capacity per well in first line)/(production capacity per well pair in both lines). $a = 660$ ft. = distance between wells and between the two lines.

appear as if they are at practically the same distance from the drive and hence should produce the same quantity of fluid. When it is recalled, however, that the shielding and leakage effects between the two lines of wells are determined only by the nature of the pressure distribution about the wells and that this is not appreciably affected by shifting the position of the line drive, once the latter is at a distance from the well array of the order of the mutual spacing, the persistence of the shielding at large distances of the drive is even to be expected. In fact, this observation at the same time explains the other feature of interest about the curves, *i.e.*, that not only does the interference persist (*i.e.*, $S > \frac{1}{2}$) for large values of d_1/a, but also that the limiting value for S is approached very rapidly and is attained to within 0.1 per cent when the line drive is distant from the first line by only 0.4 of the mutual spacing. For this behavior is again due to the fact that as the line drive is moved out beyond $d_1/a \sim 0.4$

the pressure distribution in the immediate neighborhood of the well arrays remains almost exactly the same regardless of the exact value of d_1/a. Of course, the absolute value of the pressure at any given point does depend upon d_1/a, but *the shielding and leakage is determined only by the shape of the equipressure curves and streamlines about the wells,* and these are practically independent of d_1/a, for $d_1/a > 0.5$.

Since in practical cases the distance of the line drives from the well arrays are usually of the order of several hundred feet, the above discussion shows that in the treatment of the other phases of the problem it can be supposed for convenience that d_1/a is exactly equal to unity without introducing errors of more than 0.1 per cent. Hence to study the effect of the absolute value of the mutual spacing a, we shall suppose that the two lines are separated by the value of a, and that the drive is also at a distance a from the wells, *i.e.*, that

$$d_1 = a, \qquad d_2 = 2a.$$

Equation (2) then takes the form

$$\frac{Q_1}{Q_2} = \frac{\log \dfrac{\sinh \pi r_w/a \, \sinh 3\pi}{\sinh 4\pi \, \sinh \pi}}{\log \dfrac{\sinh \pi r_w/a \, \sinh 3\pi}{\sinh 2\pi \, \sinh \pi}} \qquad (5)[1]$$

$$\cong 1 + \frac{2\pi}{\log a/2\pi r_w}.$$

The shielding ratio S corresponding to Eq. (5) is plotted in Fig. 206. It will be seen that as the spacing increases the shielding decreases, or the leakage through the first line increases. In magnitude, however, the total change is not large, as the leakage increases by only 5 per cent as the spacing is increased from 100 to 1,000 ft. This result is of interest in showing that it would hardly be practicable to attempt to appreciably diminish the leakage between two lines of wells by simply decreasing the spacing between them. It should be mentioned, however, that this conclusion refers specifically to the case where not only the bottom-hole pressures in the two lines are kept the same, but

[1] It is of interest to note that the variation of Q_1/Q_2 or of the leakage with the mutual spacing is exactly the same as that due to a variation in the well radii r_w.

where at the same time the lines of wells are separated always by the spacing within the lines. In a later section, where the problem of offsetting is treated, it will be seen that if one is willing to vary the well pressures of one of the well arrays as compared to the other, the leakage between the lines may be given, within limits, any arbitrary value.

On the other hand, when the problem is not one of offsetting, one may still keep the well pressures in the two lines the same and

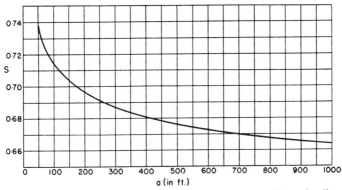

Fig. 206.—The variation of the shielding effect between two lines of wells with the well spacing a. S = (production capacity per well in first line)/(production capacity per well pair in both lines). Lines are set at distances a and $2a$ from the line drive.

yet vary the leakage by adjusting the distance between the lines. The magnitude of this effect may be seen on setting

$$d_1 = a; \quad d_2 = d_1 + \Delta d = a + \Delta d; \quad a = 660 \text{ ft.}; \quad r_w = \frac{1}{4} \text{ ft.},$$

and reducing Eq. (2) to the form

$$\frac{Q_1}{Q_2} = \frac{\log \dfrac{\sinh \dfrac{\pi r_w}{a} \sinh \pi\left(2 + \dfrac{\Delta d}{a}\right)}{\sinh 2\pi\left(1 + \dfrac{\Delta d}{a}\right) \sinh \dfrac{\pi \Delta d}{a}}}{\log \dfrac{\sinh \dfrac{\pi r_w}{a} \sinh \pi\left(2 + \dfrac{\Delta d}{a}\right)}{\sinh 2\pi \sinh \dfrac{\pi \Delta d}{a}}} \tag{6}$$

$$\cong 1 + \frac{2\pi \Delta d/a}{\log a/2\pi r_w}: \qquad \frac{\Delta d}{a} \geq 1.$$

The shielding function S for these equations is plotted in Fig. 207.

Although, as should be expected, the shielding increases as the second line of wells is removed from the first, the magnitudes of the change are only slightly larger than those caused by changing the well spacing.   It appears, therefore, that as long as the well pressures in the two lines are kept the same and none of the distances as a, d, $\Delta d/a$ become too small, one may expect the shielding to have roughly the value ⅔, *i.e.*, that roughly one third of the

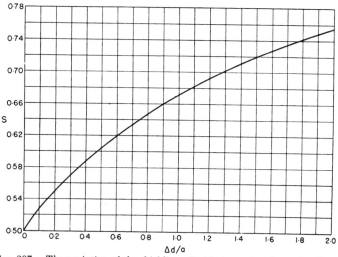

FIG. 207.—The variation of the shielding effect between two lines of wells with the distance, Δd, between them. S = (production capacity per well in first line)/(production capacity per well pair in both lines). a = 660 ft. = well spacing within the lines and distance of first from the line drive.

total flow in the system will pass by the first line of wells and enter the second line (*i.e.*, $Q_1/Q_2 \sim 2$), regardless of the exact value of a, d, or Δd.

As has already been pointed out, the general features of the leakage and shielding effects about a system of well arrays are largely determined by the pressure distribution in the immediate neighborhood of the arrays.   It is, therefore, of some importance to have at least a qualitative picture of this distribution in order to understand more clearly the basis of the effects discussed above. On the other hand, we have been unable to present the details of the distribution until now because, as Eq. (1) shows, it depends explicitly on the values of q_1 and q_2, or on the shielding effects in the system. Now, however, we know that as long as the well pressures in the two lines are the same, the shielding, or the ratio q_1/q_2, does not vary greatly with the other constants of the system. Hence a value of S near $\frac{2}{3}$ will give a representative case on which to base the pressure computation.

The specific case will, therefore, be chosen where the mutual spacing, separation between the lines, and distance of the line drive from the first line of wells are all the same. It will also be

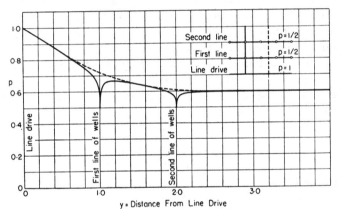

FIG. 208.—The pressure distributions along lines normal to two parallel arrays of wells supplied by a line drive. Distances between the lines and between the first line and the line drive equal the well spacing within the lines, which is taken as unity. Pressure at line drive = 1 unit; pressure at well arrays = $\frac{1}{2}$ unit. ———— pressure distribution along a line passing through a well. - - - - pressure distribution along a line passing midway between two wells.

supposed for definiteness that the well pressures are half of the drive pressure. It then follows from Fig. 207 that

$$\frac{q_1}{q_2} = 2.04; \quad p = 1: \quad y = 0; \quad p = \frac{1}{2}: \quad y = 1 \pm r_w, \quad x = 0.$$

The absolute values of q_1 and q_2 are now given by Eq. (1). with the result that $q_1 = 0.0162$. The pressure distribution computed with these constants along a normal to the arrays and passing through two of the wells, and along a normal passing midway between the well pairs is plotted in Fig. 208. The fact that the two curves of this figure practically coincide everywhere except in the immediate neighborhood of the well arrays shows again the marked localization of the effects of the individual wells about these wells and that their resultant pressure distribution rapidly merges into one that would correspond to uniform line distributions of sinks, as soon as one recedes a short distance from the arrays.

It will also be noted from Fig. 208 that for large distances behind the well arrays the pressure approaches the value of about 0.6 and remains there. Physically, this result implies that whereas the region to the left of the wells (in the figure) is at the pressure 1, the region to the right is at pressure 0.6. In an actual case, however, one might wish to assign a different or arbitrary pressure at the right of the well arrays. Although a solution which will correspond to these conditions can be readily obtained, it must be pointed out that an arbitrary predetermination of the pressures on both sides of the well arrays involves certain necessary implications with regard to the fluid flow in the system. Thus, computing the flow per well across any line $y = $ const. parallel to the well array as

$$Q = -\frac{2k}{\mu}\int_0^{\frac{a}{2}}\frac{\partial p}{\partial y}dx,$$

it is found that

$$
\begin{aligned}
Q &= \frac{4\pi k q_1}{\mu} + \frac{4\pi k q_2}{\mu}: \quad 0 \leqslant y < d_1 \\
&= \frac{4\pi k q_2}{\mu}: \quad d_1 < y < d_2 \\
&= 0: \quad d_2 < y.
\end{aligned}
\tag{7}
$$

This evidently means that the total flow, $\dfrac{4\pi k(q_1 + q_2)}{\mu}$, that leaves the high-pressure boundary $y = 0$ flows completely into the two-well arrays, leaving no net flow to pass to the lower pressure region on the other side of the well arrays. On the other hand, Eq. (7) shows that although the well pressures are lower than the pressures at $y > d_2$ (the area at some distance behind the second line), there is no net flow from this latter region into the well arrays.[1]

---

[1] It may be observed from Fig. 208 that along the lines *midway* between the wells the gradients for $y > d_2$ are *away* from the second array and thus cancel the gradients *toward* the array along the lines through the wells.

Now these results necessarily imply and are implied by the pressure distribution shown in Fig. 208, *i.e.*, that the pressure for $y > d_2$ is approximately 0.6 of the pressure at $y = 0$. Hence if one should predetermine p such that for $y > d_2$, $p \neq 0.6$, the flux relations of Eq. (7) would necessarily be destroyed. Thus as will be seen in detail in the discussion of the problem of offsetting, one may get a pressure distribution in which $p \neq 0.6$ for $y > d_2$ by adding on to Eq. (1) a term as αy and adjusting α so that at $y = d_2 + a$, for example, p has any arbitrarily assigned value. But this, on the other hand, would imply that there is a general migration flux of $a\alpha$ per well across the well-array system. Furthermore, if one should still desire that the well pressures in the two arrays should remain the same, the ratios q_1/q_2 as given by Eq. (2) will have to be modified. It follows, therefore, that the numerical features of the pressure distributions of Fig. 208 are really a necessary implication of the assumption that there is no net flux migration across the well arrays, and that only if such migration is permitted can the pressures in the low-pressure region be given arbitrarily preassigned values.

As a final point of interest with regard to the two-well-array system, we shall derive the explicit relation between the total flux, per well pair, in the system and the net pressure differential between the wells and the line drive. Assuming that d_1/a and $(d_2 - d_1)/a$ are both of the order of 1 and that $2\pi r_w/a \ll 1$, it follows from Eq. (1) that

$$\Delta p = 2q_1\left(\frac{2\pi d_1}{a} + \log\frac{a}{2\pi r_w}\right) + \frac{4\pi q_2 d_1}{a}$$

$$= \frac{Q\mu d_1}{ak} + 2q_1 \log\frac{a}{2\pi r_w}.$$

Further, by Eqs. (4) and (7),

$$Q = \frac{4\pi k q_1}{\mu S} = \frac{4\pi k q_1}{(1 - L)\mu}; \qquad L = 1 - S,$$

where L represents the leakage.

Hence,

$$Q = \frac{2\pi k \Delta p/\mu}{\dfrac{2\pi d_1}{a} + (1 - L)\log\dfrac{a}{2\pi r_w}}, \qquad (8)$$

which is evidently greater than the corresponding value of Eq.
9.8(10) for the single-well array, by virtue of the leakage factor
L. Or more explicitly, the effective resistance of the two-well
array system is

$$R = \frac{\mu\left(\dfrac{d_1}{a} + \dfrac{1}{2\pi} \log \dfrac{a}{2\pi r_w} - \dfrac{L}{2\pi} \log \dfrac{a}{2\pi r_w}\right)}{k}. \tag{9}$$

By comparing this with Eq. 9.8(12) for the single-well array, it is
seen that the decrease in resistance indicated by Eq. (9) is represented simply by the last term, which is proportional to the leakage in the second line of wells. Taking now a representative case
where

$$S = \frac{2}{3}, \qquad \frac{d_1}{a} = 1, \qquad \frac{a}{r_w} = 2640,$$

it is found that the conductivity $\dfrac{Q}{\Delta p}$ in Eq. (8) is $\dfrac{0.6094k}{\mu}$, as contrasted with the value $\dfrac{0.5098k}{\mu}$ corresponding to Eq. 9.8(10), or
for $L = 0$ in Eq. (8).

9.10. Three-line Arrays.—As the number of lines of wells is
increased the expressions giving the various shielding and leakage
characteristics of the system become increasingly more complex.
The method of the derivation of these expressions, however,
remains essentially the same as in the case of the two lines of wells.
We shall, therefore, extend the analysis no further than the three-
line system, as the additional information gained from the treatment of larger numbers of lines would hardly warrant their
discussion.

Beginning again with the expression for the pressure distribution due to the three lines of equal spacing a and placed at distances d_1, d_2, and d_3 from the y axis which is taken to be at the
pressure C (*cf.* Fig. 209), it may be written at once as

$$\begin{aligned}
p(x, y) = C &+ q_1 \log \frac{\cosh 2\pi(y - d_1)/a - \cos 2\pi x/a}{\cosh 2\pi(y + d_1)/a - \cos 2\pi x/a} \\
&+ q_2 \log \frac{\cosh 2\pi(y - d_2)/a - \cos 2\pi x/a}{\cosh 2\pi(y + d_2)/a - \cos 2\pi x/a} \\
&+ q_3 \log \frac{\cosh 2\pi(y - d_3)/a - \cos 2\pi x/a}{\cosh 2\pi(y + d_3)/a - \cos 2\pi x/a}. \tag{1}
\end{aligned}$$

As before, $4\pi kq_1/\mu$, $4\pi kq_2/\mu$, $4\pi kq_3/\mu$, represent the net fluxes, per well, entering the three lines.

Now it has already been seen from the analysis of the problems of the single and two-well arrays that unless $d_1/a < 0.5$ the variation of the shielding effects in the system with d_1 may be neglected

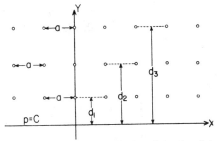

Fig. 209.—Diagrammatic representation of a line drive supplying liquid to three line arrays of wells.

and its value may be represented by that for $d_1/a = 1$.   Setting therefore,

$$\frac{d_1}{a} = 1; \qquad \frac{(d_2 - d_1)}{a} = b; \qquad \frac{(d_3 - d_2)}{a} = c; \qquad \frac{r_w}{a} = \rho$$

and assuming that the well pressures in the three lines are the same, the shielding effects in the system may be found by solving the equations

$$p(0, d_1 - r_w) = p(0, d_2 - r_w) = p(0, d_3 - r_w) \tag{2}$$

for q_2/q_1 and q_3/q_1.

This procedure gives the determinantal solutions:

$$\frac{q_2 \Delta}{q_1} = \begin{vmatrix} \log \dfrac{\sinh \pi b \sinh 2\pi}{\sinh \pi(2 + b) \sinh \pi\rho} & \log \dfrac{\sinh \pi(b + c) \sinh \pi(2 + 2b + c)}{\sinh \pi(2 + b + c) \sinh \pi c} \\[4ex] \log \dfrac{\sinh \pi(b + c) \sinh 2\pi}{\sinh \pi(2 + b + c) \sinh \pi\rho} & \log \dfrac{\sinh \pi(b + c) \sinh 2\pi(b + c + 1)}{\sinh \pi(2 + b + c) \sinh \pi\rho} \end{vmatrix} \tag{3}$$

$$\frac{q_3 \Delta}{q_1} = \begin{vmatrix} \log \dfrac{\sinh \pi b \sinh 2\pi(1 + b)}{\sinh \pi(2 + b) \sinh \pi\rho} & \\ & \log \dfrac{\sinh \pi b \sinh 2\pi}{\sinh \pi(2 + b) \sinh \pi\rho} \\ \log \dfrac{\sinh \pi b \sinh \pi(2 + 2b + c)}{\sinh \pi(2 + b) \sinh \pi c} & \\ & \log \dfrac{\sinh \pi(b + c) \sinh 2\pi}{\sinh \pi(2 + b + c) \sinh \pi\rho} \end{vmatrix} \qquad (4)$$

where

$$\Delta = \begin{vmatrix} \log \dfrac{\sinh \pi b \sinh 2\pi(1 + b)}{\sinh \pi(2 + b) \sinh \pi\rho} & \\ & \log \dfrac{\sinh \pi(b + c) \sinh \pi(2 + 2b + c)}{\sinh \pi(2 + b + c) \sinh \pi c} \\ \log \dfrac{\sinh \pi b \sinh \pi(2 + 2b + c)}{\sinh \pi(2 + b) \sinh \pi c} & \\ & \log \dfrac{\sinh \pi(b + c) \sinh 2\pi(b + c + 1)}{\sinh \pi(2 + b + c) \sinh \pi\rho} \end{vmatrix} \qquad (5)$$

One might now proceed to study in detail the variation of q_2/q_1 and q_3/q_1 with b, the distance between the first two lines, with c, the distance between the second and third lines, and with ρ which gives the effect of the mutual well spacing a. But since the detailed treatment given above of the case of two-line arrays already indicates qualitatively the nature of the effects of these variations, the quantitative discussion will be restricted to the variation of the shielding effects with c, the distance between the second and third lines, while $d_2 - d_1$, the distance between the first two lines, will be taken equal to the mutual spacing a, so that $b = 1$. For this purpose, however, it will be convenient to introduce the shielding factors S_1 and S_2, where

$$S_1 = \frac{q_1}{q_1 + q_2 + q_3}; \qquad S_2 = \frac{q_2}{q_2 + q_3}, \qquad (6)$$

so that S_1 represents the fraction of the total flow in the system that enters the first line of wells, and S_2 gives the fraction of the total flow that has leaked past the first line which flows into the second line. $1 - S_1$ will then give the total fractional leakage past the first line, and $1 - S_2$ the fraction of this leakage which passes the second line and enters the third.

The values of S_1 and S_2 computed by Eqs. (3) to (6) are plotted in Fig. 210. It will be seen that the variation of the shielding effect of the first line with the separation between the second and third lines is quite slow. On the other hand, the shielding effect of the second line changes rather rapidly with $d_2 - d_1$, and approaches a perfect shielding ($S_2 = 1$) for infinite separation between the last two lines.[1] With respect to the leakage past the first line ($1 - S_1$) it may be noted that it is higher than that in a two-line system, although in magnitude the difference is less than 5 per cent of the total flow and, of course, approaches zero as

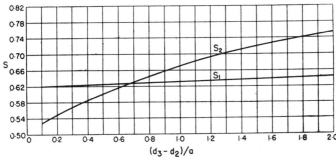

FIG. 210.—The variation of the shielding effect in a system of three parallel-line arrays with the distance, $d_3 - d_2$, between the second and third lines, the distance between the first two lines and between the first line and the line drive being equal to a, the well spacing within the lines ($= 660$ ft.). $S_1 =$ fraction of total fluid produced by all three lines which enters the first line of wells. $S_2 =$ fraction of fluid leaking past the first line which enters the second line.

the third line of wells is removed to infinity. For the case when the distance between the second and the third lines is the same as that between the first and second lines, $1 - S_1 \sim 0.37$, so that about ⅜ of the total flow in the system leaks past the first line of wells; ⅔ of this ($S_2 \cong$ ⅔)—¼ of the total flow—enters the second line, and the other ⅓—⅛ of the total flow—finally reaches the third line of wells; this corresponds to ratios of the production capacities for the three lines given by $Q_1:Q_2:Q_3 = 5:2:1$.

The fact that the shielding effect of the first line of wells changes only from the value 0.671 for a two-line system to 0.631 (for $c = 1$) in a three-line system shows that the addition of still other lines would probably give negligible changes in the shielding

---

[1] It may be noted that the curve for S_2 is, for practical purposes, identical with the shielding curve in a two-array system (*cf.* Fig. 207).

and correspondingly small increases in the leakage. And, in fact, it is not difficult to show that the upper limit to the total leakage for the case where an infinite number of well arrays would be added to the first line is 0.378. Of course, the physical basis of this result lies again in the fact that the shielding effect of and leakage past a line of wells is determined only by the pressure distribution in its immediate vicinity. Hence, since the effect on the pressure distribution about the first line due to the distant well arrays is small, the additional leakage caused by them will

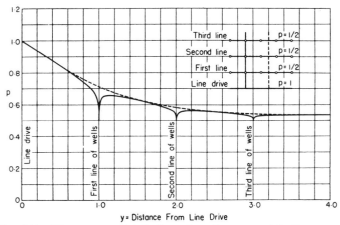

Fig. 211.—The pressure distributions along lines normal to three parallel arrays of wells; pressure at the line drive = 1 unit; pressure at the wells = ½ unit. ——— pressure distribution along a normal passing through the wells. - - - - pressure distribution along a normal passing midway between the wells; all the spacings of the wells and lines are taken equal to unity.

also be small. Evidently, then, any attempt to increase the production capacity of a one-sided system by merely adding additional lines of wells must be carried out with caution, since appreciable increases will be obtainable only by operating the more distant lines at lower and lower bottom-hole pressures.

A qualitative picture of the pressure distribution in a three-line system may be obtained from Fig. 211, which gives profiles perpendicular to the arrays on a line passing through the wells and along a line passing midway between them.

Finally it may be noted that the relation between the total flux per well triplet and the pressure differential acting on the system is given by Eq. 9.9(8), except that L represents the total leakage.

In fact, it may be readily shown that Eq. 9.9(8) holds for any number of parallel well arrays forming a rectangular network of wells.

9.11. Staggered Wells in Line Arrays.—Thus far we have dealt with multiple-well systems in which the arrays have been placed so as to form regular rectangular networks, *i.e.*, where the corresponding wells of the several arrays have been located along lines normal to the arrays. From a practical point of view, however, mere geometrical symmetry is of little moment, and the question of significance is that of the effectiveness of various well configurations with respect to their shielding, leakage, or offsetting characteristics. It is, therefore, of importance to see if staggered-

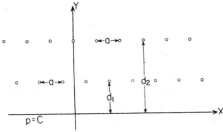

Fig. 212.—Diagrammatic representation of a line drive supplying liquid to two line arrays of wells which are mutually staggered.

well systems may have advantageous features as compared to the regular rectangular well network.

To get at once to the maximum effects of the staggering it will be supposed that the several lines are mutually shifted parallel to themselves by half of the well spacing, as is indicated in Fig. 212, and that the x axis represents a line drive at the pressure C. The pressure distribution is then, by Eq. 9.8(6),

$$p = C + q_1 \log \frac{\cosh 2\pi(y - d_1)/a - \cos 2\pi x/a}{\cosh 2\pi(y + d_1)/a - \cos 2\pi x/a}$$
$$+ q_2 \log \frac{\cosh 2\pi(y - d_2)/a + \cos 2\pi x/a}{\cosh 2\pi(y + d_2)/a + \cos 2\pi x/a}. \quad (1)$$

Imposing the condition that the well pressures in the two arrays be the same, *i.e.*, that

$$p(0, d_1 - r_w) = p\left(\frac{a}{2}, d_2 - r_w\right), \quad (2)$$

it is found that

$$\frac{q_1}{q_2} = \frac{\log \dfrac{\sinh \pi r_w/a \cosh \pi(d_2 + d_1)/a}{\sinh 2\pi d_2/a \cosh \pi(d_2 - d_1)/a}}{\log \dfrac{\sinh \pi r_w/a \cosh \pi(d_2 + d_1)/a}{\sinh 2\pi d_1/a \cosh \pi(d_2 - d_1)/a}}. \tag{3}$$

Comparing this result with Eq. 9.9(2), which applies to the case when the two lines are not staggered, it is seen that the only difference between them consists in the replacement of the ratio $\dfrac{\sinh \pi(d_2 + d_1)/a}{\sinh \pi(d_2 - d_1)/a}$ by $\dfrac{\cosh \pi(d_2 + d_1)/a}{\cosh \pi(d_2 - d_1)/a}$. Now $\cosh \pi z$ is almost exactly equal to $\sinh \pi z$ for $z \geqslant 1$. Hence, since for all practical cases $(d_2 - d_1)/a \geqslant 1$, Eq. (3) and Eq. 9.9(2) will not differ appreciably unless $(d_2 - d_1)/a$ is small. That is, staggering the well arrays will have no effect on the shielding and leakage characteristics of a system unless the distances between the lines are made appreciably less than the spacing of the wells in the lines. In particular, unless the two lines are separated by a distance less than 0.1 of the well spacing within the lines, the leakage in a staggered two-line system will exceed that in a nonstaggered system by no more than 1 per cent of the total value.

One may, therefore, conclude that for all practical purposes staggering the well system will have no appreciable affect upon the leakage or shielding characteristics of a multiple-well system.

9.12. The Theory of Offsetting. Statement of the Problem.— The discussion thus far has been concerned mainly with the natural leakage or shielding characteristics of multiple-well systems in which the well pressures have all been kept the same and the leakages have been induced by the geometrical asymmetry of the pressures acting at the boundaries of the system. The question we shall now treat is that of counteracting this leakage due to the geometrical asymmetry by an artificially imposed asymmetry in the well pressures in the several lines of wells. This is the essential question involved in the problem of offsetting, which may be formally stated as follows: Given two leases of preassigned and in general different average pressures, with two lines of wells offsetting each other across the mutual lease boundary; for a given well pressure in one of the lines, at what pressure must the other line be produced in order that there be no net fluid flow across the boundary?

In order to make clear the physical basis of the analytical treatment of this problem, a more specific meaning must be given to the term "average pressure" over a lease. Physically, the term would imply a numerical average of individual pressures measured at points distributed over various parts of the lease, and would, therefore, be a measure of the production capacity of the lease. From the point of view, however, of the interaction of the lease (or the fluid underlying it) with its neighboring leases or with any given group of wells, the above numerical average would not be immediately significant. For the determination of such effects one must rather know the average of the pressures over the lease *boundary* near the group of wells in question or the neighboring leases, *i.e.*, the average pressure at the boundary offsets. Of course, the pressure distribution in the interior of the lease partakes in determining what the boundary pressures are, but for the purpose of discussing the reaction of the lease with external flow systems the lease as a whole may be replaced by a statement of its boundary pressures.

Another point to be noted is that although the pressures in the leases will change during the course of their depletion, the changes will take place so slowly that for practical purposes steady-state conditions may be assumed. As appreciable pressure changes are observed, however, the requirements for offsetting may be varied to correspond to the new conditions.

Finally it should be observed that the theory to be developed in the following sections applies only to that component of the fluid motion which is due to a general migration into each lease from external sources, the line drives at pressures p_1 and p_2. The components representing the natural capacities of each lease to produce in virtue of the expansion of its fluid content, due either to its liquid compressibility or to the evolution and expansion of its dissolved gases, are to be considered as having been subtracted, leaving fluid velocities and pressure gradients corresponding to those in a strictly incompressible-liquid system. Thus by an extension of the following developments it would be concluded that by surrounding a given lease by properly controlled offsets the migration into the lease across its boundaries could be made to vanish, and with it the production from the wells in its interior.[1]

[1] In fact, the system of four wells in a circular reservoir discussed in Sec. 9.3, case (*d*), may be interpreted as a system in which leases in the interior

In the practical realization of such a situation it would be found, however, that even though the migration into the lease has been effectively prevented, there will nevertheless persist a flow of fluid from the wells within the lease, until its pressure has fallen to zero. Obviously this production would be due only to the expansion of the fluid content of the porous medium within the lease boundaries, and hence this production, together with the pressure gradients associated with it, must be supposed as having been subtracted from this system before applying the theory of offsetting to be developed below.

9.13. Single-line Offsetting.—From the above point of view, then, the two leases, the interaction of which is to be prevented

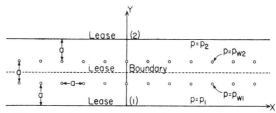

Fig. 213.—Diagrammatic representation of an offsetting system with single offset lines.

by the two lines of offset wells, will be replaced by two lines parallel to the well arrays and maintained at the "average pressures" of the leases. And for definiteness it will be supposed that each of these lines is separated from the neighboring well arrays by a distance equal to the separation between the offset lines, which in turn is to be taken equal to the well spacing a (*cf.* Fig. 213). The actual lease boundary, then, lies midway between the two lines of wells, while the two lines along which $p = p_1$ and $p = p_2$ are the analytical representations for the whole of the individual leases that lie beyond their offset lines.

Following the previous analysis, the general expression for the pressure distribution in the system may be written as

$$p = p_1 + \alpha y + q_1 \log \frac{\cosh 2\pi(y - 1) - \cos 2\pi x}{\cosh 2\pi(y + 1) - \cos 2\pi x}$$
$$+ q_2 \log \frac{\cosh 2\pi(y - 2) - \cos 2\pi x}{\cosh 2\pi(y + 2) - \cos 2\pi x}, \quad (1)$$

of the square have been so effectively offset by the four wells as to be left with no production from the interior wells.

where the mutual spacing a has been taken as the unit of distance. As it stands, Eq. (1) expresses the fact that $p = p_1$ at $y = 0$, that there are two lines of wells at $y = 1$ and $y = 2$, with fluxes $4\pi kq_1/\mu$ and $4\pi kq_2/\mu$, and that there is a *regional migration* in the system, with a magnitude α. The problem that remains is the determination of α, q_1, and q_2 such that there will be no net flux across $y = \frac{3}{2}$, the mutual lease boundary, that $p = p_{w2}$ at $y = 2$, the position of the second line of wells, and that $p = p_2$ at $y = 3$, the representation of the offset lease. Although Eq. (1) cannot give a strictly uniform value for p at $y = 3$, it has seen in the previous discussions that the variations in p parallel to an array at distances from the line of wells of the order of the mutual spacing are so small that any line parallel to the array may be taken as an exact equipressure curve.

To determine α, it need only be noted that for any general term as

$$p = q \log \frac{\cosh 2\pi(y - d) - \cos 2\pi x}{\cosh 2\pi(y + d) - \cos 2\pi x}$$

the flux Q is given by

$$Q = -\frac{k}{\mu}\int_{-\frac{1}{2}}^{\frac{1}{2}}\frac{\partial p}{\partial y}dx \quad \begin{aligned} &= \frac{4\pi kq}{\mu}: &&y < d \\ &= 0: &&y > d. \end{aligned} \tag{2}$$

It follows at once that in order that there be no net flux across the mutual lease boundary, $y = \frac{3}{2}$, α must have the value

$$\alpha = 4\pi q_2. \tag{3}$$

The values of q_1 and q_2 are now found by applying the requirements that

$$p(0, 2 - r_w) = p_{w2}; \qquad p(0, 3) = p_2, \tag{4}$$

which gives

$$\left. \begin{aligned} p_1 + 8\pi q_2 - 4\pi q_1 + 2q_2(\log 2\pi r_w - 4\pi) &= p_{w2} \\ p_1 + 12\pi q_2 - 4\pi q_1 - 8\pi q_2 &= p_2. \end{aligned} \right\} \tag{5}[1]$$

[1] The approximations are made here and below that $\frac{\cosh}{\sinh} m\pi = e^{m\pi}/2$ for $m \geqslant 1$, and that $\sinh \pi r_w = \pi r_w$.

Solving, the result is

$$q_2 = \frac{p_2 - p_{w2}}{4\pi - 2 \log 2\pi r_w}; \qquad q_1 =$$

$$\frac{4\pi(p_1 - p_{w2}) - 2(p_1 - p_2) \log 2\pi r_w}{4\pi(4\pi - 2 \log 2\pi r_w)}. \qquad (6)$$

Finally, to give the specific answer to the problem as originally stated, one must now find the pressure p_{w1} at the first line of wells, which the above values of q_1 and q_2 imply. Returning then to Eq. (1) and setting $y = 1 - r_w$, it is readily found that

$$p_{w1} = p_{w2} - 0.9614(p_1 - p_2), \qquad (7)$$

for $r_u \doteq \frac{1}{2640}$, so that

$$p_{w1} \gtreqless p_{w2} \qquad \text{as} \qquad p_1 \gtreqless p_2, \qquad (8)$$

as should be expected.

Since p_{w1} must be positive, there will be a lower limit to p_{w2} for which Eq. (7) can be satisfied. Thus for example, if the high-pressure lease has $p_1 = 1,500$ lb. and the low-pressure lease has $p_2 = 1,000$ lb., then if the offset wells of the latter have $p_{w2} = 481$ lb., p_{w1} must equal zero in order to avoid migration from the high- to the low-pressure leases. And if the low-pressure offsets are produced below 481 lb., even a vacuum on the high-pressure offsets will not completely prevent migration.

If, on the other hand, the low-pressure offsets have $p_{w2} > 481$ or if $p_1 - p_2 < 500$, p_{w1} may be kept *above* zero and still prevent migration. Or finally, if $p_1 - p_2$ becomes negative, the situation becomes reversed and p_{w1} has, in fact, to be kept *higher* than p_{w2} in order to prevent migration *into* the lease beyond $y = 0$.

These results may be conveniently summarized by replacing the factor 0.9614 in Eq. (7) by 1 and rewriting it as

$$p_{w1} - p_{w2} = p_2 - p_1, \qquad (9)$$

which can be stated in the simple rule, thus: In order to prevent migration, the difference in pressures between the offsets must be equal and opposite to the difference in pressures between the leases. If it is not possible to make the offset-pressure difference equal to the lease-pressure differential, some fluid will necessarily pass from the high to the low-pressure lease. And conversely, if the offset-pressure differential is lowered below the lease-pressure differential, there will even be a net migration into the high-pressure lease.   This rule may be given a physical interpretation on rewriting Eq. (9) as

$$p_1 - p_{w2} = p_2 - p_{w1}.$$

From this form of Eq. (9) it is readily seen that the fluid migrations from either lease to the opposite offset are equal, and hence give a *zero net* migration.

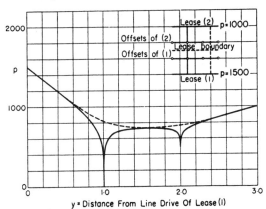

Fig. 214.—The pressure distribution along lines normal to two parallel arrays of wells with their pressures adjusted so as to prevent migration between them. ———— pressure distribution along a normal passing through the wells. - - - - pressure distribution along a normal passing midway between the wells.   Lease pressure of (1) = 1,500 lb.; lease pressure of (2) = 1,000 lb.

Applying now Eq. (7) to Eq. (6), it is seen that the production rates from the offset lines will be in the ratio

$$\frac{Q_1}{Q_2} = \frac{p_1 - p_{w1}}{p_2 - p_{w2}}.$$

The physical basis of the problem of offsetting may be more clearly seen from the pressure distribution in a system in which the pressures have been adjusted so as to have no migration between the two lines of offset wells.   Such a distribution is plotted in Fig. 214 for the numerical case treated above: lease pressures, 1,500 and 1,000 lb., and offset well pressures of 0 and 481 lb.   It may be compared with Fig. 208 which gives the pressure distribution in a two-line system without the offsetting-pressure adjustment.   It will be noted that whereas in the latter

case the pressures midway between the wells decrease uniformly
in going from the first line to the second, the pressure in Fig. 214
has a minimum between the two lines which overlies the maximum
of the pressure curve along a line *through* the well pairs.   Thus in
Fig. 208 there is a net gradient which drives liquid past the first
line into the second line, but in Fig. 214 these gradients average
out and there is no net flow between the two lines.

   Of more general significance, however, is the observation that
the pressure adjustment for offsetting is equivalent to lowering
the average effective pressures of the high-pressure lease to equal
that of the low-pressure lease.   It is only in this way that the

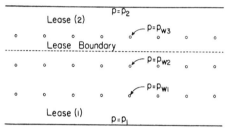

FIG. 215.—Diagrammatic representation of a multiple-line offsetting system.

general gradient between the two leases can be removed, and
with it the associated migration.   At the same time the net pressure gradient across the lease boundary will be made to vanish.
In fact, the treatments of the more general cases to be given below
are *physically* nothing more than the adjustments of the pressure
distribution in the high-pressure leases, by lowering their offset
well pressures, so as to have, on the average, pressures effectively
as low as those of their offsetting leases.

   9.14. Multiple-line Offsetting.—If in ·either or both leases
there are other well arrays parallel to the actual offset lines, the
above analysis and results will still remain strictly valid, provided
one uses for the lease pressures p_1 and p_2 the averages over the
lines in the two leases nearest to the offset wells.   In fact, a
convenient way of adjusting the effective lease pressures p_1 and p_2,
would be that of varying the well pressures in the lines parallel to
the offset lines.   To see in detail how this may be done, we may
suppose that the high-pressure lease has two offset lines instead
of one.   Then with the notation of Fig. 215 it will be supposed
that the pressure in the first line of offsets of the high-pressure

lease is p_{w1}, that nearest the lease boundary is p_{w2}, and that the offsets in the neighboring lease are produced at the pressure p_{w3}. The average lease pressures behind the offsets will be taken as p_1 and p_2. Applying the same methods as used above it follows that

$$\left. \begin{aligned}
p_{w1} - p_1 &= 4\pi q_3 + 2q_1(-2\pi + x) - 4\pi q_2 - 4\pi q_3 \\
p_{w2} - p_1 &= 8\pi q_3 - 4\pi q_1 + 2q_2(-4\pi + x) - 8\pi q_3 \\
p_{w3} - p_1 &= 12\pi q_3 - 4\pi q_1 - 8\pi q_2 + 2q_3(-6\pi + x) \\
p_2 - p_1 &= 16\pi q_3 - 4\pi q_1 - 8\pi q_2 - 12\pi q_3,
\end{aligned} \right\} \quad (1)$$

where

$$x = \log 2 \sinh \frac{\pi r_w}{a}.$$

Since there are four equations for the three unknowns q_1, q_2, q_3, the Eqs. (1) can be consistent only if

$$\begin{vmatrix}
-4\pi + 2x & -4\pi & 0 & p_1 - p_{w1} \\
-4\pi & -8\pi + 2x & 0 & p_1 - p_{w2} \\
-4\pi & -8\pi & 2x & p_1 - p_{w3} \\
-4\pi & -8\pi & 4\pi & p_1 - p_2
\end{vmatrix} = 0. \quad (2)$$

Replacing, as before, $-x/2\pi = 0.96$ (for $a/r_w = 2{,}640$) by 1, Eq. (2) reduces to

$$2p_1 + 2p_{w1} + 6p_{w2} = 5p_2 + 5p_{w3}, \quad (3)$$

which gives for the fluxes per well in the three lines the values

$$\left. \begin{aligned}
5Q_1 &= 2p_1 - 3p_{w1} + p_{w2} \\
5Q_2 &= p_1 + p_{w1} - 2p_{w2} \\
2Q_3 &= p_2 - p_{w3},
\end{aligned} \right\} \quad (4)$$

where here and in the rest of this section the ratio k/μ has been taken as 1. Hence the resultant production rates from the offset wells in the two leases will have the ratio

$$\frac{Q_1 + Q_2}{Q_3} = \frac{2}{5} \frac{3p_1 - 2p_{w1} - p_{w2}}{p_2 - p_{w3}}. \quad (5)$$

Although Eq. (3) does not permit of a simple physical interpretation such as was given to Eq. 9.13(9), its implications are, however, not difficult to see. Thus, going back to the example of Sec. 9.13, where it was assumed that $p_1 = 1,500$ and $p_2 = 1,000$, it was seen from Eq. 9.13(9) that if the low-pressure offsets were held below 500 lb.,[1] even a vacuum in the high-pressure offsets would not prevent migration across the lease boundary. Here, however, it follows at once from Eq. (3) that with the given lease pressures the offset well pressures must satisfy the relation

$$2p_{w1} + 6p_{w2} = 2,000 + 5p_{w3}.$$

Hence, even if the low-pressure offsets are kept at a vacuum ($p_{w3} = 0$), the high-pressure offsets can still prevent migration, provided only that

$$p_{w1} + 3p_{w2} = 1,000,$$

which may be satisfied not only by the combination $p_{w2} = 0$, $p_{w1} = 1,000$, but even with $p_{w2} > p_{w3}$, i.e., with $p_{w2} = 100$, $p_{w1} = 700$.

It is thus seen that simply by means of a second line of offsets in a high-pressure lease one may prevent migration with a number of different offset-pressure combinations, even though the low-pressure offsets are maintained at a vacuum. With more than two offset lines the possibilities of varying the offset pressures and yet preventing migration would be still greater. In fact, the above analysis may be readily extended to cases where the offsetting leases have any number of offsetting lines. The general results may be tabulated on introducing the following notations:

$P_1 = $ "lease pressure" in lease (1).

$P_2 = $ "lease pressure" in lease (2).

$n_1 = $ number of offset lines in lease (1) (in front of P_1).

$n_2 = $ number of offset lines in lease (2) (in front of P_2).

$p_{1k} = $ well pressure in lease (1) in kth offset line as measured from the line where the pressure is P_1.

$p_{2k} = $ well pressure in lease (2) in kth offset line as measured from the line where the pressure is P_2.

$\bar{Q}_1 = $ sum of production rates per well in offset lines of lease (1).

$\bar{Q}_2 = $ sum of production rates per well in offset lines of lease (2).

[1] This, rather than 481, would have been the limiting value for p_{w2} if x had been replaced by -2π, as has been done here.

Then the requirements for no migration and the associated lease production rates are:[1]

1. $n_1 = 1; n_2 = 1,$

$$P_1 + p_{11} = p_{21} + P_2; \frac{\bar{Q}_1}{\bar{Q}_2} = \frac{P_1 - p_{11}}{P_2 - p_{21}}; \bar{Q}_2 = \frac{1}{2}(P_2 - p_{21}).$$

2. $n_1 = 2; n_2 = 1,$

$$P_1 + p_{11} + 3p_{12} = \frac{5}{2}(p_{21} + P_2),$$

$$\frac{\bar{Q}_1}{\bar{Q}_2} = \frac{2}{5} \cdot \frac{3P_1 - 2p_{11} - p_{12}}{P_2 - p_{21}}; \bar{Q}_2 = \frac{1}{2}(P_2 - p_{21}).$$

3. $n_1 = 2; n_2 = 2,$

$$P_1 + p_{11} + 3p_{12} = 3p_{22} + p_{21} + P_2,$$

$$\frac{\bar{Q}_1}{\bar{Q}_2} = \frac{3P_1 - 2p_{11} - p_{12}}{3P_2 - 2p_{21} - p_{22}}; \bar{Q}_2 = \frac{3P_2 - 2p_{21} - p_{22}}{5}.$$

4. $n_1 = 3; n_2 = 1.$

$$P_1 + p_{11} + 3p_{12} + 8p_{13} = 13\frac{1}{2}(p_{21} + P_2),$$

$$\frac{\bar{Q}_1}{\bar{Q}_2} = \frac{2}{13} \frac{8P_1 - 5p_{11} - 2p_{12} - p_{13}}{P_2 - p_{21}}; \bar{Q}_2 = \frac{1}{2}(P_2 - p_{21}).$$

5. $n_1 = 3; n_2 = 2,$

$$P_1 + p_{11} + 3p_{12} + 8p_{13} = 13\frac{3}{5}(3p_{22} + p_{21} + P_2),$$

$$\frac{\bar{Q}_1}{\bar{Q}_2} = \frac{5}{13} \cdot \frac{8P_1 - 5p_{11} - 2p_{12} - p_{13}}{3P_2 - 2p_{21} - p_{22}}; \bar{Q}_2 = \frac{3P_2 - 2p_{21} - p_{22}}{5}.$$

6. $n_1 = 3; n_2 = 3,$

$$P_1 + p_{11} + 3p_{12} + 8p_{13} = 8p_{23} + 3p_{22} + p_{21} + P_2,$$

$$\frac{\bar{Q}_1}{\bar{Q}_2} = \frac{8P_1 - 5p_{11} - 2p_{12} - p_{13}}{8P_2 - 5p_{21} - 2p_{22} - p_{23}},$$

$$\bar{Q}_2 = \frac{8P_2 - 5p_{21} - 2p_{22} - p_{23}}{13}.$$

7. $n_1 = 4; n_2 = 1,$

$$P_1 + p_{11} + 3p_{12} + 8p_{13} + 21p_{14} = 17(p_{21} + P_2),$$

$$\frac{\bar{Q}_1}{\bar{Q}_2} = \frac{1}{17} \cdot \frac{21P_1 - 13p_{11} - 5p_{12} - 2p_{13} - p_{14}}{P_2 - p_{21}};$$

$$\bar{Q}_2 = \frac{1}{2}(P_2 - p_{21}).$$

[1] Although these relations are based on the explicit assumption that the various offset lines are spaced from each other by 660 ft., the spacing within the lines, which is equivalent to an areal well density of one well per 10 acres, they will also give at least the correct order of magnitude for the distributions of the pressures in the offset lines even when the various spacings are appreciably different from 660 ft., owing to the slow (logarithmic) variation of the quantity x, taken as -2π, with these spacings.

8. $n_1 = 4$; $n_2 = 2$,

$$P_1 + p_{11} + 3p_{12} + 8p_{13} + 21p_{14} = {}^{34}\!/_5(3p_{22} + p_{21} + P_2),$$

$$\frac{\bar{Q}_1}{\bar{Q}_2} = \frac{5}{34} \cdot \frac{21P_1 - 13p_{11} - 5p_{12} - 2p_{13} - p_{14}}{3P_2 - 2p_{21} - p_{22}};$$

$$\bar{Q}_2 = \frac{3P_2 - 2p_{21} - p_{22}}{5}.$$

9. $n_1 = 4$; $n_2 = 3$,

$$P_1 + p_{11} + 3p_{12} + 8p_{13} + 21p_{14}$$
$$= {}^{34}\!/_{13}(8p_{23} + 3p_{22} + p_{21} + P_2),$$

$$\frac{\bar{Q}_1}{\bar{Q}_2} = \frac{13}{34} \cdot \frac{21P_1 - 13p_{11} - 5p_{12} - 2p_{13} - p_{14}}{8P_2 - 5p_{21} - 2p_{22} - p_{23}};$$

$$\bar{Q}_2 = \frac{8P_2 - 5p_{21} - 2p_{22} - p_{23}}{13}.$$

10. $n_1 = 4$; $n_2 = 4$,.

$$P_1 + p_{11} + 3p_{12} + 8p_{13} + 21p_{14}$$
$$= 21p_{24} + 8p_{23} + 3p_{22} + p_{21} + P_2,$$

$$\frac{\bar{Q}_1}{\bar{Q}_2} = \frac{21P_1 - 13p_{11} - 5p_{12} - 2p_{13} - p_{14}}{21P_2 - 13p_{21} - 5p_{22} - 2p_{23} - p_{24}},$$

$$\bar{Q}_2 = \frac{21P_2 - 13p_{21} - 5p_{22} - 2p_{23} - p_{24}}{34}.$$

9.15. Numerical Example.—In applying the equations just derived for the problem of offsetting one must remember that the implicit assumption has been made that the individual well production rates, and hence their pressures, are free to be adjusted, within limits, so that an effective offsetting can be attained. For if the leases are prorated with respect to production on a per well basis, and these productions must be derived in equal amounts from the several wells, the problem of offsetting becomes, of course, meaningless. Migration across lease boundaries can then be prevented only by drilling up each lease with the same well per acre density, if the well "potentials" are the same, or with well per acre densities proportional to the potentials (lease pressures times sand thickness times sand permeability) if the latter are different but have not been taken into account in the proration. On the other hand, an attempt to prevent migration becomes equally futile if there are no proration restrictions and the offset lease is produced "wide open" (minimum obtainable bottom-hole pressures) over the extent of the whole lease. The only protection in such a case can come from drilling up the lease to be guarded against migration losses to a well per acre density equal to its neighbor and then producing these wells wide open just as the offset lease is being produced.

For the problem of offsetting to have a real meaning it must evidently be presupposed that we have at least some liberty in adjusting the production and pressure distributions in the lease in question so as to prevent migration. If the offset lease is completely depleted with regard to its own natural production capacity but is still being pumped, its neighbor is simply at its

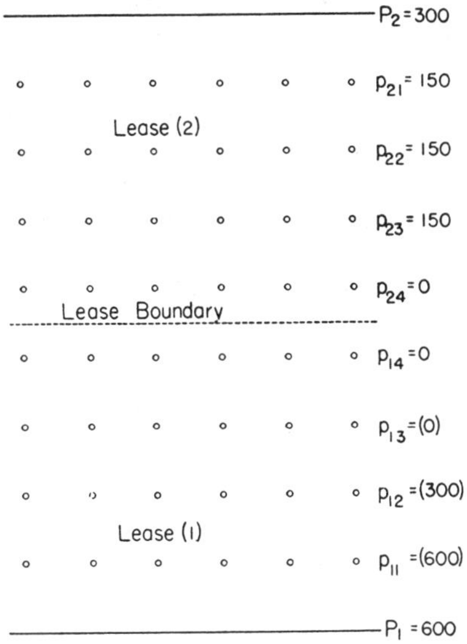

$$P_2 = 300$$

$$\circ \quad \circ \quad \circ \quad \circ \quad \circ \quad \circ \quad P_{21} = 150$$

Lease (2)

$$\circ \quad \circ \quad \circ \quad \circ \quad \circ \quad \circ \quad P_{22} = 150$$

$$\circ \quad \circ \quad \circ \quad \circ \quad \circ \quad \circ \quad P_{23} = 150$$

$$\circ \quad \circ \quad \circ \quad \circ \quad \circ \quad \circ \quad P_{24} = 0$$

Lease Boundary

$$\circ \quad \circ \quad \circ \quad \circ \quad \circ \quad \circ \quad P_{14} = 0$$

$$\circ \quad \circ \quad \circ \quad \circ \quad \circ \quad \circ \quad P_{13} = (0)$$

$$\circ \quad \circ \quad \circ \quad \circ \quad \circ \quad \circ \quad P_{12} = (300)$$

Lease (1)

$$\circ \quad \circ \quad \circ \quad \circ \quad \circ \quad \circ \quad P_{11} = (600)$$

$$P_1 = 600$$

FIG. 216.—Diagrammatic representation of an offsetting program in which four lines are used in each lease.

mercy, as it must supply all that the offset lease will produce. When, however, the neighboring lease is not yet completely depleted and its offset lines are being produced at specified bottom-hole pressures (not all of them zero), then the above formulas may be used to find a set of pressures for the offset lines in the high-pressure lease in order to prevent loss of its fluid. Moreover, one may apply the restriction that the total production rates from each lease be kept at a certain ratio, and, unless this requirement be too unfavorable for the high-pressure lease, its offset-line pressures may still be adjusted so as to prevent fluid migration across the boundary.

As a specific case a problem will be considered in which there are four lines of offset wells in each lease parallel to their mutual boundary (*cf.* Fig. 216). It will be assumed that the reservoir pressure or "effective lease pressure" of lease (1), beyond its offsets, is 300 lb. and that of lease (2) is 600 lb. These, of course, are somewhat less than the reservoir pressure at the outer extremities of the two leases owing to the general migration from these leases into their respective offsets at the mutual lease boundary. Furthermore, it will be supposed that by some proration agreement lease (2) is permitted to produce at as high a rate as lease (1), and that to obtain this allotment the offset-well pressures of lease (2) are kept at 150 lb. with the exception of the line at the boundary which is kept at a vacuum ($p_{24} = 0$). The question then is: Can the pressures in the offsets of lease (1) be adjusted so as to produce no more than its neighbor and yet prevent loss of oil across the boundary, and if so, what should these pressures be?

To find the answer one must apply the formulas of case 10 of the last section with the data[1]

$$\frac{\bar{Q}_1}{\bar{Q}_2} = 1; \quad P_1 = 600; \quad P_2 = 300; \quad p_{14} = 0 = p_{24},$$

$$p_{23} = 150; \quad p_{22} = 150; \quad p_{21} = 150.$$

From the formula of case 10, it follows that in order that $\bar{Q}_1/\bar{Q}_2 = 1$, *i.e.*, the two groups of offsets give the same production rate,

$$21P_2 - 13p_{21} - 5p_{22} - 2p_{23} - p_{24} = 21P_1 - 13p_{11} - 5p_{12} - 2p_{13} - p_{14},$$

which, with the above data, reduces to

$$13p_{11} + 5p_{12} + 2p_{13} = 9{,}300.$$

Putting the same data in the first of the formulas of case 10, the condition of no migration becomes

$$p_{11} + 3p_{12} + 8p_{13} = 1{,}500.$$

Since all the pressures must be positive but cannot exceed 600 lb., it is readily found that the only values of p_{11}, p_{12}, p_{13}

---

[1] That $p_{14} = 0$ is assumed a priori in view of the general advantages of lease (2) in inducing migration from lease (1).

which can satisfy these requirements and the above equations are

$$p_{11} = 600; \qquad p_{12} = 300; \qquad p_{13} = 0.$$

If, then, the fourth line of offsets is shut in, the second produced at 300 lb., and the third as well as the fourth are pumped at a vacuum, oil migration from lease (1) to its offset will be prevented. However, it should be mentioned that if the pressure distribution in the offset lines of lease (2) is appreciably different from that which has been chosen, *i.e.*, if the lines beyond the one adjacent to the boundary are produced at unequal pressures, one would, in such cases, be unable to prevent migration and yet avoid negative pressures or pressures higher than 600 lb. in lease (1). On the other hand, it is to be noted that this sensitivity to the pressure distribution in lease (2) is really due to the unfavorable requirement that lease (1) is to produce no more than its neighbor even though its reservoir pressure is twice as high. If this restriction is removed and lease (1) is permitted to produce at a rate twice as high as its neighbor, migration can be prevented even though the pressure distribution in lease (2) varies over a considerable range. Thus setting[1]

$$P_2 = 300; \qquad p_{21} = 200; \qquad p_{22} = 150;$$
$$p_{23} = 100; \qquad p_{24} = 0.$$

and

$$\frac{\bar{Q}_1}{\bar{Q}_2} = 2; \qquad P_1 = 600; \qquad p_{14} = 0,$$

the equations of case 10 will give

$$p_{11} + 3p_{12} + 8p_{13} = 1,150,$$
$$13p_{11} + 5p_{12} + 2p_{13} = 7,100,$$

which can be satisfied, for example, either by the set

$$p_{11} = 500; \qquad p_{12} = 103; \qquad p_{13} = 43,$$

or by

$$p_{11} = 475; \qquad p_{12} = 177: \qquad p_{13} = 18.$$

In addition to fixing the *ratio* of the production rates from the offset lines on the two leases, one will in general have to permit a preassignment of the *absolute* values of the total production rates from the sets of offset lines. This will still further restrict the possibilities of offsetting for the complete prevention of migration.

---

[1] This distribution would lead to $p_{11} = 700$ even for $p_{13} = 0$, if $\bar{Q}_1 = \bar{Q}_2$.

Nevertheless, since the total number of restrictions is three—no migration, given ratio of total offset-line production rates, and given absolute values of total production rates—a set of four offset lines should in general provide among its infinity of algebraic possibilities at least a small range of pressure distributions in these offset lines which will involve no negative well pressures or pressures higher than the lease pressures.

9.16. The Problem of Water Flooding.—One of the most favorable fields of application of the theory of the flow of "dead" liquids through sands is that in the method of artificial oil recovery known as water flooding. This method, applied with notable success in the Bradford[1] oil field of Pennsylvania, simply consists in the process of displacing the oil remaining in a sand, after the natural recovery operations have become unprofitable, by water artificially introduced into the sand.

Although the engineering technique of the process of introducing the water involves special problems of its own, the physical situation may be completely described by an interlaced network of water-injection wells—which may be either abandoned or converted oil wells, or wells drilled especially for the purpose—and oil-output wells. The water entering the sand at high pressure (owing to the weight of the water column in the well, or with added applied pressure) tends to move toward the oil wells which are maintained at lower pressure, being in fact pumped in most cases. The oil remaining in the sand owing to the depletion of the original gas dissolved in the oil may be considered as effectively "dead" and has very little or no tendency at all to move toward the output wells. The water, however, is under a high pressure and in its motion away from the injection wells floods the oil out of the sand pores and drives it forward in its path toward the output wells. Although the water neither completely washes the sand of the oil nor banks the oil in front of it with a sharp interface, these complications may be neglected without seriously affecting the validity of the conclusions of the approximate theory.

A more serious difficulty is involved in the necessary neglect of the difference in the viscosity between the injected water and the

---

[1] A discussion of the history of the water-flooding operations and their present status in this field has been given recently in a series of articles by C. R. Fettke in *Oil & Gas Jour.;* see especially p. 32, Aug. 19, 1937, and p. 48, Aug. 26, 1937.

oil to be displaced. For in view of this difference the whole problem really belongs to the class of "two-fluid systems" discussed in Chap. VIII. However, as was seen there, the treatment of two-fluid systems is extremely difficult—except possibly by graphical means—when they are other than of the simplest geometry. Indeed it appears to be almost futile to attempt an analytical discussion of multiple-well networks unless the viscosities of the two fluids be assumed the same and the treatment be developed for the equivalent homogeneous system, as outlined in Sec. 8.5. On the other hand, the application of this method to the problem of flooding has been deferred to this chapter because the treatment to be given stresses the dynamical characteristics of the infinite well networks rather more than the two-fluid phase of the problem.

The practical questions involved in the problem of water flooding concern not only the analysis and understanding of the flooding mechanism of a particular network of injection and output wells, but in addition the relative merits of various network arrangements of the injection and output wells must be considered. To arrive at the answer to these questions it will be convenient to treat separately the different phases of the problem, such as the history of the water-oil interface, the potential distributions, resistances of networks, and flooding efficiencies.

9.17. The Progress of the Water-oil Interface—Electrolytic-model Experiments.—Insofar as the viscosity difference between the flooding water and the displaced oil will be neglected, the method developed in Sec. 8.5 may be applied to trace the progress of the water as it leaves the injection wells and proceeds toward the output wells while driving the oil before it. However, this necessitates not only the explicit derivation of the equations for the equipotentials (Φ = const.) and streamlines (Ψ = const.) of the system, but the more difficult expression of $|\nabla\Phi|^2$ in terms of Φ and Ψ. And even after this has been accomplished the numerical work even in such simple cases as the line drive or elliptical drive[1] is very laborious.

Fortunately, however, this numerical work may be avoided by the use of very simple electrolytic models[2] of the various flooding

---

[1] This case is discussed by M. Muskat, *Physics*, **5**, 250, 1934.

[2] WYCKOFF, R. D., H. G. BOTSET, and M. MUSKAT, *Trans. A.I.M.E. Pet. Dev. Tech.*, **103**, 219, 1933; *cf.* R. D. Wyckoff and H. G. Botset, *Physics*, 265, 1934, for details of the technique, and also Sec. 4.17.

networks. These models are based on the observation that since the velocity of an ion in an electrolytic system is proportional to the potential gradient, just as the velocity of a fluid particle in a porous medium is proportional to the pressure gradient, the paths of the ions in the electrolytic systems must be similar to those of the fluid particles in a porous medium with the same geometry and equivalent boundary conditions.

This electrolytic model consists essentially of an electrolyte containing an ion indicator, as phenolphthalein, held in a suitable porous medium to prevent excessive velocities of normal diffusion. Ordinary blotting paper saturated with the electrolyte serves quite well. The negative electrode represents the fluid source in the corresponding flow system, the positive terminal represents the output well, and the advance of the OH ions from the negative to the positive electrode corresponds to the advancing water front in the flooding problem. The advance of the OH ions is indicated by the phenolphthalein, which is colorless in acid media but turns red in the presence of OH ions. The progress of this equivalent water-oil interface may be rather strikingly shown by photographs of the model at various time intervals after the experiments are begun.

Although the elementary considerations outlined above do lead to the conclusion that there is a real analogy between an electrolytic system and a corresponding porous medium, a closer examination suggests that there might be complications in the electrolytic models with respect to the establishment of an exact analogy. However, as the shape of the water-oil interface has been derived analytically in Chap. VIII for several simple cases, the validity of the electrolytic models may be tested by applying them to these cases.

As the first test the problem of the line drive into a single well, treated in Sec. 8.6, may be chosen. This corresponds to an idealized edge-water flood advancing into an oil field containing one well located near the edge water and produced by the water drive alone. The analytical results are given in Fig. 174. The outline of the flood for this case as traced from an electrolytic model at the instant the OH ion front first reached the positive electrode or output well is shown in Fig. 217. In this model the negative electrode or source was a long conductor contacting the electrolyte over a length great enough to be effectively infinite. The

x's indicate the calculated position of the front and show excellent agreement with the performance of the model.

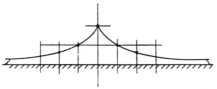

FIG. 217.—The front of a line-drive flood into a single well as it reaches the well. Solid curve is the front obtained with the electrolytic model. Crosses are points on the theoretical front derived in Sec. 8.6. (*From A.I.M.E.* **103**, 223, 1933.)

Another case which has been solved analytically in Chap. VIII is that of a single input well and a single output well located in an infinite plane. The pressure distribution, streamlines, and progress of the flood as determined analytically are shown in Fig. 176, and the outline of the flood when the first particles reach the output well is shown by the curve $t = \frac{2}{3}$. Figure 218 is a composite photograph of the model showing progressive stages of advance and the final stage when the flood first reached the output well. The x's again indicate the calculated position of the model. The fact that the flood is somewhat narrower in the model than in the calculated flood (Fig. 176) is easily explained by the fact that in the model the lateral boundaries could not be placed far enough away from the center line of the wells to simulate accurately an infinite plane in this direction.

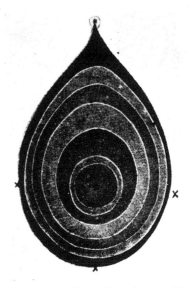

FIG. 218.—Composite photographic history of a direct-drive flood between two wells, obtained with electrolytic models. Crosses are bounding points predicted analytically (Sec. 8.7). (*From A.I.M.E.* **103**, 225, 1933.)

Having thus demonstrated that the model is competent to give not only a qualitative picture of the advancing flood but to show also quantitatively the area flooded out at any stage of the encroachment with errors of the order of only a few per cent, it is permissible to proceed with models of more complex

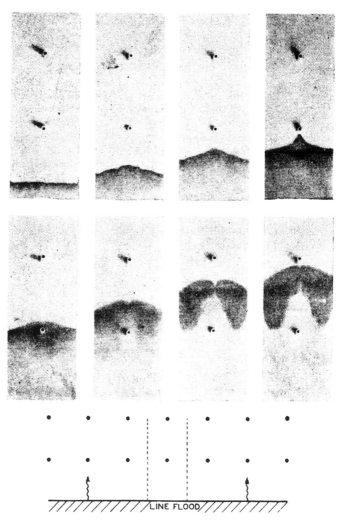

Fig. 219.—Photographic history of a line-drive flood with a double-line array in direct offsetting position, obtained with electrolytic models. (*From A.I.M.E.* **103,** 225, 1933.)

systems for which analytical results would be very difficult to obtain.

9.18. Model Experiments with Line Floods.[1]—Since many practical problems involve the encroachment of edge water which, within a reasonable distance from the edge wells, may be considered as a line flood, the application of the model to these problems will be illustrated first. As a case of practical interest we shall consider a line of edge wells running parallel to the edge water together with a second line of direct interior offsets. The assumed arrangement is shown at the base of Fig. 219, the dotted segment representing an elementary section of the front taken from an infinite line of similar sections. The model shows that the advancing front remains essentially parallel to the original line without serious distortion or "fingering" until the flood has approached the first line of wells within a distance of the order of one half the well spacing. Thereafter, the front cusps rather sharply into the first line of wells, the outline of the front just before the flood hits the wells being shown by the fourth photograph of Fig. 219. After this point was reached, the edge wells were produced at the normal rate, which in practice would, of course, involve a production of both oil and water. The flood then continued to advance as shown, until the second line of wells was reached. At this point the operation of the model was stopped. It is of interest to note the presence of "neutral" zones connecting the front-line wells with their interior offsets as indicated by the reversed cusps behind the first-line wells.

If the second line of direct offsets is replaced by a similar line of staggered offsets, a flood results, as is shown in Fig. 220. While the resulting encroachment differs from the former case in details, it is to be noted that no very pronounced change in the magnitude of the "fingering" has resulted from the use of staggered offsets.

Still another example of interest is the case of a flood acting simultaneously from opposite sides of a triple line of wells, such as represented in Fig. 221. The photographs show the course of the flood in the unit area outlined in the diagram. Being symmetrical about this unit, the appearance of the entire array at various stages is easily obtained by repetition of the single unit. One may note here the pronounced cusping into the middle-line wells, while at the same time a water-free area cusps

[1] The model photographs given here and in the following sections are reproduced from the paper by Wyckoff, Botset, and Muskat (*loc. cit.*).

sharply into the first-line wells on the side of the wells opposite
to the flood. This may be compared with **Fig. 220** where, owing

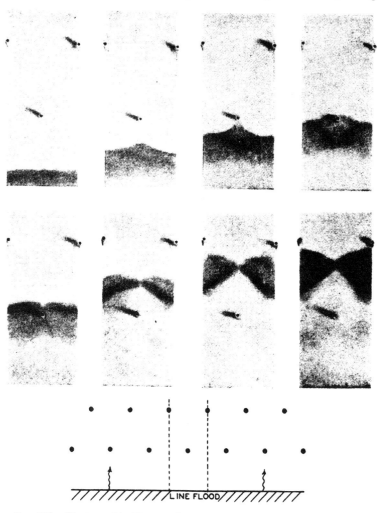

Fig. 220.—Photographic history of a line-drive flood into a line array offset
by a staggered line, as obtained with electrolytic models. (*From A.I.M.E.*
103, 226, 1933.)

to the absence of the opposite line flood, the reversed cusps about
the first line of wells are much less pronounced and almost closed
in.

For simplicity it has been assumed in these illustrations that the wells were equally spaced and were produced at the same rate. If, however, in any particular area adjacent to the water

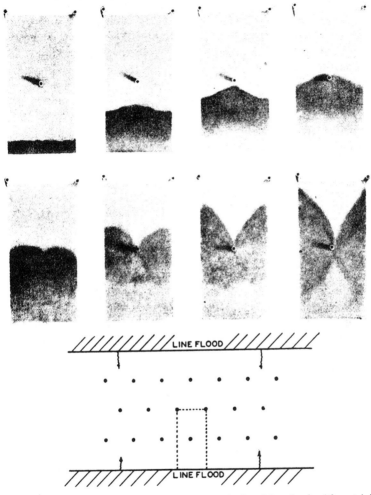

Fig. 221.—Photographic history of a double line-drive flood with a triple line of wells, as obtained with electrolytic models. (*From A.I.M.E.* **103**, 227, 1933.)

front, the wells are produced at a higher rate than elsewhere, or the equivalent of this, if the well spacing is closer, the water will necessarily advance into this area at a more rapid rate than elsewhere along the front. This situation will simply give large-scale effects of the type illustrated in Fig. 217, where the single sink or well is replaced by the area in which the production rates are high. Within this area the detailed performance will be as shown in the typical idealized line floods, but in addition there will be a "regional" cusp advancing into the area.

9.19. Experiments Corresponding to Artificial Flooding Operations.—As already pointed out, the purpose of artificial flooding operations is to obtain a final flushing out of the "depleted" oil sands by means of injected water, natural water drives being either absent or so slow as to be commercially ineffective. The aim is to cause the water to flush out the maximum possible volume of the oil horizon. Various systems have been suggested and practiced, and it is of interest to compare the performance of several such systems from the point of view of the volume swept out by the water before it first appears in the oil wells.

a. Ratio of Input to Output Wells 1 to 1.—It is clear that a continuous line drive would be the most effective from the point of view of the volume flooded out. However, since the water must be injected artificially, it is impossible to approach more closely to the line drive than with an arrangement of input wells located along a line as at a, Fig. 222. Since the drive will extend radially around each input well, the logical location of the oil wells is along the lines b and c so as to take advantage of the drive in both directions. Furthermore, in order to cover a reasonable area, the array would be extended into a network of alternate input and output lines, as shown below.

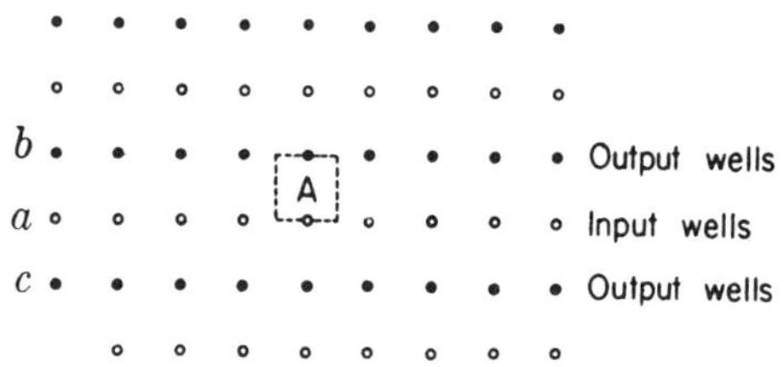

Fig. 222.—Diagrammatic representation of a direct line-drive flood.

From symmetry considerations it is clear that the system of Fig. 222 may be reproduced by the simple repetition of the section A. Using, therefore, the model for a single such unit, the shape of the advancing front at successive stages of the flood is shown in Fig. 223. The area flooded out at the time the water first reaches the output well is 57 ± 3 per cent.

An interesting variation of the simple array of Figs. 222 and 223 consists of a rearrangement of the wells so as to increase the

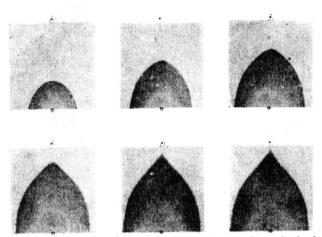

FIG. 223.—Photographic history of a direct line-drive flood, as obtained with electrolytic models. (*From A.I.M.E.* **103**, 229, 1933.)

distance between the input and output wells while leaving the distance between pairs of either type the same. Figure 224 shows such a system—the so-called five-spot array. Noting that this five-spot array not only is symmetrical about a single five-well unit A (Fig. 224) but also has perfect symmetry about the two-well section B, indicated by the dotted boundaries, such two-well units may be used in the models.

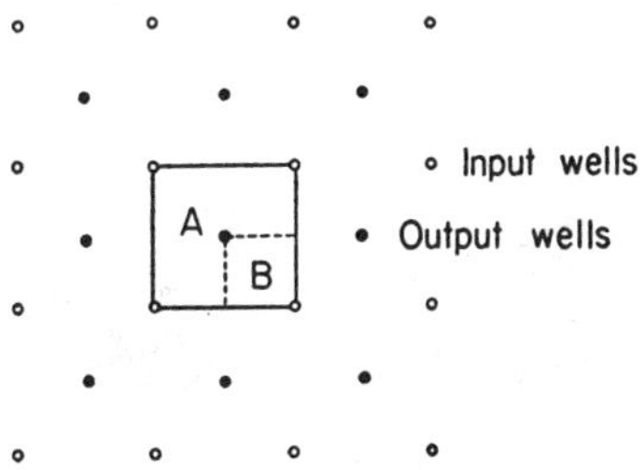

Figure 225 shows a series of photographs taken at various stages, the last picture showing the shape of the interface at the

FIG. 224.—Diagrammatic representation of a five-spot flood array.

○ Input wells
● Output wells

instant it first reaches the output well. Figure 226 is a composite picture of the complete five-spot array at various stages during the flooding process. During the early stages of the flood the advance is closely circular in character, and the first significant departure from a circular flood occurs when the front has advanced about halfway along the diagonal connecting the input and output wells.   From there on the front enters the converging

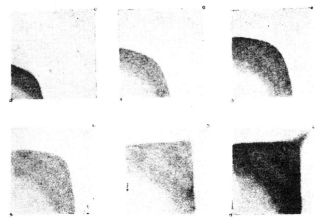

Fig. 225.—Photographic history of the five-spot flood, as obtained with electro-lytic models.   (*From A.I.M.E.* **103**, 231, 1933.)

radial field of the output well and rapid acceleration occurs, drawing out the sharp cusp into the well.   The total area flooded

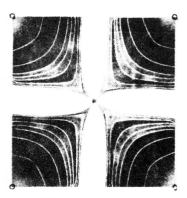

Fig. 226.—Composite photographic history of a complete five-spot flood element, as obtained with electrolytic models.   (*From A.I.M.E.* **103**, 231 1933.)

out at the instant the interface first reaches the output well is about 75 ± 3 per cent, as meas-ured on several models.

b. Ratio of Input to Output Wells 1 to 2.—The systems dis-cussed above are typical flood-ing arrays in which the numbers of input and output wells are the same.   The effect of chang-ing the ratio so as to have twice as many oil wells as there are water-injection wells will now be considered.

The system shown in Fig. 227 is built up by an extension of the simple triangular array, shown in the upper left corner, so as to form a hexagonal array. Since this array is symmetrical about both the single unit A and sector B, the two-well sector B may be used in setting up the model.

The nature of the flood obtained with this system at various stages of advance is shown in Fig. 228, where the central wells were used as input wells. The area flooded out at the time of the first appearance of water in the output well was about 80 per cent. A composite picture of this flood is shown in Fig. 229.

c. Ratio of Input to Output Wells 2 to 1.—An examination of Fig. 227 shows that to change the ratio of the number of input to output wells from ½ to 2, it is only necessary to reverse the direction of the

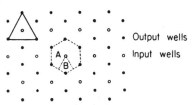

FIG. 227.—Diagrammatic representation of a hexagonal flooding array.

flood by making the well in the center of the hexagon the output well. This is the usual form of the "sevens-pot array." Again using the two-well sector B (Fig. 227) with the interchange

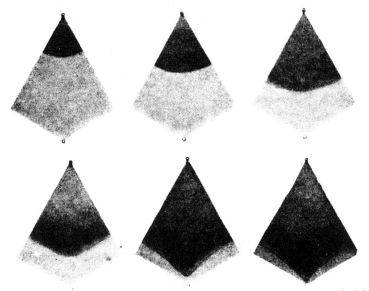

FIG. 228.—Photographic history of inverted seven-spot flood, as obtained with electrolytic models. (*From A.I.M.E.* **103**, 223, 1933.)

of the input and output wells, the resultant flood is shown in Fig. 230, with a composite picture of an entire seven-spot element in Fig. 231.

The area flooded out at the first appearance of water is 77 to 80 per cent, as measured on several models. This is to be compared with the flood of 80 per cent obtained in the "inverted" seven-spot array (Fig. 229). Since the form of the pressure distribution and the shape of the streamlines in the seven-spot array are evidently unchanged by reversing the direction of flow, it would be expected that the overall resistance of the system and the total area flooded out would also be unchanged, a result clearly verified by the models.

9.20. Effect of Barriers in the Flooding System.—The effect of impermeable barriers on the shape of the flood also offers an interesting problem in connection with encroachment or flooding phenomena. Such problems are difficult to treat analytically but may be readily solved by the use of models. Figure 232 represents the flooding history in a one-quarter section of a five-spot array with an impermeable barrier placed

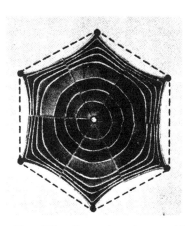

FIG. 229.—Composite photographic history of a complete inverted seven-spot flood element, as obtained with electrolytic models.

along the diagonal between the input and output wells. A comparison of the progress of the flood in Fig. 232 with that of Fig. 225 shows clearly the modifying influence of the barrier. Although this example is distinctly artificial, it is useful in indicating the general effect of barriers in any system. It is interesting to note the completeness of the flood on the input side of the barrier, and also the "neutral axis" extending from the output well to the barrier. The barrier has the effect of creating two "sources" near its end, the neutral axis being typical of two adjacent sources.

The effect of a change in the shape of the obstruction on the flood pattern may be illustrated by the model shown in Fig. 233. In this figure the rectangular obstruction has been replaced by an arc of a circle, which, in principle, covers the case of barriers having re-entrant areas or angles. By comparison with Fig. 232 it is seen that an obstruction containing

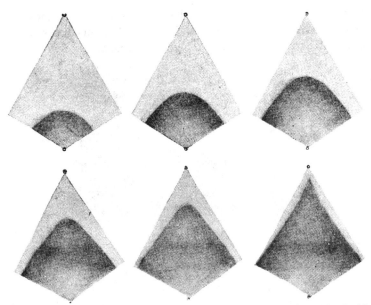

Fig. 230.—Photographic history of a normal seven-spot flood, as obtained with electrolytic models. (*From A.I.M.E.* **103**, 233, 1933.)

re-entrant angles results in very slow flooding of the re-entrant area, although it should be observed that if the flood persists long enough the "dead" areas must disappear. This may be verified by analyzing the potential distributions in such systems in the manner outlined in the next section.

9.21. Conduction-sheet Models and Potential Distributions.—Although the electrolytic models show very graphically the shape and progress of the water flood advancing from the injection water wells toward the output oil wells, the lack of strict homogeneity in the condition of the electrolytic equivalent of the sand and the difficulty of measuring the specific conductivity of the

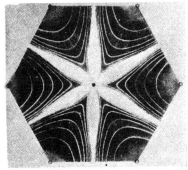

Fig. 231.—Composite photographic history of a complete normal seven-spot flood element, as obtained with electrolytic models. (*From A.I.M.E.* **103**, 234, 1933.)

electrolyte in place make it rather unsatisfactory for measurements on the total resistances of the flooding networks or the potential distributions within them.   Fortunately, however, these

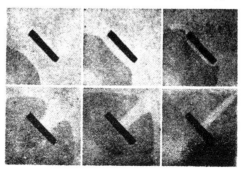

Fig. 232.—Photographic history of the flow past a rectilinear obstruction ying in a five-spot array element, as obtained with electrolytic models.   (*From A.I.M.E.* **103**, 235, 1933.)

difficulties can all be avoided by means of another very simple experimental model—the sheet-conduction model.

In Chap. III it was noted that the viscous flow of a liquid through a porous medium is exactly analogous to the current

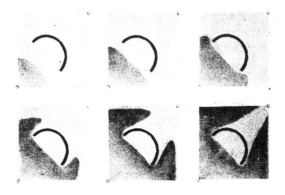

Fig. 233.—Photographic history of the flow past a curved obstruction lying in a five-spot array element, as obtained with electrolytic models.   (*After Wyckoff and Botset, Physics.*)

flow in a metallic conducting system of the same geometry.[1] For just as the fluid velocity in a porous medium is proportional to the pressure gradient, so the current density in a metallic

[1] *Cf.* also Sec. 4.17.

conducting system is proportional to the voltage or potential gradient. Hence for equivalent geometries it follows that not only will the resistances be the same (in the proper units) but that their potential distributions will also be identical.

The experimental arrangement for a two-dimensional conduction model is indicated in Fig. 234. M represents the model and is simply a thin sheet of metal of uniform thickness cut in the shape of the network element or a segment of it for the particular flood or fluid system of interest. The small electrodes W_1, W_2 correspond to the input and output wells. The potential divider r_1, r_2, connected through the galvanometer G to the

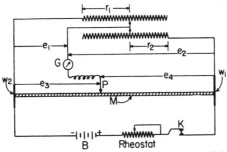

FIG. 234.—The electrical circuit for determining the potential distributions in two-dimensional flow systems by means of sheet-conduction models.

exploring electrode P, permits a null-method balance of the potential drops e_3 against e_1, and e_4 against e_2, the momentary current being supplied by the battery on tapping the key K. As under the conditions of zero galvanometer deflection

$$\frac{e_3}{e_3 + e_4} = \frac{e_1}{e_1 + e_2} = \frac{r_1}{r_1 + r_2}, \tag{1}$$

the dial reading r_1, for fixed $r_1 + r_2$, is proportional to the potential drop between the electrode W_2 and the point at P. If, in particular, $r_1 + r_2 = 100$ ohms, then the value of r_1 gives the percentage of the total drop across the system, between W_1 and W_2, occurring between P and W_2. By keeping r_1 fixed and moving P so as to maintain the null deflection at G, on closing K, an equipotential curve will be traced, and by changing r_1 and repeating the process, the whole system of equipotentials, *i.e.*, the potential distribution, in the system may be mapped.

As to the empirical precautions to be observed in the determination of the potential distributions, it should be noted that one must adjust the several electrodes in the system so as to have equivalent radii. This may, in symmetrical systems, be tested by means of points of symmetry in the potential distribution, whose positions are quite sensitive to the radii of the electrodes and will appear displaced from their proper positions if the electrodes are not of equal radii. In addition, temperature effects at the electrodes due to the high current densities there must be avoided. These are largely eliminated by keeping the key K open and hence the current zero, except during the instantaneous test of the galvanometer readings. The application of air blasts at the electrodes suffices to remove the residual heating effects.

As a final point with regard to the experimental procedure, it may be mentioned that one may determine electrically the 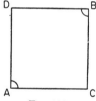 streamlines in the corresponding flow problem, as well as the equipressure curves, without essentially changing the model. For as was seen in Chap. IV, there is a mutual reciprocity between the equipotentials and streamlines in every two-dimensional potential problem. That is, if one considers the values of the potentials specified on the boundaries of a given system to be changed to equal values of the flux densities, then the equipotentials of the original system become the streamlines in the new problem and the original streamlines become the new equipotentials. Thus in the case of a network element corresponding to the five-spot flood, where the wells are placed at the corners A, B (cf. Fig. 235), it is clear that small circular arcs enclosing the wells are equipotential curves and that the boundaries ACB and ADB are limiting streamlines of the system. Hence, by the reciprocity relation referred to, if the corners at A and B are cut out along the circular arcs, and highly conducting metal strips[1] are attached along the former streamlines ACB and ADB the arcs about A and B will be limiting streamlines, while the boundary equipotentials will be the former streamlines ACB and ADB. By determining, therefore, the equipotentials of

FIG. 235.

[1] It is to be noted that the conductivity of these strips must be great enough to assure that the potential drop in them is negligible, when current is passed from the high- to the low-potential boundaries.

the new system, the streamlines will be obtained for the original
problem.

In Fig. 236 is given the pressure distribution for a quadrant of
a five-spot flood element, the flood history of which is shown in
Fig. 225. The input well at the upper right corner may be supposed to be operating at a pressure of 100 units above that of the
output well, in the lower left corner, so that the values given for
the equipressure lines represent the percentages of the total
pressure drop across the input-output system, regardless of the

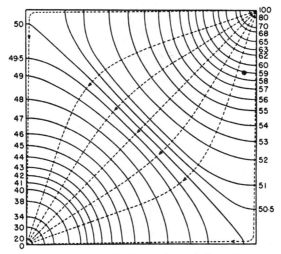

Fig. 236.—The pressure distribution and streamlines in a quadrant of a
five-spot flood network element, as obtained by sheet-conduction models. (*From
A.I.M.E.* **103**, 230, 1933.)

absolute values of the pressures. In the marked persistence of
the radial character of the pressure distributions about the two
wells is to be seen the natural explanation of the persistance of
the radial character of the flood advance in Fig. 225. Furthermore, the spreading of the equipressure lines, and hence low
gradients, near the well-free corners of the quadrant shows why
the advance along the edges of the model becomes retarded while
the diagonal region continues to form a squarelike pattern, as
in the fourth and fifth photographs of Fig 225, and finally
fingers out into a sharp cusp as the front re-enters the high
gradients about the output well.

A similar analysis and comparison with the corresponding flood advance of Fig. 230 may be made of the pressure distribution in the segment of the seven-spot flood shown in Fig. 237. Perhaps the most striking feature of this figure is the high-pressure-gradient concentration about the central well, chosen here as the output well. For, while the 50 per cent equipressure curve is midway between the input and output wells in the five-spot quadrant, it is here at a distance of only one-ninth of the total input-output distance from the central well. The reason for

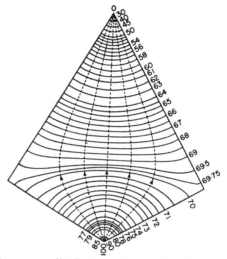

FIG. 237.—The pressure distribution and streamlines in a segment of a seven-spot flood network element, as obtained by sheet-conduction models. (*From A.I.M.E.* **107**, 67, 1934.)

this is, of course, that the flux densities at the central wells are twice those at the peripheral wells, owing to the occurrence of twice as many of the latter as of the former in the seven-spot network. The difference between the gradients at the central and peripheral wells also explains the very sharp cusps in Fig. 230, where the output well is central, and the retarded and broader cusps in Fig. 228, where the flood is driven toward the peripheral well.

Another symmetrical distribution is that of the direct line-drive flood shown in Fig. 238. The very low gradients along the corners and consequent slow advance of the flood toward the lateral boundaries show clearly why the flood concentrates itself along the line of centers, as is seen in Fig. 223, leaving almost half (43·per cent) of the network unflooded when the output well is first reached by the flood along the line of centers.

In contrast to Fig. 238 stands the pressure distribution in the staggered line-drive network shown in Fig. 239. Although the distribution about the lower well is very similar to that in the case of the direct line-drive flood, the staggering of the upper line naturally induces a preliminary broadening of the front of

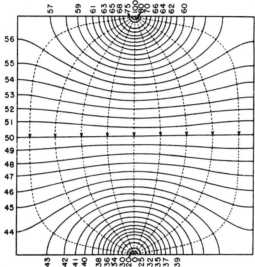

Fig. 238.—The pressure distribution and streamlines in a direct line-drive flood-network element, as obtained by conduction-sheet models. (*From A.I.M.E.* **107**, 65, 1934.)

the flood as it leaves the injection well and is then drawn out and split into the two cusps each entering symmetrically into the output wells (*cf.* Fig. 221).

In a similar manner one may readily predict at least the qualitative features of any flooding system, once the pressure distribution has been mapped and plotted.

Finally it should be observed that the electric-conduction-sheet model may be very conveniently used to determine the resistance of a network as well as the details of the potential distribution. For this purpose it is only necessary to first measure the specific resistance of the material of the model, and then the absolute value of the resistance of the model by a bridge or potentiometer null method. The absolute resistance divided by the specific resistance of the metal of the model will give the equivalent of the resistance of the flooding network for a unit permeability medium and unit viscosity fluid.

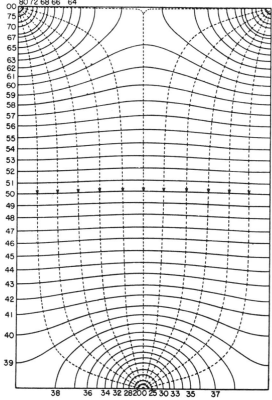

Fig. 239.—Pressure distributions and streamlines in a staggered line-drive flood-network element, as obtained by conduction-sheet models. (*From A.I.M.E.* **107**, 65, 1934.)

9.22. Analytical Calculations of the Conductivities of Flooding Networks. General Method.

—Although the sheet-conduction models discussed in the last section permit accurate measurements of the resistances of the flooding networks, they suffer the disadvantage of giving results corresponding only to the particular dimensions of the model used. To study the variation of the conductivities of the flooding networks with the well spacing in the networks it would, therefore, be necessary to construct a new model to correspond to each set of dimensions of interest. In an analytical treatment, however, the dimensions of the system may be kept throughout as parameters which need be specified only in the final numerical interpretations.

As in the case of the model experiments it will be supposed that the flooding systems consist of infinite networks of injection and output wells. As was seen in Sec. 9.8, each infinite line of wells of like sign parallel to the x axis may be represented by a contribution to the pressure distribution of the complete network of the form

$$p = q \log \left[\cosh \frac{2\pi}{a}(y - b) - \cos \frac{2\pi}{a}(x - c) \right], \quad [cf.\ \text{Eq. 9.8(2)}]$$

(1)

where a is the spacing of the wells, b is the distance of the line from the x axis, c is the distance from the y axis of the well on its right and closest to it, and q is a coefficient determining the total flux Q entering the sand from each well, per unit sand thickness, according to the relation

$$Q = \frac{4\pi k q}{\mu}. \quad [cf.\ \text{Eq. 9.8(5)}]$$

(2)

The pressure distribution corresponding to the complete network is then simply obtained by summing the terms due to the individual lines, each with its appropriate q, b, and c. However, as in the infinite network there will be an infinite number of lines of wells, both of input and output wells, the summation of the terms due to the separate lines must be made with care to avoid divergence of the series. A physical interpretation of the correct sum is that the infinite set of lines of input wells and output wells is to be added in the form of pairs of lines symmetrical about the x axis, with a constant added to the pressure term for each pair of such a magnitude that the resultant sum will be convergent.

When the expression for the pressure distribution in the network has been found, the difference in pressure between representative input and output wells will give the net pressure differential acting in the system. Dividing this into Q, the result will be the "conductivity" of the network, or production rate per network element per unit pressure differential, for a unit sand thickness, and for $k/\mu = 1$. This method will now be illustrated by treating successively the various flooding networks of practical interest.[1]

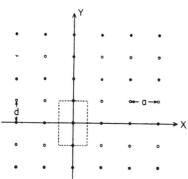

FIG. 240.—The direct line-drive flood network.

9.23. The Conductivity of the Direct Line-drive Flood.—The geometrical arrangement of the direct line-drive flood is shown in Fig. 240, where the circles indicate injection wells and dots output wells. Owing to the symmetry of the system, the whole network may be defined by the element shown as the dotted rectangle. For the analytical representation, on the other hand, it need only be noted that the terms as in Eq. 9.22(1) are to be chosen with $+q$: $b = 2nd$, $a = 0$, and $-q$: $b = (2n + 1)d$, $c = 0$.

Adding these terms and adding the constants, as outlined in the last section, the resultant pressure distribution takes the form

$$p(x, y) = q \log \left(\cosh \frac{2\pi y}{a} - \cos \frac{2\pi x}{a} \right) + q \sum_{1}^{\infty} (-1)^m \log$$

$$\frac{4 \left[\cosh \frac{2\pi(y - md)}{a} - \cos \frac{2\pi x}{a} \right]\left[\cosh \frac{2\pi(y + md)}{a} - \cos \frac{2\pi x}{a} \right]}{e^{4\pi md/a}}. \quad (1)$$

To find now the difference of pressure, implied by Eq. (1), existing between the wells in the input rows and those in the output rows, it is evidently sufficient to choose the output well at the origin and the input well immediately above it, on the y axis. Setting then, first, $x = 0$, p takes the value

$$p(x = 0) = q \log 2 \sinh^2 \frac{\pi y}{a}$$

$$+ q \sum_{1}^{\infty} (-1)^m \log \frac{16 \sinh^2 \frac{\pi(y - md)}{a} \sinh^2 \frac{\pi(y + md)}{a}}{e^{4\pi md/a}}. \quad (2)$$

---

[1] These results were given by M. Muskat and R. D. Wyckoff (*loc. cit.*)

At the output well one may set $y = r_w$, where r_w is the well radius, and is to be considered as small compared to d or a. Hence

$$p(0, r_w) = q \log 2 \sinh^2 \frac{\pi r_w}{a} + q \sum_1^\infty (-1)^m \log \frac{16 \sinh^4 \pi md/a}{e^{4\pi md/a}}.$$

(3)

At the input well, $y = d \pm r_w$, and

$$p(0, d \pm r_w) = q \log 2 \sinh^2 \pi d/a -$$
$$q \log \frac{16 \sinh^2 \pi r_w/a \sinh^2 2\pi d/a}{e^{4\pi d/a}}$$

$$+ q \sum_2^\infty (-1)^m \log \frac{16 \sinh^2 \pi d(m-1)/a \sinh^2 \pi d(m+1)/a}{e^{4\pi md/a}}. \quad (4)$$

The net pressure differential is, therefore,

$$\Delta p = 2q \log \frac{\sinh^3 \pi d/a}{\sinh^2 \pi r_w/a \sinh 2\pi d/a}$$
$$+ 2q \sum_2^\infty (-1)^m \log \frac{\sinh \pi d(m-1)/a \sinh \pi d(m+1)/a}{\sinh^2 \pi md/a}. \quad (5)$$

As in practical cases d/a will not be less than $\frac{1}{4}$, all the terms in the series beyond the first may be dropped with errors of the order of only 0.1 per cent. Δp can then be rewritten as

$$\Delta p = 2q \log \frac{\sinh^4 \pi d/a \sinh 3\pi d/a}{\sinh^2 \pi r_w/a \sinh^3 2\pi d/a}. \quad (6)$$

As already mentioned, the flux, per well, is given by[1]

$$Q = \frac{4\pi kq}{\mu}.$$

[1] Because of the assumption that k/μ has the same value both in the flooded zone and in that not reached by the water, the absolute values of the flood conductivities derived here and in the following sections may differ quite appreciably from those to be observed in actual flooding networks. However, the relative values obtained by comparing different types of networks, or different spacings for the same network, should be well approximated by the theoretical predictions. Moreover, it is probable that with a little experience in the field a suitable averaging process for k/μ might be developed with which even the absolute conductivities could be satisfactorily predicted.

The reciprocal of the resistance of the system, or its effective conductivity, is, therefore, given by

$$\frac{\mu Q}{k \Delta p} = \frac{2\pi}{\log \dfrac{\sinh^4 \pi d/a \, \sinh 3\pi d/a}{\sinh^2 \pi r_w/a \, \sinh^3 2\pi d/a}}. \tag{7}$$

For $d/a \gtrless 1$, the simplified formula

$$\frac{\mu Q}{k \Delta p} = \frac{2\pi}{\dfrac{\pi d}{a} - 2 \log 2 \sinh \dfrac{\pi r_w}{a}} \tag{8}$$

may be used to a high degree of accuracy. It is interesting to note that Eq. (8) corresponds to the value of $\mu Q/k\Delta p$ for a line

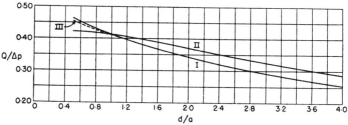

FIG. 241.—The variation of the conductivity of the line-drive flood with d/a, (the distance between the input and output lines)/(well spacing within the lines). $Q/\Delta p$ = production rate per well per unit-pressure differential in a unit permeability sand of unit thickness. I. a fixed at 660 ft., direct drive. II. d fixed at 660 ft., direct drive. III. a fixed at 660 ft., staggered drive. (*From A.I.M.E.* **107**, 68, 1934.)

drive between the *continuous* line sources and sinks with an effective separation of $d + \dfrac{2a}{\pi} \log \dfrac{a}{2\pi r_w}$. The second term evidently represents the increase in the effective distance between the input and output lines due to the fact that the flow must leave and enter the system from separated and individual well centers rather than be distributed uniformly along the input and output lines.

Equations (7) and (8) are plotted in Fig. 241 as a function of d/a; in curve I, $a/r_w = 2,640$, which corresponds to well radii of $\frac{1}{4}$ ft. and a spacing of 660 ft. within the input and output lines, and in curve II, $d/r_w = 2,640$, which corresponds to a 660 ft. spacing between the output and input lines. It will be noted that $\mu Q/k\Delta p$, or the effective conductivity of the drive, decreases rather slowly with increasing separation between the lines, owing, of course, to the fact that the main contribution to the total resistance of the system comes from the radial flow about the individual wells in the input and output arrays, so that increasing the separation of the lines affects only the relatively small contribution of the linear part of the flow between the lines. Thus even at $d/a = 4$, the effective conductivity is still only about half of what it would be if the wells were continuously spread over the input and output lines ($\mu Q/k\Delta p = 2a/d$).

For the case where the distance between the input wells is kept equal to that between the input and output wells ($d = a$)

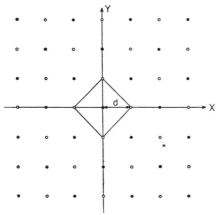

Fig. 242.—The five-spot flood network.

the effective conductivity of the direct-line flood is plotted against d in curve I of Fig. 243.

9.24. The Conductivity of the Five-spot Flood.—For the purpose of the analysis it is convenient to choose the axes in the five-spot network so as to go through the diagonals of the usual five-spot element, the network element being a square as indicated in Fig. 242, with the central well taken as an output well (with a positive flux coefficient). The five-spot network may then be represented by the following set of well groups:

$$+q: (2nd, 2md); \qquad [(2n + 1)d, (2m + 1)d],$$
$$-q: [2nd, (2m + 1)d]; \qquad [(2n + 1)d, 2md],$$

where the n places the well in its line parallel to the x axis and the m gives the distance of the line *from* the x axis. Building up the network by taking together pairs of lines symmetrically placed about the x axis, as in the case of the direct line flood, and then grouping together the input and output lines with wells just above each other, the resultant pressure distribution may be written as

$$p = q \log \frac{\cosh \pi y/d - \cos \pi x/d}{\cosh \pi y/d + \cos \pi x/d}$$

$$- q \sum_{1}^{\infty} (-1)^m \log$$

$$\frac{4\left[\cosh \frac{\pi(y - md)}{d} + \cos \frac{\pi x}{d}\right]\left[\cosh \frac{\pi(y + md)}{d} + \cos \frac{\pi x}{d}\right]}{e^{2m\pi}}$$

$$+ q \sum_{1}^{\infty} (-1)^m \log$$

$$\frac{4\left[\cosh \frac{\pi(y - md)}{d} - \cos \frac{\pi x}{d}\right]\left[\cosh \frac{\pi(y + md)}{d} - \cos \frac{\pi x}{d}\right]}{e^{2m\pi}}. \quad (1)$$

As before, Eq. (1) implies a pressure differential existing between the input and output wells which is given by
$$\Delta p = p(0, d \pm r_w) - p(0, r_w)$$

$$= 2q \log \frac{\operatorname{ctnh} \pi r_w/2d}{\operatorname{ctnh} \pi/2} + 2q \log \frac{\sinh^2 \pi/2}{\sinh \pi \sinh \pi r_w/2d}$$

$$+ 2q \sum_{2}^{\infty} (-1)^m \log \frac{\sinh \pi(m - 1)/2 \sinh \pi(m + 1)/2}{\sinh^2 \pi m/2}$$

$$+ 2q \sum_{1}^{\infty} (-1)^m \log \frac{\cosh^2 \pi m/2}{\cosh \pi(m - 1)/2 \cosh \pi(m + 1)/2}$$

$$= 2q \log \frac{\tanh^4 \pi/2 \tanh 3\pi/2 \operatorname{ctnh}^3 \pi}{\sinh^2 \pi r_w/2d}$$

$$+ 2q \sum_{3}^{\infty} (-1)^m \log \operatorname{ctnh}^2 \frac{\pi m}{2} \tanh \frac{\pi(m - 1)}{2} \tanh \frac{\pi(m + 1)}{2},$$

$$(2)$$

where cosh $\pi r_w/2d$ has been set equal to unity. As the series contributes less than 0.1 per cent of the total Δp, it may be dropped, leaving

$$\Delta p = -4q\left(\log \sinh \frac{\pi r_w}{2d} + 0.1674\right). \tag{3}$$

Hence, replacing the sinh by its argument

$$\frac{\mu Q}{k\Delta p} = \frac{\pi}{\log \dfrac{d}{r_w} - 0.6190}. \tag{4}$$

Taking again $r_w = \frac{1}{4}$ ft., Eq. (4) is plotted with $\mu Q/k\Delta p$ *versus* d in curve II of Fig. 243. It will be seen that this curve

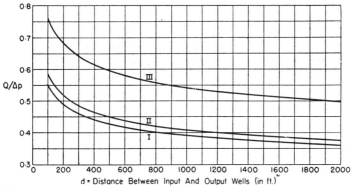

FIG. 243.—The variation of the conductivities of flooding networks with the spacing d between the input and output wells. $Q/\Delta p$ = production rate per network element per unit pressure differential in a unit-permeability sand of unit thickness. I. Regular line-drive flood ($d = a$). II. Five-spot flood. III. Seven-spot flood. (*From A.I.M.E.* **107**, 68, 1934.)

runs almost exactly parallel to but higher (4 to 6 per cent) than the conductivity curve for the direct line-drive flood, the general variation of $\mu Q/k\Delta p$ with d being very much like that of the flux in a strictly radial-flow system with the variation of the well or the external radius. Thus in doubling the spacing from $d = 1,000$ ft. to $d = 2,000$ ft., the conductivity, in either case, is decreased by only about 9 per cent, whereas in the corresponding case of exact radial flow into a single well the decrease would be 8.3 per cent.

9.25. The Conductivity of the Seven-spot Flood.—Following the procedure used in the case of the five-spot flood, the seven-

spot array may be analyzed into the following groups of wells (*cf.*
Fig. 244):

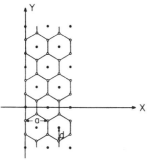

$+2q$: $(na, 3md)$;
$[(n + \frac{1}{2})a, (3m + \frac{3}{2})d]$;
$-q$: $[na, (3m + 1)d]$;
$[(n + \frac{1}{2})a, (3m + \frac{1}{2})d]$;
$-q$: $[na, (3m + 2)d]$;
$[(n + \frac{1}{2})a, (3m + \frac{5}{2})d]$

where n and m have the same significance as in the case of the five-spot flood. The factor 2 for the first group of wells enters because there are only half as many center (output) wells in

Fig. 244.—The seven-spot flood network.

the seven-spot-flood network as peripheral (input) wells. From the geometry of the hexagon it will be clear that $a = \sqrt{3}d$.

Summing the contributions of the various groups to the pressure distribution, it is found that

$$
p = 2q \log \left(\cosh \frac{2\pi y}{a} - \cos \frac{2\pi x}{a} \right)
$$

$$
+ 2q \sum_{1}^{\infty} \log
$$

$$
\frac{4\left[\cosh \dfrac{2\pi(y - 3md)}{a} - \cos \dfrac{2\pi x}{a} \right]\left[\cosh \dfrac{2\pi(y + 3md)}{a} - \cos \dfrac{2\pi x}{a} \right]}{e^{12\pi\,md/a}}
$$

$$
+ 2q \sum_{0}^{\infty} \log
$$

$$
\frac{4[\cosh 2\pi(y - 3d/2 - 3md)/a + \cos 2\pi x/a][\cosh 2\pi(y + 3d/2 + 3md)/a + \cos 2\pi x/a]}{e^{4\pi d(3m+\frac{3}{2})/a}}
$$

$$
- q \sum_{0}^{\infty} \log
$$

$$
\frac{4[\cosh 2\pi(y - d - 3md)/a - \cos 2\pi x/a][\cosh 2\pi(y + d + 3md)/a - \cos 2\pi x/a]}{e^{4\pi d(3m+1)/a}}
$$

$$(1)$$

$$- q \sum_0^\infty \log$$

$$\frac{4\left[\cosh \dfrac{2\pi(y - 2d - 3md)}{a} - \cos \dfrac{2\pi x}{a}\right]\left[\cosh \dfrac{2\pi(y + 2d + 3md)}{a} - \cos \dfrac{2\pi x}{a}\right]}{e^{4\pi d(3m+2)/a}}$$

$$- q \sum_0^\infty \log$$

$$\frac{4[\cosh 2\pi(y - d/2 - 3md)/a + \cos 2\pi x/a]}{[\cosh 2\pi(y + d/2 + 3md)/a + \cos 2\pi x/a]}{e^{4\pi d(3m+\frac{1}{2})/a}}$$

$$- q \sum_0^\infty \log$$

$$\frac{4[\cosh 2\pi(y - 5d/2 - 3md)/a + \cos 2\pi x/a][\cosh 2\pi(y + 5d/2 + 3md)/a + \cos 2\pi x/a]}{e^{4\pi d(3m+\frac{5}{2})/a}}$$

$$\tag{1}$$

Computing again the pressure differential between the input and output wells, and separating out all the terms with $m = 0$, Δp may be expressed as

$$\Delta p = p(0, d \pm r_w) - p(0, r_w) = 2q \log$$

$$\frac{\sinh^3 \pi d/a \cosh^3 \pi d/2a \cosh^4 5\pi d/2a \sinh 2\pi d/a}{\sinh^3 \pi r_w/a \cosh^6 3\pi d/2a \cosh 7\pi d/2a \sinh 3\pi d/a}$$

$$+ 4q \sum_1^\infty \log \frac{\sinh (3m - 1)\pi d/a \sinh (3m + 1)\pi d/a}{\sinh^2 3m\pi d/a}$$

$$+ 4q \sum_1^\infty \log \frac{\cosh (3m + \frac{1}{2})\pi d/a \cosh (3m + \frac{5}{2})\pi d/a}{\cosh^2 (3m + \frac{3}{2})\pi d/a}$$

$$+ 2q \sum_1^\infty \log \frac{\sinh^2 (3m + 1)\pi d/a}{\sinh 3m\pi d/a \sinh (3m + 2)\pi d/a}$$

$$\tag{2}$$

(*Equation continued on page 590.*)

$$\left. + 2q \sum_{1}^{\infty} \log \frac{\sinh^2 (3m+2)\pi d/a}{\sinh (3m+1)\pi d/a \, \sinh (3m+3)\pi d/a} \right.$$

$$+ 2q \sum_{1}^{\infty} \log \frac{\cosh^2 (3m+\frac{1}{2})\pi d/a}{\cosh (3m-\frac{1}{2})\pi d/a \, \cosh (3m+\frac{3}{2})\pi d/a}$$

$$\left. + 2q \sum_{1}^{\infty} \log \frac{\cosh^2 (3m+\frac{5}{2})\pi d/a}{\cosh (3m+\frac{3}{2})\pi d/a \, \cosh (3m+\frac{7}{2})\pi d/a} . \right\} \quad (2)$$

The infinite series may now be summed in the following manner: Taking the first series, it is readily found that

$$\sum_{1}^{\infty} \log \frac{\sinh (3m-1)\pi d/a \, \sinh (3m+1)\pi d/a}{\sinh^2 3m\pi d/a}$$

$$= \sum_{1}^{\infty} \log \left(1 - \frac{\sinh^2 \pi d/a}{\sinh^2 3m\pi d/a} \right) \cong - \sinh^2 \frac{\pi d}{a} \sum_{1}^{\infty} \frac{1}{\sinh^2 3m\pi d/a}$$

$$\cong -4 \sinh^2 \frac{\pi d}{a} \sum_{1}^{\infty} e^{-\frac{6m\pi d}{a}} = - \frac{2e^{-\frac{3\pi d}{a}} \sinh^2 \pi d/a}{\sinh 3\pi d/a},$$

the approximations evidently involving but negligible errors, since

$$\frac{\pi d}{a} = \frac{\pi}{\sqrt{3}} = 1.8138,$$

so that the replacement of $\sinh 3\pi d/a$ by $e^{\frac{3\pi d}{a}}/2$ and the neglect of terms as $\dfrac{\sinh^4 \pi d/a}{\sinh^4 3m\pi d/a}$ will give errors that are very small compared to the terms retained.

Summing the other series similarly, they may all be grouped into the term

$$\frac{-4qe^{-\frac{3\pi d}{a}} \sinh^2 \frac{\pi d}{a}}{\sinh 3\pi d/a} \left(2 + e^{-\frac{\pi d}{a}} - e^{-\frac{2\pi d}{a}} - 2e^{-\frac{3\pi d}{a}} - e^{-\frac{4\pi d}{a}} + e^{-\frac{5\pi d}{a}} \right),$$

the numerical value of which is $-0.0029q$. Evaluating now the terms in the first part of Δp, it is found that Δp may be rewritten as

$$\Delta p = 2q\left(-3 \log \sinh \frac{\pi r_w}{a} + 0.0790\right). \tag{3}$$

Noting now that the flux per output well is $8\pi kq/\mu$, and again replacing the sinh by its argument, the effective conductivity of the system is given by

$$\frac{\mu Q}{k\Delta p} = \frac{4\pi}{3 \log \dfrac{d}{r_w} - 1.7073}. \tag{4}$$

Equation (4) is plotted as curve III in Fig. 243, taking again $r_w = \frac{1}{4}$ ft. It will be seen that it is almost parallel to curves I and II, the conductivity being approximately 32 per cent higher than that of the five-spot flood and some 39 per cent higher than that of the direct line-drive flood.

This comparison refers, of course, to the three flooding systems with the same distance between the input and output wells, and does not take into account the differences in the well per acre densities in the three types of network. Thus it is easily verified that the well densities in the three cases are:

Line-drive system: 1 well per da area units.

Five-spot array: 1 well per d^2 area units.

Seven-spot array: 1 well per $0.866\ d^2$ area units.

Hence if $d = a$ in the direct line drive, as was assumed in plotting curve I of Fig. 243, this flood has the same well-per-acre density as the five-spot array, for equal distances (d) between the input and output wells, whereas the seven-spot flood has a 15 per cent higher well-per-acre density. To make them all equal, the distance between input and output wells in the seven-spot flood (d_i) must be made 1.075 times that in the five-spot flood and direct line drive. That is, if the input and output wells in the five-spot array are separated by 500 ft., the separation would have to be 537 ft. in the seven-spot flood in order to have the same number of wells per unit area in the floods. The effect, however, of this reduction to equal well-per-acre densities is very small, and still leaves the seven-spot conductivity per network element approximately 31 and 38 per cent higher than the five-spot flood and direct line drive, respectively.

9.26. Conductivity of the Staggered Line Drive.—If Fig. 242 is rotated through 45 deg., it will be noted that the five-spot network is nothing more than a staggered line-drive flood with $d/a = \frac{1}{2}$. The difference of about 3 per cent between the resistance of the five-spot array and that of the corresponding direct line drive ($d/a = \frac{1}{2}$) (*cf.* Fig. 241), and the much larger difference (130 per cent) that will be found later between their flooding efficiencies must, therefore, be due entirely to the staggering of the wells. Hence it is of interest to treat the case of a general ($d/a \neq \frac{1}{2}$) staggered line drive, or a "stretched five-spot array," and see if it offers any advantages over the usual more symmetrical arrays.

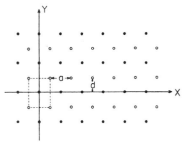

FIG. 245.—The staggered line-drive flood.

Although it would be more convenient for the calculation to be made later of the flooding efficiency of the staggered line drive to choose one of the coordinate axes so that it will pass through two neighboring input and output wells, such axes will not permit a periodic coordinate representation of all the wells in the system unless $4d^2/a^2$ is equal to an integer. For general values of d/a it is, therefore, necessary to draw the axes just as in the unstaggered drive, *i.e.*, as in Fig. 245. Following the procedure used in the previous cases, the pressure distribution in the system may be written as

$$p = q \log \left(\cosh \frac{2\pi y}{a} - \cos \frac{2\pi x}{a} \right)$$

$$+ q \sum_{1}^{\infty} \log$$

$$\frac{4 \left[\cosh \frac{2\pi(y - 2md)}{a} - \cos \frac{2\pi x}{a} \right] \left[\cosh \frac{2\pi(y + 2md)}{a} - \cos \frac{2\pi x}{a} \right]}{e^{8m\pi d/a}}$$

$$- q \sum_0^\infty \log$$

$$\frac{4 \left[\cosh \dfrac{2\pi(y - 2md - d)}{a} + \cos \dfrac{2\pi x}{a} \right]}{\left[\cosh \dfrac{2\pi(y + 2md + d)}{a} + \cos \dfrac{2\pi x}{a} \right]}{e^{4\pi d(2m+1)/a}}. \quad (1)$$

The pressure differential between the input and output wells is

$$\Delta p = p\left(\frac{a}{2}, d \pm r_w \right) - p(0, r_w)$$

$$= 2q \log \frac{\cosh^4 \pi d/a \cosh^3 3\pi d/a}{\sinh^2 \pi r_w/a \sinh^4 2\pi d/a \sinh 4\pi d/a}$$

$$+ 2q \sum_2^\infty \log \frac{\cosh (2m - 1)\pi d/a \cosh^3 (2m + 1)\pi d/a}{\sinh^3 2m\pi d/a \sinh 2(m + 1)\pi d/a}, \quad (2)$$

to which the series gives a negligible contribution if $d/a \geqslant \frac{1}{2}$. The conductivity of the system is, therefore, given by

$$\frac{\mu Q}{k \Delta p} = \frac{2\pi}{\log \dfrac{\cosh^4 \pi d/a \cosh^3 3\pi d/a}{\sinh^2 \pi r_w/a \sinh^4 2\pi d/a \sinh 4\pi d/a}} \quad (3)$$

which reduces to Eq. 9.23(8) for the unstaggered line for $d/a \geqslant 1$. In fact, a plot of Eq. (3) for $a = 660$ ft. gives curve III in Fig. 241, which merges exactly with the corresponding curve for the unstaggered network (curve I) for $d/a > 1$. It follows, then, that except for $d/a < 1$ (which includes the five-spot network) the resistance of the line drive is unaffected by staggering of the wells. This result may be considered as a generalization of the same conclusion drawn in Sec. 9.11 for finite numbers of line arrays.

9.27. The Calculation of Flooding Efficiencies.[1]—Although the "flooding efficiencies"—fractions of the total area of a net-

---

[1] The neglect in the difference in the value of k/μ on the two sides of the water-oil interface will, of course, give rise to discrepancies between the calculated absolute efficiencies and those observed in actual flooding operations, as may be inferred from Sec. 8.9. However, these should be of less serious magnitude than those in the absolute values of the flood conduc-

work flooded by the time the flooding liquid first reaches the output wells—have already been found in Sec. 9.19 by means of the electrolytic models, it is of interest to apply the pressure distributions developed for the calculation of the flood resistances to a computation of the flooding efficiencies. Since the method is quite simple it will also be used to find the variation of the flooding efficiency of the line drive as the ratio of the distance between the output and input lines to the spacing within the lines is varied.

The principle of the method is the same as applied in Secs. 8.6 and 8.7 to find the areas of flood in the problems of the line drive into a single well and the drive between two wells. Thus one simply computes the time required for a particle to travel from the input to the output well *along the streamline of highest average velocity*. This time multiplied by the flux rate into the input well gives the area (in a two-dimensional system) flooded out by the time the input liquid first reaches the output well. Dividing this area by that of the elementary unit of the flooding network (which contains just one input well), one obtains at once the fraction of the network flooded, and hence the efficiency of the flood.

9.28. The Efficiency of the Direct Line-drive Flood.—Applying this method first to the direct line drive, it is clear that the streamlines of highest average velocity are the lines of centers between the input and output wells, one of which is the y axis in Fig. 240. As the velocity along the y axis is simply the pressure gradient, it follows from Eq. 9.23(2) that

$$v_y = -\frac{k}{f\mu}\left(\frac{\partial p}{\partial y}\right)_{x=0} = -\frac{2\pi kq}{af\mu}\operatorname{ctnh}\frac{\pi y}{a}$$

$$-\frac{2\pi kq}{af\mu}\sum_{1}^{\infty}(-1)^m\left(\operatorname{ctnh}\frac{\pi(y-md)}{a}+\operatorname{ctnh}\frac{\pi(y+md)}{a}\right), \quad (1)$$

where f is the porosity of the sand.

The time for the flooding liquid to reach the output well (at the origin) will then be

---

tivities arising from this neglect (*cf.* footnote page 583), and may well be masked by effects of large-scale horizontal variations in the normal sand permeability.

$$t = \int_d^0 \frac{dy}{v_y} = \frac{f\mu}{k} \int_0^d \frac{dy}{(\partial p/\partial y)_{x=0}}. \tag{2}[1]$$

Changing the variable by means of the substitution

$$w = \cosh^2 \frac{\pi y}{a},$$

Eq. (2) may be given the form:

$$t = \frac{a^2 f\mu}{4\pi^2 kq} \int_1^{\cosh^2 \frac{\pi d}{a}} \frac{dw}{w + 2w(w-1)\sum_1^\infty (-1)^m/(w - \cosh^2 m\pi d/a)}. \tag{3}$$

Now the neglect of the terms in the series beyond the first will lead to errors in t that are of no practical significance except when d/a is considerably smaller than $\frac{1}{2}$. For $d/a \geqslant \frac{1}{2}$ Eq. (3) may be rewritten in the form

$$\left.\begin{aligned}
t &= \frac{a^2 f\mu}{4\pi^2 kq} \int_1^{\cosh^2 \frac{\pi d}{a}} \frac{(\cosh^2 \pi d/a - w)dw}{w(w + \cosh^2 \pi d/a - 2)} \\
&= \frac{a^2 f\mu}{2\pi^2 kq(\cosh^2 \pi d/a - 2)} \\
&\qquad \left(\cosh^2 \frac{\pi d}{a} \log \cosh \frac{\pi d}{a} - \sinh^2 \frac{\pi d}{a} \log 2\right).
\end{aligned}\right\} \tag{4}$$

As the area flooded out in this time is $4\pi kqt/f\mu$ per network element, of area $2da$, the fractional flood is

$$\frac{Q}{A} = \frac{1}{\pi(\cosh^2 \pi d/a - 2)d/a}\left(\cosh^2 \frac{\pi d}{a} \log \cosh \frac{\pi d}{a}\right.$$
$$\left. - 0.6932 \sinh^2 \frac{\pi d}{a}\right) \tag{5}$$

Equation (5) is plotted as curve I in Fig. 246. It will be seen that the flooding efficiency can be materially increased (the by-passing tendency decreased) by decreasing the distance a between wells of like character (output-output, input-input) or by increasing the distance d between wells of unlike character

[1] To be exact, the limits in Eq. (2) should be r_w and $d - r_w$. Since, however, $\frac{\partial p}{\partial y} \sim \infty$ at $y = 0$, d, the error in t in dropping the r_w is only $f\mu r_w^2/2kq$, which is entirely negligible as it adds to the real flooding area the areas of the two wells in each network unit.

(input-output), the effect decreasing, however, with increasing d/a. In fact, for $d/a \geqslant 1.5$, Eq. (5) may be reduced to the form

$$\frac{Q}{A} = 1 - 0.441/(d/a), \qquad (6)$$

which shows that the efficiency increases with large values of d/a in a hyperbolic manner.

It may be noted that for $d/a = 1$, Eq. (5) gives $Q/A = 0.57$, which agrees exactly with that found by the electrolytic model.

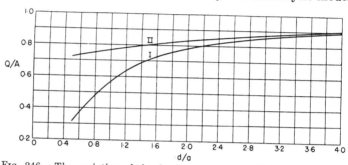

FIG. 246.—The variation of the flooding efficiency of line-drive floods with d/a (distance between input and output lines)/(well spacing within the lines). Q/A = (area of network element flooded at time water first reaches output wells)/(original area of network element). I. Direct line-drive flood. II. Staggered line-drive flood. (*From A.I.M.E.* **107**, 79, 1934.)

9.29. The Five-spot-flood Efficiency.—For the five-spot system the y axis may again be chosen as the streamline of highest average velocity, the actual value of the velocity being given by

$$v_y = -\frac{k}{f\mu}\left(\frac{\partial p}{\partial y}\right)_{x=0} = \frac{\pi k q}{df\mu}\left(\tanh\frac{\pi y}{2d} - \operatorname{ctnh}\frac{\pi y}{2d}\right)$$

$$+ \frac{\pi k q}{df\mu}\sum_1^\infty (-1)^m\left[\tanh\frac{\pi(y - md)}{2d} + \tanh\frac{\pi(y + md)}{2d}\right]$$

$$- \frac{\pi k q}{df\mu}\sum_1^\infty (-1)^m\left[\operatorname{ctnh}\frac{\pi(y - md)}{2d} + \operatorname{ctnh}\frac{\pi(y + md)}{2d}\right]$$

$$= -\frac{2\pi k q}{df\mu}\sinh\frac{\pi y}{d}\left[\frac{1}{w^2 - 1} + \frac{2\cosh\pi}{\cosh^2\pi - w^2} - \frac{2\cosh 2\pi}{\cosh^2 2\pi - w^2}\right],$$

$$(1)$$

to terms with $m = 3$, with $w = \cosh \pi y/d$.

Putting this in Eq. 9.28(2) and integrating, it is found that

$$t = \frac{2.270 f \mu d^2}{2\pi^2 k q.} \tag{2}$$

As the area flooded out will be $4\pi k q t/f\mu$, and the area of each network element is $2d^2$, the fractional flood is

$$\frac{Q}{A} = 0.7226. \tag{3}$$

The value is also in good agreement with that found by the electrolytic model (*cf.* Sec. 9.19).

9.30. The Efficiency of the Seven-spot Flood.—Here again the y axis (*cf.* Fig. 244) may be taken as the streamline of highest average velocity. The velocity along this streamline is, by Eq. 9.25(1),

$$v_y = -\frac{k}{f\mu}\left(\frac{\partial p}{\partial y}\right)_{x=0} = \frac{-4\pi k q}{af\mu} \operatorname{ctnh} \frac{\pi y}{a}$$

$$-\frac{4\pi k q}{af\mu} \sum_1^\infty \left[\operatorname{ctnh} \frac{\pi(y-3md)}{a} + \operatorname{ctnh} \frac{\pi(y+3md)}{a} \right]$$

$$-\frac{4\pi k q}{af\mu} \sum_0^\infty \left[\tanh \frac{\pi(y-3d/2-3md)}{a} + \right.$$

$$\left. \tanh \frac{\pi(y+3d/2+3md)}{a} \right]$$

$$+\frac{2\pi k q}{af\mu} \sum_0^\infty \left[\operatorname{ctnh} \frac{\pi(y-d-3md)}{a} + \operatorname{ctnh} \frac{\pi(y+d+3md)}{a} \right]$$

$$+\frac{2\pi k q}{af\mu} \sum_0^\infty \left[\operatorname{ctnh} \frac{\pi(y-2d-3md)}{a} + \right.$$

$$\left. \operatorname{ctnh} \frac{\pi(y+2d+3md)}{a} \right]$$

$$+\frac{2\pi k q}{af\mu} \sum_0^\infty \left[\tanh \frac{\pi\left(y-\dfrac{d}{2}-3md\right)}{a} + \tanh \frac{\pi\left(y+\dfrac{d}{2}+3md\right)}{a} \right]$$

(*Equation continued on page 598.*)

$$+ \frac{2\pi k q}{af\mu} \sum_{0}^{\infty} \left[\tanh \frac{\pi(y - 5d/2 - 3md)}{a} + \right.$$

$$\left. \tanh \frac{\pi(y + 5d/2 + 3md)}{a} \right]$$

$$= -\frac{4\pi k q}{af\mu} \sinh \frac{2\pi y}{a} \left[\frac{1}{w - 1} + \frac{2}{w + \cosh 3\pi d/a} \right.$$

$$- \frac{1}{w - \cosh 2\pi d/a} - \frac{1}{w - \cosh 4\pi d/a} - \frac{1}{w + \cosh \pi d/a} -$$

$$\left. \frac{1}{w + \cosh 5\pi d/a} \right] \quad (1)$$

after dropping all terms for $m > 0$, and making the substitution

$$w = \cosh \frac{2\pi y}{a}.$$

Putting this also in Eq. 9.28(2) and integrating, it is found that

$$t = \frac{2.013 f\mu a^2}{8\pi^2 k q}. \quad (2)$$

As the area of the hexagonal element is $\dfrac{\sqrt{3}a^2}{2}$, and the total flooded area is $8\pi k q t/f\mu$, the fractional flood is

$$\frac{Q}{A} = 0.740. \quad (3)$$

Although this value is about 5 per cent lower than that found by means of the electrolytic model (0.78 to 0.80), the agreement is not bad, as capillary effects at the edges of the model, in cases where the flood approaches close to its edges, tend to give an apparent flooding efficiency which is too large.

9.31. The Efficiency of the Staggered Line-drive Flood.—As already mentioned, the natural axes for a staggered line drive must be drawn through wells of the same sign only, except for special values of d/a. Hence the streamline of highest average velocity—the line of centers between input and output wells—no longer coincides with one of the axes, and both components of the pressure gradient must be computed. Thus the velocity along the actual line of centers may be expressed as

$$v_s = -\frac{k}{f\mu\sqrt{1 + 4\bar{d}^2}}\left\{\frac{\partial p}{\partial x} + 2\bar{d}\frac{\partial p}{\partial y}\right\}; \qquad \bar{d} = \frac{d}{a}, \qquad (1)$$

and the time for the flooding liquid to reach the output wells becomes

$$t = -a\sqrt{1 + 4\bar{d}^2}\int_0^{1/2}\frac{d\bar{x}}{v_s}; \qquad \bar{x} = \frac{x}{a}. \qquad (2)$$

Writing out $\frac{\partial p}{\partial x}$ and $\frac{\partial p}{\partial y}$, setting $y = 2d\bar{x}$, and finally integrating Eq. (2), graphically, for various values of $\bar{d}$, curve II of Fig. 246 was found. It will be seen that whereas the staggering had a negligible effect on the resistance, the flooding efficiency is quite materially affected by the staggering.

9.32. General Observations on Flooding Networks.—Before proceeding now with a detailed comparison of the various types of flooding systems, as would be required for the development of a practical flooding program, it will be of interest to point out several general features of the flooding problem. Thus it is to be noted first of all that the efficiency of the flood as defined here —the area flooded out in an ideal system at the time the flooding fluid first reaches the output wells—is independent of the total pressure differential acting upon the system. As may be noted from the calculation of the efficiency, the pressure differential, which is proportional to q, determines the absolute value of the time required for the flooding fluid to reach the output wells. The pressure differential cancels out, however, in the determination of the flooded area. This result indicates that the shape of the flood is also independent of the pressure differential; and, in fact, a rigorous proof of this conclusion has already been given in Sec. 8.2.

Another point to be noted is that the flooding efficiencies of the five- and seven-spot floods are independent of the dimensions of the systems. They are determined only by the geometry of the networks, which in turn controls the potential or pressure distibutions.[1] In the case of the line drive the geometry of the system is not uniquely determined by the single requirement

[1] The well radius will enter in the exact expressions for the time of complete flooding. However, the effect on the flooding efficiency will be proportional to the ratio of the area of the well bores to the area of the flooding network element, and hence will be entirely negligible.

that the input and output wells lie in alternating parallel lines of equal spacing. It may be varied both by changing the ratio of the input-output distance d to the input-input distance a, and by staggering the wells in the network. Since these geometrical changes will affect the flooding efficiency, the line drive possesses considerable flexibility.

With respect to the flood resistances, the well spacing and the physical dimensions of the systems are important, the resistance always increasing as the spacing is increased (*cf.* Fig. 243). In the case of the line drive, the resistance may be increased, for fixed well radii, not only by increasing the input-output well spacing d, but also by decreasing the input-input spacing a. However, the variation of the resistance with the dimensions of the systems is quite slow, being of an essentially logarithmic nature.

9.33. Comparison of the Flooding Networks.—In the previous sections there have been developed the separate flooding characteristics—resistance and efficiency— of the five-spot, seven-spot, and the line-drive flooding networks. With respect to their conductivities, they have been compared on the basis of equal well-per-acre densities, with the result that the seven-spot conductivity, *per network element*, is roughly 31 and 38 per cent higher than those of the five-spot and line-drive floods, respectively. However, this comparison is really not justified from a practical point of view, since the seven-spot network contains three wells per network element whereas the five-spot and line-drive networks contain only two wells per network element. Hence for the same well-per-acre density this network contains only two thirds as many elements (hexagons) as there would be in the five-spot or line-drive networks.

To develop a more practical comparison, the following question may be proposed: Given a certain area to be covered by a flooding network consisting of a definite total number n of input and ouput wells, what flooding network should be used?

For simplicity it is advisable at first to consider, among the infinity of possible direct line drives (different values of d/a), only that for which $d = a$, and determine later how the comparison will be modified by taking into account cases where d is not equal to a. Consideration of the staggered line drive will also be reserved for the later discussion. With respect to the tendencies of the systems to induce regional bypassing of water (flooding efficiencies), a definite answer to the question is given by the results already presented. Thus the seven-spot network would be better than the five-spot array, and the latter better than the direct line drive. In fact, the by-passing tendencies of the seven-spot, five-spot, and line-drive arrays may be represented by the ratios 1:1.02:1.30.

However, with respect to the production rates to be derived from the flooded area for a given pressure differential between input and output wells, the answer is quite different. Although the conductivity or production rate *per network element* is roughly 31 and 38 per cent higher for the seven-spot than that for the five-spot or direct line-drive network, for equal well-per-acre densities, the seven-spot network, containing three wells per element as compared to two for the other networks, will contain only two thirds as many elements as the other networks contain. Therefore, the ordinates of curve III, Fig. 243, must be given a weight of two-thirds with respect to those of curves I and II. Hence for a given pressure differential the production rates for the total flooded area will be lowest for the seven-spot network, and the ratios for the five-spot, direct line-drive, and seven-spot arrays will be approximately

$$Q_5:Q_d:Q_7 = 1.00:0.95:0.88.$$

This may be seen more clearly from Fig. 247, in which the data of Fig. 243 have been recomputed so as to give the conductivities per acre as functions of the well densities (acres per well).

Since the production capacity of the five-spot array is thus 12 per cent higher than that of the seven-spot array and its efficiency is only 2 per cent lower, it may be concluded that the five-spot flood should be definitely preferable to the seven-spot flood. The latter, however, should be preferred to the simple, direct line-drive flood, since the 30 per cent higher by-passing tendency of the line drive will evidently outweigh its advantage of giving a 7 per cent higher production rate than the seven-spot network.

On the other hand, the line drive may be made more favorable if advantage is taken of the possibility of increasing its efficiency by making the well spacing closer within the input and output lines. For in order to have the same well-per-acre density—and

hence the same total number of wells—it is only necessary that
the product da equal the $d_5{}^2$ of the five-spot or $0.866d_7{}^2$ of the
seven-spot network, while the ratio d/a, which determines the
efficiency and resistance of the flood, remains perfectly arbitrary.
Thus by Fig. 246 it is clear that if $d/a = 1.59$ the flooding
efficiency of the direct line drive will be raised to that of the five-
spot array. Supposing then that the input-output distance in
the five-spot system is 660 ft., the dimensions of the correspond-

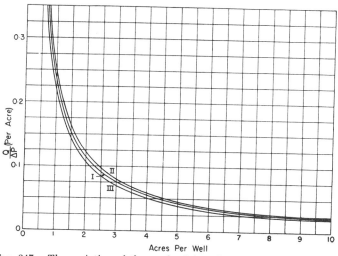

Fig. 247.—The variation of the conductivity of flooding networks with the
well density. $Q/\Delta p$ = production rate obtainable per acre per unit pressure
differential in a unit-permeability sand of unit thickness. I. Regular line-drive
flood $(d = a)$. II. Five-spot flood. III. Seven-spot flood. (*From A.I.M.E*
107, 71, 1934.)

ing line drive (for $d/a = 1.59$) would be $a = 523$ ft. and
$d = 832$ ft., so that by Eq. 9.23(8) the conductivity would
be 0.379. From Fig. 243, on the other hand, the five-spot
conductivity would be 0.433 and the seven-spot conductivity
0.378.

Thus it is to be noted that in increasing the efficiency of the
direct line drive to be equivalent to that of the five-spot network
by varying d/a, its conductivity has been lowered so as to be
only 87 per cent of that of the five-spot array. The latter must,
therefore, be considered as definitely superior to the unstaggered
line drive. With respect to the seven-spot network, however,
it is seen that for practical purposes the direct line drive and

seven-spot array can be made essentially equivalent, for by taking $d/a = 1.59$, the conductivities of the two become almost exactly the same and the efficiencies differ by only 2 per cent in absolute value.

Thus far, modifications of the line drive have involved variations of d/a alone. If now use is made of the additional possibility of staggering the line drive, the latter will be found even more efficient than the five-spot network. Leaving $d/a = 1.59$ and staggering the drive, the efficiency is raised to 0.81, which represents a 12 per cent increase relative to the five-spot efficiency of 0.723 (see Fig. 246). Since the lower conductivity of this line drive (0.87 of that of the five-spot system) can be compensated for by a 13 per cent increase in applied pressure differential, while the flooding efficiency cannot be improved once the geometry of the system has been chosen, it is clear that the 12 per cent gain in flooding efficiency will more than counterbalance the 13 per cent loss in conductivity. This staggered line drive, therefore, must be considered as even better than the five-spot system for the same well-per-acre density.

For practical purposes it may be more convenient to fit into an actual flooding program the ratio $d/a = 1.5$, rather than the value 1.59 used above. With this ratio the various flooding networks may be compared on the basis of a density of 1 output well to 10 acres ($d_5 = 660$ ft.) as follows:

TABLE 18.—THE RELATIVE EFFICIENCIES OF VARIOUS FLOODING NETWORKS

Type of network	Relative production rate (conductivity for equal well/acre densities)	Flood efficiency, per cent
Staggered line drive		
$d/a = 1.50$; $a = 539$ ft.; $d = 808$ ft.	0.383	80.0
Five-spot array, $d_5 = 660$ ft.	0.433	72.3
Regular line drive, $d/a = 1.50$; $a = 539$ ft.;		
$d = 808$ ft.	0.383	70.6
Seven-spot array, $d_7 = 709$ ft.	0.378	74.0

Here again there can be little question but that the staggered line drive would make the best flooding arrangement. On the other hand, it may be observed that from a purely economic point of view the seven-spot flood possesses an advantage over the five-spot and line-drive floods which is not involved in the physical comparisons developed above. In these comparisons the total number of wells over a given area was assumed to be the same for all systems. This implies that the numbers of input and output wells are equal for the line-drive or five-spot floods, but in the seven-spot flood only one-third of all the wells need be output wells if they are placed at the centers of the hexagons. If the output wells are pumpers and, therefore, more expensive to operate than the input wells, there will evidently be a distinct advantage in distributing the total number of wells among two-thirds input wells and one-third pumpers rather than using a 1:1 distribution, as would be required in the other two flooding networks. If it should happen that input wells are more expensive to operate, they may be put at the hexagon centers and the advantage will be retained, as the efficiency and the resistance of the system are unaffected by interchanging the output and input wells.

As a practical illustration of these comparisons a specific numerical example approximately corresponding to a Bradford-field flooding problem will be considered. Thus it will be supposed that there are four 1,000-acre tracts, on each of which shall be put 400 wells, including injection and output wells. They will be arranged respectively in the staggered drive with $d/a = 1.5$, direct line drive, five-spot, and seven-spot arrays. The sand will be assumed to be 40 ft. thick, with a porosity of 10 per cent and a permeability of 0.01 darcys. Table 19 gives the significant data for comparison of the various floods. The superiority of the staggered line drive will be quite evident.

It should be remembered, of course, that no detailed account has been taken of the end effects involved in fitting the networks within the boundaries of the 1,000-acre tracts. Rather it has been assumed that the tracts are simply sections of infinite networks. Furthermore, the initial transients due to the compression of the gas have been neglected, and, in fact, the supposition has been made that, to begin with, the total porosity of 10 per cent was filled with oil and is moreover completely flushed out

TABLE 19.—COMPARISON OF VARIOUS FLOODING NETWORKS FOR A SPECIFIC FLOODING PROGRAM

Total area of tracts: 1,000 acres; total number of wells: 400; total pressure differential: 1,000 lb.; sand thickness: 40 ft.; sand permeability: 0.01 darcy; sand porosity: 10 per cent; effective oil-water viscosity: 10 centipoises.

	No. input wells	No. output wells	Input-input spacing (ft.)	Input-output spacing (ft.)	Rate of fluid input or output (bbl./day)	Time to reach output wells (yrs.)	Flood efficiency (%)	Total water-free oil prod. (10⁶ bbl.)
Staggered line drive $d/a = 1.50$..	200	200	270	426	3776	18.0	80.0	2.48
Direct line drive, $d = a$........	200	200	330	330	4094	11.8	57.0	1.77
Five spot.....................	200	200	467	330	4305	14.3	72.3	2.25
Seven spot....................	267	133	355	355	3765	16.7	74.0	2.30

Staggered line drive — Direct line drive — Five-spot — Seven-spot

by the advancing water. If it be supposed that there will be only a 50 per cent recovery, the times given in the third column from the right in the table will be cut by a factor of approximately ½.

Even with this factor of ½ the times required for the water to reach the output wells are still admittedly larger than the times observed in actual flooding programs by a factor of about 10. Although this may appear to cast doubt upon the validity of the theory presented here, it is rather felt that this discrepancy throws considerable light upon the sand conditions under which such floods as those in the Bradford field have been operating. For it simply shows that there is a more pronounced water by-passing in the Bradford field than would be possible in an ideal homogeneous sand. That is, it indicates that there must be considerable channeling in the floods, due either to high-permeability beds or to considerable streaking of high-permeability zones along the main body of sand of permeability of the order of 0.01 darcy. Another factor which may enhance this channeling is a by-passing of the water through the oil even under uniform sand conditions, much as gas is known to by-pass oil in homogeneous sands. The main effect, however, is probably due to the high permeabilities, streaks or beds, for the actual existence of permeability variations as high as 100-fold have been found in measurements of cores from the producing sands.[1]

Furthermore, it may be pointed out that regardless of the detailed explanations invoked in the analysis of the field data, the term efficiency as used here may be taken as a quantitative measure of the effectiveness of a flooding program. And this may be obtained simply by dividing the water-free oil output by the volume of oil in the sand at the beginning of the flooding operations. This procedure will give a specific measure of the extent of the water by-passing. In fact, we may already conclude from the general comparison between the theoretically ideal example given above and the results of practical field experience that perhaps the questions of well spacing and arrangement are after all of relatively minor importance in determining the success of a flooding program, and that attention should rather be concentrated upon preventing high-permeability by-passing as the major problem to be solved before the efficiency of artificial water-flooding programs can be materially improved.

Returning finally to the question of the absolute well spacing, one need only recall the earlier result obtained in previous sections that it has no effect on the flood efficiency and only a minor effect on the conductivity. The problem of well spacing, therefore, must be relegated to the economic side of the question in which one must balance the value of high production rates with a close spacing and large investment against lower production rates with wider spacing and proportionately lower investment.

9.34. Summary.—While many problems of interest can be idealized and reduced to that of a single well draining a porous medium, there are many others, such as that of offsetting at the boundaries of leases over oil reservoirs and that of water flooding, which must be explicitly treated as multiple-well systems. Furthermore, since the "external boundaries" always entering in the specification of single-well systems are usually, in practical situations, nothing more than *effective* boundaries

[1] *Cf.* Sec. 2.12.

formed by the presence of other wells drilled in the neighborhood of the one in question, it is of interest to see in detail just how such boundaries are set up.

In the treatment of multiple-well systems it is convenient to consider separately those containing finite and small numbers of wells distributed over areas which are small compared to the total area of the liquid-bearing sand, and those consisting of large or effectively infinite numbers of wells. In the former case each well may be defined by the average of the pressure over its surface, the mutual separations being small as compared to the radius of the effective external boundary. The "external-boundary pressure" may be defined by the average over the boundary of the logarithmic terms representing the individual contributions of the several wells to the resultant pressure distribution. In this way a series of linear equations are obtained relating the individual well pressures to their associated fluxes and to the external boundary pressure [cf. Eqs. 9.2(5) and 9.2(6)].

The mutual interference and interactions between the wells of small groups may be studied by means of these equations, by supposing their well pressures to be equal and computing their associated fluxes. For two, three, four, or any number of wells situated symmetrically, as on the circumference of a circle, the fluxes of the several wells are necessarily equal. However, there is still some interference between them, as is shown by the fact that these common fluxes decrease as the wells are brought closer together [cf. Eqs. 9.3(3), (4), (5), (18)]. Furthermore, the fluxes per well decrease as the total number of wells in the group increases [cf. Eqs. 9.3(18), 9.4(2)]. Thus each well of a group of four will have a flux equal to only 51.5 per cent of that which a single well would have in the same reservoir and flowing under the same pressure differential. In fact, this interference grows so rapidly with the number of wells that a group of 16 wells placed in a rectangular pattern (cf. Fig. 196) will produce at a rate only 3.33 times as large as that of a single well under the same conditions [cf. Eq. 9.4(1)].

The interference effects within small groups are still more clearly shown by comparing the fluxes of wells in the interior of well groups with those along the outer ring of wells. Thus a well placed at the center of a square pattern of four has a flux of only 78 per cent of that of each of those surrounding it. The central well of a group of 9 as in Fig. 195 will produce at a rate which is only 61.5 per cent as high as that of those on the corners, even if it is produced at the same bottom-hole pressure (for $d/r_w = 800$). Similarly, the central four of the group of 16 of Fig. 196 will produce at rates equal to but 44.5 per cent of those of the four corner wells, if the mutual spacings are 200 ft., and only 36.2 per cent if their mutual spacing is 50 ft. [cf. Eq. 9.3(16)].

Although these results are essentially independent of the absolute position of the group of wells with respect to the external boundary, their validity does depend upon the assumption that none of the wells is very close to it. A detailed examination of the pressure distribution over the external boundary which is implied by the analysis, shows that for the theory to be valid it is sufficient, to a first approximation, that the center of the external boundary be not too distant from the center of gravity of the well system when weighted with respect to their fluxes.

Another type of small group of wells of interest is that in which several wells in a line are supplied with fluid by a neighboring and parallel line drive. In such systems the external boundary is treated explicitly as an infinite linear source, the appropriate pressure distributions being readily derived by applying the method of images. Here again, the mutual interference between the wells is shown in the result that the fluxes per well in a group decrease as the number of wells increases. Thus each of two wells separated by 100 ft. and 100 ft. distant from the line drive will have a flux of only 89.26 per cent of that which a single well alone in the sand would have. If there are three wells in the group, the exterior two will have fluxes of 86.0 per cent and the middle well only 79.3 per cent of a single-well flux. When the number of wells becomes infinite the fluxes per well are reduced to 64.1 per cent of that of a single well. As to the effect of the mutual well spacings, it is found here, as before, that the interference increases with decreasing well separations, this variation, in both types of well groups, being essentially of a logarithmic nature.

When the number of wells becomes so large as to traverse completely the fluid-bearing sand, they may be treated as infinite arrays or networks. Supposing all the wells are at the same pressure and are uniformly spaced in a linear array, the pressure distribution due to the individual wells may be summed and the resultant treated as a unit, being represented by the single term

$$p(x, y) = q \, \log\left[\cosh \frac{2\pi(y - d)}{a} - \cos \frac{2\pi x}{a}\right], \qquad \text{[cf. Eq. 9.8(2)]}$$

where the line array lies parallel to and at a distance d from the x axis, with a well spacing a and individual production rates of $4\pi kq/\mu$. This expression may be treated as that for a single well, and may be added to others when there are several arrays in the system. It gives a pressure distribution symmetrical about the array axis and about each well (*cf.* Fig. 201), its most significant feature, however, being the fact that the oscillations of the equipressure lines rapidly die out as one recedes from the array, so that at a distance equal to the mutual well spacing the equipressure lines become, for all practical purposes, strictly straight and parallel to the axis of the array.

For further study of the properties of linear arrays of wells, it is necessary to specify more precisely the source of the fluid which enters the wells. This may be done by supposing the fluid to be supplied by a "line drive" or infinite line source, held at a uniform pressure above that maintained at the wells. The pressure distribution is then no longer symmetrical about the axis of the array, but rather shows a regional gradient from the line drive toward the array (*cf.* Fig. 203).

The interference properties of the linear array may be investigated by an analysis of the pressure distribution and fluxes in two or more such arrays placed parallel to, and on one side of, a line drive. Here, evidently, the flux in the more distant arrays will be simply that which leaks past the first line, and, owing to the interference of the latter, will be smaller than that entering the first line. Specifically it is found that for a system containing two arrays the leakage past the first line decreases as the well spacing within the lines decreases and as the two arrays—of fixed separation—are withdrawn from the line drive. However, the minimum leakage is attained already when the distance of the first line from the line drive is only 0.4 of the well spacing within the lines, further increases in the separation of the lines from the line drive or decreases in the well spacing decreasing the leakage only inappreciably. The minimum value of this leakage is 33 per cent for a well spacing of 660 ft. With respect to the absolute value of the well spacing, the calculations show that although the leakage increases as the well spacing increases, the effect is quite small. Thus, if the first line is distant from the line drive and from the second line by the value of the well spacing, the leakage past the first line decreases only from 33.6 to 28.5 per cent of the total flow in the system as the well spacing decreases from 1,000 to 100 ft. Hence it would not be practicable to diminish appreciably the leakage past a linear array of wells by simply decreasing the well spacing.

Similarly, for fixed well spacing, the leakage past a line array decreases as the distance between the two lines increases. Here again, however, the effect is relatively small, and for a well spacing of 660 ft. the leakage decreases from 46 to 28 per cent as the distance between the lines increases from 100 to 1,000 ft. Hence except when the various distances in the system are very large or very small, roughly one third of the total flux in a system of two parallel arrays with the same well pressures will pass by the first line and enter the second line.

A similar analysis can be made of a system consisting of three parallel lines on one side of a line drive, and all flowing at the same well pressures. Although the leakage past the first line will now clearly be greater than if the third line were not present, the actual magnitude of this increase is only of the order of 5 per cent. Thus when the separations between the lines and the distance of the first line from the line drive are of the order of the well spacing within the lines, the total leakage past the first line is three-eighths of the total flux in the system; two thirds of this—one-fourth of the total flux—enters the second line, and the remaining third—one-eighth of the total flux—finally reaches the third line. The production rates from the three lines will, therefore, be in the ratio 5:2:1. As still more lines of wells are added, the additional leakages induced by them, and hence the flow into them, very rapidly decrease to vanishingly small values, thus showing that unless these distant lines are kept at much lower pressures than those near the line drive—as a river or canal —they would be of no practical value in providing drainage into the region past the first two or three lines.

A quite interesting result with respect to the shielding and leakage characteristics past infinite linear arrays of wells is found when the effect of mutually staggering adjacent lines is analyzed. For the analysis shows that unless the distance between the lines is appreciably less than the well spacing within them, the shielding and leakage effects will be, for all practical purposes, the same as if the lines are not staggered. Thus unless the separation of the two lines is less than 0.1 of the well spacing, the leakage in the staggered two-line system will exceed that in the nonstaggered system by no more than 1 per cent of the total leakage value.

As long as one assumes that the well pressures in the several lines are the same, the interference and leakage characteristics of the systems must be due to the geometrical structure of the well systems. Thus the fact that the first line in a group of several line arrays has been found to have higher fluxes then the others is the direct result of the assumption that the fluid source supplying the lines is placed unsymmetrically on one side of the group. In the problem of offsetting, however, an attempt is made to counteract the leakage induced by the geometrical asymmetry in the system by an artificially imposed asymmetry in the well pressures in the several lines.

The problem of offsetting in its simplest form may be specifically formulated as follows: Given two leases of preassigned and, in general, different average pressures with one or more lines of wells offsetting each other across the mutual lease boundary; for given pressures maintained at the offsets of one of the leases, what pressures should be maintained at the offsets of the other lease in order that there be no net fluid flow across the lease boundary? These "average" pressures must be considered as the averages of the pressures near the offset lines, rather than as averages over the whole of each of the leases, as the former alone determine the details of the interaction of the two leases.

If each lease has only a single line of offset wells, the analysis shows that fluid migration across the lease boundary can be prevented if the difference in operating well pressures between the offset lines is equal and opposite to the difference between the "average" pressures of the two leases, as defined above. When there are two or more offset lines in either or both of the leases, the corresponding restrictions upon the pressures to be maintained at the offsets can be derived without difficulty (cf. Sec. 9.14), although the interpretation of the analytical expressions is not so obvious as in the case of single line offsets. In general, however, the pressure adjustment for offsetting is essentially nothing more than a procedure of lowering the effective pressures of the high-pressure lease to equal that of the low-pressure lease, thus removing the regional gradient between them at the lease boundary, and with it the associated migration. On the other hand, in the application of the analytical results to a practical situation, it must be remembered that the analysis applies only to the dead-liquid component of the total flow in the systems. The contribution due to the compressibility of the fluid—evolution and expansion of the dissolved gases—must be first subtracted, as well as the pressure gradients associated with it, before applying the analytical results based on the incompressible-liquid hypothesis.

. A final problem involving multiple-well systems, and one which is of considerable practical interest, is that of water flooding. This is a method of artificial oil recovery in which water is introduced into an oil sand so as to displace the oil from the pores and drive it to output wells which, together with the water-injection wells, are usually arranged in regular geometrical patterns. Although the water and oil are of different viscosity and hence form a two-fluid system, the viscosity difference must be neglected in order to make the treatment reasonably tractable. Under this assumption several methods of treatment, experimental and theoretical, may be applied to the problem, and together furnish a fairly complete picture of all the significant features involved.

For the purpose of tracing the history of the water-oil interface as it travels from the injection to the output wells, a very simple electrolytic model serves as a very convenient tool. Here, the advance of OH ions in a porous medium, as saturated blotting paper, from electrodes representing the injection wells takes the place of the water-oil interface, the motion of the ions being revealed by the red coloration of the medium as they advance and react with a suitable indicator permeating the medium. The equivalence of this model to that of *fluid* flow in a porous medium can be readily proved by comparison of the model traces with those derived analytically for several simple cases where the analysis is not too laborious. The advance of the OH ions may be permanently and strikingly recorded by a series of photographs of the model at various time intervals. It is to be noted that here, as well as in the case of the electrical-conduction-sheet model and the analytical treatment to be discussed below, one need study only a single element or cell of symmetry of the flooding system rather than the whole of the infinite network.

In this way a detailed history of the flood is obtained in which the input fluid advances at first radially outward from the injection wells, then assumes a distorted front appropriate to the geometry of the network, and then finally develops rather sharp "fingers" or cusps as the interfaces reach the vicinity of the output wells and are drawn in rapidly by the high gradients there (*cf*. Figs. 219–231). In addition to giving the detailed pictures of the various stages of the flood in such a simple manner, the electrolytic model is particularly suited for determining the "efficiency" of the various flooding networks. The "efficiency" may be defined as the fraction of the total area of the network which is swept out by the advancing interface as it just reaches the output wells, thus giving, for the ideal conditions assumed here, the volume of oil produced by the output wells by the time the water first reaches them. This quantity may be obtained from the model photographs by simply measuring the area of the colored portion of the flood at the instant the ion front first touches the output-well electrodes, and dividing by the area of the network element. In this way it is readily found that the efficiency of the direct line-drive flood, in which the input and output lines are placed in alternate parallel lines, is 57 ± 3 per cent; that of the regular five-spot flood, in which the output lines are staggered with respect to the input lines, the distance between the lines of the same kind being equal to the well spacing within the lines, is 75 ± 3 per cent; and that of the seven-spot flood, in which each output or input well is surrounded by six input or output wells placed at the corners of a hexagon, is 77 to 80 per cent. The range of measured values is due to the effects of diffusion and capillarity in the model which in general tend to make the observed area of flood somewhat larger than in the ideal flood.

Another particularly interesting application of the electrolytic model is that of studying the effect of impermeable barriers which may be present in the sand and tend to distort the advance of the flood. Such systems are practically impossible of treatment by any other means, and yet may be investigated with the electrolytic model by simply cutting out of the porous medium sections of the same shape as the assumed barrier. (*cf.* Figs. 232 and 233.)

The differences in the behavior of the various flooding networks are clearly due to the differences in the pressure distribution in the networks as caused by their varying geometrical forms. Because of the lack of strict homogeneity in the condition of the electrolyte in the electrolytic model it cannot be used to determine the pressure distributions. This phase of the problem can, however, be studied by means of the equally simple electrical-conduction-sheet model. Here a thin sheet of a homogeneous metallic conductor is cut to the shape of the element of the network of interest, with electrodes attached to simulate the input and output wells, and the potential distribution is measured by a bridge or potentiometer null method. The exact equivalence of the electrical-conduction system to that of liquid flow in a porous medium has already been pointed out in Sec. 3.6.

By means of the pressure distributions obtained by the sheet-conduction models (*cf.* Figs. 236–239), one can readily interpret the details of the flood history as derived from the electrolytic models. In fact, one can predict both the qualitative and quantitative features of any flooding system once the pressure distribution has been mapped and plotted.

In addition to these pressure distributions, the sheet-conduction model is suited for the measurement of the resistance of any particular flooding network. This resistance gives the pressure differential required to induce a unit production rate in the flooding network for a unit-permeability sand and unit-viscosity fluid.

Although the history of the flood advance as shown by the electrolytic models and the pressure distributions and resistances given by the sheet-conduction models suffice to give a complete description of any particular flooding arrangement, they suffer from the disadvantage that, especially with respect to the latter features, they must be repeated every time the relative dimensions of the network elements are changed. In the analytical treatment, on the other hand, the absolute dimensions are kept throughout as parameters which need be specified only in the final numerical interpretation. To compute the resistance, or its reciprocal, the "conductivity," of a given network, one simply adds up the individual logarithmic pressure contributions of the various wells, and then finds the pressure differential acting on the system by computing the difference of the value of this sum at a representative input and output well. By dividing this difference into the flux rate per network element, the conductivity per network element is obtained.

Such calculations show that the flood conductivity per network element of the seven-spot flood is approximately 39 per cent higher than that of the five-spot flood. In all cases the conductivity decreases slowly with increasing well spacing, an increase of the input-output well spacing from 400 to 800 ft. decreasing the conductivity by only about 10 per cent (*cf.* Fig. 243).

Although the five- and seven-spot floods are completely defined by their geometry, the line-drive flood may be modified from its regular pattern both by staggering the wells in the alternate input and output lines and varying the distance between the lines as compared to the well spacing within the lines. The analysis shows that when the distance between the lines is appreciably less than the well spacing within the lines, the staggering will slightly lower the conductivity, but that for line separations of the order of, or greater than, the well spacing, the effect is entirely negligible. Stretching the line drive gives larger decreases of the conductivity, although the effect is still quite small (*cf.* Fig. 241). Thus for a fixed well spacing the conductivity decreases by only 17 per cent as the line spacing is increased from a value equal to the well spacing to one twice as great.

When it is observed that each network element of a seven-spot flood contains three wells, while those of the five-spot or line-drive floods contain only two wells, it is seen that a comparison of their conductivities, *per network element*, will give the seven-spot flood an apparent advantage, since in the same number of network elements it will have 50 per cent more wells than the other systems. In fact, when the more practical comparison on the basis of equal well-per-acre densities is made, it is found that the seven-spot flood has even a lower conductivity than the regular line-drive flood, while the five-spot flood has the highest conductivity.

The analytical computation of the efficiency of a flooding network is even simpler than that of its resistance. For one has only to compute the time required for the water first to reach the output wells. This time, multiplied by the rate of water input per network element, gives the area flooded by the time the water first reaches the output wells, and this area divided by that of the network element gives at once the flood efficiency.

In this way the efficiencies of the seven and five-spot floods are found to be 74.0 and 72.3 per cent, respectively, which agree, within the experimental errors, with those derived from the electrolytic models. While these values are fixed by the geometry of these two networks and are independent of the absolute well spacings, pressure differentials, sand permeability, or liquid viscosity (neglecting the difference between the oil and water), the efficiency of the line-drive flood may be varied both by stretching and staggering the drive. In contrast to what was found in the case of the flood conductivity where the staggering had practically no effect, the efficiency of the line-drive flood is very materially increased by staggering, especially when the line separations are less than twice the well spacings within the lines. Thus when the line spacing is one half of that within the lines—which corresponds to the five-spot flood—the unstaggered flood would have an efficiency of only 31.4 per cent; this is increased to the five-spot efficiency of 72.3 per cent simply by staggering (*cf.* Fig. 246).

Stretching the line-drive flood also has a marked effect on the efficiency, and changing the ratio of the line spacing to the well-spacing within the lines from 1:2 to 2:1 will increase the efficiency from 31.4 to 77.9 per cent. If now the stretched flood is staggered, the efficiency will be still further increased—to 83 per cent in the last example. Thus by a combination of stretching and staggering the efficiency of the regular line drive can be raised from the low value of 57 per cent to appreciably exceed the value of 74 per cent of the seven-spot flood.

Noting that a low flood conductivity can be compensated for by increasing the pressure differential driving the flood, whereas the flood efficiency cannot be changed once the geometry of the network has been chosen, it is readily seen that a suitably stretched and staggered line drive possesses the most favorable physical features. Choosing for definiteness a staggered line drive in which the line separations are 1.50 times the well spacing within the lines, it is found that it has a conductivity of 12 per cent less than the five-spot flood, but equal to that of the seven-spot flood, for the same well-per-acre densities. On the other hand, its efficiency will be 80 per cent in comparison with those of 72.3 and 74 per cent of the five- and seven-spot floods. Hence the seven-spot array is certainly the least favorable network, and the relative efficiency excess of 11 per cent of the line drive over the five-spot array should more than counterbalance its higher resistance.

Although it has thus been found by a strictly theoretical investigation that the stretched and staggered line drive should be preferable to the more frequently used five-spot or seven-spot floods, there are other phases of the practical flooding problem which are even more serious than that of the particular value of the well spacing or network arrangement. For if the time required for the water to travel from the input to the output wells is computed for a practical situation corresponding approximately to the Bradford field, it is found that these times are of the order of 10 times those actually observed. This can only mean that there is much more by-passing of water in the Bradford sand than would be possible in an ideal homogeneous sand. This in turn shows that there is considerable channeling in the sands, owing either to high-permeability beds or streaks of high permeability in the main body of the sand. As such channeling can far more than offset any advantages obtainable by variations of the network arrangement, the most important practical problem for the effective development of a flooding program is that of plugging off high-permeability beds or streaks. Likewise, in view of the slow variation of the conductivity with the well separation and the fact that the efficiency is strictly independent of the well spacing, the question of well spacing in flooding problems is of minor physical significance and must be solved primarily on the basis of economic considerations relevant to the particular flooding program of interest.

PART III

THE NONSTEADY-STATE FLOW OF LIQUIDS

CHAPTER X

THE FLOW OF COMPRESSIBLE LIQUIDS THROUGH POROUS MEDIA

10.1. Introduction.—The discussions of the various problems of fluid flow developed in Part II have all been based upon the assumption that the flow systems either were 'permanently in a "steady state" or were experiencing a "continuous succession of steady states." The time variable, when it entered at all, played the role of a parameter rather than an independent variable, and did not occur in the fundamental differential equation—Laplace's equation for the potential or pressure distribution. Thus while the analytical details were carried through with the explicit assumption of time-independent boundary conditions, this was done with the understanding that if the boundary conditions did vary with the time, the potential distribution would also vary, but its variation would be such that each instantaneous state would be a *steady-state* distribution appropriate to the corresponding instantaneous values of the boundary conditions (*cf.* Sec. 3.4). In the developments of Chap. VIII, where two-fluid systems were discussed, the time variable did play a more significant role, since the boundaries of the regions of interest were changing with the time, and the *nature* of the potential distributions, as well as the absolute values of the potentials, were varying continually. However, even there the histories of these flow systems were described in terms of continuous sequences of states, each of which being the *steady-state* distribution appropriate to the corresponding instantaneous shape and position of the boundaries of the system and the instantaneous values of the boundary variables.

The strict physical implications of the assumption that a "continuous succession of steady states" will give the correct representation of the physical time variations in a system that is actually in an unsteady state of flow, are that the velocity of propagation of disturbances in the porous medium is infinite, and that the variations in the mass fluid content of the system associated with those in the boundary conditions are small as compared to the steady-state flux through the system. The infinite velocity of the propagation of disturbances in the porous medium would have as a consequence that any change in the boundary conditions or even shape of the boundaries would be immediately conveyed to all interior points of the system, and, if there were no "attenuation" in the propagation of such changes, it would result in instantaneous adjustments of the pressure distributions to ones appropriate to the new boundary conditions or bounding surfaces. However, insofar as the porous medium and its fluid form a continuous and homogeneous distribution of matter, the propagation of disturbances (pressure variations) through it must be equivalent to that in continuous elastic media. Such elastic media, on the other hand, are characterized—except for dispersion effects—by definite maximal velocities of propagation given by the velocity of longitudinal waves in the medium, namely, $v = \sqrt{E/\gamma}$, where E is its bulk modulus and γ its density. As γ is always of the order of 1, the assumption of infinite v therefore implies an infinitely high E, and hence a vanishing compressibility.

Under laboratory conditions the compressibilities of such liquids[1] as are found in underground reservoirs are known to have magnitudes of the order of 10^{-4} to 10^{-5}/atm. It would appear, therefore, that the velocity of propagation is *not* infinite, but has a definite upper limit. However, it is clear that *physically* it is not the absolute magnitude of the velocity of propagation which is of primary importance, but rather its magnitude *relative* to the fluid velocity in the medium. Now the maximum fluid velocities that may be expected in ordinary reservoirs will be of the order of 1 cm./sec. (*cf.* Sec. 2.2)—except within the first foot of sand about the well bore—whereas the longitudinal-wave velocity will be of the order of 10^5 cm./sec. Evi-

---

[1] The compressibility of the medium itself is only of second-order significance as long as the *fluid* pressures are less than the mean internal pressure within the grains of the medium, as is practically always the case. A change in the *fluid* pressure will then affect only the details of the pressure distribution within the individual grains, leaving the mean internal pressure the same, and hence the amplitude of the "disturbance" propagated through the medium itself will be of an order smaller than that transmitted through the liquid in the medium.

dently, then, the numerically high velocity of longitudinal waves is also equivalent to a very high *relative* velocity, when compared with the actual particle velocities in the medium. The assumption of the fluid incompressibility or infinitely high velocities of propagation of disturbances should, therefore, be quite valid under normal conditions, and the developments of Part II may be considered as resting upon a physically sound basis, with respect to this implication of the assumption of the "continuous succession of steady states."

The second implication that the variations in the mass fluid content of the flow system, associated with changes in the boundary conditions, should be small as compared to the steady-state flux through the system, is a more severe condition for the justification of the assumption of the "continuous succession of steady states." For unless this is satisfied, the amplitudes propagated into the interior of the porous medium of the pressure variations at the boundary will be so attenuated that the readjusted internal distribution will not correspond at all to the steady-state distribution appropriate to the new boundary conditions. Thus the changes in the pressure distribution will at the same time involve changes in the density distribution. Now if the flow system is of large dimensions or if the compressibility of the liquid is high, the mass of fluid involved in this readjustment may become so large as to require many days for its absorption, if the average pressure has been raised, or for its removal if the average pressures have been lowered, and there will be a corresponding delay in the reestablishment of the steady-state distribution appropriate to the new boundary pressures.

A somewhat more specific formulation of this condition is given by considering a radial system of external radius r_e, in which the rate of change in the average pressure due to those at the boundaries is denoted by $\dfrac{dp}{dt}$. The corresponding rate of change in the mass content of the system, per unit thickness, is then approximately

$$\frac{dM}{dt} = \pi r_e{}^2 f \beta \bar{\gamma} \frac{dp}{dt}, \tag{1}$$

where $\bar{\gamma}$ is the average fluid density, β the fluid compressibility, and f the porosity of the medium. Now the steady-state mass flux passing through the system for a difference in density $\Delta\gamma$ between r_e and the internal (well) radius r_w, will be, for a liquid of viscosity μ,

$$Q_0 = \frac{2\pi k \Delta\gamma}{\beta\mu \log r_e/r_w}, \qquad [cf.\ Eq.\ (15)] \qquad (2)$$

k being the permeability of the medium. The ratio will therefore have the value

$$\frac{\dfrac{dM}{dt}}{Q_0} = \frac{\mu f \bar{\gamma} \beta^2 r_e{}^2 \dfrac{dp}{dt} \log \dfrac{r_e}{r_w}}{2k\Delta\gamma} \cong 100\beta^2 r_e{}^2 \frac{dp}{dt}, \qquad (3)$$

for $f = 0.2$, $k/\mu = 0.1$, $(\bar{\gamma}/\Delta\gamma) \log r_e/r_w \sim 100$. Hence for $\beta = 5 \cdot 10^{-5}$/atm., and $r_e = 500$ ft. $= 15.24 \cdot 10^3$ cm., and for a pressure variation as rapid as 1 atm./day, the above ratio will still equal only $0.67 \cdot 10^{-3}$. In systems of moderate dimensions flowing liquids of normal compressibility, one is, therefore, fully justified in using the steady-state solutions of Part II, with a formal neglect of the effects of the liquid compressibility, and in synthesizing continuous sequences of such steady-state solutions to give the physically significant variations with time in the pressure distributions or fluxes if the systems are not strictly in the steady state.

A similar justification of the use of the concept of the continuous succession of steady states is obtained by computing the time required for the establishment of the steady state in the same case of radial flow. Thus it may be readily verified that the mass content of a radial system bounded by radii r_e, $r_w(r_w \ll r_e)$, which is initially of a uniform density, γ_e, will be lowered by the quantity

$$Q = \frac{\pi r_e{}^2 f(\gamma_e - \gamma_w)}{2 \log r_e/r_w}, \qquad (4)$$

by the time the assumed uniform density has readjusted itself to the steady-state pressure distribution in which the density at r_w is γ_w, that at r_e being maintained at γ_e. Hence assuming that the early stage in the flow history simply consists in a steady-state removal of the excess mass content of the reservoir between r_w and r_e, it would take a time of the order of[1]

$$t \sim \frac{Q}{Q_0} = \frac{f\beta\mu r_e^2}{4k}.$$ (5)

For a system with the same dimensions as those used above, t is found to have the value of only 1.6 hr. Thus here, again, it is seen that the analysis of the various problems presented in Part II on the basis of the steady-state Laplace's equation may be considered as sufficiently exact for all practical purposes, as long as the liquids possess normal compressibilities and the geometrical dimensions of the flow systems are of moderate magnitude.

When, however, the flow system is of very large dimensions, or when the liquid has an abnormally high compressibility, or both, the assumption of the "continuous succession of steady states" will no longer be valid, and one must use a theory which explicitly takes into account the transient states in the pressure or density distribution, which in general will be quite different from those of the steady state corresponding to the instantaneous values of the boundary pressure or fluxes.[2] This is clear from the fact that both the ratio in Eq. (3) and the time t in Eq. (5) increase with β and r_e. Hence if, for example, the effective "reservoir radius" is 5 miles rather than 500 ft., the time t of Eq. (5) will be increased to 186 days, and the ratio of Eq. (3) will increase to 1.87. And if the compressibilities are also higher than normal, these ratios will become still larger and the possibility of a satisfactory steady-state description of the details of the flow characteristics will be even more remote.

---

[1] Although the flux through the system would be infinitely large at the initial instant and would approach the steady-state value only exponentially, the value of Eq. (5) should give the correct order of magnitude of t. The accurate theory of this problem will be given in Sec. 10.4.

[2] An important practical example of a case where the time transients play the predominant role in the production characteristics of the flow system is that of the "East Texas" oil field, where practically all of the production of the first $3\frac{1}{2}$ years—520 million barrels—has been due to the expansion of the water in the Woodbine sand within a radius of 20 miles about the field (*cf.* Sec. 10.8). Indeed, in this case the compressibility must be given an abnormally high value such as may be attributed to distributed free-gas masses.

Such abnormally high liquid compressibilities as just referred to will occur when there is free gas dispersed in the liquid in the porous medium. By this is not meant the free gas that will always be found—except for supersaturation effects—in the pores of a porous medium carrying a gas-saturated liquid wherever the liquid pressure has fallen below the saturation pressure. For then the fluid takes on the characteristics of a gas-liquid "mixture," and although it will exhibit many of the features of a compressible liquid, it can no longer be treated in detail as a *homogeneous* fluid obeying Darcy's law. What is meant is rather a dispersion of free gas in the form of localized gas "pockets," of dimensions small compared to those of the whole fluid reservoir and distributed within the reservoir with some degree of uniformity. While the main body of liquid may still be considered as essentially incompressible, the presence of the gas pockets will evidently give an elastic capacity to the whole of the fluid system which will result in a large-scale behavior much as if the liquid itself were appreciably compressible.

The volume of such dispersed gas masses required to give a resultant effective compressibility of $\bar{\beta}$ to a liquid of original compressibility β_0 is easily shown to be given by

$$x = \frac{\bar{\beta} - \beta_0}{\dfrac{1}{p} - \beta_0}, \qquad (6)$$

where x is the fraction of the total fluid volume occupied by free gas, and p is the fluid pressure, assuming an ideal gas and neglecting its solubility in the liquid. Thus if $\beta_0 = 5 \cdot 10^{-5}$/atm. and $p = 100$ atm., a distributed free-gas content of only $4\frac{1}{2}$ per cent would give a resultant compressibility 10 times as large as β_0. x itself as well as $\bar{\beta}$ will vary with the pressure, but it will usually suffice to use average values of x and p in estimating the effective resultant compressibility.

Although the chief interest in taking into account the compressibility of the liquids flowing in a porous medium is in the study of the *time* variations in the pressure and flux distributions in a flow system consequent upon variations—continuous or discontinuous—in the boundary conditions, it must be remembered that the explicit assumption of a nonvanishing compressibility does not exclude the possibility of steady-state conditions of flow for compressible liquids. On the contrary, every compressible-fluid system will tend to approach a steady-state condition of flow regardless of its previous history, if and when the boundary values of the flux or pressure assume a permanent distribution. The compressibility of the fluid will determine the details and *rate* of the approach—always of infinite duration in a strict mathematical sense—to the steady-state condition, but once the latter has been attained, the compressibility no longer remains in the problem except in the translation of the fluid-density distribution in the system into its equivalent pressure distribution.

Thus it will be recalled from Sec. 3.4 that for a general non-steady state of flow of a compressible liquid in a homogeneous porous medium (effect of gravity being neglected) the governing equation is

$$\frac{\partial^2\gamma}{\partial x^2} + \frac{\partial^2\gamma}{\partial y^2} + \frac{\partial^2\gamma}{\partial z^2} = \frac{f\beta\mu}{k}\frac{\partial\gamma}{\partial t}, \qquad (7)$$

where γ is the density of the liquid, β is its compressibility, μ is its viscosity, and k and f are the permeability and porosity of the medium, respectively, and the density is related to the fluid pressure by the equation

$$\gamma = \gamma_0 e^{\beta p}, \qquad (8)$$

γ_0 being the density at zero pressure. It is clear that if the boundary conditions do not vary with the time, the right side of Eq. (7) may be equated to zero, leaving as the equation for the *steady state*

$$\frac{\partial^2\gamma}{\partial x^2} + \frac{\partial^2\gamma}{\partial y^2} + \frac{\partial^2\gamma}{\partial z^2} = 0. \qquad (9)$$

This equation is nothing more than Laplace's equation, and is identical with that for the pressure distribution in an incompressible-liquid system. In fact, it follows at once from this observation that all the solutions for the problems treated in Part II, except those in Chaps. VI and VIII, can be immediately reinterpreted as those for corresponding compressible-liquid systems by simply replacing the term "pressure" by "density" with respect to both the distributions within the flow systems and those on their boundaries. For example, when the system is linear, with densities γ_1, γ_2 at the boundaries, the density distribution is

$$\gamma = \gamma_2 + \frac{\gamma_1 - \gamma_2}{L}x, \tag{10}$$

where L is the length of the linear channel. In terms of the pressure distribution this is equivalent to the more complicated expression

$$p = \frac{1}{\beta} \log\left[\frac{(e^{\beta p_1} - e^{\beta p_2})}{L}x + e^{\beta p_2} \right]. \tag{11}$$

The mass flux through the system is then

$$Q = -\frac{k\gamma}{\mu}\frac{\partial p}{\partial x} = -\frac{k}{\beta\mu}\frac{\partial \gamma}{\partial x} = \frac{k}{\beta\mu}\frac{\gamma_2 - \gamma_1}{L} = Q_0(1 + \beta\bar{p}) \tag{12}$$

to terms in β^2, where Q_0 is the flux for the case of an incompressible liquid of density γ_0 and viscosity μ, and $\bar{p}$ is the algebraic mean pressure.

Similarly for a radial-flow system, an analysis similar to that given in Sec. 4.2 gives at once

$$\gamma = \frac{\gamma_e - \gamma_w}{\log r_e/r_w} \log \frac{r}{r_w} + \gamma_w, \tag{13}$$

with a pressure distribution of the form

$$p = \frac{1}{\beta} \log\left\{ \frac{(e^{\beta p_e} - e^{\beta p_w}) \log r/r_w}{\log r_e/r_w} + e^{\beta p_w} \right\}, \tag{14}$$

and a mass flux

$$Q = \frac{2\pi k(\gamma_e - \gamma_w)}{\beta\mu \log r_e/r_w} = Q_0(1 + \beta\bar{p}) \tag{15}$$

to terms in β^2, Q_0 being the radial mass flux for an incompressible liquid of density γ_0.

As $\gamma_0(1 + \beta\bar{p})$ is, to terms in β^2, the average density of the liquid in the linear- and radial-flow systems, it follows from Eqs. (12) and (15) that the mass flux in the steady-state compressible-liquid systems may be computed, to terms in β^2, by taking for the liquid density the algebraic mean of those at the boundaries, but otherwise using the ordinary formulas where the compressibility is neglected. This result also holds for all steady-state-flow systems with uniform pressures or densities over the boundaries, as the replacement of Δp by $\Delta \gamma$ will always lead to the factor $(1 + \beta \bar{p})$ multiplying Q_0.

The details of the pressure distribution in a compressible-liquid system as compared to that in which the compressibility is neglected may be studied by means of equations as (11) and (14). These will evidently differ by terms proportional to β. Furthermore it may be shown that for uniform boundary pressures the pressure gradient, and hence velocity, at any point in a steady-state compressible-liquid system will be less or greater—to terms in β^2—than that in the same incompressible-liquid system with the same boundary pressures as the density at the point is greater or less than the algebraic mean density.

As to the detailed treatment of nonsteady-state-flow systems, one could attempt to study the nonsteady-state behavior of all the various systems discussed in Part II. However, as has already been pointed out, the assumption of a steady state or of a continuous succession of steady states may be justifiably used under practically all conditions, except when the liquid has an abnormally high compressibility or when the dimensions of the flow system are quite large. The problems to be treated in Part III will, therefore, be confined to those in which the effect of the compressibility may have some real physical significance, and to such others as illustrate types of transient behavior that may be of interest for the qualitative discussion of gas-liquid-mixture systems. Of course, in the study of the first class of problems the actual dimensions will enter the analysis only as parameters which may be ultimately chosen as being either large or small. However, systems with numerical dimensions of intermediate magnitude will possess physical interest only when they are associated with rather high fluid compressibilities.[1]

Finally, it should be noted that for analytical convenience all the boundary conditions and values will be expressed in terms of

---

[1] In fact, as will be seen below, or as follows directly by transforming the coordinate and time variables in Eq. 10.2(1) so they become dimensionless, the fundamental variables in the analytical treatment will involve the product of the compressibility and the square of one of the dimensions of the system, and the time transients will be of importance only when this product is of appreciable magnitude [*cf.*, for example, Eq. (5)].

the density γ rather than the pressure p. In the final physical interpretation the translation into the equivalent values of p will be made by means of Eq. (8).

10.2.[1] Radial-flow Systems. Some Preliminary Analytical Formulas.—When the flow system is radial, the general Eq. 10.1(7), governing the motion of compressible liquids in porous media, may be rewritten in plane-polar coordinates as

$$\kappa\left\{\frac{\partial^2\gamma}{\partial r^2} + \frac{1}{r}\frac{\partial\gamma}{\partial r}\right\} = \frac{\partial\gamma}{\partial t}; \qquad \kappa = \frac{k}{f\beta\mu}. \tag{1}$$

This equation has four essentially different elementary solutions, which are

$$\gamma = \log r \tag{2}$$

$$\gamma = r^2 + 4\kappa t \tag{3}$$

$$\gamma = \left.\begin{matrix} J_0(\alpha r) \\ Y_0(\alpha r) \end{matrix}\right\} e^{-\kappa\alpha^2 t} \tag{4}$$

$$\gamma = \frac{1}{t} e^{-r^2/4\kappa t}, \tag{5}$$

where α is any real number[2] and J_0, Y_0 are the zero-order Bessel functions of the first and second kind, respectively. As will be seen in the following sections, the various solutions corresponding to different types of physical radial systems may be synthesized by forming appropriate linear combinations of the first three of the above solutions. The solution of Eq. (5) will be applied to a somewhat different type of problem in Sec. 10.15.

In the synthesis of these solutions it will be convenient to make use of the fact that if $\gamma_0(r, t)$ is a solution of Eq. (1) which vanishes at $t = 0$ and equals 1 at $r = r_0$, then

---

[1] The analyses of this and the following sections, up to Sec. 10.15, is taken from M. Muskat, *Physics*, **5**, 71, 1934.

[2] Equation (4) will also be a solution if α is complex. However, unless the real part of α^2 is positive, it would be of no physical significance with respect to the problems of interest here. These solutions of Eq. (4) may be derived by the method of the separation of variables illustrated in Sec. 7.9. The detailed properties of the functions Y_0, as well as J_0, will be found in "Modern Analysis," Chap. XVII, and in the treatise by G. N. Watson which is entirely devoted to the study of Bessel functions, namely, "The Theory of Bessel Functions," 1922.

$$\gamma = \int_0^t f(\lambda)\frac{\partial}{\partial t}\gamma_0(r, t - \lambda)d\lambda \qquad (6)^1$$

is a solution of Eq. (1) which also vanishes at $t = 0$, but has the value $f(t)$ at $r = r_0$. By means of this formula a solution with variable boundary conditions may be derived from one with constant boundary values.

Frequent use will also be made of the following integrals. Denoting by $U(\alpha r)$ the function

$$U(\alpha r) = AJ_0(\alpha r) + BY_0(\alpha r), \qquad (7)$$

where A and B are any two arbitrary constants which may depend on α but not on r, then

$$\int_a^b rU(\alpha r)dr = \frac{-1}{\alpha^2}\left[r\frac{dU}{dr}\right]_a^b, \qquad (8)$$

$$\int_a^b rU(\alpha r)\log r dr = \frac{1}{\alpha^2}\left[U - r\frac{dU}{dr}\log r\right]_a^b, \qquad (9)$$

$$\int_a^b rU(\alpha r)U(\alpha'r)dr = \frac{1}{\alpha^2 - \alpha'^2}\left[rU(\alpha r)\frac{dU(\alpha'r)}{dr} - rU(\alpha'r)\frac{dU(\alpha r)}{dr}\right]_a^b, \qquad (10)$$

$$\int_a^b rU^2(\alpha r)dr = \frac{1}{2}\left\{r^2U^2(\alpha r) + \frac{1}{\alpha^2}\left[r\frac{dU(\alpha r)}{dr}\right]^2\right\}_a^b. \qquad (11)$$

These may be verified without difficulty on making use of the fact that the $U(\alpha r)$ satisfy the differential equation of the Bessel functions, namely,

$$\frac{1}{r}\frac{d}{dr}\left[r\frac{dU(\alpha r)}{dr}\right] = -\alpha^2 U(\alpha r). \qquad [cf.\ \text{Eq. 7.9(6)}] \qquad (12)$$

Finally it will be convenient to use the relation

$$J_0(\alpha r)Y_1(\alpha r) - Y_0(\alpha r)J_1(\alpha r) = \frac{-2}{\pi\alpha r}, \qquad (13)^2$$

where J_1, Y_1 are the Bessel functions of the first order, and are equal to the negative derivatives of J_0, Y_0.

---

[1] This formula is a special case of Duhamel's theorem, a discussion and proof of which will be found in Carslaw's "Theory of the Conduction of Heat," 2d ed., p. 16, 1921.

[2] *Cf.* Watson, *loc. cit.*, p. 76.

As in the case of problems of steady-state flow, a radial system will be defined by two concentric circular boundaries of radii r_1, r_2, over each of which the density (and hence pressure) or flux will be distributed with uniform fixed or time-varying values. When one of the boundaries is a well of practical dimensions—small as compared to all others in the system—with a specified flux rate, the internal boundary will be replaced by a mathematical sink.

10.3. Radial Systems in Which the Density Is Specified over Both Boundaries.—The general case of a radial system in which the density is specified over both boundaries may be defined by the conditions

$$\begin{aligned}
\gamma = f_1(t): \quad & r = r_1 \\
= f_2(t): \quad & r = r_2 \\
= g(r): \quad & t = 0,
\end{aligned} \tag{1}$$

where $g(r)$ gives the initial density (or pressure) distribution. The solution to this problem will give not only the pressure distribution at any later instant, but also the fluxes entering or leaving the system.

To derive a solution satisfying all the conditions of Eq. (1) simultaneously, it is convenient first to set up the preliminary and simpler solutions γ_1, γ_2, γ_3, defined by

$$\begin{aligned}
(\gamma_1, \gamma_2, \gamma_3) = (1, 0, 0): \quad & r = r_1 \\
= (0, 1, 0): \quad & r = r_2 \\
= [0, 0, g(r)]: \quad & t = 0.
\end{aligned} \tag{2}$$

For if from these the solution

$$\gamma = \gamma_3(r, t) + \int^t \left[f_1(\lambda) \frac{\partial \gamma_1(r, t - \lambda)}{\partial t} + f_2(\lambda) \frac{\partial \gamma_2(r, t - \lambda)}{\partial t} \right] d\lambda \tag{3}$$

is constructed, all the conditions of Eq. (1) will be satisfied simultaneously, as a consequence of Eq. 10.2(6). The individual solutions γ_1, γ_2, γ_3 may now be developed as follows: Introducing the function $U(\alpha_n r)$,

$$U(\alpha_n r) = Y_0(\alpha_n r_2) J_0(\alpha_n r) - J_0(\alpha_n r_2) Y_0(\alpha_n r), \tag{4}$$

and choosing α_n such that

$$U(\alpha_n r_1) = 0, \tag{5}$$

γ_1, γ_2, and γ_3 may be expanded as

$$\gamma_1 = \frac{\log r/r_2}{\log r_1/r_2} + \sum_{\alpha_n} A_n U(\alpha_n r) e^{-\kappa \alpha_n^2 t}, \tag{6]1}$$

$$\gamma_2 = \frac{-\log r/r_1}{\log r_1/r_2} + \sum B_n U(\alpha_n r) e^{-\kappa \alpha_n^2 t}, \tag{7}$$

$$\gamma_3 = \sum C_n U(\alpha_n r) e^{-\kappa \alpha_n^2 t}, \tag{8}$$

from which it is immediately clear that γ_1, γ_2, γ_3 satisfy the conditions of Eq. (2) at $r = r_1$, r_2, regardless of the values of the constants A_n, B_n, and C_n. These are to be so chosen that the γ_1, γ_2, γ_3 also satisfy the requirements of Eq. (2) at the initial instant, $t = 0$. Thus setting $t = 0$ in Eqs. (6), (7), and (8), it is clear that if Eq. (2) is to be satisfied, the A_n, B_n, C_n must satisfy the equations

$$\sum A_n U(\alpha_n r) = \frac{-\log r/r_2}{\log r_1/r_2}, \tag{9}$$

$$\sum B_n U(\alpha_n r) = \frac{\log r/r_1}{\log r_1/r_2}, \tag{10}$$

$$\sum C_n U(\alpha_n r) = g(r). \tag{11}$$

Such series as on the left are known as Fourier-Bessel series, and are completely analogous to the Fourier series discussed in Sec. 4.3, the only essential difference being the replacement of the trigonometric sine and cosine terms of the latter by the linear combinations of Bessel functions $U(\alpha_n r)$. Formally, however, the important fact is that, subject to conditions similar to those required in the case of Fourier-series expansions, and which will always be satisfied in problems of physical interest, any arbitrary function as $g(r)$ can be expanded in the form of the series of Eqs. (9), (10), and (11) such that the sum of the series at any point where $g(r)$ is continuous equals the value of $g(r)$, and the algebraic mean of the right and left limits of $g(r)$ at any point of discontinuity.[2]

---

[1] The summations in the series here and in all the similar series in this chapter are to be extended only over the positive roots α_n of Eq. (5), or its analogue in the case of the other similar series.

[2] *Cf.* Watson, *loc. cit.* Chap. XVIII.

The determination of the coefficients C_n, as in Eq. (11), may now be determined by a procedure quite similar to that giving the coefficients in a Fourier series. Thus multiplying through both sides of Eq. (11) by $rU(\alpha_m r)dr$, and integrating each side between r_1 and r_2, it is seen by Eq. 10.2(10) and Eqs. (4) and (5), that the integrals with coefficients C_n will all vanish. For that with $n = m$, however, Eq. 10.2(11) gives a nonvanishing value, with the result that

$$\frac{C_m}{2\alpha_m^2}[r_2^2 U'^2(\alpha_m r_2) - r_1^2 U'^2(\alpha_m r_1)] = \int_{r_1}^{r_2} rg(r)U(\alpha_m r)dr, \quad (12)$$

where the prime denotes differentiation with respect to r. There is thus obtained an explicit expression for each coefficient C_m. Furthermore the A_m and B_m will evidently also be given by Eq. (12) if for $g(r)$ in Eq. (12) are substituted the right sides of Eqs. (9) and (10), respectively. Carrying out the resulting integrations with the aid of Eqs. 10.2(8) and 10.2(9), then simplifying by means of Eq. 10.2(13), which shows that

$$U'(\alpha_n r_2) = \frac{-2}{\pi r_2}; \qquad U'(\alpha_n r_1) = \frac{-2J_0(\alpha_n r_2)}{\pi r_1 J_0(\alpha_n r_1)}, \quad (13)$$

and finally putting in the values of A_n, B_n, C_n in Eqs. (6), (7), and (8), it is readily found that γ_1, γ_2, γ_3 assume the forms

$$\gamma_1 = \frac{\log r/r_2}{\log r_1/r_2} + \pi \sum \frac{J_0(\alpha_n r_1)J_0(\alpha_n r_2)U(\alpha_n r)e^{-\kappa \alpha_n^2 t}}{J_0^2(\alpha_n r_1) - J_0^2(\alpha_n r_2)}, \quad (14)$$

$$\gamma_2 = \frac{-\log r/r_1}{\log r_1/r_2} - \pi \sum \frac{J_0^2(\alpha_n r_1)U(\alpha_n r)e^{-\kappa \alpha_n^2 t}}{J_0^2(\alpha_n r_1) - J_0^2(\alpha_n r_2)}, \quad (15)$$

$$\gamma_3 = \frac{\pi^2}{2} \sum \frac{\alpha_n^2 J_0^2(\alpha_n r_1)U(\alpha_n r)e^{-\kappa \alpha_n^2 t}}{J_0^2(\alpha_n r_1) - J_0^2(\alpha_n r_2)} \int_{r_1}^{r_2} rg(r)U(\alpha_n r)dr. \quad (16)$$

Adding these solutions as indicated in Eq. (3), the final solution satisfying the conditions of Eq. (1) may be expressed as

$$\gamma = \pi \sum \frac{\alpha_n^2 J_0(\alpha_n r_1)U(\alpha_n r)e^{-\kappa \alpha_n^2 t}}{J_0^2(\alpha_n r_1) - J_0^2(\alpha_n r_2)} \left[\frac{\pi}{2} J_0(\alpha_n r_1) \int_{r_1}^{r_2} rg(r)U(\alpha_n r)dr \right.$$
$$\left. - \kappa J_0(\alpha_n r_2) \int_0^t f_1(\lambda)e^{\kappa \alpha_n^2 \lambda}d\lambda + \kappa J_0(\alpha_n r_1) \int_0^t f_2(\lambda)e^{\kappa \alpha_n^2 \lambda}d\lambda \right]. \quad (17)$$

It should be observed in using this equation that to check for the boundary values one must not replace r directly by $r = r_1, r_2$, for then, by virtue of Eqs. (4) and (5), γ will appear to vanish identically. One should rather sum the series first and then make r approach the limiting values of r_1 and r_2. This apparent difficulty arises from the fact that the series of Eqs. (9), (10), and (11), from which γ_1, γ_2, and γ_3 were derived, are discontinuous at r_1 and r_2. However, if the series are summed first and the boundary values for r are substituted in the sums no difficulties will arise.

10.4. Production Decline in a Field Produced by a Water Drive with Variable Field Pressures.—Equation 10.3(17) gives the density distribution at any time t within any system bounded

by the radii r_1, r_2, at which the densities are $f_1(t)$ and $f_2(t)$, and which initially had a radial density distribution $g(r)$. By differentiation one can also obtain the flux through the system as a function of the time. To illustrate the use of Eq. 10.3(17) we shall consider the case of an oil field producing by a water drive in which the whole field is lumped together as a single circular well of radius r_w equal to a dimension of the field (*cf.* Fig. 248). The flux through this radius r_w is then to be considered as the

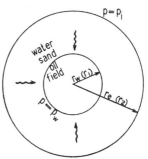

FIG. 248.—Diagrammatic representation of an oil field produced by a radial water drive.

combined production from the various wells in the field.

Supposing that the initial instant is that just before the field is tapped, the initial density distribution may be taken as a uniform density γ_i. The radius r_2 will correspond to the external radius r_e, the pressure at which—and hence the density—will be taken as constant, and equal to the initial pressure. It will be furthermore supposed that at the boundary of the field, defined by the "well radius" $r_w(r_1)$, the pressure, in one case, is suddenly lowered to $p_w(\gamma = \gamma_w)$, and in an other case that the pressure declines continuously so that the density falls linearly to γ_0.

Analytically, these assumptions imply that

$$\left.\begin{array}{l} \gamma = g(r) = \gamma_i; \quad (p = p_i): \quad t = 0 \\ \gamma = \gamma_e = \gamma_i; \quad (p_e = p_i): \quad r = r_e(r_2) \\ (a): \gamma = \gamma_w; \quad \left(p_w = \dfrac{1}{\beta}\log\dfrac{\gamma_w}{\gamma_0}\right) \\ (b): \gamma = \gamma_i - \epsilon t; \quad \left(p_w = \dfrac{1}{\beta}\log\dfrac{\gamma_i - \epsilon t}{\gamma_0}\right) \end{array}\right\} r = r_w(r_1). \tag{1}$$

Putting these into Eq. 10.3(17), and summing the series not containing the time, making use of the expansions of Eqs. 10.3(9) and 10.3(10), it is found that

$$(a): \gamma = \frac{\gamma_i \log \dfrac{r}{r_w} + \gamma_u \log \dfrac{r_e}{r}}{\log r_e/r_w} -$$

$$\pi(\gamma_i - \gamma_w) \sum \frac{J_0(\alpha_n r_e) J_0(\alpha_n r_w) U(\alpha_n r) e^{-\kappa \alpha_n{}^2 t}}{J_0{}^2(\alpha_n r_w) - J_0{}^2(\alpha_n r_e)}, \quad (2)$$

$$(b): \gamma = \frac{\gamma_i \log \dfrac{r}{r_w} + (\gamma_i - \epsilon t) \log \dfrac{r_e}{r}}{\log r_e/r_w} -$$

$$\frac{\pi \epsilon}{\kappa} \sum \frac{J_0(\alpha_n r_e) J_0(\alpha_n r_w) U(\alpha_n r)}{\alpha_n{}^2 [J_0{}^2(\alpha_n r_w) - J_0{}^2(\alpha_n r_e)]}$$

$$+ \frac{\pi \epsilon}{\kappa} \sum \frac{J_0(\alpha_n r_e) J_0(\alpha_n r_w) U(\alpha_n r) e^{-\kappa \alpha_n{}^2 t}}{\alpha_n{}^2 [J_0{}^2(\alpha_n r_w) - J_0{}^2(\alpha_n r_e)]}. \quad (3)$$

That these equations satisfy the boundary and initial conditions of Eq. (1) may be verified by direct substitution.

Of particular practical interest is the flux or production rate from the field that would be obtained under the above conditions. This will evidently be given by

$$Q = \frac{2\pi k}{\mu} \left(r\gamma \frac{\partial p}{\partial r} \right)_{r_w} = \frac{2\pi k}{\mu \beta} r_w \left(\frac{\partial \gamma}{\partial r} \right)_{r_w}. \quad (4)$$

Applying this to Eqs. (2) and (3), it is found that

$$(a): Q = \frac{2\pi k}{\mu \beta} (\gamma_i - \gamma_w) \left[\frac{1}{\log r_e/r_w} - \right.$$

$$\left. \pi r_w \sum \frac{J_0(\alpha_n r_e) J_0(\alpha_n r_w) U'(\alpha_n r_w) e^{-\kappa \alpha_n{}^2 t}}{J_0{}^2(\alpha_n r_w) - J_0{}^2(\alpha_n r_e)} \right], \quad (5)$$

$$(b): Q = \frac{2\pi k \epsilon}{\mu \beta} \left\{ \frac{t}{\log r_e/r_w} - \frac{\pi r_w}{\kappa} \sum \frac{J_0(\alpha_n r_e) J_0(\alpha_n r_w) U'(\alpha_n r_w)}{\alpha_n{}^2 [J_0{}^2(\alpha_n r_w) - J_0{}^2(\alpha_n r_e)]} \right.$$

$$\left. + \frac{\pi r_w}{\kappa} \sum \frac{J_0(\alpha_n r_e) J_0(\alpha_n r_w) U'(\alpha_n r_w) e^{-\kappa \alpha_n{}^2 t}}{\alpha_n{}^2 [J_0{}^2(\alpha_n r_w) - J_0{}^2(\alpha_n r_e)]} \right\}. \quad (6)$$

Applying Eq. 10.3(13) to Eq. (5), and summing the constant series of Eq. (6) by means of Eq. 10.2(8), these may be finally rewritten as

$$(a): Q = \frac{2\pi k(\gamma_i - \gamma_w)}{\mu\beta}\left[\frac{1}{\log \rho} + 2\sum\frac{J_0^2(x_n\rho)e^{-x_n^2\bar{t}}}{J_0^2(x_n) - J_0^2(x_n\rho)}\right], \quad (7)$$

$$(b): Q = 2\pi f\epsilon r_w^2\left\{\frac{\bar{t}}{\log \rho} + \frac{\rho^2}{2(\log \rho)^2}\left[\frac{1}{2} - \frac{1}{2\rho^2} - \frac{1}{\rho^2}\log \rho\right.\right.$$

$$\left.\left. - \left(\frac{\log \rho}{\rho}\right)^2\right] - 2\sum\frac{J_0^2(x_n\rho)e^{-x_n^2\bar{t}}}{x_n^2[J_0^2(x_n) - J_0^2(x_n\rho)]}\right\}, \quad (8)$$

upon introducing the dimensionless notation

$$\frac{r_e}{r_w} = \rho; \qquad \alpha_n r_w = x_n; \qquad \frac{\kappa t}{r_w^2} = \bar{t}. \quad (9)$$

For the purposes of numerical illustration, the physical dimensions of the system will be supposed to have the values

$$r_w \text{ ("well" radius)} = 20 \text{ miles,}$$
$$r_e \text{ (external reservoir radius)} = 100 \text{ miles.}$$

In evaluating now Eqs. (7) and (8), one must determine first the roots α_n of Eq. 10.3(5). In the notation of Eq. (9) the first seven roots for the case $\rho = 5$ are:

n	1	2	3	4	5	6	7
x_n	0.7632	1.5575	2.3470	3.1352	3.9210	4.7073	5.4933

The values of the Bessel functions may then be read off from tables such as those of Hayashi or Watson.[1] The final results are given in Fig. 249, where the quantities $\dfrac{Q\beta\mu}{2\pi k(\gamma_i - \gamma_w)}$ and $\dfrac{Q}{2\pi f\epsilon r_w^2}$, denoted by $\bar{Q}$, are plotted against $\bar{t}$ as curves I and II, respectively.

As should be expected physically, curve I begins with infinitely large values and asymptotically approaches the constant $\dfrac{1}{\log 5} = 0.62$. This is the steady-state flux that will pass through the system after the transient effects represented by the series in Eq. (7), due to the compressibility, have disappeared. Taking $k/\mu = 2$, and β so small that $\gamma_i - \gamma_w \sim \gamma_0(p_e - p_w)$, the

[1] HAYASHI, K., "Tafeln der Besselschen, Theta-, Kugel- und anderer Funktionen," 1930; Watson, *loc. cit.*

value $\bar{Q} = 1$, for curve I, corresponds to about 200 bbl./day/ft. of sand per atmosphere pressure differential ($p_e - p_w = 1$ atm.).

To get the numerical equivalent of the $\bar{t}$ scale, values of the compressibility β and porosity f must be chosen. Taking $f = 0.2$ and $\beta \sim 4.5 \cdot 10^{-4}$ atm.$^{-1}$ (10 times that of water), one has $\bar{t} \sim 1.85 \cdot 10^{-4}t$ (days), so that $\bar{t} = 0.1$ corresponds to approximately $1\frac{1}{2}$ years. This system will, therefore, possess very long period transients, for even after six years the flux rate would still be more than twice that in the steady state.

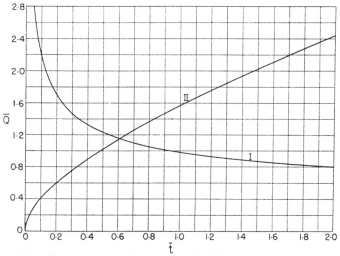

Fig. 249.—The production history of an oil field produced by a water drive maintained at constant pressure. I. Field pressure suddenly dropped to and maintained at p_w; $\bar{Q} = Q\mu/2\pi k\gamma_0\Delta p$, $\bar{Q} = 1 \sim 200$ bbl./day/ft. sand/atm. differential. II. Field pressure decreasing continuously so that the fluid density drops at a constant rate (ϵ); $\bar{Q} = Q\mu/2\pi f\epsilon r_w{}^2$, $\bar{Q} = 1 \sim 150,000$ bbl./day/ft. sand for a pressure decline of 2 lb./day. (*From Physics*, **5**, 80, 1934.)

In the case of curve II, one unit on the scale of ordinates has the numerical equivalent of $4.0\epsilon \cdot 10^{14}$/ft. sand, or approximately 153,000 bbl./day/ft. sand for a pressure decline at r_w of 2 lb./day. However, if the initial pressure is 1,600 lb., the rate of decline can be maintained for only 800 days, or to $\bar{t} \sim 0.15$, at which $\bar{Q} \sim 0.5$. Hence the production rate by the time the field pressure will have dropped to zero will have risen to 75,000 bbl./day/ft. of sand. At the same time, $\bar{Q} \sim 1.9$ for curve I, so that if $\Delta p \sim 1,600$ lb. the system in which the pressure has been suddenly dropped to zero will be producing at the rate of approximately 41,000 bbl./day/ft. of sand.

With regard to the cumulative production, it may be noted that

$$P(a) = \int_0^t Q(a)dt = \frac{\Delta\gamma}{\epsilon}Q(b) \cong \frac{\Delta p}{(dp/dt)}Q(b) = \frac{\Delta p}{2}Q(b), \quad (10)$$

where $Q(a)$ and $Q(b)$ represent the expressions of Eqs. (7) and (8), respectively. Hence the cumulative production for case (a) at $\bar{t} = 0.15$ is

$$P(a) = 800Q(b) = 6 \times 10^7 \text{ bbl./ft. sand.}$$

$P(b)$, on the other hand, may be most readily obtained from the area under curve II, the result being 4×10^7 bbl./ft. of sand. The general variation of $P(b)$ with time will be essentially of the form of t^2, since after an initial short-period transient, $Q(b)$ increases approximately linearly with t. For the same reason, $P(a)$, in virtue of Eq. (10), will increase approximately linearly in t after the initial transients have passed.

Although the actual constants used in the above discussion of Fig. 249 roughly correspond to the "East Texas" oil field,[1] it should be observed that the curves of this figure are independent of the absolute dimensions or numerical values for the physical constants of the flow system. The brackets of Eqs. (7), (8), and the roots x_n, depend only on the ratio $r_e/r_w = \rho$. The absolute values of r_e, r_w, and the actual values of β, f, k, μ, $\Delta\gamma$, enter only in determining the physical equivalents of the ordinate and abscissa scales of Fig. 249. Unless a different value of ρ is chosen, curves I and II may be used for any combination of geometrical and physical constants.

10.5. The Limiting Case of Vanishing Internal Radius.— Although in the derivation of Eq. 10.3(17) the numerical values of r_w and r_e were left perfectly arbitrary, one may not proceed to the limit $\rho \sim \infty$, $r_w \sim 0$. For if the flux should be nonvanishing with a well of vanishing radius, the well must be replaced by a mathematical sink, with a negative infinite pressure. The specification of finite fixed or variable pressures, therefore, becomes meaningless as $r_w \to 0$. If, however, one does not require a nonvanishing flux, the pressure or density at the well ($r_w \sim 0$)

[1] The detailed analysis of the production history of this field will be given in Sec. 10.8.

becomes uniquely determined by that on the external boundary. In a manner similar to that developed in Sec. 10.3 it may be shown that the distribution in γ is then given by

$$\gamma = \frac{2}{r_e{}^2} \sum \frac{J_0(\alpha_n r)}{J_1{}^2(\alpha_n r_e)} e^{-\kappa\alpha_n{}^2 t} \left[\int_0^{r_e} r g(r) J_0(\alpha_n r) dr \right.$$
$$\left. + \kappa r_e \alpha_n J_1(\alpha_n r_e) \int_0^t f_2(\lambda) e^{\kappa\alpha_n{}^2\lambda} d\lambda \right], \quad (1)$$

where

$$J_0(\alpha_n r_e) = 0,$$

$f_2(t)$ is the value of γ maintained at the external boundary $r = r_e$, and $g(r)$ is the initial distribution.

10.6. The Rise of the Bottom-hole Pressures in Closed-in Wells.—An application of Eq. 10.5(1) of some practical interest may be found in the problem of taking closed-in bottom-hole pressures in a producing oil field. The usual procedure is to send down to the bottom of a producing well a recording or maximum reading bottom-hole-pressure bomb, close in the well, and after a certain period remove the bomb and record the reading. This value is taken as the reservoir pressure in the field about the well chosen, and the question arises as to the time that is required after closing in for the pressure at the well to rise to a value appropriate to its "vicinity." The following solution to this problem will not strictly apply to cases in which there is free gas in the sand about the well, owing to the variation in the effective permeability of the sand as the pressure builds up and the gas is compressed. However, it will be valid in fields producing entirely by water drive—as is the case in the "East Texas" field—in which no gas is evolved from the oil until it reaches the well bore; and it will give a *qualitative* description of the pressure rise even in sands flowing both gas and oil, the curve of Fig. 250 falling more steeply in these cases, owing to the rise in the effective permeability as the pressure about the well rises.

Supposing, then, that the well, before closing in, has been flowing in a steady-state condition, its density distribution at the initial instant,[1] $g(r)$ in Eq. 10.5(1), will be given by

---

[1] It is to be understood, of course, that at the "initial instant" for the pressure rise the well bore is filled with liquid.

$$g(r) = \frac{(\gamma_e - \gamma_{wi}) \log r/r_w}{\log r_e/r_w} + \gamma_{wi} \qquad [cf. \text{ Eq. } 10.1(13)] \quad (1)$$

where γ_e is the "reservoir density" to be determined, and γ_{wi} is the flowing bottom-hole density just before the test is made, r_w being the well radius. Since the field as a whole continues flowing, $f_2(t)$ may be given the value γ_e.

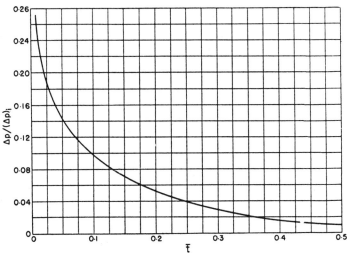

FIG. 250.—The establishment of the "reservoir pressure" on shutting in a well producing a compressible liquid. $(\Delta p)/(\Delta p_i)_- =$ (pressure differential after shutting in)/(flowing pressure differential). $\bar{t} = kt/\mu f \beta r_e^2 =$ dimensionless time. (*From Physics,* **5,** 84, 1934.)

Under these conditions, Eq. 10.5(1) takes the form

$$\gamma = \gamma_e - \frac{2(\gamma_e - \gamma_{wi})}{r_e^2 \log r_e/r_w} \sum \frac{J_0(\alpha_n r)e^{-\kappa \alpha_n^2 t}}{\alpha_n^2 J_1^2(\alpha_n r_e)}. \quad (2)$$

The pressure differential between the well (at $r = r_w$) and the radius r_e is then given by

$$\frac{p_e - p_w}{p_e - p_{wi}} \simeq \frac{\gamma_e - \gamma_w}{\gamma_e - \gamma_{wi}} = \frac{2}{\log \rho} \sum \frac{e^{-x_n^2 \bar{t}}}{x_n^2 J_1^2(x_n)}, \quad (3)$$

where $\rho = r_e/r_w$, $x_n = \alpha_n r_e$ and $\bar{t} = \kappa t/r_e^2$, with p_e, p_w corresponding to γ_e, γ_w. Thus the fractional rise in the pressure differential at any time is independent of the absolute values of the pressures and is determined only by the dimensionless constants in the series of Eq. (3). Equation (3) is plotted in Fig. 250 for

$\rho = 2,000$, and, except for the factor $\dfrac{1}{\log \rho}$, is a universal curve. Thus for $k/\mu = 0.2, f = 0.2, \beta = 4.5 \times 10^{-5}$ atm.$^{-1}$, $r_e = 500$ ft., $\bar{t} = 0.1 \sim t = 0.3$ hr., so that 90 per cent of the original pressure differential would be destroyed in about $\frac{1}{3}$ hr. and 99 per cent in $1\frac{1}{2}$ hr. These values are of the same order of magnitude as those observed in the "East Texas" field where the constants within the field proper correspond approximately to those chosen above.[1]

10.7. Radial Systems in Which the Density Is Specified over One Boundary and the Flux over the Other.—The analytical definition of radial systems in which the density is specified over one boundary (r_1) and the flux over the other (r_2) may be expressed by the conditions that

[1] A somewhat more practical representation of the pressure rise in closed-in wells with, however, an analytical treatment much more approximate than the above, is obtained by assuming the shutting-in process to be one in which the fluid head simply rises from its initial flowing value, thereby continually increasing the back pressure upon the sand face until equilibrium is reached (M. Muskat *A.I.M.E.*, **103**, 44, 1937). Thus if A is the effective cross section of the well bore, h_w the height of the fluid level above the level of the sand face, γ_0 the density of the fluid, g the acceleration of gravity, p_e the reservoir pressure, h the sand thickness, and p_w, μ, r_e have their usual significance, it readily follows on assuming instantaneous steady-state-flow conditions that

$$A\frac{\partial h_w}{\partial t} = \frac{2\pi kh(p_e - p_w)}{\mu \log r_e/r_w}; \qquad p_w = \gamma_0 g h_w,$$

with the solutions

$$\frac{h_e - h_w}{h_e - h_i} = \frac{p_e - p_w}{p_e - p_{wi}} = e^{-Ct}; \qquad C = \frac{2\pi kh\gamma_0 g}{\mu A \log r_e/r_w},$$

where p_{wi}, h_i are the initial well pressures and fluid heads, and h_e is the fluid-head equivalent of the reservoir pressure p_e. If a plot of this equation on semilogarithmic paper gives a straight line, and thus confirms the assumption of "dead" liquid flow upon which the derivation is based, the slope will give the constant C and hence a means of determining the sand permeability k. At the same time the adjustment of the constant p_e so that the plot is straight, gives a very sensitive method for determining the reservoir pressure without waiting for strict equilibrium to be established, while the determination of C itself will permit the prediction of the maximum production capacity of the well.

$$\left. \begin{array}{ll} \gamma = f_1(t): & r = r_1 \\ r\dfrac{\partial \gamma}{\partial r} = f_2(t): & r = r_2 \\ \gamma = g(r): & t = 0, \end{array} \right\} \tag{1}$$

where $g(r)$ is again the initial density distribution.

Following the procedure of Sec. 10.3, the solution satisfying simultaneously all the conditions of Eq. (1) will be synthesized from the simpler solutions $(\gamma_1, \gamma_2, \gamma_3)$ which satisfy the conditions

$$\left. \begin{array}{ll} (\gamma_1, \gamma_2, \gamma_3) = (1, 0, 0): & r = r_1 \\ r_2\dfrac{\partial}{\partial r}(\gamma_1, \gamma_2, \gamma_3) = (0, 1, 0): & r = r_2 \\ (\gamma_1, \gamma_2, \gamma_3) = [0, 0, g(r)]: & t = 0. \end{array} \right\} \tag{2}$$

Then, similarly to Eq. 10.3(3), the final solution satisfying Eq. (1) may be written as

$$\gamma = \gamma_3(r, t) + \int_0^t \left[f_1(\lambda)\frac{\partial \gamma_1(r, t - \lambda)}{\partial t} + f_2(\lambda)\frac{\partial \gamma_2(r, t - \lambda)}{\partial t} \right] d\lambda. \tag{3}$$

Introducing now the new function

$$U(\alpha_n r) = Y_1(\alpha_n r_2)J_0(\alpha_n r) - J_1(\alpha_n r_2)Y_0(\alpha_n r), \tag{4}$$

so that

$$U'(\alpha_n r_2) = 0 \tag{5}$$

and choosing α_n such that

$$U(\alpha_n r_1) = 0, \tag{6}$$

and then applying again Eqs. 10.2(8) to 10.2(11), the values of $\gamma_1, \gamma_2, \gamma_3$ may readily be shown to be given by the expansions

$$\gamma_1 = 1 + \pi \sum \frac{J_0(\alpha_n r_1)J_1(\alpha_n r_2)U(\alpha_n r)e^{-\kappa\alpha_n^2 t}}{J_0^2(\alpha_n r_1) - J_1^2(\alpha_n r_2)}, \tag{7}$$

$$\gamma_2 = \log \frac{r}{r_1} + \frac{\pi}{r_2} \sum \frac{J_0^2(\alpha_n r_1)U(\alpha_n r)e^{-\kappa\alpha_n^2 t}}{\alpha_n[J_0^2(\alpha_n r_1) - J_1^2(\alpha_n r_2)]}, \tag{8}$$

$$\gamma_3 = \frac{\pi^2}{2} \sum \frac{\alpha_n^2 J_0^2(\alpha_n r_1)U(\alpha_n r)e^{-\kappa\alpha_n^2 t}}{J_0^2(\alpha_n r_1) - J_1^2(\alpha_n r_2)} \int_{r_1}^{r_2} rg(r)U(\alpha_n r)dr. \tag{9}$$

The resultant solution is then, by Eq. (3),

$$\gamma = \pi \sum \frac{\alpha_n^2 J_0(\alpha_n r_1)U(\alpha_n r)e^{-\kappa\alpha_n^2 t}}{J_0^2(\alpha_n r_1) - J_1^2(\alpha_n r_2)} \left[\frac{\pi}{2}J_0(\alpha_n r_1)\int_{r_1}^{r_2} rg(r)U(\alpha_n r)dr \right.$$
$$\left. - \kappa J_1(\alpha_n r_2)\int_0^t f_1(\lambda)e^{\kappa\alpha_n^2\lambda}d\lambda - \kappa\frac{J_0(\alpha_n r_1)}{\alpha_n r_2}\int_0^t f_2(\lambda)e^{\kappa\alpha_n^2\lambda}d\lambda \right]. \tag{10}$$

10.8. The Pressure Decline in the "East Texas" Oil Field.—
Perhaps the most striking practical application of the theory of
the flow of homogeneous fluids through porous media, and the
nonsteady-state flow of liquids in particular, is that of Eq. 10.7
(10) to the question of the pressure decline in the "East Texas"
oil field.   Whereas the original reservoir pressure in the field was
1,620 lb., tests[1] of the oil have shown that it was saturated with

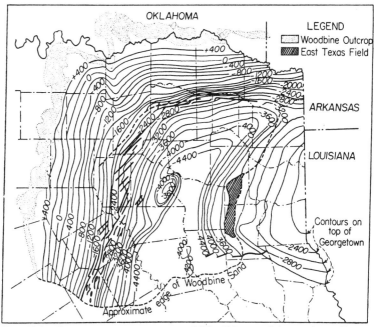

Fig. 251.—Structure contours on top of Woodbine sand in northeast Texas.
(After Schilthius and Hurst, Oil Weekly.)

gas to a pressure of only 755 lb.   Evidently, then, until the pressure within the sand has fallen to 755 lb., there will be no free
gas evolved, and the oil will necessarily flow as a homogeneous
"dead" liquid.   Now in order that a porous medium carrying a
homogeneous liquid discharge that liquid into outlet wells it is
necessary that either the liquid itself expand on lowering the
reservoir pressure, or that it be "driven" into the outlet wells
by means of some external mobile agency.   In the case of the
"East Texas" field, the observed performance has been that

[1] Lindsly, B. E., U. S. Bur. Mines, Rept. Investigations, 3212, 1933.

the reservoir pressure has dropped approximately 375 lb. during the course of some 500 million bbl. of oil production. Using for the fluid content of the field the estimated value of $7 \cdot 10^9$ bbl., it is readily seen that this fluid content would have to have a compressibility of 0.0028/atm., which is 20 times that which has actually been measured for the oil of the "East Texas" field with its dissolved gas,[1] if one is to attribute the oil produced from the field to the expansion of the fluid within the field itself. Although even the theory given below will require the assumption of an

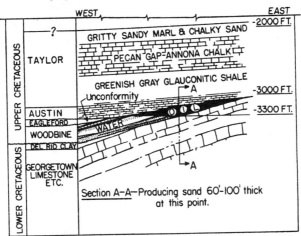

Fig. 252.—West-east cross section of the "East Texas" oil field. (*After Reistle,* "*Drilling and Production Practice,*" 1934, *Amer. Pet. Inst.*)

abnormally high—by a factor of 12—compressibility for the fluid within the adjacent water sand, it is impossible to justify the high value for the oil within the field, since it is known to be undersaturated and hence cannot have dispersed through it any free gas masses.

It is therefore necessary to suppose that the oil is "driven" into the outlet wells by an external mobile agency or "drive." This supposition is, however, well supported by the fact that the west edge of the field, which produces from the Woodbine sand, is flanked by a water sand extending westward for more than 100 miles before it outcrops, as shown in the map of **Fig. 251.** A vertical east-west section through the field is shown in **Fig. 252.** It is natural, then, to attribute the production from the

[1] LINDSLY, *loc. cit.*

field to the drive of the water in the Woodbine sand, which expands as the pressure is lowered. In fact, in comparison to the expansion of the tremendous volume of this water reservoir, the expansion of the oil within the field—it would contribute only about 20 million barrels during a pressure drop of 20 atm.— can be neglected, and the wells within the field may be lumped together into the equivalent of a single large well with a flux equal to the sum of those of the individual wells. In particular, the field will be replaced by an arc of 120 deg. along a circular sink or well of radius 20 miles. It will furthermore be supposed that at a radius of 100 miles the pressure is maintained at its original value of 1,620 lb., the sand between the two boundaries having an average thickness of 130 ft.[1] These conditions may be expressed analytically as

$$\left.\begin{array}{llll} \gamma = \gamma_e = \gamma_i; & (p_e = p_i): & r = r_e(r_1) = 100 \text{ miles} \\ \gamma = g(r) = \gamma_i; & (p = p_i): & t = 0. \end{array}\right\} \quad (1)$$

When these two conditions are applied to Eq. 10.7(10), the density difference between the two boundaries takes the form

$$\Delta\gamma = \gamma_i - \gamma_w = -\frac{2\kappa}{r_w^2}\sum\frac{J_0{}^2(\alpha_n r_e)e^{-\kappa\alpha_n{}^2 t}}{J_0{}^2(\alpha_n r_e) - J_1{}^2(\alpha_n r_w)}\int_0^t f_2(\lambda)e^{\kappa\alpha_n{}^2\lambda}d\lambda,$$
$$. (2)$$

where $f_2(t)$ is proportional to the production rate from the field at the time t. While this production rate is, in general, to be considered as a function composed of continuously varying segments, it is sufficient for practical purposes to take $f_2(t)$ as composed of segments of constant value over finite time intervals. That is, it may be expressed as

$$\begin{aligned} f_2(t) = r_w\left(\frac{\partial\gamma}{\partial r}\right)_{r_w} &= a: & 0 \leqslant t \leqslant t_1 \\ &= b: & t_1 \leqslant t \leqslant t_2 \\ &= c: & t_2 \leqslant t \leqslant t_3, \text{ etc.} \end{aligned} \quad (3)$$

Introducing the dimensionless notation of Eq. 10.4(9) and the function $F(t)$ defined by

[1] These values are, of course, very much idealized. However, they are actually effective averages of data derived from a detailed analysis of well logs in the water reservoir adjacent to the "East Texas" field.

$$F(\bar{t}) = 2 \sum \frac{J_0^2(x_n\rho)e^{-x_n^2 t}}{x_n^2[J_0^2(x_n\rho) - J_1^2(x_n)]}, \tag{4}$$

Eq. (2) can be written out more explicitly as

$$
\left.
\begin{aligned}
0 \leqslant \bar{t} \leqslant \bar{t}_1: \quad & \Delta\gamma = a \log \rho + aF(\bar{t}), \\
\bar{t}_1 \leqslant \bar{t} \leqslant \bar{t}_2: \quad & \Delta\gamma = b \log \rho + aF(\bar{t}) + (b - a)F(\bar{t} - \bar{t}_1), \\
\bar{t}_2 \leqslant \bar{t} \leqslant \bar{t}_3: \quad & \Delta\gamma = c \log \rho + aF(\bar{t}) + (b - a)F(\bar{t} - \bar{t}_1) + \\
& \qquad\qquad (c - b)F(\bar{t} - \bar{t}_2), \text{ etc.}
\end{aligned}
\right\} \tag{5}
$$

To get the numerical values of $\Delta\gamma$, specific values must be chosen for the various physical constants of the system. These are

$$
\left.
\begin{aligned}
& k/\mu = 2.12; \quad f = 0.2; \quad \beta = 5.3 \cdot 10^{-4}/\text{atm.}; \\
& \qquad\qquad r_w = 20 \text{ miles}; \quad r_e = 100 \text{ miles}, \\
\text{so that} & \\
& \qquad \rho = 5; \quad \bar{t} = 1.6686 \cdot 10^{-4}t \text{ (days)}.
\end{aligned}
\right\} \tag{6}[1]
$$

Observing now that

$$\Delta\gamma \cong \gamma_0\beta\Delta p; \qquad a = r_w\left(\frac{\partial\gamma}{\partial r}\right)_{r_w} = \frac{\mu\beta Q}{2\pi kh}, \tag{7}$$

where h is the thickness of the water sand (130 ft.), and Q is the flux that would enter the field if it substended an angle of 2π, Eq. (5) may be used to compute directly the pressure decline in the field as a function of the production rates a, b, etc. The function $F(\bar{t})$ may be calculated from Eq. (4), after finding the roots of Eq. 10.7(6) for $\rho = 5$.

The decline of the pressure *versus* the age of the field and the cumulative productions, as calculated by means of these equations, are plotted in Figs. 253 and 254, the production data used being averages of monthy values published in the oil-industry trade journals;[2] these are indicated in the lower discontinuous

---

[1] The value of k/μ used here was obtained from an average of the permeabilities of cores from "East Texas" wells, combined with an estimated effective viscosity for the water in the Woodbine sand taking into account its probable temperature at the depth of the sand.

[2] The cumulative production shown in Fig. 254 for Jan. 1, 1937, is some 50 million barrels less than the value which is now generally accepted. This discrepancy arises from changes that have been made, since the calculation of the curves of Figs. 253 and 254 was completed, in the

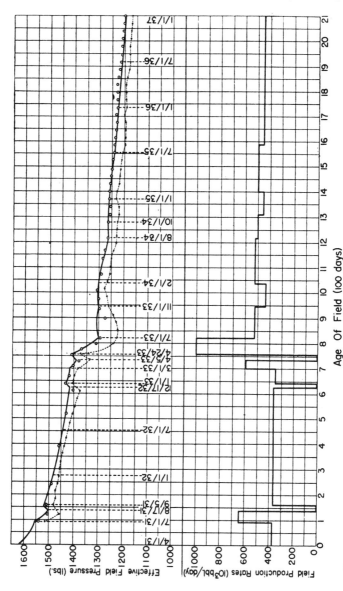

Fig. 253.—The pressure decline in the "East Texas" oil field. ——— calculated effective pressures; – – – – – observed average pressures; circles: observed pressures corrected for the east-west gradient (7 lb./100,000 bbl./day). Lower curve gives averages of field production rates.

curves of the figures. These production data have been multi-plied by 1.18, before applying Eq. (7), to take account of the shrinkage of the oil from its volume within the sand to that measured at the surface after the release of its dissolved gases.

It is to be noted that the pressures computed here must be considered as "effective" field pressures, due to the fact that the whole field of some 4 to 6 miles width has been lumped together as a circular sink concentrated over the 120-deg. arc of a circle of 20 miles radius. While the details of the actual disposal within any circular boundary of the flux that passes through it can have no effect upon the pressure variation on the boundary or in the medium beyond it, provided the time variation of the flux is the same as that originally preassigned, the theoretically calculated pressure variation will only be comparable with that observed in the immediate neighborhood of the boundary. Thus since the eastern edge of the "East Texas" field derives its production by virtue of the migration of the oil across the field from the west, the pressures observed within the field will decrease in going from west to east. Furthermore, in addition to the gradient across the field associated with the migration of the fluid from west to east, there will be a pressure gradient arising from the fact that the producing sand of the field is wedge-shaped and tapers off toward the east (*cf.* Fig. 252), so that the effective resistance of the sand as a whole to migration increases toward the east. It follows, therefore, that the calculated pressures will correspond only to those observed on the western edge of the field,[1] and hence will be higher than the pressures which are averaged over the whole field and which are universally quoted as the "field pressure." And when changes in the flux rate are

---

estimates of the actual withdrawals from the field in its early life. It should be noted further that since these calculations were made in the summer of 1934, it was assumed that the production rates after Aug. 1, 1934, would be maintained at 450,000 barrels per day, although the actual rates turned out to correspond approximately to those indicated in the lower halves of Figs. 253 and 254.

[1] The effective internal boundary representing the field cannot be placed somewhere along the interior of the field, as this system would involve a removal of some of the fluid in the sand between the two boundaries—at 20 and 100 miles—whereas the theory given here explicitly implies that no fluid leaves the sand until it has passed through the internal boundary.

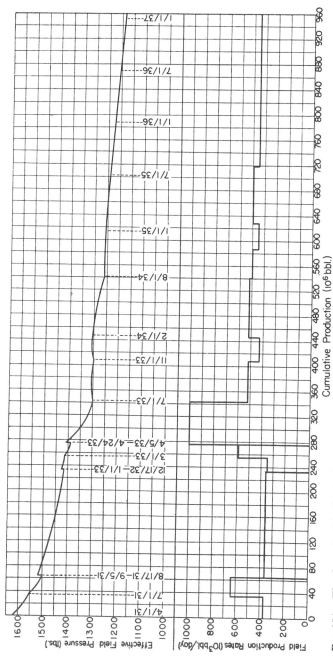

FIG. 254.—The theoretical decline in the effective pressures of the "East Texas" oil field vs. the cumulative production from the field.

made, the eastern edge of the field will show much more pronounced pressure changes than will the western edge, so that
the averages taken over the whole field will show larger fluctuations than those calculated theoretically. These will also involve
time transients in the establishment of the pressure distributions
over the field to correspond to the changed total flux rate at
the western boundary, which again are not taken into account
in the theoretical decline curve calculated above. Finally, it
is to be noted that the field-production rates are not known with
any great accuracy, owing essentially to the fact that a considerable amount of "hot" oil—oil produced illegally in violation
of the proration allocations, and hence not officially reported
and on record—has been produced at various times in the field
and in varying amounts. The data used here, which are those
gathered from trade-journal reports, appear to be the most
reliable, although it is difficult to estimate precisely their
accuracy.

It is for these reasons that the theoretically predicted pressure-decline curve does not, and should not be expected to, agree
exactly with the reported pressure-decline data, denoted by the
broken line in Fig. 253, which are averages referring to the field as
a whole. The agreement to within some 50 lb. over most of the
plotted history of the field and the close parallelism between the
computed and observed pressures are, however, significant, and do
confirm the theory unambiguously. For practical purposes of
predicting average pressures to be observed in the future, one may
approximately correct for the effects discussed above by assuming
that the pressure distribution across the field is essentially linear,
with a total west-east pressure differential proportional to the production rate from the field[1] so that the predicted western pressures
should equal the observed averages plus half of the west-east
differentials across the field. Taking as an average of the differentials indicated by the field pressure contour maps, a total
differential of 7 lb. per 100,000 bbl. daily-production rate, the
"corrected" observed data are indicated by the circles in Fig.
253. The agreement now is clearly as good as could be desired.

---

[1] In fact, this linear variation can be derived theoretically if one assumes
the velocity over the vertical sections of the sand to be uniform, and the
production from the field to be uniformly distributed over its width.

It is to be noted that this agreement has been attained with a calculation involving only a single essentially arbitrary constant, the compressibility, its value having been so chosen as to make the effective field pressure calculated for April 1, 1933, agree approximately with that observed at that time along the western edge of the field. While this value—some 12 times that of gas-free water—is indeed abnormally high, it may be explained on the assumption that there are gas pockets dispersed through the water horizon to the extent of 4.9 per cent of the total pore volume of the sand [*cf*. Eq. 10.1(6)]. Although there is no evidence at present either for or against such an explanation, it

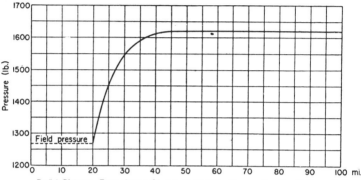

Fig. 255.—Calculated radial pressure distribution for July 1, 1934, in the water reservoir adjacent to the "East Texas" oil field.

may be noted that it does not seem to be inherently impossible, in view of the fact that there are other known oil fields in the Woodbine sand west of the "East Texas" field; it may therefore well be that the Woodbine basin actually does contain small gas fields which are as yet undiscovered.

The pressure distribution in the Woodbine sand as of July 1, 1934, computed by means of Eq. 10.7(10) together with Eqs. (1) and (3) is given in Fig. 255. It will be seen that although 520 million barrels of oil had been produced from the field— occupying a space of 613.6 million barrels in the reservoir— the pressure in the water reservoir has been appreciably affected only within a radius of 20 miles from the field. Beyond this radius the original pressure of 1,620 lb. may be considered for practical purposes as unaffected. While this result is at first thought surprising, an elementary calculation readily shows that the whole production could be accounted for by a uniform drop in pressure of 20 atm. extending for only 8.8 miles beyond the western edge of the field. The accurate theory shows indeed that such a rough estimate is quite close to the truth, and that the production until July 1, 1934 was due entirely to the expansion of the water within a relatively short distance from the field, the component of mass fluid motion from the distant parts of the reservoir being for practical purposes entirely negligible.

The pressure distribution given in Fig. 255 is also of interest in showing that for practical purposes the Woodbine sand reservoir can be considered as effectively infinite for what will probably be the complete life history of the field, and certainly until the internal gas drive due to the evolution of gas within the sand becomes of appreciable magnitude. In fact, from the above type of elementary computation it is readily seen that if the pressure drop extends over the whole of the reservoir to the radius of 100 miles, the expansion alone will produce a displacement of 580 million barrels per atmosphere of average pressure decline.

10.9. A Single Well in a Closed Reservoir.—Although it is not of great practical interest, Eq. 10.7(10) can be used to find the decline in flux from a single well in a closed reservoir. We shall briefly derive the appropriate formulas, omitting, however, a numerical discussion of the problem. The specific case treated will be that in which the flux at the external boundary, $r = r_e$, is zero—a closed reservoir—and the internal boundary radius is kept at a fixed pressure (density). As Eq. 10.7(10) is valid for any problem in which a flux is specified on one boundary and the density on the other, it may be adapted to the present case by setting

$$\left.\begin{array}{l} r_2 = r_e \text{ (reservoir radius)}; \qquad f_2(t) = 0 \\ r_1 = r_w \text{ (well radius)}; \qquad f_1(t) = \gamma_w \\ g(r) = \gamma_i = \text{const.} \end{array}\right\} \tag{1}$$

Equation 10.7(10) then takes the form

$$\gamma = \gamma_w - \pi(\gamma_i - \gamma_w) \sum \frac{J_0(\alpha_n r_w) J_1(\alpha_n r_e) U(\alpha_n r) e^{-\kappa \alpha_n^2 t}}{J_0^2(\alpha_n r_w) - J_1^2(\alpha_n r_e)} \tag{2}$$

for the density distribution within the sand at any time t.

The density decline at the outer reservoir boundary, $\gamma(r_e)$, and the decline in the flux from the well will, therefore, be given in the notation of Eq. 10.4(9) by

$$\gamma_e = \gamma_w + 2\frac{(\gamma_i - \gamma_w)}{\rho}\sum\frac{J_0(x_n)J_1(x_n\rho)e^{-x_n^2\bar{t}}}{x_n[J_0^2(x_n) - J_1^2(x_n\rho)]} \tag{3}$$

and:

$$Q = \frac{4\pi k h(\gamma_i - \gamma_w)}{\mu\beta}\sum\frac{J_1^2(x_n\rho)e^{-x_n^2\bar{t}}}{J_0^2(x_n) - J_1^2(x_n\rho)}, \tag{4}$$

where the x_n are to be determined as the roots of

$$Y_1(x_n\rho)J_0(x_n) - J_1(x_n\rho)Y_0(x_n) = 0, \tag{5}$$

ρ having a value of the order of 2,000 to 3,000.

10.10. A Well of Infinitesimal Radius.—When the flux at a well of practical dimensions is specified, one may replace the well by a mathematical sink and thereby simplify the analysis. Restating the problem in the form

$$\left.\begin{aligned}\lim_{r\to 0}\left(r\frac{\partial\gamma}{\partial r}\right) &= f_w(t)\\\gamma = f_e(t)\!: \quad & r = r_e\\\gamma = g(r)\!: \quad & t = 0,\end{aligned}\right\} \tag{1}$$

the effect of the sink can be represented by a single term logarithmic in r, so that the Bessel functions of the second kind which also vary in a logarithmic manner for small r can be omitted from the series. The general solution is then easily shown to be given by

$$\gamma = \frac{2}{r_e^2}\sum\frac{J_0(\alpha_n r)e^{-\kappa\alpha_n^2 t}}{J_1^2(\alpha_n r_e)}\left[\int_0^{r_e} rg(r)J_0(\alpha_n r)dr\right.$$
$$\left. - \kappa\int_0^t f_w(\lambda)e^{\kappa\alpha_n^2\lambda}d\lambda + \kappa\alpha_n r_e J_1(\alpha_n r_e)\int_0^t f_e(\lambda)e^{\kappa\alpha_n^2\lambda}d\lambda\right], \tag{2}$$

where

$$J_0(\alpha_n r_e) = 0. \tag{3}$$

For the special case where

$$f_w(t) = \text{const.} = q_0; \quad f_e(t) = \text{const.} = \gamma_i; \quad g(r) = \text{const.} = \gamma_i,$$

Eq. (2) reduces to

$$\gamma = \gamma_i + q_0 \log \frac{r}{r_c} + \frac{2q_0}{r_e^2} \sum \frac{J_0(\alpha_n r) e^{-\kappa \alpha_n^2 t}}{\alpha_n^2 J_1^2(\alpha_n r_e)}. \tag{4}$$

Although the roots of Eq. (3) may be readily found from available tabulations, the present case is not of great practical interest, and hence will not be treated here numerically.

10.11. Radial Systems in Which the Flux Is Specified over Both Boundaries.—A final type of radial-flow problem is that in which the flux rate is specified over both boundaries of the system. It may be generally defined by the conditions

$$\left.\begin{array}{ll} r\dfrac{\partial \gamma}{\partial r} = f_1(t): & r = r_1 \\[2mm] \qquad\quad = f_2(t): & r = r_2 \\[2mm] \quad \gamma = g(r): & t = 0. \end{array}\right\} \tag{1}$$

Once more it is convenient first to derive preliminary and simpler solutions corresponding to constant boundary conditions. Thus in terms of the solutions $\gamma_1, \gamma_2, \gamma_3$, satisfying the conditions

$$\left.\begin{array}{ll} r\dfrac{\partial}{\partial r}(\gamma_1, \gamma_2, \gamma_3) = (1, 0, 0): & r = r_1 \\[2mm] \qquad\qquad\qquad\quad = (0, 1, 0): & r = r_2 \\[2mm] \quad (\gamma_1, \gamma_2, \gamma_3) = (0, 0, g(r)): & t = 0, \end{array}\right\} \tag{2}$$

that corresponding to Eq. (1) will be given by

$$\gamma = \gamma_3(r, t) + \int_0^t \left[f_1(\lambda)\frac{\partial \gamma_1(r, t - \lambda)}{\partial t} + f_2(\lambda)\frac{\partial \gamma_2(r, t - \lambda)}{\partial t} \right] d\lambda. \tag{3}$$

For the Bessel-function expansions, the same form for $U(\alpha_n r)$ will be taken as in Eq. 10.7(4), but it will be required that the α_n are the roots of

$$U'(\alpha_n r_1) = 0 \tag{4}$$

rather than of Eq. 10.7(6). By using this definition, and applying Eqs. 10.2(8) to (11) once more to find the expansion coefficients, it is found that

$$\gamma_1 = -\frac{(r^2 + 4\kappa t)}{2(r_2{}^2 - r_1{}^2)} + \frac{r_2{}^2}{r_2{}^2 - r_1{}^2} \log r + \frac{1}{(r_2{}^2 - r_1{}^2)^2}$$

$$\left[\frac{(r_2{}^2 - r_1{}^2)(r_1{}^2 + 3r_2{}^2)}{4} + r_2{}^2(r_1{}^2 \log r_1 - r_2{}^2 \log r_2)\right]$$

$$-\frac{\pi}{r_1}\sum \frac{J_1(\alpha_n r_1) J_1(\alpha_n r_2) U(\alpha_n r) e^{-\kappa\alpha_n{}^2 t}}{\alpha_n[J_1{}^2(\alpha_n r_1) - J_1{}^2(\alpha_n r_2)]}, \quad (5)^1$$

$$\gamma_2 = \frac{(r^2 + 4\kappa t)}{2(r_2{}^2 - r_1{}^2)} - \frac{r_1{}^2}{r_2{}^2 - r_1{}^2} \log r + \frac{1}{(r_2{}^2 - r_1{}^2)^2}$$

$$\left[\frac{(r_1{}^2 - r_2{}^2)(r_2{}^2 + 3r_1{}^2)}{4} + r_1{}^2(r_2{}^2 \log r_2 - r_1{}^2 \log r_1)\right]$$

$$+\frac{\pi}{r_2}\sum \frac{J_1{}^2(\alpha_n r_1) U(\alpha_n r) e^{-\kappa\alpha_n{}^2 t}}{\alpha_n[J_1{}^2(\alpha_n r_1) - J_1{}^2(\alpha_n r_2)]}, \quad (6)$$

$$\gamma_3 = \frac{2}{(r_2{}^2 - r_1{}^2)}\int_{r_1}^{r_2} rg(r)dr +$$

$$\frac{\pi^2}{2}\sum \frac{\alpha_n{}^2 J_1{}^2(\alpha_n r_1) U(\alpha_n r) e^{-\kappa\alpha_n{}^2 t}}{J_1{}^2(\alpha_n r_1) - J_1{}^2(\alpha_n r_2)} \int_{r_1}^{r_2} rg(r) U(\alpha_n r)dr, \quad (7)$$

so that finally

$$\gamma = \frac{2\kappa}{r_2{}^2 - r_1{}^2}\int_0^t [f_2(\lambda) - f_1(\lambda)]d\lambda + \frac{2}{(r_2{}^2 - r_1{}^2)}\int_{r_1}^{r_2} rg(r)dr$$

$$+ \pi \sum \frac{\alpha_n{}^2 J_1(\alpha_n r_1) U(\alpha_n r) e^{-\kappa\alpha_n{}^2 t}}{J_1{}^2(\alpha_n r_1) - J_1{}^2(\alpha_n r_2)}\left[\frac{\pi}{2}J_1(\alpha_n r_1)\int_{r_1}^{r_2} rg(r) U(\alpha_n r)dr\right.$$

$$\left. + \kappa\frac{J_1(\alpha_n r_2)}{\alpha_n r_1}\int_0^t f_1(\lambda) e^{\kappa\alpha_n{}^2\lambda}d\lambda - \kappa\frac{J_1(\alpha_n r_1)}{\alpha_n r_2}\int_0^t f_2(\lambda) e^{\kappa\alpha_n{}^2\lambda}d\lambda\right]. \quad (8)$$

10.12. The Limiting Case of Vanishing Internal Radius.—

Equation 10.11(8) might be applied to the problem of the pressure
decline in a field producing artesian water or an oil field driven by
a water drive in which the flux into the water reservoir, such as

---

[1] The constant terms in Eqs. (5) to (8) and 10.12(2) to (6) enter here
because of the vanishing of $U'(\alpha r)$ on both boundaries. For Eqs. 10.12(2)
to (6), the expansions are Dini-series expansions which take on a constant
term for the zero-order Bessel functions when the derivatives at the limiting
radius vanishes. Equations (5) to (8) are generalizations of the Dini series
expansions and take on constant terms when the derivatives of the Bessel
functions vanish at both boundaries. This phenomenon is similar to the
entry of the constant term in the Fourier series expansions (*cf.* Sec. 4.3 and
Watson, "Theory of Bessel Functions," p. 597).

that due to rainfall, is known or assumed. Of more practical interest, however, is the case where the sand is closed off at the external boundary, and the internal boundary represents an actual well of radius small as compared to that of the external boundary. Here again the well may be replaced by a mathematical sink, and for the more general case where the external boundary flux is not taken to be zero, *i.e.*, where

$$\lim_{r \to 0} \left(r\frac{\partial \gamma}{\partial r} \right) = f_w(t): \quad \left(r\frac{\partial \gamma}{\partial r} \right)_{r_e} = f_e(t); \quad \gamma = g(r): \quad t = 0, \quad (1)$$

the functions $(\gamma_1, \gamma_2, \gamma_3)$ take the form

$$\gamma_1 = \frac{3}{4} - \frac{(r^2 + 4\kappa t)}{2r_e^2} + \log \frac{r}{r_e} + \frac{2}{r_e^2} \sum \frac{J_0(\alpha_n r)e^{-\kappa\alpha_n^2 t}}{\alpha_n^2 J_0^2(\alpha_n r_e)}, \quad (2)$$

$$\gamma_2 = -\frac{1}{4} + \frac{r^2 + 4\kappa t}{2r_e^2} - \frac{2}{r_e^2} \sum \frac{J_0(\alpha_n r)e^{-\kappa\alpha_n^2 t}}{\alpha_n^2 J_0(\alpha_n r_e)}, \quad (3)$$

$$\gamma_3 = \frac{2}{r_e^2} \int_0^{r_e} rg(r)dr + \frac{2}{r_e^2} \sum \frac{J_0(\alpha_n r)e^{-\kappa\alpha_n^2 t}}{J_0^2(\alpha_n r_e)} \int_0^{r_e} rg(r)J_0(\alpha_n r)dr, \quad (4)$$

where

$$J_1(\alpha_n r_e) = 0, \quad (5)$$

so that

$$\gamma = \frac{2\kappa}{r_e^2} \int_0^t [f_e(\lambda) - f_w(\lambda)]d\lambda + \frac{2}{r_e^2} \int_0^{r_e} rg(r)dr + \frac{2}{r_e^2} \sum \frac{J_0(\alpha_n r)e^{-\kappa\alpha_n^2 t}}{J_0^2(\alpha_n r_e)}$$
$$\left[\int_0^{r_e} rg(r)J_0(\alpha_n r)dr - \kappa \int_0^t f_w(\lambda)e^{\kappa\alpha_n^2\lambda}d\lambda + \kappa J_0(\alpha_n r_e)\int_0^t f_e(\lambda)e^{\kappa\alpha_n^2\lambda}d\lambda \right]. \quad (6)$$

10.13. A Well in a Closed Sand.—If the sand is closed off $(f_e = 0)$, and the pressure, and hence density, is initially uniform $[g(r) = \gamma_i]$, Eq. 10.12(6) reduces to the form

$$\gamma = \gamma_i - \frac{2\kappa}{r_e^2}\left[\int_0^t f_w(\lambda)d\lambda + \sum \frac{J_0(\alpha_n r)e^{-\kappa\alpha_n^2 t}}{J_0^2(\alpha_n r_e)} \int_0^t f_w(\lambda)e^{\kappa\alpha_n^2\lambda}d\lambda \right]. \quad (1)$$

If now the flux at the well, $f_w(t)$, has the constant value q, γ is given by

$$\gamma = \gamma_i + q\left[\frac{3}{4} + \log \bar{r} - \frac{1}{2}(\bar{r}^2 + 4\bar{t}) + 2\sum \frac{J_0(x_n\bar{r})e^{-x_n^2\bar{t}}}{x_n^2 J_0^2(x_n)} \right], \quad (2)$$

where

$$\bar{r} = \frac{r}{r_e}; \qquad x_n = \alpha_n r_e; \qquad \bar{t} = \frac{\kappa t}{r_e{}^2}. \tag{3}$$

Even without any numerical assumptions it is clear that Eq. (2) involves two distinct types of transient. The first, represented by the series, will disappear at an exponential rate as the time increases. The second, however, represented by the term linear in $\bar{t}$, will persist indefinitely (since q has been assumed to be maintained indefinitely), and gives a constant rate of variation of the density with the time after the first transient has become of negligible magnitude. Furthermore, this linear term is independent of r, so that it will not affect the density, and hence pressure *distribution*. That is, the density gradients will remain fixed, except for the series transient, although the absolute value of γ will decrease linearly with the time.

From a numerical point of view, however, the series transient turns out to be of such short duration, for cases of practical interest, as to be of no importance. Thus, assuming that

$$\frac{k}{\mu} = 1; \qquad f = 0.2; \qquad \beta = 4.5 \times 10^{-4} \text{ atm.}^{-1}; \qquad r_e = \left. \begin{array}{c} \\ 500 \text{ ft.,} \\ \\ \end{array} \right\} \quad (4)^1$$

so that

$$\bar{t} = 4.133t \text{ (days)},$$

and applying Eq. (2), the drop in pressure at the well ($r_w = \frac{1}{4}$ ft.) from its initial value, for a production rate of 1 bbl./day/ft. of sand, will be given by the expression

$$p_i - p_w = 0.1412\left(6.8509 + 2\bar{t} - 2\sum \frac{e^{-x_n{}^2\bar{t}}}{x_n{}^2 J_0{}^2(x_n)}\right), \tag{5}$$

where $J_0(x_n \bar{r}_w)$ has been replaced by 1, owing to the small value of $\bar{r}_w$ ($\frac{1}{2000}$), and $p_i - p_w$ has the units of pounds.

For the pressure decline at the external boundary of the sand, $r = r_e(\bar{r} = 1)$, Eq. (2) gives

---

[1] The compressibility is taken here as abnormally high so as to give what may be considered as an upper limit to the length of the initial transient in single-well systems of moderate dimensions. For a liquid of normal compressibility this transient will evidently be of still smaller duration.

$$p_s - p_e = 0.1412\left(-0.25 + 2\bar{t} - 2\sum\frac{e^{-x_n^2\bar{t}}}{x_n^2 J_0(x_n)}\right) \quad (6)$$

Equations (5) and (6) are plotted in Fig. 256. Because of the extremely short life of the series transient, the time scale has been taken as hours. For as appears from the curves, after a period of only 1 hr. the system settles down to an effectively steady-state drainage by the well, and hence a uniform pressure decline over the whole reservoir, always maintaining a pressure differential of 1.00 lb. between the reservoir boundary and the well. The

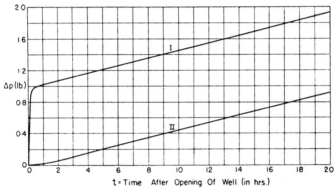

Fig. 256.—The pressure decline due to a single well at the center of a closed reservoir producing 1 bbl./day/ft. sand of a liquid of compressibility $4.5 \cdot 10^{-4}$/atm. I. Pressure drop at the well ($\frac{1}{4}$ ft. radius). II. Pressure drop at the external closed boundary (500 ft. radius). $k/\mu = 1$; sand porosity $= 0.20$. (*From Physics*, **5**, 88, 1934.)

very rapid pressure decline at the well and the lag in the decline at the external boundary are, of course, to be expected.

Although the pressure distribution after the initial transient is clearly not a strictly steady-state distribution, the logarithmic term, characteristic of the steady state, does predominate everywhere except very near the external boundary. Using the same constants as in Eq. (4), the pressure variation after the series transient has disappeared may be expressed as

$$p(r) - p_w = 0.1412\left(7.6009 - \frac{r^2}{2r_e^2} - \log\frac{r_e}{r}\right). \quad (7)$$

Equation (7) is plotted in Fig. 257. In general appearance it resembles the strictly steady-state logarithmic distribution and, in fact, approximates it very closely as the well is approached.

At the external boundary, however, the gradients vanish in Fig.
257, since the reservoir is closed, whereas in the steady-state
system they are of the order of $1/r_e$. As the result of these
smaller gradients, the total pressure drop $p_e - p_w = 1.00$ lb.
is less than the corresponding value of 1.07 lb. required to give
the same production. rate in a strictly incompressible steady-
state system.

One may nevertheless derive from these results a rather
simple physical approximation of the description of the decline
in the system. For as has been seen above, after the series

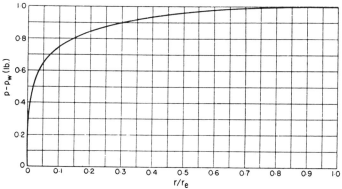

Fig. 257.—The pressure distribution about the well with the decline curve
of Fig. 256, after the passing of the initial transient. (*From Physics*, **5**, 88,
1934.)

transient has passed, the density decreases uniformly over the
whole reservoir and at a rate proportional to q. And, in fact,
if Q is the actual production rate, it is readily seen from Eq.
(2) that

$$Q = \frac{2\pi k}{\mu}\left(r\gamma\frac{\partial p}{\partial r}\right)_{r=0} = \frac{2\pi kq}{\beta\mu} = -f\pi r_e^2\left(\frac{\partial\gamma}{\partial t}\right) = -f\pi r_e^2\gamma\beta\left(\frac{\partial p}{\partial t}\right). \quad (8)$$

Thus the production from the reservoir is supplied by the equivalent density or pressure decline distributed uniformly over the
reservoir, the instantaneous pressure distribution differing only
slightly from the steady-state type. One is therefore led quite
naturally to the representation of the dynamical behavior of
the system as a "continuous succession of steady states." We
are thus provided with at least an indirect analytical justification

for the basic hypothesis, underlying the whole treatment of Part II, that continuous sequences of the various steady-state solutions, derived there, would give very close approximations to the actual time variations of normal liquid-flow systems of moderate dimensions (*cf.* Sec. 10.1). Furthermore, it will be convenient explicitly to use such a hypothesis in the theory of gas flow where the rigorous solutions cannot be obtained in analytic form (*cf.* Part IV).

10.14. Nonradial Flow. Well Interference. Green's Function.—Thus far we have treated problems involving single wells placed at the centers of their reservoirs. In practical cases, however, the reservoir will be covered by a number of wells, and the question arises as to the extent of their mutual interaction and their interference relations. Although only a simple case of such interactions will be treated in detail, we shall present the fundamental element by means of which the solutions for more general problems may be synthesized. This element is

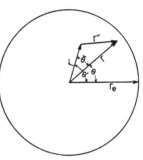

FIG. 258.

essentially the Green's function of the circle which, for the present purposes, may be defined as the function γ having a unit logarithmic singularity at (r', θ'), with a vanishing normal derivative at the circle $r = r_e$, and which is identically zero at $t = 0$. Physically, this Green's function gives the density distribution within a closed circular reservoir, initially at "zero density," containing a well of unit "strength" at (r', θ').

With the notation of Fig. 258, the constant-strength well or sink may be represented by the term

$$\gamma_0 = \log \frac{r''}{r_e} = \frac{1}{2} \log \left[\frac{r^2}{r_e^2} - \left(\frac{2r'r}{r_e^2} \right) \cos \bar{\theta} + \frac{r'^2}{r_e^2} \right]. \qquad (1)$$

As is readily verified, a solution of Eq. 10.1(7)[1] with a vanishing normal derivative (at $r = r_e$) will be obtained by adding

---

[1] While γ_0 itself may be considered as a solution of the radial-flow Eq. 10.2(1), the problem considered here is essentially of a nonradial character, governed by Eq. 10.1(7), the elementary solutions of which, in polar coordinates, are the terms in the series of Eq. (3).

the function

$$\gamma_a = \frac{1}{2} \log\left[1 - \left(\frac{2r'r}{r_e^2}\right) \cos \bar{\theta} + \frac{r'^2 r^2}{r_e^4} \right] - \left(\frac{1}{2r_e^2}\right)(r^2 + 4\kappa t). \quad (2)$$

For the final solution, or the Green's function, one may, therefore, write

$$\gamma = \gamma_0 + \gamma_a + \sum_{nm} A_{nm} J_n(\alpha_{nm} r) \cos n\bar{\theta}\, e^{-\kappa \alpha_{nm}^2 t} + \text{const.}, \quad (3)$$

where

$$J_n'(\alpha_{nm} r_e) = 0, \quad (4)$$

and

$$\gamma_0 + \gamma_a(t = 0) + \sum_{nm} A_{nm} J_n(\alpha_{nm} r) \cos n\bar{\theta} + \text{const.} = 0, \quad (5)$$

so that Eq. (4) determines the parameters α_{nm} and Eq. (5) the coefficients A_{nm}. Expanding γ_0 and γ_a as Fourier series, and equating first the coefficients of $\cos n\bar{\theta}$, it is readily found that

$$\left.\begin{aligned}
\sum_m A_{nm} J_n(\alpha_{nm} r) &= \frac{(r'r/r_e^2)^n}{n} + b_n \\
b_n &= \frac{(r/r')^n}{n}: \qquad r \leqslant r' \\
&= \frac{(r'/r)^n}{n}: \qquad r \geqslant r'.
\end{aligned}\right\} n > 0 \qquad (6)$$

$$\left.\begin{aligned}
\sum_m A_{0m} J_0(\alpha_{0m} r) &= \frac{r^2}{2r_e^2} + b_0 - \text{const.} \\
b_0 &= \log \frac{r_e}{r'}: \qquad r \leqslant r' \\
&= \log \frac{r_e}{r}: \qquad r \geqslant r'.
\end{aligned}\right\} \qquad (7)$$

In virtue of Eq. (4), the Bessel equation, and recurrence relations, which the $J_n(\alpha_{nm} r)$ satisfy, namely,[1]

$$\left.\begin{aligned}
\frac{d^2 J_n(z)}{dz^2} + \frac{1}{z}\frac{dJ_n(z)}{dz} + \left(1 - \frac{n^2}{z^2}\right) J_n(z) &= 0 \\
z^{n+1} J_n(z) = \frac{d}{dz}[z^{n+1} J_{n+1}(z)]; \qquad z^{-n} J_{n+1}(z) &= -\frac{d}{dz}[z^{-n} J_n(z)] \\
J_n'(z) = \frac{n}{z} J_n(z) - J_{n+1}(z) &= -\frac{n}{z} J_n(z) + J_{n-1}(z),
\end{aligned}\right\} \quad (8)$$

[1] Whitaker and Watson, "Modern Analysis," Chap. XVII.

it may be shown that

$$
\left.
\begin{aligned}
\int_0^{r_e} r J_n{}^2(\alpha_{nm}r)dr &= \left[\frac{(\alpha_{nm}{}^2 r_e{}^2 - n^2)}{2\alpha_{nm}{}^2}\right] J_n{}^2(\alpha_{nm}r_e) \\
\int_0^{r_e} r^{n+1} J_n(\alpha_{nm}r)dr &= \left(\frac{nr_e{}^n}{\alpha_{nm}{}^2}\right) J_n(\alpha_{nm}r_e) \\
\int_0^{r_e} b_n r J_n(\alpha_{nm}r)dr &= \left(\frac{1}{\alpha_{nm}{}^2}\right)\left[2J_n(\alpha_{nm}r') - \left(\frac{r'}{r_e}\right)^n J_n(\alpha_{nm}r_e)\right] \\
\int_0^{r_e} r^3 J_0(\alpha_{0m}r)dr &= \left(\frac{2r_e{}^2}{\alpha_{0m}{}^2}\right) J_0(\alpha_{0m}r_e) \\
\int_0^{r_e} b_0 r J_0(\alpha_{0m}r)dr &= \left[\frac{J_0(\alpha_{0m}r') - J_0(\alpha_{0m}r_e)}{\alpha_{0m}{}^2}\right].
\end{aligned}
\right\} \quad (9)
$$

Setting now

$$
\alpha_{nm}r_e = x_{nm}; \qquad \frac{\kappa t}{r_e{}^2} = \bar{t}; \qquad \frac{r}{r_e} = \bar{r}; \qquad \frac{r'}{r_e} = \rho, \qquad (10)
$$

Eq. (3) may be finally expressed as

$$
\begin{aligned}
\gamma = {} & \frac{1}{2}\log\left[\bar{r}^2 - 2\bar{r}\rho\cos(\theta' - \theta) + \rho^2\right] + \frac{1}{2}\left(\frac{3}{2} - \rho^2\right) + \\
& \frac{1}{2}\log\left[1 - 2\bar{r}\rho\cos(\theta' - \theta) + \bar{r}^2\rho^2\right] \\
& - \frac{1}{2}(\bar{r}^2 + 4\bar{t}) + 2\sum\frac{J_0(x_{0m}\rho)J_0(x_{0m}\bar{r})e^{-x_{0m}{}^2\bar{t}}}{x_{0m}{}^2 J_0{}^2(x_{0m})} \\
& + 4\sum_1^\infty n\sum m\frac{J_n(x_{nm}\rho)J_n(x_{nm}\bar{r})}{(x_{nm}{}^2 - n^2)J_n{}^2(x_{nm})}\cos n(\theta' - \theta)e^{-x_{nm}{}^2\bar{t}}. \quad (11)
\end{aligned}
$$

This expression gives the density distribution in a closed unit-radius [in the units of Eq. (10)] reservoir initially at density 0, which is drained by a unit strength well at (ρ, θ'). As might be expected, this Green's function is symmetrical in $(\bar{r}, \theta)$ and (ρ, θ'); *i.e.*, the density at $(\bar{r}, \theta)$ at a time $\bar{t}$, due to a source at (ρ, θ'), is the same as that at the same time at (ρ, θ') due to a source at $(\bar{r}, \theta)$.

If the initial density has a distribution given by $g(r, \theta)$, the function

$$\gamma_i = \frac{1}{\pi}\int_0^1 \bar{r}d\bar{r}\int_0^{2\pi} g(\bar{r},\,\theta)d\theta \;+$$

$$\frac{1}{\pi}\sum \frac{J_0(x_{0m}\bar{r})e^{-x_{0m}^2 t}}{J_0^2(x_{0m})}\int_0^1 \bar{r}J_0(x_{0m}\bar{r})d\bar{r}\int_0^{2\pi} g(\bar{r},\,\theta)d\theta \;+$$

$$\frac{2}{\pi}\sum_{n>0} nm \frac{x_{nm}^2 J_n(x_{nm}\bar{r})\; \cos n\theta e^{-x_{nm}^2 t}}{(x_{nm}^2 - n^2)J_n^2(x_{nm})}\int_0^1 \bar{r}J_n(x_{nm}\bar{r})d\bar{r}\int_0^{2\pi} g(\bar{r},\,\theta)$$

$$\cos \frac{n\theta}{2}d\theta \quad (12)$$

should be added to the right side of Eq. (11).

If, finally, it be supposed that the well has a flux strength[1] $q(t)$ instead of the permanent unit value, the resultant density distribution will be

$$\gamma = \gamma_i(\bar{r},\,\bar{t}) + \int_0^t q(\lambda)\frac{\partial \gamma_1(\bar{r},\, t - \lambda)}{\partial t}\partial\lambda, \quad (13)$$

where γ_1 is the function of Eq. (11).

By adding together a number of functions as in Eq. (13) with different values of r', θ', and $q(t)$, the density and therefore pressure distribution due to the corresponding group or network of wells in a closed reservoir will be obtained. A study of such distributions will give the interference effects and mutual interactions among the wells due to their simultaneous drainage of the same reservoir.

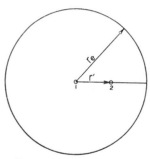

Fig. 259.—Diagrammatic representation of a two-well interference system.

As an example of such an interference problem indicating in more detail the method of treatment, we shall consider the simplest case of interference, that of two wells in a closed circular reservoir. As is shown in Fig. 259, well No. 1 will be placed at the center, and well No. 2 at a radius r', this radius being taken as the polar axis. It also will be supposed that No. 1 has been producing at a uniform rate of q_1 units since the initial instant, and that No. 2 is opened after a time $\bar{t}_1$, and is produced

[1] The "strength" is here taken as the value of the actual flux per radian.

thereafter at the rate of q_2 units, the flux of well No. 1 being maintained at q_1.

Up to the time $\bar{t}_1$, the pressure at well No. 1 will evidently decline without any interference, and according to curve I of Fig. 256 which corresponds to this case. The opening of No. 2, however, will clearly accelerate this decline, and it is this reaction of well No. 2 on No. 1 that shall be computed. One could also compute the interaction of well No. 2 on itself, *i.e.*, the effect on the normal decline at the position of No. 2 as caused by the production of the central well, owing to the drilling of No. 2; but to show the method the reaction of No. 2 on No. 1 will suffice.

As just mentioned, for $\bar{t} < \bar{t}_1$ the pressure at No. 1 (at $r = r_w$) will decline as in a radial system. Hence, by Eq. 10.13(2), to which Eq. (11) reduces for this case, the density at (1) will be given by

$$\bar{t} \leqslant \bar{t}_1: \qquad \gamma(1) = \gamma_i + q_1\left\{\frac{3}{4} - \log\frac{r_e}{r_w} - 2\bar{t} + 2\sum\frac{e^{-x_n^2\bar{t}}}{x_n^2 J_0^2(x_n)}\right\},$$
(14)

where $(r_w/r_e)^2$ has been dropped, and $J_0(x_n r_w/r_e)$ has been replaced by unity, γ_i being the initial uniform density.

This contribution will continue for $\bar{t} > \bar{t}_1$, but upon it will now be superposed the effect due to well No. 2, which by Eq. (11) is

$$\gamma_2(1) = q_2\log\rho + \frac{q_2}{2}\left(\frac{3}{2} - \rho^2\right) - 2q_2(\bar{t} - \bar{t}_1) +$$
$$2q_2\sum\frac{J_0(x_{0m}\rho)e^{-x_{0m}^2(\bar{t}-\bar{t}_1)}}{x_{0m}^2 J_0^2(x_{0m})}. \quad (15)$$

This is to be added to the $\gamma(1)$ of Eq. (14). For the case where $q_1 = q_2 = q$, $\rho = \frac{1}{2}$, and $r_e/r_w = 2,000$, the density drop at well No. 1 may be expressed numerically as

$$\bar{t} \leqslant \bar{t}_1: \qquad \frac{[\gamma_i - \gamma(1)]}{q} = 6.851 + 2\bar{t} - 2\sum\frac{e^{-x_n^2\bar{t}}}{x_n^2 J_0^2(x_n)}, \qquad (16)$$

$$\bar{t} \geqslant \bar{t}_1: \qquad \frac{[\gamma_i - \gamma(1)]}{q} = 6.919 + 2(2\bar{t} - \bar{t}_1) -$$
$$2\sum\left[\frac{e^{-x_n^2\bar{t}}}{x_n^2 J_0^2(x_n)}\right]\left[1 + J_0\left(\frac{x_n}{2}\right)e^{x_n^2\bar{t}_1}\right]. \quad (17)$$

These equations are plotted in Fig. 260 with t_1 chosen as 0.3. The dotted curve indicates the pressure decline if No. 2 well had not been drilled.   It will be seen that after the initial transient, in which the original decline is altered, the rate of decline assumes a value twice as large as its original or extrapolated value.   This is, of course, to be expected since for $\bar{t} > 0.3$ twice as much fluid is being taken from the reservoir as before.

10.15. The Use of Sources and Sinks in the Solution of Problems Involving the Nonsteady State Flow of Compressible Liquids through Porous Media.—We shall close this chapter with a brief outline of the method of sources and sinks for treat-

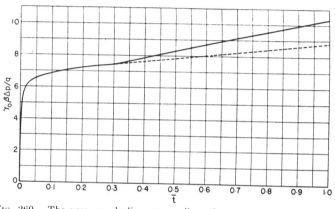

Fig. 260.—The pressure decline at a well at the center of a closed circular reservoir which produces at a constant rate, as affected by the drilling of another well at $\bar{t} = 0.3$ at a radius $\frac{1}{2}$ of the reservoir radius, which is produced at the same rate as the first. Δp = pressure drop at the central well from its initial pressure. $\bar{t}$ = dimensionless time = $kt/\mu f\beta r_e^2$. q = flux rate from each well (per radian). (*From Physics,* **5**, 91, 1934.)

ing transient-flow problems.   This method is in principle analogous to that frequently used in Part II in the representation of wells by permanent sources or sinks.   It is of particular interest in the treatment of systems in which the porous medium may be considered to be infinitely extended, at least along one direction, although by synthesizing independent solutions it can also be applied to systems of finite dimensions.   Only the case of most practical interest in which the system may be taken as two-dimensional will be discussed explicitly, as the modifications required for one- and three-dimensional systems can be made without difficulty.

The principle element in the method is the instantaneous "line-source" solution of Eq. 10.2(1), namely,

$$\gamma = \frac{q}{4\pi f \kappa t} e^{-\frac{r^2}{4\kappa t}} \qquad [cf. \text{ Eq. } 10.2(5)]. \qquad (1)$$

This solution vanishes everywhere at $t = 0$ and represents an instantaneous emission of q mass units of fluid at the origin, $r = 0$, at the initial instant, $t = 0$. If the strength of the source is "permanent" and of magnitude $q(t)$, the corresponding solution is

$$\gamma = \frac{1}{4\pi f \kappa} \int_0^t \frac{q(\tau)}{t - \tau} e^{-\frac{r^2}{4\kappa(t-\tau)}} d\tau. \qquad (2)$$

Finally, if there is an initial density distribution given by $g(x, y)$, the complete solution will be

$$\gamma = \frac{1}{4\pi\kappa} \left\{ \frac{1}{t} \int_{-\infty}^{+\infty} d\xi \int_{-\infty}^{+\infty} d\eta g(\xi, \eta) e^{-\frac{[(x-\xi)^2 + (y-\eta)^2]}{4\kappa t}} + \frac{1}{f} \int_0^t \frac{q(\tau)}{t-\tau} e^{-\frac{r^2}{4\kappa(t-\tau)}} d\tau \right\}. \qquad (3)$$

If the initial density (and hence pressure) distribution has the uniform value γ_i, the first term reduces to γ_i, and Eq. (3) will give the density distribution at any later time due to a well at the origin with a production rate of $-q(t)$ units. When the actual production rate has the constant value Q, Eq. (3) reduces to

$$\gamma = \gamma_i - \frac{Q}{4\pi f \kappa} \int_{\frac{r^2}{4\kappa t}}^{\infty} \frac{e^{-w} dw}{w} = \gamma_i + \frac{Q}{4\pi f \kappa} Ei\left(\frac{-r^2}{4\kappa t}\right), \qquad (4)$$

where the function Ei can be read from tables such as those of Jahnke and Emde.[1] The density (and hence pressure) decline at the well can be readily obtained by evaluating Eq. (4), after setting $r = r_w$, the well radius. Physically, of course, Eq. (4) becomes meaningless when the density γ falls below the "zero pressure value" γ_0. This clearly means that the production rate Q can no longer be obtained, so that the assumption of a permanent flux rate of Q breaks down.

Just as in the case of the steady-state flow of incompressible liquids the individual logarithmic terms due to a number of separate wells can be added together so as to give the resultant pressure distribution appropriate to the several wells in the

[1] Jahnke, E., and F. Emde "Funktionentafeln," 1928.

groups, so here to find the density distribution in a multiple-well system, one need only add together terms as in Eq. (2) with the appropriate values of Q and r. Thus if the wells, of fluxes Q_i, are located at the points (x_i, y_i), the resultant density distribution, for an initial uniform density, will be

$$\gamma = \gamma_i - \frac{1}{4\pi f \kappa} \int_0^t \frac{d\tau}{t - \tau} \sum_i Q_i(\tau) e^{-\frac{[(x-x_i)^2 + (y-y_i)^2]}{4\kappa(t-\tau)}} \tag{5}$$

These wells of flux Q_i may be real wells or only "images" set at appropriate points (x_i, y_i) in order to satisfy particular boundary conditions. Or, the wells may be distributed continuously over an area or curve to give a continuous flux distribution. Thus for a finite line sink of length L along the x axis, of flux density $q(t)$ per unit length, draining an infinite reservoir of initial uniform density γ_i, the density distribution at any later instant will be given by

$$\gamma = \gamma_i - \frac{1}{4\pi f \kappa} \int_0^t \frac{q(\tau)}{t - \tau} e^{-\frac{y^2}{4\kappa(t-\tau)}} d\tau \int_{-\frac{L}{2}}^{\frac{L}{2}} e^{-\frac{(x-\xi)^2}{4\kappa(t-\tau)}} d\xi. \tag{6}[1]$$

In a similar manner the flux density $q(t)$ may be distributed over a circular boundary to give a representation of a single well or a field lumped together as a single well. For this case, the density distribution at the time t and distance r from the center of the well of radius a will take the form

$$\gamma = \gamma_i - \frac{1}{4\pi f \kappa} \int_0^t \frac{q(\tau)}{t - \tau} d\tau \int_0^{2\pi} e^{-\frac{r^2 + a^2 - 2ar \cos\theta}{4\kappa(t-\tau)}} d\theta$$

$$= \gamma_i - \frac{1}{2 f \kappa} \int_0^t \frac{q(\tau)}{t - \tau} e^{-\frac{(r^2 + a^2)}{4\kappa(t-\tau)}} I_0\left(\frac{ar}{2\kappa(t-\tau)}\right) d\tau, \tag{7}[2]$$

where I_0 is the zero-order Bessel function of the third kind.[3]

[1] This is equivalent to the representation used by Schilthuis and Hurst, (*Oil Weekly*, Oct. 18, 1934) for computing the pressure decline of the "East Texas" field, the results corresponding closely to those derived by another method in Sec. 10.8.

[2] Equation (7) could also be used to give the pressure decline in such a field as "East Texas," and would correspond exactly to the theory of Sec. 10.8 except for the assumption that the reservoir is effectively infinite, whereas there a finite external boundary radius was used.

[3] Watson "Theory of Bessel Functions," p. 79.

Although we shall not enter here into a numerical study of the above equations, it should be observed that the assumption— on which they are based—that the flow system is of infinite extent is of no serious consequence from a practical point of view, except in the later stages of the pressure decline in the system. For until the effective "radius of drainage" has actually receded to the real boundaries of the system, it may be treated as effectively infinite without appreciably affecting the computed course of the decline. On the other hand, most of the problems that appear now to be of practical interest can be treated just as easily by means of the theory developed in the previous sections, the method of sources and sinks being given here essentially for completeness[1] in illustrating the more familiar analytical procedures for solving problems governed by Eq. 10.1(7).

10.16. Summary.—Although almost all problems of the flow of homogeneous fluids through porous media of practical interest are inherently of a time-varying character, owing either to natural or artificial variations in the boundary conditions, it is important to distinguish carefully between those in which the time transients play only a minor role in determining the physical behavior of the system and those in which the variations in the system with time are the predominating features of significance. Such a distinction can be made without difficulty once the exact manner in which time variations can physically enter a problem is clearly understood. Now the velocity of propagation of disturbances within a fluid medium, which evidently determines the time required for variations in the boundary conditions to be transmitted to the internal points of the system, will in general be very high as compared to the velocity of the fluid particles, as the former, being equal to the velocity of sound in the fluid, will be of the order of 10^5 cm./sec., while the latter will usually be limited to values of the order of 1 cm./sec. The transmission of pressure disturbances in a liquid-bearing porous medium may, therefore, be considered as effectively instantaneous. In

---

[1] It may also be observed that quite analogous to the use of source or sink distributions in representing boundary flux values, one may develop a theory of "doublets" by means of which arbitrary boundary density distributions may be attained (*cf.* Byerly "Fourier Series," p. 94). For still other methods of finding solutions of Eq. 10.1(7), see "The Conduction of Heat," by H. S. Carslaw, (1921).

fact, the analytical theory of the steady-state behavior of flow systems in porous media implicitly assumes that the velocity of transmission of the pressure variations is really infinitely great.

A far more significant question with respect to the transmission of pressure disturbances in a fluid-bearing porous medium, however, is that of the time required for the internal points to adjust themselves to the new boundary conditions. Thus if the complete readjustment of the internal pressures should require a finite change in the fluid content of the system, it will clearly take a finite time for the re-establishment of steady-state internal conditions appropriate to the new values of the pressures or fluxes at the boundaries of the porous medium. Now the actual change in the fluid content of a system necessary to bring it in equilibrium with the new pressures at its boundaries will be proportional to the compressibility of the fluid and the area (porous) of the system, whereas the rate at which this fluid mass can be removed or absorbed (for decreases or increases in the boundary pressures) is directly proportional to the permeability of the medium and inversely proportional to the viscosity of the fluid. The resultant times of readjustment of the pressures within the porous medium will, therefore, be proportional to the quantity $f\beta\mu r_e^2/k$, where f and k are the porosity and permeability of the medium, β and μ the compressibility and viscosity of the fluid, and r_e one of the significant dimensions of the system.[1]

Another representation of the time lags involved in the readjustment of the internal-pressure distribution is obtained by a consideration of the rate of change of mass content of the reservoir due to the variations in the boundary conditions as compared to the steady-state carrying capacity of the system. If the former is small as compared to the latter, it is clear that the times of readjustment of the internal-pressure distribution, so as to follow the instantaneous conditions at the boundaries, will be small, and conversely.

[1] These considerations can also be interpreted in terms of the attenuation of the pressure wave as it is transmitted through the system, as the time lag in the readjustment of the internal pressures may be considered as the result of the attenuation of the pressure wave originating at the boundaries, so that its initial amplitude at interior points is but a small fraction of the exciting boundary-pressure variation.

Assuming now that a given system which is in a steady state—in static or dynamic equilibrium—is subjected to a change in the pressure at its boundaries, it is clear that if the time of readjustment of the internal pressures is small, there will be quickly established another steady-state distribution appropriate to the new boundary conditions.   The history of such a system may then be described by a continuous succession of steady-state distributions following the variations in boundary conditions without lag.   It is this type of treatment which has been applied to the various problems discussed in Part II, the justification being that the fluids under consideration have been taken as normal liquids for which β is very small (of the order of 5×10^{-5}/atm.), and that the problems have in all cases, except those of Chap. IX, referred to systems of small or moderate dimensions[1] (of the order of 500 ft.).   And even in Chap. IX, where the porous media were taken as of large or infinite dimensions, the features of particular interest were the mutual interactions of wells which were individually separated by relatively small distances, so that the areas of porous medium associated with each well were in most cases of the same order as those for the single-well systems.   The results derived in Part II may, therefore, be considered as physically applicable to the practical equivalents of the systems discussed there as long as they carry normal liquids, with the understanding that the significant transients of these systems are to be obtained by simply replacing in the steady-state solutions the constant boundary values by their appropriate variable values.

When, however, the liquid has a particularly high compressibility, or when the porous medium is of large dimensions, the times of readjustment will become correspondingly high, and the pressure distributions within the medium will lag behind the changes that may take place at the boundaries.   While a high compressibility is indeed in itself an unreasonable hypothesis when considering a normal liquid, an equivalent effect in the case of extended media may be obtained if there is a dispersion throughout the medium of a small amount of free gas in the form

---

[1] Of course, the assumption in the case of two-dimensional systems that they are infinitely extended along one direction does not invalidate their representation as systems of small dimensions, since they involve no transfers of fluid across the planes defining the fluid motion.

of gas masses of moderate dimensions.   Thus a normal compressibility of 5×10^{-5}/atm. can be raised to an effective value 10 times as great by a distributed gas volume of only $4\frac{1}{2}$ per cent, at 100 atm.   Furthermore, when the area of the system is very large, the times of readjustment even for a liquid of normal compressibility may become so large as to lend practical interest to the time transients of the system, as given by a rigorous treatment of the nonsteady-state problem.

The types of problems involving compressible liquids in nonsteady states of flow may be conveniently classified by means of their boundary conditions.   For radial systems these are: (1) Those in which the pressure (analytically, the density) is specified over both the internal and external radial boundaries; (2) those in which the flux is preassigned for one boundary and the pressure over the other; and (3) those in which the fluxes are given for both boundaries.   In all cases the solutions take the form of infinite series of Bessel functions with the radial coordinate as their arguments, multiplied by exponentials in the time and constants so adjusted that the initial state of the system is reproduced by the solution.

The first case, in which the pressures are specified over both boundaries, will give solutions from which one may derive not only the internal pressure distribution at any time after the initial instant, but also the fluxes passing through the system under the given boundary conditions.   As a particular example one may consider an oil field, lumped together as a single equivalent well, produced by a water drive, and thus study the production history of the field to be expected for various types of variable field pressures (assuming the pressure at the effective external boundary or the water-oil interface to have known values).   Examples of such systems in which the field pressures are either suddenly dropped from their initial uniform values or decrease approximately linearly (the densities decrease in an exactly linear manner) with the time are treated in detail in Sec. 10.4.   Another example of this type of problem is an idealized representation of that of the pressure rise in a well upon shut-down.   Thus, using constants corresponding to the "East Texas" field, it is found from the solution developed for this problem (Sec. 10.6) that only about 6 per cent of the original pressure differential in an "East Texas" well should remain after

½ hr. of shutting in, which is of the same order of magnitude as has been observed in that field. While such solutions will not apply if there is an appreciable amount of free gas in the sand about the well—due to the simultaneous change of the effective sand permeability with the rise in pressure—it should give at least the qualitative features of the effect of the compressibility even in the case when free gas is present in the sand.

The second type of radial-flow problem involves the situation in which are preassigned the flux rate at one of the boundaries and the pressure at the other. Thus if the sand forms a closed system the flux at the external boundary will be zero, and if the pressure at the internal boundary, defining a well bore for example, is known, the solution to the problem will give the fluxes passing through the internal boundary. Or, the pressures at the external boundary might be preassigned together with the fluxes at the internal boundary, the solution then giving the variation with time of the pressure at the latter.

A very interesting practical application of this latter type of problem arises in the interpretation of the pressure variations of the "East Texas" oil field. Analyses of the oil having shown that the oil of this field is saturated with gas only to the pressure of 755 lb., the production from the field until its pressures have fallen to 755 lb. must evidently be attributed to the "drive" exerted on the oil by the water from the adjacent Woodbine sand. The tremendous area of this water reservoir immediately suggests that the time transients in the system must play a significant role in determining the instantaneous pressures in the field, and indeed the very fact that the pressures in the field have shown declines even while the production rates have been kept constant proves that the water in the Woodbine sand is really a compressible liquid. Furthermore, the total pressure drops in the water reservoir are considerably too low to induce a flow of water into the field sufficient to replace the oil removed from it, if the pressure distribution in the reservoir were simply the steady-state distribution.

By lumping together the individual wells of the field into an equivalent well of a radius (20 miles) approximating the general contour of the field, and representing the water reservoir of the Woodbine sand as a radial sector of 120 deg. extending from a constant (original) pressure external boundary at a 100-mile radius and converging into the well, the pressure decline at the internal boundary—the western edge of the field—becomes determinate, once the production history from the field is preassigned. Choosing for the latter the observed production rates for the field since its discovery, the computed pressure decline is almost exactly parallel to and higher than the recorded pressures as averaged over the individual wells of the field. The difference between the computed and observed pressures are readily explained—and indeed to be expected—when it is noted that the theoretical calculation gives the pressures to be found at the western edge of the field, whereas the reported field pressures are averages taken over the whole width of the field and include not only the pressure drop over the field inherently associated with the fluid migration from west to east, but also the accentuation of this pressure drop due to the fact that the producing section of the sand is wedge-shaped, pinching out toward the east. When the observed average pressures are corrected by the addition of a term proportional to the field production rate— 7 lb. per 100,000 bbl./day—the resultant pressures agree with the calculated values as closely as can be expected—5–10 lb.— in view of the uncertainties in the real production rates from the field due to the unknown extent of the illegally produced and unreported oil taken from the field. This agreement, over a period of six years during which the production rates have fluctuated from extremely high values to zero values during shut-down periods, is particularly significant when it is noted that all the physical constants chosen for the system, excepting only the water compressibility, were actual averages of data derived from well logs in the Woodbine sand.

The single adjustable constant, the effective compressibility of the water, had to be taken as 12 times that of gas-free water. While this may at first appear surprising, such an *effective* value can be explained by supposing that there are gas pockets distributed throughout the water horizon to an extent of only 4.9 per cent of the total pore volume of the sand, an assumption contradictory to nothing at present known of the Woodbine reservoir.

The fundamental role played by the nonsteady-state character of the flow in the Woodbine sand is shown in a striking manner by the computed pressure distribution in the sand for July 1,

1934, some $3\frac{1}{2}$ years after the discovery of the field. For although 520 million barrels of oil had been removed from the field, the calculations show that the pressure distribution at that time was still far from that of the steady state.   In fact, instead of the pressure rising uniformly in a logarithmic manner to the original field pressure of 1,620 lb. as the external boundary at 100 miles is approached, the actual pressure distribution, of that date, rises rapidly to the pressure of 1,620 lb. within a distance of 25 miles from the field.   Thus the whole production of the field to that time had actually been replaced by the expansion of the water in the Woodbine sand within the 45 miles radius, rather than by the bodily movement of the water throughout the whole of the water reservoir, as would be implied by a steady-state representation of the flow system.

This concentration of the pressure drop within a short distance of the field even after $3\frac{1}{2}$ years of production shows that the placing of the external boundary at a 100-mile radius will be of no significance with respect to the pressure decline at the field until the latter has been entirely depleted.   In fact, the whole estimated oil content of the field of 7 billion barrels could be replaced by a simple expansion of the water (assuming the high compressibility) involving a drop in pressure over the whole reservoir up to the radius of 100 miles of only 177 lb.

The same analytical procedure applied to the calculation of the pressure decline in the "East Texas" field can be used to compute the variation of the flux from a well or field draining a closed reservoir (*cf*. Sec. 10.9) for preassigned well or field pressures. However, this problem does not at present seem to be of great practical interest except in the case of actual oil production from wells penetrating lenses of limited volume, when the internal drive due to the evolution and expansion of the dissolved gases will in general be the significant feature.   Such a treatment, however, is beyond the scope of this work.

This case of a well producing from a closed reservoir may, however, be used as a convenient illustration of the final type of radial-flow problem, in which the flux is specified over both boundaries.   The solution of this problem, in which the initial pressure in the sand is taken as uniform and the production rate from the well is preassigned to have a constant value, shows two types of transient in the pressure decline both at the well and the external boundary. The first, of extremely short life—about an hour for the constants chosen in the numerical example—disappears exponentially with the time. It is then followed by a linear decline in both the well and external-boundary pressures in which the system assumes an effectively steady-state drainage with the pressures declining uniformly over the whole reservoir. The actual pressure distribution is almost exactly logarithmic except at the external boundary, where the gradient is necessarily zero owing to the closure of the reservoir. The decline of the system here naturally assumes a continuous succession of steady states in which the flux through the outflow well is supplied by a uniform depletion of the fluid content at all points of the reservoir.

In addition to strictly radial-flow systems, those involving nonradial flow can also be treated analytically. Such problems arise in the consideration of the interference between wells draining the same reservoir. For in closed systems the pressure decline must evidently be approximately proportional to the total flux from the system. A specific calculation on the effect on the pressure decline at a well producing at a uniform rate at the center of a closed sand, owing to drilling another with the same rate midway between the center and the external circular boundary, shows indeed that the decline at the center becomes doubled after a short-period transient. From a practical point of view, however, it should suffice for most purposes to study the interference effects in nonsteady-state systems by neglecting the local short-period transients and replacing the nonsteady-state pressure distributions by continuous sequences of steady-state distributions with density declines distributed uniformly over the system in such a manner as will supply the total flux from it.

In addition to the method of solving directly the fundamental partial differential equation [*cf.* Eq. 10.1(7)] for the nonsteady-state flow of liquids, solutions can be obtained by the synthesis of particular solutions representing fluid sources or sinks giving instantaneous or permanent fluxes entering or leaving the flow system. These are particularly suited to the analysis of problems in which the reservoir may be taken as effectively infinite, although one may apply the method of images, outlined in the discussion of steady-state flow systems, to construct solutions corresponding also to regions of finite extent.

PART IV

THE FLOW OF GASES THROUGH POROUS MEDIA

.

CHAPTER XI

THE FLOW OF GASES THROUGH POROUS MEDIA

11.1. Introduction.—It was seen in Chap. II that Darcy's law that the velocity of a homogeneous fluid at any point in a porous medium is, under viscous-flow conditions, proportional to the pressure gradient at the point, and which was originally established for the case of liquids, holds also for the flow of gases. Hence the dynamical law of motion can here, too, be written as

$$\bar{v} = -\frac{k}{\mu}\nabla p, \tag{1}$$

where k is the permeability of the medium and μ is the viscosity of the fluid. Furthermore, it was shown in Sec. 3.4 that on applying the equation of continuity to Eq. (1), the density γ of a gas flowing in a homogeneous porous medium must obey the fundamental differential equation

$$\nabla^2\gamma^{\frac{1+m}{m}} = \frac{(1+m)\mu f \gamma_0^{\frac{1}{m}}}{k}\frac{\partial\gamma}{\partial t}, \tag{2}$$

where the equation of state of the gas has been defined by

$$\gamma = \gamma_0 p^m, \tag{3}$$

and f is the porosity of the medium. The exponent m determines the thermodynamic character of the expansion of the gas as it moves from the high- to the low-pressure regions; in particular, $m = 1$ corresponds to isothermal expansion, while the case of adiabatic expansion is given when

$$m = \frac{\text{(specific heat at constant volume)}}{\text{(specific heat at constant pressure)}}.$$

Equation (2) governs both the steady- and unsteady-state conditions of flow of a gas in a homogeneous porous medium, in complete analogy to Eq. 10.1(7) for the case of a compressible liquid. However, whereas the latter is linear and permits explicit

679

solutions, Eq. (2) involves powers of γ—in derivative form —other than the first, and hence is nonlinear. Furthermore, for the type involved in Eq. (2), it has not been thus far possible to derive explicit solutions for general boundary and initial conditions, thus necessitating the use of an approximate method of analysis, as will be developed later.

11.2. The Steady-state Flow of Gases. Linear Systems.— For steady-state conditions of flow, on the other hand, the analysis can be carried through rigorously. For then, setting the right side of Eq. 11.1(2) equal to zero, the equation reduces to Laplace's equation in the dependent variable $\gamma^{\frac{1+m}{m}}$ or p^{1+m}, namely,

$$\nabla^2\gamma^{\frac{1+m}{m}} = \nabla^2 p^{1+m} = 0. \tag{1}$$

As this is the same equation as that in the dependent variable p, governing the flow of incompressible liquids, the various solutions derived in Part II for the latter case may be directly applied to the solution of the corresponding problems in the case of gases. In particular, the pressure distribution in a linear system is given by

$$p^{1+m} = (p_1{}^{1+m} - p_2{}^{1+m})\frac{x}{L} + p_2{}^{1+m}, \tag{2}$$

where p_2, p_1 are the boundary values of the pressure at $x = 0, L$. The rate of mass flow for the system per unit cross section is, therefore,

$$Q = -\frac{k\gamma}{\mu}\frac{\partial p}{\partial x} = -\frac{k\gamma_0}{(1 + m)\mu}\frac{\partial p^{1+m}}{\partial x}, \tag{3}\text{[1]}$$

which, applied to Eq. (2), takes the form

$$Q = \frac{k\gamma_0}{\mu(1 + m)L}(p_2{}^{1+m} - p_1{}^{1+m}). \tag{4}$$

Denoting by Q_1 the value of Q for isothermal flow ($m = 1$), with the same density at the pressure p_2—so that $\gamma_0 \sim p_2{}^{-m}$ in Eq. (4)—Eq. (4) may be expressed as

---

[1] It may be noted that here and in all the formulae for Q to be given below, Q/γ_0 will give the *volume* flux as measured at atmospheric pressure.

$$Q = \frac{2Q_1}{1+m} \frac{1 - (p_1/p_2)^{1+m}}{1 - (p_1/p_2)^2}. \tag{5}$$

It follows that the flux rate for nonisothermal flow ($m < 1$) exceeds that for isothermal flow, Q/Q_1 increasing as m decreases. Furthermore, for a given type of flow, fixed m, Q/Q_1 increases as p_1/p_2 decreases—as the pressure differential and the flow rates increase.

As a specific example, one may choose $p_1/p_2 = 0.1$ and $m = 0.71$, which, for air, corresponds to adiabatic expansion. It follows then from Eq. (5) that $Q/Q_1 = 1.16$, so that 16 per cent

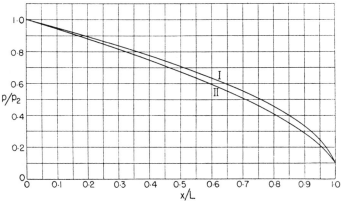

FIG. 261.—The pressure distributions in a linear channel flowing an ideal gas under isothermal conditions (curve I) and adiabatic conditions (for air, curve II). x/L = fractional distance along channel. p/p_2 = (pressure at x)/(pressure at $x = 0$).

more air will pass through a linear system if it expands adiabatically (and $p_1/p_2 = 0.1$) than if its expansion is isothermal. It is to be noted that this excess is due essentially to the higher outflow density, for equal inflow densities, for the case of adiabatic expansion than for the isothermal flow. In fact, this difference in densities more than compensates for the lower outlet-pressure gradients in the nonisothermal flow system. This may be seen from the pressure-distribution curves plotted in Fig. 261 for a linear system for $m = 1$ and $m = 0.71$.

11.3. Two-dimensional Systems in the Steady State.—For a radial flow system, the pressure distribution, by analogy with Eqs. 4.2(8) and 10.1(13), will be given by

$$p^{1+m} = \frac{(p_e{}^{1+m} - p_w{}^{1+m}) \log r/r_u}{\log r_e/r_w} + p_w{}^{1+m}. \tag{1}$$

The associated mass flux through the system, of thickness h, will be

$$
\begin{aligned}
Q &= \frac{2\pi r h k \gamma}{\mu} \frac{\partial p}{\partial r} = \frac{2\pi r h k \gamma_0}{(1+m)\mu} \frac{\partial p^{1+m}}{\partial r} \\
&= \frac{2\pi h k \gamma_0 (p_e{}^{1+m} - p_w{}^{1+m})}{(1+m)\mu \log r_e/r_w} = \frac{2Q_1}{1+m} \frac{1 - (p_w/p_e)^{1+m}}{1 - (p_w/p_e)^2},
\end{aligned} \tag{2}
$$

where Q_1 is again the flux for isothermal flow ($m = 1$), namely,

$$Q_1 = \frac{\pi h k \gamma_0 \Delta p^2}{\mu \log r_e/r_w}, \tag{3}[1]$$

In a similar manner one may write down the solutions to other steady-state problems in the flow of gases in porous media corresponding to those treated in Part II for incompressible liquids by simply replacing the pressures occurring there in the pressure distributions by p^{1+m} to get the pressure distributions for the case of gas flow, and the Δp occurring in the expressions for the flux by[2] $\gamma_0 \Delta p^{1+m}/(1 + m)$. Thus it may be shown that, even though the pressure distribution over the circular boundaries of a radial system is not uniform, the simple radial-flow formula for the mass flux may still be used, provided for the boundary values of p the average values of p^{1+m} are substituted. In such cases, therefore,

$$Q = \frac{2\pi h k \gamma_0 (\overline{p_e{}^{1+m}} - \overline{p_w{}^{1+m}})}{(1+m)\mu \log r_e/r_w}. \qquad [cf.\ \text{Eq. } 4.5(12)] \tag{4}$$

On the other hand, if the external boundary enclosing a well is not circular, the flux may in general be expressed in the form

$$Q = \frac{2\pi h k \gamma_0 \Delta p^{1+m}}{(1+m)\mu \log c/r_w}, \qquad [cf.\ \text{Eq. } 4.16(6)] \tag{5}$$

[1] This may also be evidently written as $Q_1 = \dfrac{2\pi h k \overline{\gamma} \Delta p}{\mu \log r_e/r_w}$, where $\overline{\gamma}$ is the algebraic mean density in the system, so that $Q_1/\overline{\gamma}$, the *volume* flux at the mean pressure in the system, is given by the same formula as that applicable to a liquid-flow system.

[2] For isothermal flow one need only multiply the expression for the flux (volume) as given in Part II by the algebraic mean density $\overline{\gamma}$ to find the *mass* flux in the corresponding gas-flow system (*cf.* preceding footnote).

where c is a constant depending on the shape of the external boundary, and which may be approximated by an average distance of the well from the external boundary.

11.4. Three-dimensional Systems in the Steady State.—The solutions for the three-dimensional systems presented in Chap. V may also be transposed without further analysis into their equivalents for porous media carrying gaseous fluids. Thus the potential distributions for partially penetrating wells as are shown in Figs. 81 and 82 will also apply for wells partially penetrating

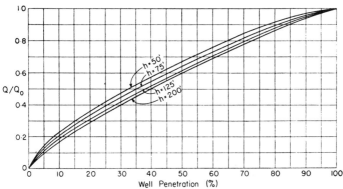

Fig. 262.—The relative production capacities of partially penetrating wells producing either gases or liquids. Q/Q_0 = (production capacity of partially penetrating well)/(production capacity of completely penetrating well in same sand). h = sand thickness; well radius = $\frac{1}{4}$ ft.; reservoir boundary radius = 500 ft.

gas horizons, of only the pressures or potentials there are interpreted as p^{1+m}. The production capacities of partially penetrating wells as functions of the well penetration or sand thickness, plotted in Figs. 83 and 84, cannot be used directly, as the production capacities of these figures are expressed in barrels per day per atmosphere. However, the ratios of these production capacities for partially penetrating wells to those for completely penetrating wells will be identical for the flow of liquids and gases. These are plotted in Figs. 262 and 263. The absolute values of the mass-production rates of the gas wells, which may be obtained from these figures by multiplying the ratios by the absolute mass fluxes for a completely penetrating well, as given by Eq. 11.3(2), depend not only on the type of flow—value of m—but also on the absolute pressures. Thus for the isothermal flow ($m = 1$) of methane ($\mu = 0.012$ centipoise) into a $\frac{1}{4}$-ft. well at atmospheric pressure, from a 50-ft. sand with a permeability of 1 darcy and "reservoir" pressure of 7 atm. ($\sim$103 lb.) at a distance of 500 ft. from the well, Eq. 11.3(3) gives a flux rate of 2.52×10^6 cc./sec. (at atmospheric pressure) or 7.68×10^6 ft.^{3}/day. From Fig. 262 it is seen that if the well penetration is 75 per cent the flux rate will be only 87.3 per cent of that of the completely penetrating well, and 66.9 per cent if the well penetration is 50 per cent. Hence the absolute production rates for the wells

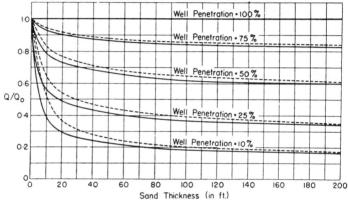

Fig. 263.—The variation of the relative production capacities of partially penetrating wells with the sand thickness. Q/Q_0 = (production capacity of partially penetrating well)/(production capacity of completely penetrating well in same sand). ——————————: well radius = $\frac{1}{4}$ ft.; ....: well radius = $\frac{1}{2}$ ft.; reservoir boundary radius = 500 ft.

of 75 and 50 per cent penetration will be 6.71×10^6 and 5.14×10^6 cu. ft./day, respectively.

For gas wells partially penetrating an anisotropic sand, one may again take over the results derived in Part II for incompressible liquids (*cf.* Sec. 5.5). Thus from Fig. 87 one may get directly the production rates (in volume at atmospheric pressure) from a 125-ft. sand for various well penetrations and vertical sand permeabilities by simply multiplying the ordinates by $\dfrac{hk\Delta p^{1+m}}{(1 + m)\mu}$, h being 125 ft. Figure 88 remains strictly valid without any change whatever.

11.5. The Effect of Gravity on the Flow of Gases through Porous Media.—Owing to the very low density of gases, even when under such pressures as occur inc underground gas reservoirs, the effect of gravity may be entirely neglected in the discussion of the flow of gases through porous media, insofar as direct effects on the gas itself are concerned. Furthermore, as a gas will fill completely any space into which it can enter at all, there will be no free surfaces in a gas-flow system, even though gravity is permitted to act freely upon the gas.[1] The problems of gravity flow of liquids treated in Chap. VI are, therefore, of no interest in the study of the flow of gases through porous media.

11.6. The Steady-state Flow of Gases within Systems of Nonuniform Permeability.—When a porous medium through which a gas is flowing in the steady state is not strictly homogeneous, the procedure of simply solving the Laplace's Eq. 11.2(1) as outlined in the previous sections must be modified. If the permeability of the medium varies continuously, Eq. 11.2(1) must be replaced by

$$\frac{\partial}{\partial x}\left(k\frac{\partial p^{1+m}}{\partial x}\right) + \frac{\partial}{\partial y}\left(k\frac{\partial p^{1+m}}{\partial y}\right) + \frac{\partial}{\partial z}\left(k\frac{\partial p^{1+m}}{\partial z}\right) = 0, \qquad (1)$$

neglecting the effect of gravity, and may be treated in the same manner as the corresponding Eq. 7.2(2) for the case of liquids.

If the permeability varies discontinuously, one may again apply the methods of Chap. VII. The Laplace's equation in p^{1+m}, Eq. 11.2(1), must then be solved for each region of uniform permeability, and the individual solutions connected at the "surfaces of discontinuity" so that the pressures and normal velocities are continuous there. To get the actual solutions, one need only replace p in the pressure distributions of Chap. VII by p^{1+m}. In particular, the effects on the production rates of local inhomogeneities are, for gas wells, exactly the same as given in Chap. VII. Thus for radial systems Eq. 7.3(11) and Fig. 150 are valid without change. For limestone-fracture systems producing gas, the analysis of Secs. 7.4 and 7.5 may be carried over by simply changing p to p^{1+m} as already indicated.

The theory of acid treatment of limestone wells producing oil, developed in Secs. 7.6 to 7.8, is equally valid for the case of

---

[1] To be exact, there will be some density variation with height owing to the action of gravity, such as exists in the atmosphere; however, this will be of infinitesimal magnitude as compared to the density variations associated with the *dynamical-pressure* variations in any system of practical interest.

limestone reservoirs producing gas, Figs. 157 to 159 remaining true without change.

The theory of partially penetrating gas wells in stratified horizons may also be taken over directly from that given in Chap. VII. Figure 162 giving the production capacity of a well just tapping a sand underlain, at a depth of 25 ft., by an infinitely thick sand of different permeability, will also apply to a gas well in giving the production capacities relative to that in an infinitely extended homogeneous sand, if the ordinates of Fig. 162 are reduced so that for $k_2/k_1 = 1$ becomes unity.

Finally if the well bore of a gas well becomes partly filled with sand, the relative effects on the production capacity of the well will be exactly the same as if the well produced liquid. Equations 7.10(8) and 7.10(19) and Fig. 164 remain valid without change, while the ordinates of Fig. 165 must be taken as those of $(p^{1+m} - p_w{}^{1+m})/\Delta p^{1+m}$.

11.7. Two-fluid Systems. Water Coning.—Although gas is almost universally associated with the oil in underground oil reservoirs, this association does not necessarily imply that the flow of the gas and oil is that of a two-fluid system of the type considered in Chap VIII. For as long as the gas stays in solution, the oil will flow as a homogeneous liquid, the effect of the gas coming into play only in the reduction of the viscosity of the oil. And this is indeed the basic implicit assumption underlying the whole treatment in this work of the flow of liquids in porous media as that of a homogeneous fluid, insofar as the liquid is taken to be an oil flowing through its underground reservoir. On the other hand, if the gas has come out of solution and is more or less uniformly disseminated through the oil, it is still not to be considered as a two-fluid system of the type discussed in Chap. VIII, but rather the gas and oil must be treated as a mixture or single heterogeneous fluid, the nature of which—as defined by its equation of state—will in general vary from point to point, and the discussion of which is beyond the scope of this work.

The two-fluid systems treated in Chap. VIII up to Sec. 8.10 involved an approximately vertical separation between the two liquids, so that one fluid was considered as "encroaching" on the other. While this situation will never be strictly realized in isotropic media as long as the two fluids are of unequal density, it may be taken as an approximation to the physical problem if the density difference between the two fluids is small, as is the case in the encroachment of water into an oil sand. However, if the water is encroaching into a gas zone, the assumption of a vertical interface will clearly be entirely inadmissible. For the water will evidently flow to the bottom of the porous medium, and will form an interface with the gas that may be more closely described as a horizontal than a vertical interface.

If the water does come into a gas sand, along the bottom of the pay, and if the sand is not completely penetrated by the well, there will arise a problem of water coning quite similar to that when the upper part of the pay is carrying oil. The physical analysis of the problem will be identical with that given in Sec. 8.10 for the coning of water into an oil sand. However, in the analytical treatment account must be taken of the fact that the unperturbed potential distributions derived in Sec. 5.3 and used in Sec. 8.11 give the values of p^{1+m}, rather than p, in the gas-flow system. The pressure distribution itself will, therefore, show more concentrated gradients about the well in the present case than in that of Chap. VIII, and as the fundamental condition for equilibrium, Eq. 8.10(1), involves p linearly, it is clear that water coning will be more difficult for a gas well than for one producing oil. Furthermore, as the total pressure differentials in the system required to bring the water into the well are proportional to the density difference between the two fluids, the critical-pressure differentials for water coning will, on this account alone, be some three to four times as high for a gas well as for an equivalent oil well. It follows, therefore, that water coning will be much more readily suppressed and will involve less serious difficulties for wells producing from gas zones than for wells producing oil. If the water does cone into the well, the methods of suppression by pinching in or plugging back the well as discussed in Chap. VIII should be equally successful for the gas well as for an oil well. Finally the general nature of the curves of Fig. 187 giving the critical-pressure differentials for coning will also apply to gas wells, although the absolute values of the pressure differentials will probably be greater than those shown in Fig. 187 by a factor of at least 4.

11.8. Gas-oil Ratios in Porous Media Carrying Gas and Oil as Homogeneous Fluids.—While the problem of analyzing theoretically the gas-oil ratios to be expected of wells producing both gas and oil must ultimately be treated from the point of view that the gas and oil within the underground reservoir form a single heterogeneous fluid, the limiting cases where the gas and oil zones are essentially separated due to the segregation and migration of the gas to the upper strata of the pay, or where the gas and oil flow in two distinct noncommunicating but parallel zones penetrated by the same well, may be discussed at least approximately by the laws of flow of homogeneous fluids. The significance of the gas-oil ratio—defined as the volume of gas, measured under atmospheric conditions, produced for each volume of oil—is that it is a direct measure of the efficiency of the production, and determines the ultimate recovery of oil to be expected from the oil reservoir. For if R_i is the initial reservoir gas-oil ratio, p_i the initial reservoir pressure, and if $\bar{R}$ is the average gas-oil ratio until the average field pressure has fallen to $\bar{p}$, the fractional recovery of oil P from the sand at the time corresponding to $\bar{p}$ is given by the relation

$$P \gtrless \frac{R_i(p_i - \bar{p})}{\bar{R}p_i}. \tag{1}$$

The equality applies to cases where there are no free-gas sands, whether or not in communication with the oil sand, which are exposed by the well bore and are contributing to the observed average gas-oil ratio $\bar{R}$. If there are such free-gas sands producing gas through the tubing, the inequality in Eq. (1) is to be applied.[1] Except for this qualification, Eq. (1) is true regardless of the details or method of the production. No assumption is implied as to the presence or absence of a water drive, that the gas does or does not obey Henry's law, that the field is produced wide open or with back-pressure control, that the well spacing is wide or close, that the sand is thick or thin, or that the permeability is high or low. A consistently high gas-oil ratio will, therefore, necessarily imply a low ultimate recovery, and any method which will, in a given situation, give the lowest gas-oil ratio must represent, for *that case*, the most efficient method of production. In view of this fact, the importance of noting carefully and controlling the gas-oil ratio during the course of the production becomes self-evident.

---

[1] If R_i is also taken to include the gas in any such free-gas sands, then the equality will, of course, still be valid.

As the only control, available at the surface, upon the details of the flow within the reservoir is that of varying the bottom-hole pressure by controlling the production rates from the wells, the essential question with respect to the gas-oil ratio is its dependence upon the back pressure or bottom-hole pressure maintained at the sand face.   It is this variation which will be investigated here.

11.9. Gas-oil Ratios for Communicating Gas and Oil Zones.— As is indicated in the diagram, Fig. 264, it will be supposed that the gas was originally separated or has been segregated in the course of the production in the upper part of the producing horizon, so that the gas and oil flow individually as homogeneous fluids with a common interface. Although there will be a tendency for the gas to cone down and induce a gravity flow and

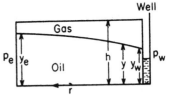

FIG. 264.—Diagrammatic representation of communicating oil and free-gas zones exposed by the same well bore.

vertical velocities in both the gas and oil zones, we shall, as a first approximation neglect them, and take the horizontal velocity components as uniform over the cross section of the flow channel. Furthermore, since the oil and gas have a common interface, the pressures in the two fluids must be the same on the two sides of the interface, and hence by the above approximation the fluid heads will be the same over any vertical cylinder in the sand (coaxial with the well) whether one refers to the gas or oil zone. Denoting now by $y(r)$ the height of the interface at r, and by Q_g, Q_o, the rates of gas and oil production from the two zones, it follows from the fundamental laws of gas and liquid flow through sands, that of the former being under isothermal conditions, that under steady-state conditions

$$Q_g = \frac{\pi k r (h - y)}{\mu_g} \frac{\partial p^2}{\partial r} \qquad [cf.\ \text{Eq. 11.3(2)}] \qquad (1)$$

$$Q_o = \frac{2\pi k r y}{\mu_o} \frac{\partial p}{\partial r}, \qquad (2)$$

where h is the total sand thickness, μ_g, μ_o are the viscosities of the gas and oil, respectively, p is the pressure at r, and Q_g gives the gas volume at atmospheric pressure.

Although in addition to the flow, Q_o, due to the radial pressure gradients, there will also be a gravity component, it can be shown by a separate calculation that its effect on the gas-oil ratio will be entirely negligible. One may, therefore, take Q_g/Q_o for the gas-oil ratio, or

$$R = \frac{Q_g}{Q_o} = \frac{\mu_o(h - y)p}{\mu_g y}. \tag{3}$$

As this ratio, under steady-state conditions, must be evidently the same for all values of r, R may be computed from the value of p and y at the back of the sand, *i.e.*, at a large distance from the well. Hence,

$$R = \frac{\mu_o p_e(1 - y_e/h)}{\mu_g y_e/h} \tag{4}$$

where p_e is the reservoir pressure, and y_e/h is the fraction of the total sand thickness saturated with oil at a large distance from the well. As Eq. (4) does not involve the well pressure p_w, it is seen that for the present case the gas-oil ratio is directly proportional to the reservoir pressure but is *independent of the back pressure*.

As the sand is depleted, however, p_e and y_e will decrease and R will change, although at any given time R will remain fixed regardless of variations in the back pressure or production rates. If it is assumed that y_e decreases linearly or even more rapidly with decreasing p_e, R will increase as the sand becomes depleted.

To complete the physical picture for this type of production, we shall compute the shape of the gas-oil interface and pressure distribution under the approximations already indicated. Thus multiplying Eq. (1) $hy\mu_g$ and Eq. (2) by $\mu_o p$ and adding, it is found that

$$r\frac{\partial p^2}{\partial r} = \frac{Q_o\mu_o}{\pi kh}p + \frac{Q_g\mu_g}{\pi kh},$$

the solution of which is

$$\log \frac{r_e}{r} = \frac{1}{b}\left[p_e - p - \frac{a}{b}\log \frac{p_e + a/b}{p + a/b} \right]. \tag{5}$$

where

$$a = \frac{Q_g\mu_g}{2\pi kh}; \qquad b = \frac{Q_o\mu_o}{2\pi kh}. \tag{6}$$

The ratio a/b may be found from the relation

$$\frac{a}{b} = \frac{1 - y_e/h}{y_e/h} p_e,$$

which follows at once from Eqs. (4) and (6); and b may be computed by setting $r = r_w$, the well radius (where $p = p_w$), in Eq. (5).

Assuming for a practical example that

$$\frac{y_e}{h} = 0.9; \qquad p_e = 70 \text{ atm.}; \qquad p_w = 10 \text{ atm.}; \qquad \frac{r_e}{r_w} = 2,000,$$

and computing the pressure distribution from Eq. (5), and then the shape of the gas-oil interface by means of Eq. (3), the results

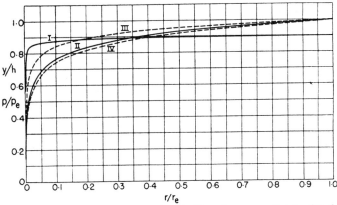

FIG. 265.—The shape of the interface (y/h) and pressure distribution (p/p_e) in communicating oil and free-gas zones. y = height of gas-oil interface above the bottom of the sand. h = total thickness of the sand. I: y/h; II: p/p_e; III: p/p_e for gas zone if isolated; IV: p/p_e for oil zone if isolated. p_e = reservoir pressure (at r_e) = 70 atm. $y_e/h = 0.9$; well pressure = 10 atm.

are those shown in Fig. 265. It will be seen that the coning effect of the gas is very localized about the well, although the oil surface at the face of the well is depressed to 62 per cent of its height at the back of the sand. This pressure distribution in the composite system may be compared with those of the dotted curves giving the distributions that would exist in the gas and oil zones if they were not connected.

These curves show the physical reason for the coning of the interface when the gas and oil zones are interconnected. For owing to the higher pressures in the gas zone it would naturally tend to depress the oil surface if it were permitted to react with it freely. In the equilibrium state the resultant distribution should evidently be an average between those for the isolated gas and oil zones, as curve II actually proves. The fact that this resultant pressure distribution lies closer to that of the liquid is clearly due to the assumption that the oil zone is nine times as thick as the gas zone ($y_c/h = 0.9$), so that the effect of the latter is essentially like that of a small perturbation upon the oil zone.

11.10. Gas-oil Ratios for Noncommunicating Gas and Oil Sands.—As in the present case the gas zones and oil zones may lie in entirely unrelated sands, though penetrated by the same well, the reservoir pressures for the two zones need not neces-

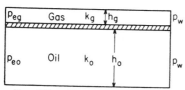

FIG. 266.—Diagrammatic representation of noncommunicating oil and free gas zones exposed by the same well bore.

sarily be the same. Denoting the reservoir pressure for the gas by p_{eg} and that for the oil p_{eo}, one may distinguish between three possible cases, namely, $p_{eg} > p_{eo}$; $p_{eg} = p_{eo}$; and $p_{eg} < p_{eo}$. In all cases, of course, the well pressures[1] will be taken the same, and will be

denoted by the symbol p_w. The thickness of the gas and oil zones will be denoted by h_g and h_o, and their permeabilities by k_g and k_o (cf. Fig. 266).

Applying, then, the formulas for steady-state gas and liquid radial flow, the gas and liquid volume-production rates will be given by

$$Q_g = \frac{\pi k_g h_g}{\mu_g} \frac{(p_{eg}{}^2 - p_w{}^2)}{\log r_e/r_w} + R_0 Q_o, \qquad (1)$$

$$Q_o\dot{} = \frac{2\pi k_o h_o}{\mu_o} \frac{(p_{eo} - p_w)}{\log r_e/r_w}, \qquad (2)$$

where R_0 represents the gas-oil ratio due to any other gas in the system besides that in the noncommunicating zone, assuming that this gas does not destroy the validity of Eq. (2) for the flow of the oil, and that the gas-oil ratio R_0 is independent of the back pressure.

---

[1] To be exact, the well pressures minus the fluid heads from the top of the sand will be the same in both zones, but since the reservoir pressures also involve these corrections there will be no net error in using simply the pressures.

The resultant gas-oil ratio will, therefore, be

$$R = \frac{Q_g}{Q_o} = R_0 + \frac{c(p_{eg}^2 - p_w^2)}{p_{co} - p_w}, \tag{3}$$

where c is the constant: $k_g \mu_o h_g / 2 k_o \mu_g h_o$.

Since, however, R_0 has been assumed to be independent of the back pressure, it may be omitted in the discussion of R as a function of p_w; only the last term need, therefore, be considered as

$$R = \frac{c(p_{eg}^2 - p_w^2)}{p_{co} - p_w}. \tag{4}$$

The detailed variation of R with p_w may be seen from the slope of the R versus p_w curves, which is given by

$$\frac{\partial R}{\partial p_w} = c\left[1 + \frac{p_{eg}^2 - p_{eo}^2}{(p_{eo} - p_w)^2}\right]. \tag{5}$$

It follows at once that if

$$(a) \; p_{eg} > p_{co},$$

$\dfrac{\partial R}{\partial p_w} > 0$ and increases as the back pressure is increased to the value of the reservoir pressure in the oil zone. The gas-oil ratio curve will be of the form given in Fig. 267.

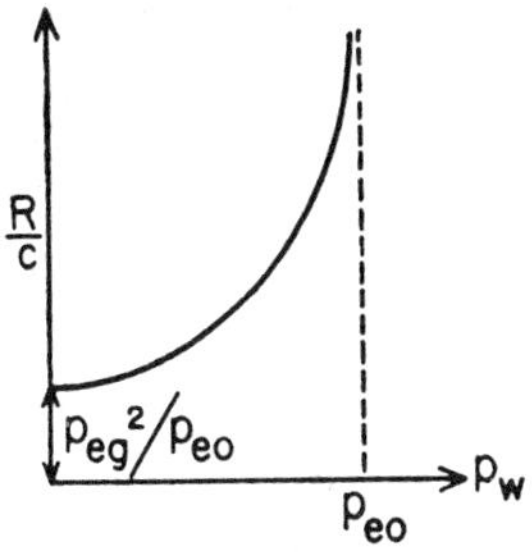

Fig. 267.—Diagrammatic representation of the variation of the gas-oil ratio (R) with the back pressure (p_w) if the reservoir pressure in the free-gas zone (p_{eg}) exceeds that in the oil zone (p_{eo}). $c = k_g \mu_o h_g / 2 k_o \mu_g h_o$, where the subscripts g, o refer to the gas and oil zones of thickness h_g, h_o.

$$(b) \; p_{eg} = p_{eo},$$
$$\frac{\partial R}{\partial p_w} = c = \text{const},$$
$$R = c(p_e + p_w).$$

The gas-oil ratio here, therefore, increases linearly with the back pressure, as shown in Fig. 268.

$$(c) \; p_{eg} < p_{co},$$

$\dfrac{\partial R}{\partial p_w} > 0$, for $p_w \sim 0$, but changes sign before p_w is raised to the gas-reservoir pressure, the gas-oil ratio due to the free gas zone finally going to zero as p_{eg} is reached. The shape of the curve is given in Fig. 269.

Under no circumstances will there be a minimum in the gas-oil ratio.

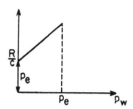

Fig. 268.—Diagrammatic representation of the variation of the gas-oil ratio (R) with the back pressure (p_w) if the reservoir pressure in the free-gas zone (p_{eg}) equals that in the oil zone (p_{eo}), the notation being the same as for Fig. 267.

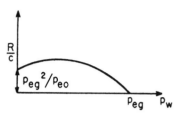

Fig. 269.—Diagrammatic representation of the variation of the gas-oil ratio (R) with the back pressure (p_w) if the reservoir pressure in the free-gas zone (p_{eg}) is less than that in the oil zone (p_{eo}), the notation being the same as for Fig. 267.

With respect to the variation of the gas-oil ratio with the rates of oil production, one need only reverse the scale of p_w. The above results will then take the form:

(a) $p_{eg} > p_{eo}$: R increases (more rapidly than linearly) as the production rate decreases. It becomes infinitely large as the back pressure is increased to the oil-reservoir pressure, corresponding to a bleeding off of the gas while the oil is not being produced at all.

(b) $p_{eg} = p_{eo}$: R increases linearly with decreasing production rates.

(c) $p_{eg} < p_{eo}$: As the production rates decrease the gas-oil ratio slowly rises to a maximum, and then decreases to a value lower than that for wide-open flow as the production rate is still further decreased.

It is to be understood, of course, that the assumed constant gas-oil ratios R_0 due to the other gas in the system are to be added to the ratios indicated above. Furthermore, it is to be noted that since the pressures in free-gas zones will in general tend to decline more rapidly than those in oil zones, gas-oil ratio curves which are initially of the form of Fig. 267 will tend to change to that of Fig. 268 and possibly also to that of Fig. 269 as the sand as a whole becomes depleted.[1] The details of this transformation will depend essentially on the relative permeabilities of the free gas and oil sands, their relative reservoir volumes, and on the differences in original reservoir pressures.

It may be also noted that insofar as Eq. (4) remains valid, the numerical value of the gas-oil ratio will decrease with decreasing reservoir pressures as the sand is depleted, provided the ratio p_{eg}/p_{eo}, if $\leqslant 1$, remains fixed or decreases. This means that for corresponding bottom-hole pressures the gas-oil ratios should ultimately decrease with the life of the field. Since the only change that can take place in the flow in the free-gas zone is that due to the decline in its reservoir pressure, a rise in the gas-oil ratio as the oil production declines must be attributed to changes in the character of the oil sand. These will involve either or both the decreasing of the liquid permeability due to the development of gas bubbles in the sand and the development of a free-gas horizon within the oil sand. These effects will evidently increase the gas-oil ratio contribution due to the oil sand, and if the increase more than counterbalances the decline in the contribution due to the free-gas sand, the resultant gas-oil ratio will rise as the sands are depleted. In fact, it is usually observed that the gas-oil ratio of a field does increase during both its natural flowing life and later pumping stage. This characteristic must, therefore, be due to the effects just mentioned, and without doubt the most important single factor is the formation of the free-gas zone overlying the oil horizon, *i.e.*, the by-pass zone.[2] This zone is characterized not by a complete depletion of the oil and the formation of a "dry"-gas channel, but rather by a higher proportion of free gas than exists at lower levels in the sand so that it will permit by-passing much as if it were such a channel.

[1] The development of a large differential between p_{eg} and p_{eo} will, however, be counteracted by the tendency of the oil zone to "repressure" the gas zone through the well bore, if p_{eg} should fall appreciably below p_{eo}.

[2] In fact, recent studies of the flow of heterogeneous fluids through sands (M. Muskat and M. W. Meres, *Physics*, **7**, 346, 1936) show definitely that both the empirically observed minima in the gas-oil ratio versus production-rate curves and the monotonic rise in the gas-oil ratio with the age of a field can be explained as consequences of the variations in the permeability of the sand with changes in its liquid and free-gas contents.

11.11. The Effect of the Tubing on the Gas-oil Ratio.—The analysis of the last two sections was based on the assumption that the gas-oil ratio was determined only by the bottom-hole pressure and not by the apparatus in the well bore. Further, it was explicitly assumed that the bottom-hole pressures p_w were always the same for the oil and gas sands. Although the results obtained on this assumption will still remain valid, for all

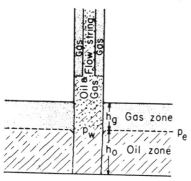

practical purposes, when the sands are thin and the production rates are high, they do need some modification when applied to thick sands producing at low-pressure differentials and when the tubing is set *below* the top of the producing sand.

To see the nature and magnitude of the modification, we shall consider again in some detail the calculation of the gas-oil ratio for a sand producing gas and oil from adjacent gas and oil zones.

FIG. 270.—Diagrammatic representation of adjacent oil and free-gas zones producing through tubing set above the top of the gas zone.

Thus if the pressures at the interface between the oil and gas zones be denoted by p_w and p_e, the differential pressure acting on the oil zone will be $p_e - p_w$, whereas in the gas zone it will have

an average value of $p_e - \left(p_w - \dfrac{\gamma g h_g}{2} \right)$, where γ is the density of the oil-gas mixture and h_g is the gas-zone thickness (*cf.* Fig. 270).[1] Now if the bottom-hole pressure be kept fixed, the above pressure differentials, and hence the oil and gas-production rates and their ratio, will be independent of the position of the tubing as long as it is kept *above* the top of the sand.

If, however, the tubing is lowered below the top of the sand, even though p_w is kept fixed, there will be a change introduced in the relative pressure differentials acting on the gas and oil zones. For simplicity it may be supposed that the tubing is lowered just to the top of the oil zone (*cf.* Fig. 271). The total pressure differential acting on the oil will then still be

[1] Here, as in all the discussions of this work, the dynamical-friction drop along the producing sand face due to the flow of the fluids into the flow string is neglected.

$p_e - p_w$. But in the gas zone there will now be the *same* differential of $p_e - p_w$, since the sand face opposite the gas zone must now be exposed to gas (of negligible density) at pressure p_w. The rate of gas flow, Q_g', with this lowered tubing will, therefore, be *less* than that with the higher tubing, Q_g. And, in fact, their ratio will be

$$\frac{Q_g'}{Q_g} = \frac{p_e^2 - p_w^2}{p_e^2 - \left(p_w - \frac{\gamma g h_g}{2}\right)^2} = \frac{R'}{R} < 1, \qquad (1)$$

which will also represent the ratio between the gas-oil ratios after and before lowering the tubing.

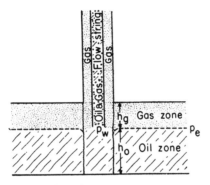

FIG. 271.—Diagrammatic representation of adjacent oil and free-gas zones producing through tubing set at the bottom of the gas zone.

Lowering the tubing will, therefore, lower the gas-oil ratio by effectively increasing the back pressure on the gas zone as compared to that in the oil zone. The quantitative features of the effect of lowering the tubing may be seen from Fig. 272, which gives R'/R, of Eq. (1), plotted as a function of the pressure differential $p_e - p_w$, the reservoir pressure being taken as 1,000 lb. and the gas zone thickness as 25 ft. As should be expected, the effect decreases as the pressure differential increases, since the fluid head of 25 ft., which gives the increased back pressure on the gas zone when the tubing is lowered, becomes a small fraction of the total pressure differential when the latter exceeds about 100 lb. As to the effect of the absolute value of the reservoir pressure, R'/R remains practically constant down to values of p_e of 100 lb., when it begins to increase appreciably over the values of Fig. 272.

When the tubing is not set exactly opposite the gas-oil zone interface, the expression for R' becomes more complex. As long as it does not reach below the top of the oil zone, the back pressure on the latter will be unaffected and only that on the gas zone will be raised. When, however, the bottom of the tubing is opposite the oil zone, there will be effective back-pressure increases on both the oil and gas zones. Denoting by h_g and h_o the thicknesses of the gas and oil zones, respectively, by $\bar{p}_e$ and $\bar{p}_w$ the reservoir pressure and bottom-hole pressures, as reduced to the level of the gas-oil interface with the tubing set above the

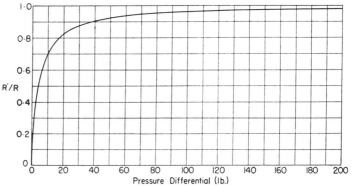

Fig. 272.—The effect on the gas-oil ratio of lowering the tubing as a function of the flowing pressure differential. R'/R = (gas-oil ratio after lowering the tubing to the bottom of the free-gas zone)/(gas-oil ratio with tubing set above the free-gas zone); gas-zone thickness = 25 ft.; reservoir pressure = 1,000 lb.; oil density is taken as 0.8 g./cc.

gas zone,[1] and by h_t the total depth of the bottom of the tubing below the top of the gas zone, and neglecting the drop in pressure along the gas column within the well bore, it may be readily shown that

for $h_t \leqslant h_g$:

$$\frac{R'}{R} = \frac{(h_g - h_t)\left\{\bar{p}_e^2 - \left[\bar{p}_w - \dfrac{\gamma g(h_g - h_t)}{2}\right]^2\right\} + h_t\{\bar{p}_e^2 - [\bar{p}_w - \gamma g(h_g - h_t)]^2\}}{h_g\left[\bar{p}_e^2 - \left(\bar{p}_w - \dfrac{\gamma g h_g}{2}\right)^2\right]};$$

(2)

[1] $\bar{p}_e$, $\bar{p}_w$ are therefore the true pressures p_e, p_w at the base of the oil zone minus the oil-zone head $\gamma g h_o$, neglecting the difference between the density of the fluid in the flow string and the oil within the sand body.

for $h_t \geqslant h_g$:

$$\frac{R'}{R} = \frac{h_o(\bar{p}_e - \bar{p}_w)\{\bar{p}_e^2 - [\bar{p}_w + \gamma g(h_t - h_g)]^2\}}{\left[h_o(\bar{p}_e - \bar{p}_w) - \frac{\gamma g}{2}(h_t - h_g)^2\right]\left[\bar{p}_e^2 - \left(\bar{p}_w - \frac{\gamma g h_g}{2}\right)^2\right]},$$ (3)[1]

where R' and R are again the gas-oil ratios after and before the lowering of the tubing, it being assumed throughout that the flow takes place in horizontal planes.

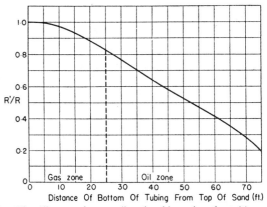

FIG. 273.—The effect on the gas-oil ratio of lowering the tubing as a function of the depth of the tubing below the top of the free-gas zone. R'/R = (gas-oil ratio after lowering the tubing)/(gas-oil ratio with tubing set above the free-gas zone); reservoir pressure = 1,000 lb.; bottom-hole pressure = 980 lb.; free-gas-zone thickness = 25 ft.; oil-zone thickness = 50 ft.; oil density is taken as 0.8 g./cc.

The effect of lowering the tubing as a function of the depth below the top of the gas zone is shown graphically in Fig. 273, as computed from Eqs. (2) and (3) with the constants $\bar{p}_e = 1,000$ lb., $\bar{p}_w = 980$ lb., $h_g = 25$ ft., $h_o = 50$ ft., and $\gamma = 0.8$ gm./cc. It will be seen that if the tubing is lowered to the bottom of the sand, the gas-oil ratio under the above conditions will be lowered to less than one-fifth of its original value. And it may be noted that since the gravity flow in the oil zone that will be induced

---

[1] For this case where the fluid level in the well bore is depressed to the bottom of the tubing, and hence below the top of the oil zone, the flow in the latter will be a composite radial and gravity flow and should strictly be computed by the formulas of Chap. VI; to show the general nature of the effect, however, the simpler formula of Eq. (3), based on two-dimensional flow in both the oil and gas zones, should be adequate.

when the tubing is set below the top of the oil zone was neglected in deriving Eqs. (2) and (3), the actual lowering of the gas-oil ratio will be even somewhat greater than that given by Fig. 273. As a final point concerning the effect of the tubing on the gas-oil ratio, which may be studied by means of Eqs. (2) and (3), it is of interest to examine the relation of R'/R to the total gas- and oil-zone thicknesses. For this purpose one may suppose that

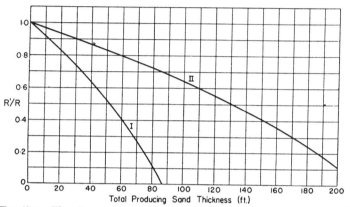

Fig. 274.—The effect on the gas-oil ratio of tubing lowered to the bottom of the producing sand as a function of the total sand thickness. R'/R = (gas-oil ratio after lowering the tubing)/(gas-oil ratio with tubing set above the free-gas zone); ¡reservoir pressure = 1,000 lb.; oil-zone thickness = 2 gas-zone thickness. I. Flowing pressure differential at the level of the gas-oil-zone interface = 20 lb. II. Flowing pressure differential at the level of the gas-oil-zone interface = 50 lb.; oil density is taken as 0.8 g./cc.

the tubing has been set to the bottom of the producing sand, *i.e.*, $h_t = h_o + h_g$. Equation (3) then reduces to

$$\frac{R'}{R} = \frac{(\bar{p}_e - \bar{p}_w)[\bar{p}_e{}^2 - (\bar{p}_w + \gamma g h_o)^2]}{\left(\bar{p}_e - \bar{p}_w - \dfrac{\gamma g h_o}{2}\right)\left[\bar{p}_e{}^2 - \left(\bar{p}_w - \dfrac{\gamma g h_g}{2}\right)^2\right]}. \tag{4}$$

Taking again $\bar{p}_e = 1,000$ lb. and supposing that the gas-zone thickness h_g is always half of the oil-zone thickness h_0, Eq. (4) is plotted in Fig. 274 as a function of the total sand thickness, for pressure differentials of 20 lb. ($\bar{p}_w = 980$ lb., curve I) and 50 lb. ($\bar{p}_w = 950$ lb., curve II). These curves show clearly the increasing effect of lowering the tubing for the thicker sands. Curve I further brings out the fact that if the sand is so thick that its equivalent fluid head equals the pressure differential on the sand, the added back pressure due to this fluid head may completely stop the flow of gas from the gas zone and reduce the gas-oil ratio to zero. Of course, this vanishing of the gas-oil ratio, as well as the ordinates of all the plots of this section, refers only to the free gas coming from the free-gas zone immediately above the oil zone, for it is only this gas flow which can be affected by the tubing position.

It should also be observed that all the above discussion has been based on the assumption that the gas has no avenue of escape from the sand except through the tubing, so that as the latter is lowered below the sand face the fluid level between the tubing and casing is depressed to the bottom of the tubing as illustrated in Fig. 271. If, however, the casing is not gastight and permits the gas to escape, the fluid head in it will not necessarily be depressed even as low as the top of the sand. In such cases no effect upon the gas-oil ratio on lowering the tubing should be observed. As a corollary to these considerations, it follows that if the casing is gastight the casing-head pressure should always be equal, except for the weight of the gas and variations due to surging, to the pressure at the bottom of the tubing, even if it is above the sand face; hence it should increase as the tubing is lowered. If this is not the case, and the casing-head pressure corresponds to a fluid head *above* the sand face even when the tubing is lowered below the top of the sand, any changes—for the *same* oil-production rate—that may be observed in the gas-oil ratio must be regarded as spurious and attributed to a redistribution in the same total amount of gas leaving the gas zone between the components leaking out through the casing and that leaving the well bore through the tubing. There appears to be no other alternative, if one is to retain the fundamental principle that the flow of oil or gas from producing sands of given geometry, permeability, and reservoir pressure is determined only by the back pressure maintained at the sand face exposed by the well bore.

11.12. Gas Coning in Tubed Wells.—The discussion of the last section, leading to Eqs. 11.11(2) and 11.11(3), implied that even though the tubing was set below the bottom of the gas zone the gas would flow down against the differential action of gravity and escape from the sand at the bottom of the tubing. Such a condition of flow is evidently equivalent to that in the case of water coning discussed in Secs. 8.10 and 11.7 in which the water rises against the differential action of gravity. While the analysis for the problem of water coning was rather involved owing to the necessity of taking into account the pressure distribution immediately below and surrounding the well bore, the fact that the well bore penetrates also the zone of the coning fluid considerably simplifies the analysis of the problem of gas coning, so that one can give not only the critical conditions for the suppression of the cone but also a good approximation to the flow conditions after the cone has broken through.

Thus by neglecting the pressure drop along the gas column in the well bore, the pressure along the sand face opposite the gas zone may be considered as uniform, so that the flow in the gas zone after the cone has broken through is still essentially radial, except for the depression in the oil-gas interface, which is highly localized about the well bore (*cf.* Fig. 265). Similarly, the flow in the oil zone may be approximated by that in two adjacent radial systems (except for the correction indicated in the footnote on page 699). Indeed, it is by means of these representations that Eqs. 11.11(2) to (4) have been derived. The gas-oil ratio (the gas being that only from the gas zone) under the condition of gas coning therefore will be, by Eq. 11.11(3),

$$R' = \frac{\mu_o k_g h_g \{\bar{p}_e^2 - [\bar{p}_w + \gamma g(h_t - h_g)]^2\}}{2\mu_g k_o \left[h_o(\bar{p}_e - \bar{p}_w) - \frac{\gamma g}{2}(h_t - h_g)^2 \right]}, \tag{1}$$

in the notations of Eq. 11.11(3), k_g and k_o being the permeabilities of the gas and oil zones, respectively.

The vanishing of R' clearly means that gas coning is not taking place, and that the fluid level between the tubing and sand face lies statically above the bottom of the tubing. This condition will occur when

$$p_e - p_w = \bar{p}_e - \bar{p}_w \leqq \gamma g(h_t - h_g), \tag{2}$$

where the equality gives the critical condition, and the inequality implies the existence of stable gas cones lying statically above the oil zone. Since p_e and p_w are the reservoir and well pressures, as measured as the bottom of the oil zone, it is seen that there will be no gas coning as long as the total pressure differential in the system $(p_e - p_w)$ does not exceed the equivalent of the fluid

SEC. 11.12] *FLOW OF GASES THROUGH POROUS MEDIA* 703

head between the top of the oil zone and the bottom of the tubing. When this fluid head is exceeded, the gas zone will contribute to the total gas-oil ratio of the well by an amount approximately given by Eq. (1).

As the distance between the bottom of the tubing and the bottom of the well bore $(h_o + h_g - h_t)$ represents the effective well penetration with respect to the production of the gas from the upper gas zone, Eq. (2) shows that the critical-pressure differential for gas coning decreases linearly with increasing effective well penetration, assuming that the real penetration through the oil zone is complete. This may be compared with the case of water coning in which this pressure differential falls much more rapidly for high and low penetrations than for moderate penetrations (*cf.* Fig. 187). Thus while the critical-pressure differentials for water coning may be some ten times as large as the fluid head of an oil column of height equal to the thickness of the oil zone—the maximal differential possible without gas coning—this differential will be much more sensitive to the effective well penetration in the case of water coning than for gas coning.

These differences are clearly due to those in the details of the pressure distribution in the immediate vicinity of the well bore. For in a real partially penetrating well underlain by water the pressure distribution below the well is approximately spherical, and the pressure gradients are highly concentrated about the well bore, thus leaving only a small fraction of the total pressure differential to lift the water into the well. In the present case, however, where the well actually penetrates both the gas and oil zones, the pressure distribution is almost exactly radial, and, except for the fluid head of the oil column, the same total differential pressure acts on the gas zone as on the oil zone. Hence it is only necessary for the total pressure differential to exceed this fluid head to make the gas cone down and enter the well through the bottom of the tubing. If, on the other hand, the sand face is packed off to the bottom of the tubing,[1] so that the gas of the gas zone must force itself through the oil zone to enter the tubing, the problem becomes essentially identical with that of

---

[1] If the packer is set only opposite the gas-oil interface, there will be but little effect in suppressing the cone, as a short packer will not appreciably concentrate the pressure gradients about the well bore.

water coning studied in Chap. VIII, and all the results for the latter problem will be applicable to the case of gas coning, with the only change that the critical-pressure differentials and production rates of Figs. 187 and 188 must be multiplied by the density of the oil divided by 0.3.

11.13. Multiple-well Systems.—Just as in the case of artesian or oil wells, so also in the study of wells producing gas, one cannot always restrict oneself to the consideration of single wells, but must also frequently take into account the mutual interactions between the various wells distributed over a given area. For the steady state these interactions will in general have exactly the same effect upon the individual well capacities as in the case of wells producing from water- or oil-bearing sands. Thus for small groups of wells, the theoretical results are those already derived in Secs. 9.2 to 9.6, where again the *relative* production capacities, such as given in Eqs. 9.3(7), (11), or (15), are strictly applicable also for gas wells, while the pressure distributions as given in Eqs. 9.2(1) or 9.2(5) must be modified by the substitution of p^{1+m} for p.   In a similar manner one may take over the results derived in Chap. IX for problems involving linear arrays, including the theory of offsetting, again noting that in the pressure requirements for the latter—as given in Sec. 9.14—the pressures must be replaced by the corresponding values of p^{1+m}.

While the theory of the regular well networks developed in Chap. IX may be applied to gas wells as well as to those producing liquids, there is, of course, no practical interest in the problem of water flooding.   For although the pressure in an oil sand may decline to the atmospheric value after only 15 to 35 per cent of the original oil content has been taken from the sand, the complete depletion of the pressure in a gas sand necessarily means a recovery of all but $1/p$ of the gas originally in the reservoir, p being the original reservoir pressure in atmospheres.   Thus if the original reservoir pressure is 50 atm.—corresponding to a depth of some 1,700 ft.—the recovery will be 98 per cent when the sand is depleted.   Artificial recovery methods as that of water flooding are, therefore, of no interest whatever in the production from gas fields.

11.14. The Nonsteady-state Flow of Gases through Porous Media.—Strictly speaking, there are probably no cases of actual field production from gas reservoirs in which the conditions of flow are exactly those of the steady state. For if the gas reservoir is limited in extent, the removal of the gas through the producing wells will deplete the sand pressure so that the fluxes from the wells will continually decrease if the pressures at the wells are kept "steady." And if the reservoir is bounded by a water sand, it will still be unlikely, in most cases, that the encroachment of the water will suffice to maintain the reservoir pressure. For assuming a gas-production rate per well of 10^5 bbl. (at atmospheric pressure)/day ($\sim 5.6 \cdot 10^5$ ft.^{3}/day) it would require a water encroachment rate of 100,000 bbl./day to maintain the pressure (~ 100 atm.) in a field drained by only 100 wells. Such encroachment rates, however, are probably considerably higher than those usually encountered in practice, and, as is actually observed, the pressures in most gas fields do decline quite rapidly.

Of course, the same considerations apply in principle to the production of liquids from porous reservoirs. There, however, the transients in the system (due to the fluid compressibility) are inherently of longer life, and furthermore the rates of fluid depletion are considerably slower owing to the small production capacities—of the order of 1,000 bbl./day—compared to those of gas wells. In fact, the relative rates of decline in transients in gas and liquid systems may be readily estimated as follows: Taking the particular case of isothermal gas flow ($m = 1$), it is seen from Eq. 11.1(2) that the time will enter in the expression for γ^2 essentially as $kt/f\gamma_0\mu$. Now for a given sand, f and k will be independent of the fluid, and for ideal gases $\gamma_0 = w/RT$, where R is the gas constant per mol, w the molecular weight, and T the absolute temperature. Hence, for isothermal gas flow, t will enter in γ^2 essentially as $tRT/w\mu$, so that the *time rate of change* in the system will be proportional to $RT/w\mu$. For compressible liquids, on the other hand, it follows similarly from Eq. 10.1(7) that the time rate of change will be proportional to $1/\beta\mu$ [*cf.* also Eq. 10.1(5)].

Now for a gas such as methane, $RT/w \sim 1,500$, whereas β has a value for water and mineral oils of the order of $5 \cdot 10^{-5}$. The viscosity, however, for water or mineral oils is of the order of $10^2 - 10^3$ times that of methane. Hence from the resultant of these two factors, it follows that the rate of decline in a **gas** reservoir under isothermal conditions will be of the order of 10–100 times as large as in one containing a gas-free but compressible liquid. Apparently, then, the low viscosity of gases more than counterbalances their high elasticity in giving rise to very rapid transients.[1]

Thus while one may justifiably approximate many of the problems of liquid flow by those of the steady state, this approximation will in general be invalid in the case of gas-flow systems. For these the transient behavior will in all practical cases be the predominating feature. Unfortunately, however, the exact treatment of such transient states necessitates the solution of the nonlinear Eq. 11.1(2), for which it has as yet not been possible to construct analytic solutions. We shall, therefore, introduce some physical approximations which will simplify the analytical problem and make it tractable.

Returning to Sec. 10.13, it will be recalled that if a closed sand containing a compressible liquid is pierced by a well of a given (uniform) flux, the pressure distribution in the system first undergoes a very short-period transient, during which the pressure about the well develops an essentially logarithmic distribution and that at the external boundary assumes a linear rate of decline (*cf.* Fig. 256). After this period—of about an hour in Fig. 256, where the constants are such as to give relatively long transients—the system enters a transient state in which the pressure decreases uniformly over the whole sand at a rate proportional to the flux rate from the well, the distribution within the sand being, for practical purposes, identical with the steady-state distribution except very near the external boundary. In fact, the total pressure drop between the closed external boundary and the well is only 7 per cent less than what it would be in a steady state-flow system with the same flux. Thus as pointed out in Sec. 10.13, the production from the reservoir after the first transient has passed may be considered as supplied by an equivalent density decline distributed uniformly over the reservoir, the instantaneous pressure distribution differing only slightly from

---

[1] The fact that Eq. 11.1(2) involves γ^2 (for $m = 1$) whereas Eq. 10.1(7) is linear in γ does not invalidate this comparison. For if Eq. 10.1(8) is approximated by $\gamma = \gamma_0(1 + \beta p)$, it is found that the equation for γ for compressible liquids takes exactly the same form as for gases, *i.e.*, it also involves $\nabla^2\gamma^2$ rather than $\nabla^2\gamma$.

that of the steady state. Formulated more precisely, the real system may be approximated by one which undergoes a "continuous succession of steady states," the time entering only as a parameter in the boundary conditions which are taken to vary in such a manner as to give the preassigned flux in the system.

Although there are available no rigorous solutions of the nonlinear Eq. 11.1(2) to use as a test for this approximation in the case of gas-flow systems, there seems to be no reason why the approximation of the continuous succession of steady states should not give as good a representation of the actual flow conditions in the case of gas flow as for the flow of a compressible liquid. On the contrary, owing to the higher pressure gradients in the gas-flow system about the output well, and correspondingly lower gradients at the external boundaries, than in a compressible-liquid system—for equal total pressure differentials—the error due to the extrapolation of the logarithmic steady-state distribution to a closed external boundary should, therefore, be even less in the former case than in the latter.[1] Furthermore, this long-period fundamental transient of the succession of steady states will be more quickly established for gas-flow systems— the short-period transient immediately following a change in the boundary conditions dying off more rapidly—than for compressible-liquid systems, thus giving relatively smaller errors when neglecting the finer details of the early short-period transients.

In view of these considerations, the problem of the nonsteady-state flow of gases through porous media may be reduced—approximately, of course—to that of finding the steady-state pressure distributions, and then the time variation of the boundary conditions complementary to those originally preassigned in such a way that the latter are satisfied. It is because of this reduction of the rigorous problem of solving Eq. 11.1(2) that we have outlined in the preceding sections of this chapter the nature of

---

[1] Even so, the approximation will probably be unsatisfactory if applied to linear systems, as then the gradients in the steady state fall off much more slowly with increasing distance from the outlet surfaces, so that it will afford a poor approximation at an external closed boundary where the actual boundary requirement would be that the gradient be zero. In most practical problems, however, the outflow surfaces are of appreciably smaller area than the external boundaries, so that the flow systems may be represented geometrically as radial rather than linear systems.

the steady-state solutions corresponding to the various types of problems treated in Part II, although these steady-state solutions are not of great interest in themselves. Now that they are available, the problem that remains is that of forming the proper continuous sequence of these steady-state solutions so as to correspond to the particular transient problem of interest.

Because of the essentially transient character of all gas-flow problems of practical interest, the only external boundary condition of practical significance is that of vanishing flux, corresponding to a closed reservoir. With respect to the outlet wells, the conditions may be either that of preassigned flux or pressure. In the following we shall present the solutions for the case where the gas reservoir is drained by a single well, either at a constant pressure or with a constant production rate. The methods given for solving these problems may be applied to the more complex cases involving the other steady-state solutions discussed in Secs. 11.2 to 11.13.

11.15. A Closed Gas Reservoir Drained by a Well Producing at Constant Pressure.—Although, as pointed out in the last section, the initial transient during which the system adjusts itself to a uniform drainage in a continuous succession of steady states is of very short duration—less than an hour in the example discussed numerically below—we shall show for the present case how even this transient can be treated to essentially the same approximation as the later and fundamental transient decline of the system. For this purpose it is convenient to introduce the concept of a "radius of drainage," which recedes to the boundary of the reservoir as the steady-state condition is established. That is, the "radius of drainage" gives the distance from the well to which the approximately steady-state distribution has been established, and the production during the initial transient is simply the rate at which gas is removed from the system in the establishment of this steady-state distribution, it being assumed that no gas is removed from any point until the radius of drainage has passed that point.

Considering, then, the specific problem of a closed reservoir of radius r_e and a central well of radius r_w, and assuming that the initial reservoir pressure is p_i and that the well pressure is permanently maintained at p_w, the production rate per unit sand thickness at a time t when the "radius of drainage" has receded to the radius r_0 will have the value [*cf.* Eqs. 11.3(2), 11.1(3) and Fig. 275]

$$Q = \frac{\pi k \Delta \gamma^2}{\mu \gamma_0 \log r_0/r_w} = -2\pi f \frac{\partial}{\partial t}\left[\int_{r_w}^{r_0} \gamma r dr + \int_{r_0}^{r_e} \gamma_i r dr\right], \quad (1)$$

where

$$\gamma_0 = \gamma(p = 1); \quad \Delta\gamma^2 = \gamma_i{}^2 - \gamma_w{}^2; \quad \gamma_i = \gamma(t = 0), \quad (2)$$

and where for simplicity m has been taken as 1, the value for isothermal flow.

By introducing the notation

$$2 \log \frac{r_0}{r_w} = z; \quad 2 \log \frac{r}{r_w} = y \quad (3)$$

and applying Eqs. 11.3(1) and 11.1(3), Eq. (1) may be rewritten as

$$t = \frac{\gamma_0 \mu f r_w{}^2}{4k} \int_0^z \frac{dz}{z} \int_0^z \frac{y e^y dy}{[(\Delta\gamma^2/z)y + \gamma_w{}^2]^{1/2}}. \quad (4)$$

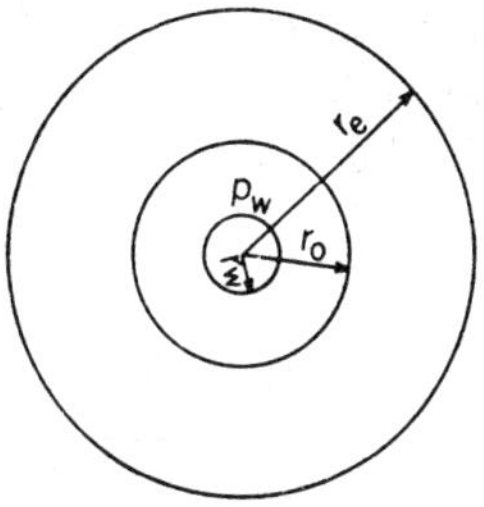

FIG. 275.

In principle, Eq. (4) permits the determination of t as a function of z, or of r_0 as a function of t, and hence the time variation of Q, on the application of Eq. (1), until r_0 recedes to the reservoir radius r_e. To show the nature of the process, it will be sufficient to choose the case where $p_w = 0(= \gamma_w)$, when Eq. (4) can be reduced, by partial integration, to the form

$$t = \frac{\mu \gamma_0 f r_w{}^2}{2\gamma_e k} \int^z (e^y - e^{\frac{y^2}{z}})dy, \quad (5)$$

which is rather convenient for graphical integration, giving, when evaluated graphically,

$$e^{z/2} = \frac{r_0}{r_w} \cong 1 + 1.068\left(\frac{4kp_e t}{fr_w{}^2\mu}\right)^{0.4967} \cong \frac{2}{r_w}\left(\frac{kp_e t}{\mu f}\right)^{1/2}, \quad (6)$$

except when $t \sim 0$.

With the constants

$$f = 0.2; \quad k = 1 \text{ darcy}; \quad \mu = 0.012 \text{ centipoise};$$
$$p_e = 100 \text{ atm.}; \quad r_w = \frac{1}{4} \text{ ft.}, \quad (7)$$

Eq. (6) is plotted in Fig. 276 as curve I. It will be observed that the "radius of drainage" recedes extremely rapidly, growing to a radius of 100 ft. in about 1 min., and reaching the effective reservoir boundary ($r_e \sim 500$ ft.) in less than half an hour, thus confirming the conclusion stated in the last section that the initial transient is of very short duration.

The production rate, which is given during this initial transient by

$$Q = \frac{2\pi k \gamma_0 p_e^2}{\mu z} = \frac{3.50 \cdot 10^3}{z} \text{ gm./sec./cm.} =$$

$$(4.87 \cdot 10^8/z)\text{cu. ft./day/ft. (sand)}, \quad (8)$$

is plotted as curve II, with the constants as in Eq. (7) and γ_0 taken as $6.68 \cdot 10^{-4}$ gm./cc., the atmospheric density of methane (at 20°C.).

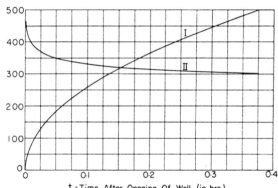

t = Time After Opening Of Well (in hrs.)

Fig. 276.—The production rates and recession of the "radius of drainage" during the initial transient about a gas well under isothermal-flow conditions. I. Extent of the "radius of drainage" in feet. II. Production rate in 10^5 cu. ft./day/ft. sand. $k = 1$ darcy; $\mu = 0.012$ centipoises; $f = 0.20$; initial reservoir pressure = 100 atm.; well radius = $\frac{1}{4}$ ft.

The total production during this period is evidently given by the difference between the initial mass of gas in the reservoir and that still left in the reservoir when the radius of drainage r_0 has receded to the external boundary r_e. Denoting this cumulative production per unit sand thickness by $\bar{Q}$, it follows that

$$\bar{Q} = \pi \gamma_i (r_e^2 - r_w^2) - 2\pi \gamma_i \int_{r_w}^{r_e} r \left\{ \frac{\log r/r_w}{\log r_e/r_w} \right\}^{\frac{1}{2}} dr. \quad (9)$$

If $\bar{Q}_i$ is the initial mass content of the reservoir, $\bar{Q}/\bar{Q}_i$ can be expressed as

$$\frac{\bar{Q}}{\bar{Q}_i} = \frac{e^{-z}}{2\sqrt{z}}\int_0^z \frac{e^y}{\sqrt{y}}dy = 0.0338, \tag{10}$$

where, here and in the following, $z = \log \dfrac{r_e{}^2}{r_w{}^2}$, and $\dfrac{r_e}{r_w} = 2,000$.
This shows again that the initial transients following a discontinuous change in the boundary conditions represent only a very minor part of the whole decline history of a gas reservoir.

After the initial transient has passed, *i.e.*, after the "radius of drainage" has reached the reservoir boundary, we shall assume, as explained in Sec. 11.14, that the reservoir pressure will decline uniformly throughout, its instantaneous values following a steady-state distribution. Hence Q will now be given by

$$Q = \frac{2\pi k\Delta\gamma^2}{\mu z\gamma_0} = -2\pi f\frac{\partial}{\partial t}\int_{r_w}^{r_e}\gamma r\,dr$$

$$= \frac{-\pi r_w{}^2 f\gamma_e}{z}\frac{\partial\gamma_e}{\partial t}\int_0^z \frac{ye^y dy}{\left[\Delta\gamma^2\frac{y}{z} + \gamma_w{}^2\right]^{1/2}}. \tag{11}$$

Again choosing the case $\gamma_w = 0 = p_w$ this gives

$$\left.\begin{aligned}\frac{1}{\gamma_e{}^2}\frac{\partial\gamma_e}{\partial t} &= \frac{-2kz^{-1/2}}{\mu f r_w{}^2\gamma_0\int_0^z y^{1/2}e^y dy}\\[2mm] &\cong \frac{-2k}{f\mu\gamma_0 zr_e{}^2}; \qquad \frac{r_e}{r_w} = 2,000,\end{aligned}\right\} \tag{12}$$

so that, with the constants of Eq. (7), and recalling that $\gamma = \gamma_0 p$,

$$p_e = \frac{1}{\left[2.039\cdot 10^{-2}(t-t_0)+\dfrac{1}{p_i}\right]}, \tag{13}$$

where $t = t_0$ is the time (in days) when the radius of drainage reaches the reservoir boundary.

The production rate for the present case ($p_w = 0$) is given by

$$Q = \frac{\pi k\gamma_0 p_e{}^2}{\mu \log r_e/r_w} = 2.301\cdot 10^{-2}p_e{}^2 \text{ gm./sec./cm.} =$$
$$3.201\cdot 10^3 p_e{}^2 \text{ cu. ft./day/ft. sand.} \tag{14}$$

Equations (13) and (14) are plotted in Fig. 277. Here again, the decline is very rapid, the pressure dropping to a tenth of its initial value in about 4¼ days. As to the relation of these decline curves to those which are observed in practice, it should be remembered that not only is the assumed permeability of 1 darcy considerably higher than that of most gas sands, but also the effective reservoirs drained by each well in a field will in general be appreciably higher than that of 500 ft. assumed here.[1] Both these factors will tend to increase the lives of the wells over that indicated by Fig. 277.

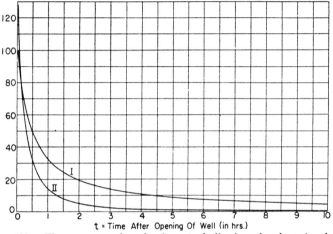

t = Time After Opening Of Well (in hrs.)

Fig. 277.—The pressure and production rate decline in a closed gas (methane) reservoir drained by a single well. I. Pressure (atm.) at the reservoir boundary (500 ft.). II. Production rate in $2.5 \cdot 10^5$ cu. ft./day/ft. sand. Initial reservoir pressure = 100 atm.; sand permeability = 1 darcy.

11.16. A Closed Gas Reservoir Drained by a Well Producing at a Uniform Rate.—For this case we shall neglect the initial transient following immediately upon opening the well and thus disturbing the initial uniform pressure distribution. Rather, it will be supposed that the "radius of drainage" has already receded to the external boundary of the system, so that the pressure distribution is about to begin a continuous succession of steady states. Each of these states will correspond to different pressures at the well and the external boundary, and the problem to be solved will be that of finding the decline in these boundary

[1] The average spacing in the Appalachian gas fields is of the order of one well to 80 acres, which corresponds to a drainage area per well 4.4 times as great as that assumed above.

pressures which will correspond to a uniform flux rate from the well.

Denoting the flux from the well by Q, it will clearly be given, just as in Sec. 11.15 by

$$Q = \frac{2\pi k \Delta \gamma^2}{\mu z \gamma_0} = -2\pi f \frac{\partial}{\partial t} \int_{r_w}^{r_e} r\gamma dr, \tag{1}$$

where again for simplicity the assumption of isothermal flow has been made ($m = 1$), the notation being that used in Sec. 11.15. Integrating, it is seen that

$$\int_0^z (\gamma_a - \gamma)e^y dy = \frac{Qt}{\pi f r_w^2}, \tag{2}[1]$$

where γ_a is the density distribution at the initial instant—after the passage of the initial transient—and is given by

$$\gamma_a^2 = \frac{(\gamma_i^2 - \gamma_w^2)y}{z} + \gamma_{wa}^2; \qquad Q = \frac{2\pi k(\gamma_i^2 - \gamma_{wa}^2)}{\mu z \gamma_0}, \tag{3}$$

γ_i being the initial uniform density before the well is opened to production, and γ_{wa} the density at the well at $t = 0$. Expressing now γ by the relation

$$\gamma^2 = \frac{\mu \gamma_0 Q y}{2\pi k} + \gamma_w^2, \tag{4}$$

.Eq. (2) can be integrated (graphically) for various values of γ_w, giving, when inverted, γ_w as a function of t. Setting $y = z$ in Eq. (4), the corresponding values of γ_e will be obtained. When this process is carried out for the constants

$$f = 0.2; \qquad k = 1 \text{ darcy}; \qquad \mu = 0.012 \text{ centipoise};$$
$$r_w = \tfrac{1}{4} \text{ ft.}; \qquad r_e = 500 \text{ ft.}; \qquad p_i = 100 \text{ atm.}; \left.\begin{array}{l}\\\\\\\end{array}\right\} \tag{5}$$
$$\gamma_0 = 6.68 \cdot 10^{-4} \text{ gm./cc.}; \qquad Q = 5 \cdot 10^4 \text{gm./day/cm. sand} \sim$$
$$8.0 \cdot 10^4 \text{ cu. ft./day/ft. sand,}$$

and the results are converted into equivalent pressure declines, the curves of Fig. 278 are obtained.

---

[1] If Q is not constant, one need only replace Qt by $\int_0^t Q dt$, which can be considered as a known function of time.

It will be seen that for practical purposes the pressures decline linearly with the time. However, it is to be noted that this result is due not only to the constancy assumed for Q, but also to the high values of p_w over most of the life of the well, so that the total pressure drop in the system is very small and the internal pressures are essentially equal to p_w plus small correction terms. Furthermore, it is for this latter reason that the two curves for the pressure p_e at the external boundary and that at the well p_w practically coincide until the very last part of the decline history of the well. For as the flux in a gas-flow system is proportional to the differences in the squares of the boundary pressures, the pressure differential required to produce a given flux is inversely proportional to the average pressure in the system.

In a similar manner more complex cases can be treated both with respect to the pressure declines for preassigned flux rates and, as in Sec. 11.15, for the production-rate decline for preassigned well pressures.

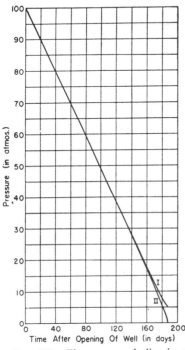

FIG. 278.—The pressure decline in a closed gas (methane) reservoir drained by a single well at the uniform rate of $8 \cdot 10^4$ cu. ft./day/ft. sand   I. Pressure at the external boundary (500 ft. radius).   II. Pressure at the well (¼ ft. radius).   Initial reservoir pressure = 100 atm.; sand permeability = 1 darcy.

11.17. Summary.—From a physical point of view all problems of gas flow of practical interest are fundamentally of the nonsteady state, in which the gas content of, and pressures in, a closed reservoir continually decline as the wells draining the reservoir are produced. A strictly accurate treatment of the gas-flow systems would, therefore, involve the solution of the nonlinear partial differential equation governing the density decline in such systems [*cf*. Eq. 11.1(2)], and which is analogous to the linear equation in the density [*cf*. Eq. 10.1(7)] for the nonsteady flow of liquids, which was solved rigorously for a number of specific cases in Part III. Fortunately, however, such a program which would require a numerical treatment of the nonlinear equation—since there are as yet no general analytical solutions of this equation—can be reduced to a more tractable form upon a closer analysis of the physical nature of the transients in gas-flow systems.

In analogy to the results derived by a strict analysis of the pressure decline in a closed liquid-bearing reservoir drained by a single well (*cf.* Sec. 10.13), it is to be expected that the transients in a gas-flow system will in general consist of two types. The first will be of a very short period, following immediately any artificial change in the boundary conditions, such as the opening of a well or a variation in its production rate. Such transients will precede or be superposed upon the second type, which is that giving the general decline of pressure in the reservoir or production from it, due to its depletion. That the first type will actually be of extremely short duration follows from the fact that, owing to the relatively low viscosity of gases, rates of change in gas-flow systems will in general be many times greater than in liquid-bearing sands, and for the latter it was seen that the duration of the initial short-period transients will be only of the order of an hour. Furthermore, the change in fluid content of a reservoir associated with this transient will in general be no greater than about 3 per cent, even if upon opening a well it is suddenly reduced to atmospheric pressure, the reservoir pressure being 100 atm. For practical purposes, therefore, these short-period transients can be entirely neglected.

The second type of transient, on the other hand, does represent the fundamental decline history of a gas-flow system. By analogy with the result found by a rigorous calculation in the case of liquid flow, it is reasonable to suppose that it may be represented by a continuous succession of steady states, in which the density declines everywhere uniformly at such a rate as to provide for the instantaneous total flow from the system. In fact, it is to be expected that such an approximate treatment should be still more accurate for gas-flow systems than for those producing liquids. For not only will the readjustments to the approximately steady-state conditions take place more rapidly in gas-bearing media, but, moreover, the actual approximation of the steady-state pressure distributions, after the passage of the initial transients, should involve smaller errors in gas-flow systems, owing to the naturally smaller gradients in the latter at points distant from outlet surfaces which should strictly vanish at the outer closed boundaries of the system.

The analytical treatment of gas-flow systems may, therefore, be divided into a preliminary analysis of the steady-state behavior of the system, and then a derivation of the continuous sequence of these steady states appropriate to the particular boundary conditions that are preassigned. While the steady-state characteristics are thus involved only as a preliminary step in the analysis, they should give even in themselves a fair representation of the main features of the flow, especially if the reservoir is large and the total flow from it not excessive, so that the general decline in the system is not too rapid. Furthermore, they should provide a good approximation for the comparative study of various types of system.

In view of the fact that the steady-state flow of gases is governed by Laplace's equation in the dependent variable p^{1+m} [*cf.* Eq. 11.2(1)], where m determines the thermodynamic character of the flow, the analytical expressions for the solutions for various problems can be simply taken as those already derived for the corresponding systems of steady-state liquid flow, also governed by Laplace's equation, in Part II. The only change that needs to be made with respect to the pressure distributions is the replacement of the pressure p in the expressions for the cases of steady-state liquid flow by p^{1+m}. And the values of the mass fluxes for the steady-state solutions for gas flow are to be obtained from those for the volume flux for steady-state liquid systems by replacing the $p_e - p_w$ of the latter by $\dfrac{\gamma_0(p_e^{1+m} - p_w^{1+m})}{(1 + m)}$, γ_0 being the gas density at unit pressure. In the case of most practical interest, that of isothermal flow ($m = 1$), the above change is equivalent to that of multiplying the volume fluxes of the steady-state liquid flow by the mean density in the gas-flow system to get the mass flux in the latter.

Reviewing briefly the types of steady-state liquid-flow systems discussed in Parts I and II from the point of view of the above considerations, it may be noted first that for linear gas flow the pressure distribution will be given simply by a linear variation of p^{1+m}. As to the effect of the thermodynamic character of the flow, the value of m, on the flow of a gas, it is found that the isothermal flow ($m = 1$) will give the least flow, the flux increasing as the value of m decreases, so that, for example, 16 per cent more air will pass adiabatically ($m = 0.71$) through a linear system (if $p_1/p_2 = 0.1$) than if it flows isothermally. This result is due to the higher outflow densities when $m < 1$, and more than compensates for the lower outflow-pressure gradients in the latter case.

For radial-flow systems the quantity p^{1+m} varies logarithmically with the radial distance [cf. Eq. 11.3(1)]. When the flow is not exactly radial, the flux will still be given by the expression for radial flow, provided only that the quantity ($p_e^{1+m} - p_w^{1+m}$) in the latter—p_e, p_w being the external boundary and well pressures—is replaced by its average $(\overline{p_e^{1+m}} - \overline{p_w^{1+m}})$ [cf. Eq. 11.3(4)]. Moreover, the effect of a noncircular shape of the external boundary on the flow from the system may be corrected, as in the case of liquid flow, by simply using for the radius of the external boundary in the expression for the flux a reasonable average distance of the well from the external boundary.

The same changes may be applied to give the results for three-dimensional steady-state gas-flow systems from those developed in Chap. V for liquids. Thus the pressure distributions for partially penetrating wells derived in Sec. 5.3 may be applied directly to wells partially penetrating gas reservoirs by simply replacing everywhere p in the former by p^{1+m}. The ratios of the production capacities for various partially penetrating wells will remain exactly the same as for wells producing liquids in the steady state. The effect of an anisotropy in the sand permeability for gas wells may be taken over directly from that derived in Sec. 5.5 for liquid-producing wells.

Because of the low density of gases and the fact that gases always completely fill any available space, the effect of gravity is of no practical significance for gas-flow systems.

The behavior of gas wells draining sands of nonuniform permeability will be formally identical with that for wells producing liquids. The methods of Chap. VII may be applied to gas wells both when the sand permeability varies continuously and when its variation is discontinuous. Limestone-fracture systems bearing gas will be governed by the same theory as those bearing liquids and will react to acid treatment in the same way. Likewise the effect of sand in the bore of a gas well on its production capacity will be the same as that calculated in Sec. 7.10 for wells producing liquids.

While "two-fluid" liquid systems may involve, with reasonable approximation, either essentially vertical or horizontal interfaces, a gas in contact with a liquid in the same porous medium and flowing as a homogeneous fluid will always be separated from the liquid by an effectively horizontal interface, the gas occupying the upper layers of the sand.[1] When the liquid lying below the gas zone is water, and the well only partially penetrates the gas zone, there will arise a problem of coning, in which the water in the lower strata may be elevated into a static cone as the gas is produced from the upper strata. The formal theory of this coning problem will be exactly the same as that developed for the case of water coning in oil production (*cf.* Sec. 8.10), the only changes being the recalculation of the pressure distribution about the partially penetrating wells used for liquid-bearing sands by replacing the pressures there by p^{1+m}, and the correction for the larger density contrast in the gas-flow system. The results of these corrections will be that the critical-pressure differentials for water coning for gas wells will be some three to four times as great as those for oil wells (*cf.* Fig. 187), so that water coning will be more easily suppressed than in the latter case. The general features of the coning problem, however, will be the same as for wells producing oil.

A two-fluid problem in the flow of gases that has no analogy of practical interest in the systems carrying only liquids, arises in the treatment of the gas-oil ratios—volume of gas, measured under atmospheric conditions, produced per unit volume of oil— to be expected of wells producing oil from sands overlain by gas zones. Idealizing the problem so as to neglect the gas dispersed in, and flowing with, the oil, and treating the flow of the latter as that of a homogeneous liquid and that of the gas in the gas zone as a homogeneous gas, the flow characteristics of the combined system can be derived without difficulty. When the gas zone is contiguous with the oil zone so that the two have a

---

[1] An exception is to be noted in the case of large-scale regional migrations of gas up and down the flanks of an oil-bearing structure into, or out of, a gas cap.

common interface, the resultant gas-oil ratio will be directly proportional to the reservoir pressure but independent of the back pressure at the face of the sand.   When the gas and oil zones are noncommunicating, lying in different strata,[1] the gas-oil ratios will depend on the relative reservoir pressures in the two zones. Thus if the reservoir pressure in the gas zone exceeds that in the oil zone, the gas-oil ratio will increase—more rapidly than linearly—as the production rate decreases, until gas alone will be bled off when the back pressure equals the reservoir pressure in the oil zone.   When the two zones have equal reservoir pressures, the gas-oil ratio will increase linearly as the oil-production rate is decreased, and when the reservoir pressure of the gas zone is less than that of the oil zone, the gas-oil ratio will rise to a maximum and then decrease as the oil-production rate is decreased.[2]

The same type of analysis can be applied to give the effect of the position of the tubing on the gas-oil ratio.   Although it has been a fundamental tenet throughout the preceding chapters that the behavior of the fluids in a fluid-bearing sand is uniquely determined by the pressures or fluxes maintained at the face of the sand exposed at the well bore, and is entirely independent of the equipment in the well bore, one is here concerned with the question of the effect of the equipment on these sand-face pressures.   The answer to this question is the following: As long as the tubing is set above the top of the gas zone, its exact position does not affect the sand-face pressures, and the gas-oil ratios will be governed by the results quoted above.   If, however, the bottom of the tubing is set below the top of the gas zone, the sand-face pressure opposite the gas zone and above the bottom of the tubing will be increased to that at the bottom of the tubing.   This will result in a decreased flow from the gas zone and hence in a lower gas-oil ratio.   This effect will increase as the

---

[1] Here the gas zone may, of course, lie between two oil-producing strata.

[2] While it would be beyond the scope of the present treatment to enter into details, it may be mentioned that in comparing the above conclusions with actual field observations, account must be taken not only of the gas segregation and development of upper gas zones and an associated rise in gas-oil ratio as the oil sand becomes depleted, but also of the effect of variations in the oil velocity (production rate) on the gas by-passing within the oil. When these effects are superposed upon the above results for *homogeneous* fluids, they may entirely mask the latter.

depth of the tubing is increased and may become of a very appreciable magnitude if the pressure differential over the sand is small and if the sands are thick. Thus if the gas zone is 25 ft. thick, the oil zone is 50 ft. thick, and the well is produced with a 20-lb. pressure differential, the reservoir pressure being 1,000 lb., tubing set at the top of the oil zone will cut the gas-oil ratio to 83 per cent of its value when the bottom of the tubing is above the gas zone. If the tubing be lowered to the bottom of the oil zone, the gas-oil ratio will be reduced to only 20 per cent of its initial value. If, therefore, the casing is gastight so that the gas can escape from the gas zone only by flowing down around the bottom of and through the tubing, one may very effectively vary the gas-oil ratio by adjusting the depth of the tubing.

An extreme case of the effect of the tubing on the gas-oil ratio is that when the gas-oil ratio (the contribution from the free-gas zone) is simply cut to zero. This condition will obtain when the tubing is set below the top of the oil zone, and the total pressure differential in the system does not exceed the equivalent of the fluid head between the top of the oil zone and the bottom of the tubing. The gas will then evidently form an inverted static cone overlying the oil zone. The nonvanishing gas-oil ratios when the tubing is set below the top of the oil zone correspond to the breaking of the static gas cone in exactly the same way as a static water cone will break and rise into the bore of a gas or oil well if the pressure differential over the sand exceeds the critical value given by the coning theory. Because of the differences in the pressure distribution immediately surrounding the well bore the details of the gas-coning phenomena will be different from those for water coning. The general physical basis of the problem is, however, the same in the two cases.

The mutual interactions of gas wells can also be found simply by reinterpreting the corresponding results for wells producing liquids, derived in Chap. IX. With respect to the practical significance of such problems, however, it is to be noted that regular networks of gas wells of the type used in water-flooding programs are of no practical interest, as gas reservoirs will always naturally deplete themselves completely, so that artificial methods of recovery are of no importance for gas production.

As mentioned at the beginning of this section, the various steady-state solutions must be synthesized into sequences appropriate to the real boundary conditions in order to obtain the decline history of the corresponding gas-flow system. Thus in particular, account must be taken of the fact that not only will the fluxes or pressures (or both) at the wells decline, but at the same time the external-boundary pressures occurring in the steady-state formulas will also decline owing to the closure of the reservoir at its external boundaries and its depletion. As previously explained, this succession of steady states is to be obtained by imposing a decline in density distributed uniformly over the whole reservoir, the decline proceeding at such a rate as to provide for the production from the outlet wells.[1]

This procedure may be illustrated by the problems of closed reservoirs drained by single wells produced either at constant pressure or constant flux. In such problems, in fact, one may represent the initial transient following the opening of the well as the period during which the steady-state condition of flow spreads out from the well bore and recedes to the external boundary. Such a calculation for isothermal flow with the well pressure suddenly dropped from 100 atm. to 0, leads to a life of the initial transient of less than 0.4 hr. and a total flux during this period of 3.3 per cent of the reservoir-gas content, thus confirming the conclusion stated at the beginning of this section that for practical purposes the short-period transients in gas-flow systems can be entirely neglected. The fundamental transient giving the major decline history of the reservoir involves a hyperbolic decline of the reservoir pressure with time, and a corresponding decline in the flux. In the case where the flux is kept uniform, a similar analysis leads to an essentially linear decline of both the pressures at the well and external boundary until the reservoir is entirely depleted.

[1] If the system is a multiple-well system with the wells producing at different rates, the density declines should be adjusted locally about the various wells at rates proportional to their respective fluxes.

APPENDIX I

CONVERSION FACTORS

LENGTH

1 cm. = 0.39370 in.

= 0.03281 ft.

= 6.2137 × 10^{-6} mile

1 in. = 2.5400 cm.

1 ft. = 30.480 cm.

1 mile = 160,934 cm.

AREA

1 sq. cm. = 0.15500 sq. in.

1 sq. m. = 10.764 sq. ft.

= 3.8610 × 10^{-7} mile²

= 2.4710 × 10^{-4} acre

1 sq. in. = 6.4516 sq. cm.

1 sq. ft. = 929.03 sq. cm.

= 2.2957 × 10^{-5} acre

1 sq. mile = 2.5900 × 10^6 sq. m.

1 acre = 4.0469 × 10^3 sq. m.

= 43,560 sq. ft.

VOLUME

1 cc. = 0.99997 ml.

= 3.5315 × 10^{-5} cu. ft.

= 0.26417 × 10^{-3} gal.

= 6.2900 × 10^{-6} bbl.

= 8.1073 × 10^{-10} acre-ft.

1 cu. ft. = 28.317 × 10^3 cc.

= 7.4806 gal.

= 0.17811 bbl.

= 2.2957 × 10^{-5} acre-ft.

1 gal. = 3.7854 × 10^3 cc.

= 0.13368 cu. ft.

= 2.3810 × 10^{-2} bbl.

= 3.0688 × 10^{-6} acre-ft.

1 bbl. = 158.98 × 10^3 cc.

= 5.6146 cu. ft.

= 42 gal.

= 1.2889 × 10^{-4} acre-ft.

1 acre-ft. = 1.2335 × 10^9 cc.

= 43,560 cu. ft.

= 7758.5 bbl.

$$\frac{\mu}{d} = 0.226t - \frac{195}{t} \text{ (for oils of 100 sec. Saybolt or less)}[1]$$

$$= 0.220t - \frac{135}{t} \text{ (for oils of 100 sec. Saybolt or greater)},$$

where μ = absolute viscosity in centipoises.

d = density of oil at same temperature.

t = "Saybolt seconds" (Universal).

PRESSURE

1 atm. = 76.0 cm. Hg at 0°C. (of density of 13.5951 gm./cc. under normal gravity, which is 980.655 cm./sec.²)

[1] An improvement over these conversion formulas, which are those recommended by the A.S.T.M., has recently been proposed by S. Erk and H. Eck (*Phys. Zeits*, **38**, 469, 1937). It is $\frac{\mu}{d} = \frac{t}{29.0} \cdot 6.163^{1 - \left(\frac{29.0}{t}\right)^3}$

$$1 \text{ atm.} = 1.0133 \times 10^6 \text{ dynes/cm.}^2$$
$$= 14.696 \text{ lb./sq. in.}$$
$$= 33.899 \text{ ft. } H_2O \text{ at } 4°C. \text{ under normal gravity}$$
$$= 1033.2 \text{ gm./cm.}^2$$
$$1 \text{ dyne/cm.}^2 = 0.98692 \times 10^{-6} \text{ atm.}$$
$$= 14.504 \times 10^{-6} \text{ lb./sq. in.}$$
$$= 33.455 \times 10^{-6} \text{ ft. } H_2O$$
$$1 \text{ lb./in.}^2 = 0.06805 \text{ atm.}$$
$$= 6.8948 \times 10^4 \text{ dynes/cm.}^2$$
$$= 2.3067 \text{ ft. } H_2O$$
$$1 \text{ cm. } H_2O \ (4°C.) = 9.6781 \times 10^{-4} \text{ atm.}$$
$$= 980.64 \text{ dynes/cm.}^2$$
$$= 0.99998 \text{ gm./cm.}$$
$$= 0.073554 \text{ cm. Hg (at } 0°C.)$$
$$1 \text{ ft. } H_2O = 0.029500 \text{ atm.}$$
$$= 29.891 \times 10^3 \text{ dynes/cm.}^2$$
$$= 0.43353 \text{ lb./in.}^2$$

RATE

$$1 \text{ cc./sec.} = 2.1189 \times 10^{-3} \text{ cu. ft./min.}$$
$$= 1.5850 \times 10^{-2} \text{ gal./min.}$$
$$= 0.54335 \text{ bbl./day}$$
$$1 \text{ gal./min.} = 63.090 \text{ cc./sec.}$$
$$= 0.13368 \text{ cu. ft./min.}$$
$$= 34.286 \text{ bbl./day}$$
$$1 \text{ bbl./day} = 1.8404 \text{ cc./sec.}$$
$$= 0.0038990 \text{ cu. ft./min.}$$
$$= 0.029167 \text{ gal./min.}$$

APPENDIX II

LAPLACE'S EQUATION IN CURVILINEAR COORDINATES

Because of the fundamental role played by Laplace's equation in the study of the flow of homogeneous fluids through porous media, we shall present a derivation of this equation in generalized curvilinear coordinates. While more rigorous developments might be attained by the procedure of a direct transformation of the partial derivatives $\dfrac{\partial}{\partial x}, \dfrac{\partial}{\partial y}, \dfrac{\partial}{\partial z}$ [1] or by an application of the calculus of variations,[2] or by the use of the divergence integral theorem,[3] a more physical deriva-tion[4] based on the development of the expressions for the divergence of a vector in curvilinear coordinates, similar to that used in Sec. 3.1, might be more appropriate here.

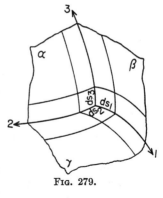

FIG. 279.

For this purpose we suppose a set of orthogonal curvilinear coordinates:

$$\alpha(x, y, z) = \text{const}; \beta(x, y, z) = \text{const}; \gamma(x, y, z) = \text{const}.$$

to be defined for every point of the region of interest. Their inter-sections may be considered to define three mutually orthog-onal directions as indicated in Fig. 279. The net flux cor-responding to a vector velocity $\bar{v}$ out of a differential volume element of sides ds_1, ds_2, ds_3, or the divergence of the vector $\bar{v}$, can then be shown, by the method of Sec. 3.1, to be

[1] GOURSAT, E., "Cours d'Analyse," vol. I, 153, 1927.

[2] R. COURANT and D. HILBERT, "Methoden der Mathematischen Physik," vol. I, p. 194, 1924.

[3] WEBSTER, A. W., "Partial Differential Equations of Mathematical Physics," p. 301, 1927.

[4] This derivation follows Riemann-Weber's, "Differential-Gleichungen der Physik," vol. I, p. 74, 1925.

$$ds_1 ds_2 ds_3 \ \operatorname{div} \ \bar{v} = \frac{\partial}{\partial s_1}(v_1 ds_2 ds_3)ds_1 + \frac{\partial}{\partial s_2}(v_2 ds_3 ds_1)ds_2 +$$

$$\frac{\partial}{\partial s_3}(v_3 ds_1 ds_2)ds_3,$$

where v_1, v_2, v_3 are the components of $\bar{v}$ parallel to the directions 1, 2, 3. Denoting the ratios of the differential changes in the coordinates α, β, γ, to the actual differential elements of length, ds_1, ds_2, ds_3, by h_1, h_2, h_3, so that

$$d\alpha = h_1 ds_1; \qquad d\beta = h_2 ds_2; \qquad d\gamma = h_3 ds_3,$$

and dividing by the volume element $ds_1 ds_2 ds_3$, the above expression for the divergence becomes

$$\operatorname{div} \bar{v} = h_1 h_2 h_3 \left[\frac{\partial}{\partial \alpha}\left(\frac{v_1}{h_2 h_3}\right) + \frac{\partial}{\partial \beta}\left(\frac{v_2}{h_3 h_1}\right) + \frac{\partial}{\partial \gamma}\left(\frac{v_3}{h_1 h_2}\right) \right].$$

If now $\bar{v}$ be considered as the gradient of a potential function Φ,

so that $\bar{v} = \nabla\Phi$, we get

$$\operatorname{div} \bar{v} = \nabla^2\Phi = h_1 h_2 h_3 \left[\frac{\partial}{\partial \alpha}\left(\frac{h_1}{h_2 h_3}\frac{\partial\Phi}{\partial\alpha}\right) + \frac{\partial}{\partial \beta}\left(\frac{h_2}{h_3 h_1}\frac{\partial\Phi}{\partial\beta}\right) + \right.$$

$$\left. \frac{\partial}{\partial \gamma}\left(\frac{h_3}{h_1 h_2}\frac{\partial\Phi}{\partial\gamma}\right) \right].$$

This expression set equal to zero represents the curvilinear-coordinate transformation of the Laplace equation $\nabla^2\Phi = 0$.

Thus for the Cartesian system (x, y, z) it is clear that $h_1 = h_2 = h_3 = 1$, so that the normal Laplacian form of Eq. 3.4(4) is obtained. For the cylindrical-coordinate system (r, θ, z) (*cf.* Fig. 22), $(h_1, h_2, h_3) = (1, 1/r, 1)$, leading to Eq. 3.7(3). And for the spherical-coordinate system (r, θ, χ) (*cf.* Fig. 23), $(h_1, h_2, h_3) = (1, 1/r, 1/r \sin \theta)$, from which Eq. 3.7(6) follows at once. In a similar manner the corresponding equation for any other coordinate system may be derived, once the appropriate (h_1, h_2, h_3) have been found.

APPENDIX III

˙SOME TWO-DIMENSIONAL GREEN'S FUNCTIONS

To facilitate the application of the method of the Green's function to two-dimensional problems other than that discussed in Sec. 4.6, the following additional Green's functions are listed:
Infinite half plane: $y \geqslant 0$. (*Cf.* Fig. 280)

$$G(x, y; x', y') = \frac{1}{2} \log \frac{(x - x')^2 + (y + y')^2}{(x - x')^2 + (y - y')^2}.$$

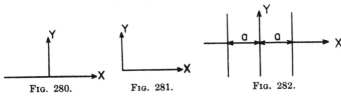

FIG. 280.   FIG. 281.   FIG. 282.

Infinite quadrant: $y \geqslant 0; x \geqslant 0$. (*Cf.* Fig. 281)

$$G(x, y; x', y') = \frac{1}{2} \log$$
$$\frac{[(x - x')^2 + (y + y')^2][(x + x')^2 + (y - y')^2]}{[(x - x')^2 + (y - y')^2][(x + x')^2 + (y + y')^2]}.$$

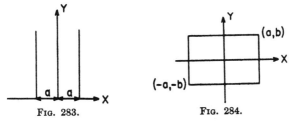

FIG. 283.   FIG. 284.

Infinite strip: $-\infty < y < +\infty; -a \leqslant x \leqslant +a$. (*Cf.* Fig. 282)

$$G(x, y; x', y') = \frac{1}{2} \log \frac{\cosh \dfrac{\pi(y - y')}{2a} + \cos \dfrac{\pi(x + x')}{2a}}{\cosh \dfrac{\pi(y - y')}{2a} - \cos \dfrac{\pi(x - x')}{2a}}.$$

Infinite half strip: $y \geqslant 0$; $-a \leqslant x \leqslant +a$.   (*Cf.* Fig. 283)

$$G(x, y; x', y') = \frac{1}{2} \log$$

$$\frac{\left[\cosh\dfrac{\pi(y - y')}{2a} + \cos\dfrac{\pi(x + x')}{2a}\right]\left[\cosh\dfrac{\pi(y + y')}{2a} - \cos\dfrac{\pi(x - x')}{2a}\right]}{\left[\cosh\dfrac{\pi(y - y')}{2a} - \cos\dfrac{\pi(x - x')}{2a}\right]\left[\cosh\dfrac{\pi(y + y')}{2a} + \cos\dfrac{\pi(x + x')}{2a}\right]}.$$

Rectangle: $-b \leqslant y \leqslant +b$; $-a \leqslant x \leqslant +a$.   (*Cf.* Fig. 284)

$$G(x, y; x', y') = \frac{8}{\pi ab} \sum_{nm = 1}^{\infty} \frac{1}{k_{nm}} \sin\frac{n\pi(x + a)}{2a}$$

$$\sin\frac{n\pi(x' + a)}{2a} \sin\frac{m\pi(y + b)}{2b} \sin\frac{m\pi(y' + b)}{2b},$$

where

$$k_{nm} = \frac{n^2}{a^2} + \frac{m^2}{b^2}.$$

THE TRANSFORMATION PROPERTIES OF THE
MODULAR ELLIPTIC FUNCTION $\Theta(q) = k^{*2}$

It is shown in Secs. 33 and 54 of "Formeln und Lehrsätze zum Gebrauche der elliptischen Functionen," by H. A. Schwarz, that if the modular elliptic function $\Theta(q) = k^{*2}$ of modulus $k^*(\leqslant 1)$ is defined by

$$\Theta(0) = 1, \qquad \Theta(1) = \infty, \qquad \Theta(i\infty) = 0, \qquad \Theta(1 + i\infty) = 0,$$

the function $\Theta\left(\dfrac{m' + n'q}{m + nq}\right)$ will have the values

I	II	III	IV	V	VI
$\Theta(q)$	$\dfrac{\Theta(q)}{\Theta(q) - 1}$	$\dfrac{1}{\Theta(q)}$	$\dfrac{1}{1 - \Theta(q)}$	$\dfrac{\Theta(q) - 1}{\Theta(q)}$	$1 - \Theta(q)$

when $(m, n; m', n')$ are given by the table

	m	n	m'	n'
I	1	0	0	1
II	1	0	1	1
III	1	1	0	1
IV	1	1	1	0
V	0	1	1	1
VI	0	1	1	0

where the values of m, n; m', n' are expressed with respect to the modulus 2, so as to equal 1 if the true values are odd, and 0 if even.

The inversion of the function $\Theta(q) = k^{*2}$ is given by Eq. 6.5(15).

The use of these tables may be illustrated by their application to the function $\lambda(q)$ defined by Eq. 6.4(14). From this it follows that

$$\lambda(q) = \frac{\Theta(q) - 1}{\Theta(q)} = \Theta\!\left(\frac{q+1}{q}\right); \qquad \frac{1}{\lambda(q)} = \frac{\Theta(q)}{\Theta(q) - 1} = \Theta(q+1);$$

$$1 - \lambda(q) = \frac{1}{\Theta(q)} = \Theta\!\left(\frac{q}{q+1}\right); \qquad \frac{1}{1 - \lambda(q)} = \Theta(q);$$

$$\frac{\lambda(q)}{\lambda(q) - 1} = 1 - \Theta(q) = \Theta\!\left(\frac{1}{q}\right); \qquad \frac{\lambda(q) - 1}{\lambda(q)} = \frac{1}{1 - \Theta(q)} =$$

$$\Theta\!\left(\frac{1}{q+1}\right).$$

Hence to find the arguments q for which

$$-\infty < \lambda \leqslant 0,$$

one may set, since $k^{*2} \leqslant 1$,

$$k^{*2} = \Theta(q) = \frac{1}{1 - \lambda}, \qquad \text{or} \qquad k^{*2} = \Theta\!\left(\frac{-1}{q}\right) = \frac{\lambda}{\lambda - 1},$$

so that

$$iq = -\frac{K'}{K}\!\left(\frac{1}{1 - \lambda}\right), \qquad \text{or} \qquad \frac{i}{q} = \frac{K'}{K}\!\left(\frac{\lambda}{\lambda - 1}\right),$$

where the arguments of K'/K represent the values of k^{*2}.

Likewise, for

$$0 \leqslant \lambda \leqslant 1,$$

one may set

$$k^{*2} = \Theta\!\left(\frac{q}{1 - q}\right) = 1 - \lambda, \qquad \text{or} \qquad k^{*2} = \Theta\!\left(\frac{q - 1}{q}\right) = \lambda,$$

so that

$$\frac{iq}{1 - q} = -\frac{K'}{K}(1 - \lambda), \qquad \text{or} \qquad i\!\left(\frac{q - 1}{q}\right) = -\frac{K'}{K}(\lambda),$$

and for

$$1 \leqslant \lambda < \infty,$$

one may set

$$k^{*2} = \Theta(q - 1) = \frac{1}{\lambda}, \qquad \text{or} \qquad k^{*2} = \Theta\!\left(\frac{1}{1 - q}\right) = \frac{\lambda - 1}{\lambda},$$

so that

$$i(q - 1) = -\frac{K'}{K}\left(\frac{1}{\lambda}\right), \quad \text{or} \quad \frac{i}{1 - q} = -\frac{K'}{K}\left(\frac{\lambda - 1}{\lambda}\right).$$

The values of K'/K as a function of k^{*2} are tabulated in K. Hayashi, "Tafeln der Besselschen, Theta, Kugel- und anderer Funktionen," 1930.

APPENDIX V

PROOF OF THE GENERALIZED POISSON FORMULA
EQ. 6.4(15)

This formula states that the complex potential function $\theta - i\tau$ (θ, τ being conjugate functions) which satisfies the condition that $\theta(x, y = 0) = \theta(x)$ is given by

$$\theta(x, y) - i\tau(x, y) = -i\tau_0 + \frac{1}{\pi i}\int_{-\infty}^{+\infty}\frac{(1 + z\epsilon)\theta(\epsilon)}{(\epsilon - z)(1 + \epsilon^2)}d\epsilon,$$

where $z = x + iy$.

We begin with the Fourier integral solution of the potential problem referring to $\theta(x, y)$ alone, which may be expressed as[1]

$$\theta(x, y) = \frac{1}{\pi}\int_0^\infty d\alpha \int_{-\infty}^{+\infty} e^{-\alpha y}\theta(\epsilon)\cos\alpha(\epsilon - x)d\epsilon$$

$$= \frac{1}{\pi}\int_{-\infty}^{+\infty}\frac{y\theta(\epsilon)d\epsilon}{y^2 + (x - \epsilon)^2}.$$

Splitting the integrand into two conjugate terms, so that

$$\theta(x, y) = \frac{-i}{2\pi}\int_{-\infty}^{+\infty}\theta(\epsilon)d\epsilon\left[\frac{1}{\epsilon - z} - \frac{1}{\epsilon - \bar{z}}\right],$$

where the bar denotes the complex conjugate, and noting from the theory of conjugate functions (*cf.* Sec. 4.8) that

$$\tau = -\int\frac{\partial\theta}{\partial x}dy,$$

it follows that

$$\tau = \frac{i}{2\pi}\int_{-\infty}^{+\infty}\theta(\epsilon)d\epsilon\int\left[\frac{1}{(\epsilon - z)^2} - \frac{1}{(\epsilon - \bar{z})^2}\right]dy$$

$$= \frac{1}{2\pi}\int_{-\infty}^{+\infty}\theta(\epsilon)d\epsilon\left[\frac{1}{\epsilon - z} + \frac{1}{\epsilon - \bar{z}} - C\right],$$

where the constant of integration C may be set equal to $2\epsilon/(1 + \epsilon^2)$ to make the integrals convergent. Adding now $-i\tau$ to the above value of θ, the Poisson formula is the immediate result.

[1] *Cf.* W. E. Byerly, "Fourier's Series and Spherical Harmonics," pp. 70, 73.

APPENDIX VI

TABULATION OF THE SPECIFIC QUANTITATIVE RESULTS DEVELOPED IN THIS WORK, IN FORMULA OR GRAPHICAL FORM

1. Darcy's law: 2.3(3).

$$v = \frac{k}{\mu} \frac{dp}{dx} = \text{volume rate of fluid per unit area of porous medium.}$$

k = permeability of medium; μ = viscosity of fluid.

$\frac{dp}{dx}$ = pressure gradient where pressure is p and velocity v.

2. Liquid flow through a horizontal linear channel, and permeability formula for horizontal linear liquid flow experiments: 2.5(4).

$$Q = \frac{Ak(P_1 - P_2)}{\mu L}; \qquad k = \frac{\mu QL}{A(P_1 - P_2)};$$

Q = liquid flux (vol./time) through a linear channel of area A and length L, at the termini of which the pressures are P_1, P_2.

3. Gas flow through a linear channel, and permeability formula for linear-gas-flow experiments: 2.5(7), (8).

$$Q_m = \frac{\gamma_0 Ak(P_1{}^2 - P_2{}^2)}{2\mu L}; \qquad k = \frac{2\mu Q_m L}{\gamma_0 A(P_1{}^2 - P_2{}^2)};$$

$$\bar{Q} = \frac{Ak(P_1 - P_2)}{\mu L}; \qquad k = \frac{\mu \bar{Q} L}{A(P_1 - P_2)};$$

Q_m = mass flux of gas, of density γ_0 at atmospheric pressure.

$\bar{Q}$ = volume gas flux reduced to the algebraic mean pressure in column, $(P_1 + P_2)/2$.

4. Permeability formula for field experiments with a well producing liquid by artesian flow: 2.11(1).

733

$$k = \frac{\mu Q \log r_e/r_w}{2\pi h(P_e - P_w)}.$$

r_e = "external boundary" radius ($\sim$ 500 ft.) where pressure is P_e.

r_w = well radius, where pressure is P_w.

h = sand thickness.

5. Permeability formulas for field experiments with a gas well: 2.11(2).

$$k = \frac{\mu \bar{Q} \log r_e/r_w}{2\pi h(P_e - P_w)} = \frac{\mu Q_m \log r_e/r_u}{\pi h \gamma_0(P_e^2 - P_w^2)}.$$

6. Correction factors for well penetration of both pressure-drive and gravity-flow wells. Figure 17.

7. Permeability formula for field experiments with a well producing entirely by the action of gravity: 2.11(3).

$$k = \frac{\mu Q \log r_e/r_w}{\pi \gamma g(h_e^2 - h_w^2)}.$$

γ = liquid density.

g = acceleration of gravity (980 cm./sec.^{2}).

h_e = fluid head in sand at r_e.

h_w = fluid head in well (at r_w).

8. Permeability formula for field experiments with a well producing under combined effect of gravity and pressure drive: 2.11(4).

$$k = \frac{\mu Q \log r_e/r_w}{\pi \gamma g(2hh_e - h^2 - h_w^2)}.$$

h_e = fluid head ($> h$) at r_e in sand of thickness h.

h_w = fluid head ($< h$) in well (at r_w).

9. Permeability formula for field experiments with artesian wells in which the bottom-hole pressures are varied: 2.11(5).

$$k = \frac{\mu \Delta Q \log r_e/r_w}{2\pi h \Delta P_w}.$$

ΔQ = change in liquid flux caused by a change ΔP_w in the bottom-hole pressure.

10. Permeability formula for field experiments with gas wells in which the bottom-hole pressures are varied: 2.11(6).

$$k = \frac{\mu \Delta \bar{Q} \log r_e/r_w}{2\pi h \Delta P_w}.$$

11. Permeability formula for field experiments with wells producing by simple gravity or composite gravity and pressure drives, in which the fluid level in the well is varied: 2.11(7).

$$k = \frac{\mu \Delta Q \log r_e/r_w}{\pi \gamma g \Delta h_w^2}.$$

If in formulas (1) to (11), the various lengths are expressed in centimeters, areas in centimeters squared, volume fluxes in cubic centimeters per second, densities in grams per centimeters cubed, μ in centipoises, g in centimeters per second squared (980), and pressures in atmospheres, k will be given as darcys.

12. Formula for determining total porosities by volumetric or gravimetric methods: 2.13(1).

$$f = 100\left(1 - \frac{V_g}{V_b}\right) = 100\left(1 - \frac{\gamma_b}{\gamma_g}\right) = \text{porosity in per cent.}$$

V_g, γ_g = volume and density of the sand grains of the sample.
V_b, γ_b = bulk volume and bulk density of the sample.

13. General form of Darcy's law: 3.3(3) − (5).

$$\bar{v} = -\nabla\Phi; \qquad v_x = -\frac{\partial\Phi}{\partial x}; \qquad v_y = -\frac{\partial\Phi}{\partial y}; \qquad v_z = -\frac{\partial\Phi}{\partial z};$$

$$\Phi = \frac{k}{\mu}(p - V).$$

$\bar{v}$ = vector fluid macroscopic velocity (of components v_x, v_y, v_z).

∇ = differential vector operator of components $\frac{\partial}{\partial x}, \frac{\partial}{\partial y}, \frac{\partial}{\partial z}$.

Φ = velocity potential; V = potential of body forces acting on fluid, $= \pm\gamma gz$ if the body force is that of gravity, and the vertical coordinate $+z$ is directed upward $(+)$ or downward $(-)$.

14. Fundamental differential equation governing the flow of incompressible liquids: 3.4(4).

$$\nabla^2\Phi = \frac{\partial^2\Phi}{\partial x^2} + \frac{\partial^2\Phi}{\partial y^2} + \frac{\partial^2\Phi}{\partial z^2} = 0 = \nabla^2 p = \frac{\partial^2 p}{\partial x^2} + \frac{\partial^2 p}{\partial y^2} + \frac{\partial^2 p}{\partial z^2}.$$

15. Fundamental differential equation governing the flow of compressible liquids: 3.4(6).

$$\nabla^2\gamma = \frac{\partial^2\gamma}{\partial x^2} + \frac{\partial^2\gamma}{\partial y^2} + \frac{\partial^2\gamma}{\partial z^2} = \frac{f\beta\mu}{k}\frac{\partial\gamma}{\partial t}.$$

t = time variable.

β = liquid compressibility, defined by: $\gamma = \gamma_0 e^{\beta p}$, or:

$$\beta = \frac{1}{\gamma}\frac{\partial\gamma}{\partial p}.$$

16. Fundamental differential equation governing the flow of gases: 3.4(7).

$$\nabla^2\gamma^{\frac{1+m}{m}} = \frac{\partial^2\gamma^{\frac{1+m}{m}}}{\partial x^2} + \frac{\partial^2\gamma^{\frac{1+m}{m}}}{\partial y^2} + \frac{\partial^2\gamma^{\frac{1+m}{m}}}{\partial z^2} = \frac{(1+m)f\mu\gamma_0^{\frac{1}{m}}}{k}\frac{\partial\gamma}{\partial t},$$

the nature of the gas and the character of its flow being defined by: $\gamma = \gamma_0 p^m$.

17. Analogies of fluid flow through porous media with other physical problems: Table 13.

STEADY-STATE LIQUID FLOW SYSTEMS

18. Strictly radial flow into a well: 4.2(10).

$$Q = \frac{2\pi kh(p_e - p_w)}{\mu \log r_e/r_w}.$$

Notation as in 4.

19. Flow into a well with variable boundary pressures: 4.5(12).

$$Q = \frac{2\pi kh(\bar{p}_e - \bar{p}_w)}{\mu \log r_e/r_w}.$$

$\bar{p}_e$ = average pressure over external boundary (of radius r_e).
$\bar{p}_w$ = average pressure over well boundary (of radius r_w).

20. Flow into a well displaced from the center of the external boundary: 4.6(10), Fig. 34.

$$Q = \frac{2\pi kh(p_e - p_w)}{\mu \log \frac{r_e^2 - \delta^2}{r_e r_w}}.$$

δ = displacement of well center from center of external boundary.

21. Flow from an infinite line source ("line drive") into a well: 4.7(8).

$$Q = \frac{2\pi k h (p_e - p_w)}{\mu \log 2d/r_w}.$$

d = distance of well from the line source.

22. Flow from a finite line source into a well: 4.9(16).

$$Q = \frac{2\pi k h (p_e - p_w)}{\mu \log \frac{4y_0}{r_w} \sqrt{\dfrac{(c^2 - r_0^2)^2 + 4y_0^2 c^2}{c^2 - r_0^2 + \sqrt{(c^2 - r_0^2)^2 + 4y_0^2 c^2}}}}.$$

$2c$ = width of the finite line source.
r_0 = distance of well from center of line source.
y_0 = perpendicular distance of well from line source.

23. Pressure distribution underneath a dam without sheet piling: 4.10(1), Fig. 44.

$$p = \frac{p_2 - p_1}{\pi} \cos^{-1} \frac{2x}{w} + p_1.$$

p_2, p_1 = upstream and downstream pressures.
x = distance along base measured from its center.
w = width of base.

24. Upward force on a dam without sheet piling: 4.10(2).

$$F = \frac{(p_2 + p_1)w}{2} = \text{total upward force per unit length of dam.}$$

25. Total uplift moment with respect to the heel of a dam without sheet piling: 4.10(3).

$$M = \frac{w^2(3p_2 + 5p_1)}{16} = \text{total uplift moment per unit length}$$

of dam.

26. Pressure distribution underneath a dam with sheet piling: 4.11(11), Figs. 50, 51.

$$p = \frac{p_2 - p_1}{\pi} \cos^{-1}$$

$$\left[\frac{2\sqrt{d^2 + (x' - \bar{x})^2} - \sqrt{d^2 + (w - \bar{x})^2} + \sqrt{d^2 + \bar{x}^2}}{\sqrt{d^2 + (w - \bar{x}_1)^2} + \sqrt{d^2 + \bar{x}_1^2}} \right] + p_1.$$

d = depth of piling; x' = distance along base from heel.
$\bar{x}$ = distance of piling from heel; sign of first radical is $+$ for $x' > \bar{x}$, and $-$ for $x' < \bar{x}$.

27. Total uplift forces on dams with sheet piling: Fig. 52.

28. Total uplift moments with respect to the heel of a dam with sheet piling: Fig. 53.

29. Pressure drops over the piling: 4.11(17).

$$\delta p = \frac{p_2 - p_1}{\pi} \cos^{-1} \frac{1}{[\sqrt{d^2 + (w - \bar{x})^2} + \sqrt{d^2 + \bar{x}^2}]^2}$$
$$[\{\sqrt{d^2 + (w - \bar{x})^2} - \sqrt{d^2 + \bar{x}^2}\}^2 - 4 + 4\bar{x}(w - \bar{x})].$$

30. Seepage flux under dams without sheet piling: 4.12(10) and Fig. 61.

31. Seepage flux under dams with sheet piling: 4.13(5) and Figs. 66, 67.

32. Pressure drops over piling if permeable bed is of finite thickness 4.13(7) and Fig. 68.

33. Seepage flux under coffer dams: 4.14(9) and Fig. 71.

34. Differential equation giving the pressure distribution in homogeneous anisotropic media: 4.15(2).

$$k_x \frac{\partial^2 p}{\partial x^2} + k_y \frac{\partial^2 p}{\partial y^2} + k_z \frac{\partial^2 p}{\partial z^2} = 0.$$

35. General formula for the flow of a liquid into a well: 4.16(6).

$$Q = \frac{2\pi k h \Delta p}{\mu \log c/r_w}.$$

c = an "effective" average of the distance from the well center to the external boundary.

36. Formula for the flux in a flow system as given by a graphically constructed square network of the equipotentials and streamlines: 4.17(9).

$$Q = \frac{m}{n} \Delta \Phi.$$

m = number of squares lying between two neighboring equipotentials extending from one bounding streamline surface to another.

n = number of squares lying between two neighboring streamlines extending from the high to the low-potential boundaries.

$\Delta \Phi$ = total potential drop over the system.

37. Flow into a spherical surface ("nonpenetrating" well): 5.2(8).

$$Q = \frac{4\pi(\Phi_e - \Phi_w)}{\dfrac{1}{r_w} - \dfrac{1}{r_e}}.$$

Φ_e, Φ_w = potentials $\left[\dfrac{k}{\mu}(p - \gamma gz)\right]$ at spherical bounding surfaces of radii r_e, r_w.

38. Production capacities of partially penetrating wells: 5.4(6) and Figs. 83 to 85.

$$Q = \frac{2\pi kh\Delta p/\mu}{\dfrac{1}{2\bar{h}}\left[2\log\dfrac{4h}{r_w} - \log\dfrac{\Gamma(0.875\bar{h})\Gamma(0.125\bar{h})}{\Gamma(1 - 0.875\bar{h})\Gamma(1 - 0.125\bar{h})}\right] - \log\dfrac{4h}{r_e}}.$$

$$Q = \frac{2\pi kh\bar{h}\Delta p/\mu}{\log r_e/r_w}\left(1 + 7\sqrt{\dfrac{r_w}{2h\bar{h}}}\cos\dfrac{\pi\bar{h}}{2}\right).$$

h = sand thickness.

$\bar{h}$ = well penetration expressed as a fraction of the sand thickness.
Γ are gamma functions.

39. Production capacities of partially penetrating wells in anisotropic sands: Fig. 87.

40. Fluid caught by a ditch on a sloping sand: 6.2(8).

$$\frac{Q}{Q_0} = \frac{H}{h_1}.$$

Q = fluid caught by the ditch.
Q_0 = normal flow through the sand (without the ditch).
H = drawdown of free water surface at the ditch.
h_1 = thickness of undisturbed layer of water-saturated sand.

41. Seepage flux through a dam with vertical faces: 6.5(16).

$$Q = \frac{k\gamma g(h_e{}^2 - h_w{}^2)}{2\mu L} = \text{seepage flux per unit length of dam.}$$

h_e = fluid head on upstream face.
h_w = fluid head on downstream face.
L = width of dam.

42. The seepage of water out of ditches or canals when the seepage stream does not merge with the water table: 6.7(20); 6.8(6); 6.9(5); Figs. 120 and 124.

$$Q = \frac{k\gamma g}{\mu}\left[B + \frac{2HK(k^*)}{K'(k^*)} \right] = \text{seepage flux per unit length of ditch}$$

or canal.

B = total width of the ditch or canal profile at the top of the free-water surface.

H = depth of the free water at its deepest point.

K, K' are complete elliptic integrals of the first kind with moduli k^* and $\sqrt{1 - k^{*2}}$, the value of k^* being determined by the exact shape of the ditch (*cf.* Fig. 121) and depth of the highly permeable bed carrying the water table.

43. The extent of the lateral drainage out of a canal into a coarse sand before seeping into an underlying bed: 6.13(2).

$$x_1 = \sqrt{\frac{2hh_0k_g}{k_c}} = \text{length of lateral drainage.}$$

h_0 = fluid head at the face of the canal at the level of the coarse sand (gravel).

h = thickness of coarse sand, of permeability k_g.

k_c = permeability of tight beds (clay) underlying the coarse sand.

44. Angle of dip of top soil on a hillside required to carry away the rainfall without flood erosion: 6.15(4) and Fig. 137.

$$\cos \theta_1 = \frac{\cos \theta_2}{1 - \dfrac{q\mu}{k\gamma g}}.$$

θ_1 = inclination of top face of soil bed.

θ_2 = inclination of hillside.

q = rainfall intensity, per unit horizontal area.

45. Drainage spacing required to avoid water logging: 6.16(6) and Fig. 139.

46. Artesian flow into a well under the action of gravity alone: 6.18(6).

$$Q = \frac{\pi k\gamma g(h_e{}^2 - h_w{}^2)}{\mu \log r_e/r_w}.$$

Notation as in 7.

47. Flow into a well producing under the combined action of gravity and pressure heads: 6.19(2).

$$Q = \frac{\pi k \gamma g (2hh_e - h^2 - h_u{}^2)}{\mu \log r_e/r_w}.$$

Notation as in 8.

48. Equivalent nongravity potential drop over a sand producing by gravity flow: 6.20(9).

$$\Delta\bar{\Phi} = \frac{k\gamma g(h_e{}^2 - h_w{}^2)}{2\mu h_e}.$$

h_e = actual fluid head at inflow (external) boundary.
h_w = actual fluid head at outflow (well) surface.

49. Equivalent nongravity potential drop over a sand producing under the combined action of gravity and pressure heads: 6.20(10).

$$\Delta\bar{\Phi} = \frac{k\gamma g(2hh_e - h^2 - h_w{}^2)}{2\mu h}.$$

h = sand thickness.
h_e = fluid head ($> h$) at inflow (external) boundary.
h_w = fluid head ($< h$) at outflow (well) surface.

50. Production capacity of a well in a nonuniform sand: 7.3(10) and Fig. 150.

$$Q = \frac{2\pi hk_1(p_e - p_w)/\mu}{\log \dfrac{r_0}{r_w} + \dfrac{k_1}{k_2} \log \dfrac{r_e}{r_0}}.$$

k_1 = permeability in the annulus of radius r_0 surrounding the well bore (r_w).
k_2 = permeability in the remainder of the sand (up to r_e).

51. The effect of acid treatment on radial flow systems: Figs. 157 and 158.

52. Effective permeability of a fracture: 7.8(2).

$$k = \frac{10^8 w^2}{12} \text{ darcys.}$$

w = fracture width in centimeters.

53. The effect of acid treatment in highly fractured limestones: Fig. 159.

54. The production capacities of wells in stratified horizons: Fig. 162.

55. Production capacities of wells with sanded liners: 7.10(7). 7.10(17), 7.10(19), and Fig. 164.

56. The history of the water encroachment in a linear system: 8.3(5) and Fig. 169.

57. The history of the water encroachment in a radial system: 8.4(5) and Fig. 172.

58. The history of a line drive water encroachment into a single well: 8.6(9) and Fig. 174.

59. Maximal pressure differentials over a sand without water coning: Fig. 187.

60. Maximal production rates from partially penetrating wells without water coning: Fig. 188.

61. Production capacities of wells in a well pair within a given area: 9.3(3).

$$Q = \frac{2\pi kh(p_e - p_u)}{\mu \log R^2/r_w d} = \text{production capacity per well.}$$

R = radius of external boundary.

d = separation between wells ($< R$).

62. Production capacities of wells in a group of three wells in a triangular pattern: 9.3(4).

$$Q = \frac{2\pi kh(p_e - p_w)}{\mu \log R^3/r_w d^3}.$$

63. Production capacity of wells in a group of four wells in a square pattern: 9.3(5).

$$Q = \frac{2\pi kh(p_e - p_w)}{\mu \log \dfrac{R^4}{\sqrt{2}r_w d^3}}$$

64. Production capacities of wells in a circular battery: 9.3(18) and Fig. 198.

$$Q = \frac{2\pi kh(p_e - p_w)}{\mu \left[\log \dfrac{R^n}{r^{n-1}r_w} - \sum_1^{n-1} \log 2 \sin \dfrac{\pi m}{n} \right]} = \text{production capacity per well.}$$

n = number of wells in battery of radius r.

65. Production capacities of linear groups of wells supplied by infinite line drives: 9.6(2).

66. Production capacities of wells on an infinite linear array supplied by a parallel line drive: 9.8(9).

$$Q = \frac{2\pi kh\Delta p}{\mu \log \sinh \dfrac{2\pi d/a}{\pi r_w/a}} = \text{production capacity per well.}$$

Δp = pressure differential between the line drive and well surfaces.

d = distance from line drive to the linear array of wells.

a = well spacing within the array.

67. Relative production capacities of wells in two infinite parallel arrays supplied by a line drive: 9.9(2).

$$\frac{Q_1}{Q_2} = \frac{\log \dfrac{\sinh \pi r_w/a \sinh \pi(d_2 + d_1)/a}{\sinh 2\pi d_2/a \sinh \pi(d_2 - d_1)/a}}{\log \dfrac{\sinh \pi r_w/a \sinh \pi(d_2 + d_1)/a}{\sinh 2\pi d_1/a \sinh \pi(d_2 - d_1)/a}}.$$

Q_1 = production capacity per well in line array at distance d_1 from the line drive.

Q_2 = production capacity per well in line array at distance $d_2(> d_1)$ from the line drive.

a = well spacings within the line arrays.

68. Shielding and leakage characteristics of two-line arrays: Figs. 205, 206, and 207.

69. Shielding and leakage characteristics of three-line arrays: Fig. 210.

70. Pressure requirements for single-line offsetting: 9.13(9).

$p_{w1} - p_{w2} = p_2 - p_1.$

p_{w1} = bottom-hole pressures in offset wells of line 1.

p_{w2} = bottom-hole pressures in offset wells of line 2.

p_1 = average "lease pressure" of lease 1.

p_2 = average "lease pressure" of lease 2.

71. Pressure requirements for multiple-line offsetting: p. 555, 556.

72. The flooding history of the direct line-drive flood: Fig. 223.

73. The flooding history of the five-spot flood: Fig. 225.

74. The flooding histories of the seven-spot floods: Figs. 228 and 230.

75. Conductivity of direct line-drive flooding networks as a function of the line and well spacings: 9.23(7) and Fig. 241.

$$Q = \frac{2\pi kh\Delta p}{\mu \log \dfrac{\sinh^4 \pi d/a \, \sinh 3\pi d/a}{\sinh^2 \pi r_u/a \, \sinh^3 2\pi d/a}} = \text{production capacity per}$$

well or per network element.

d = spacing between the input and output lines.

a = well spacing within the lines.

76. Conductivity of the five-spot flooding network: 9.24(4) and Fig. 243.

$$Q = \frac{\pi kh\Delta p}{\mu \left(\log \dfrac{d}{r_w} - 0.6190\right)} = \text{production capacity per network}$$

element.

d = distance between neighboring input and output wells.

77. Conductivity of the seven-spot flooding network: 9.25(4) and Fig. 243.

$$Q = \frac{4\pi kh\Delta p}{\mu \left(3 \log \dfrac{d}{r_w} - 1.7073\right)} = \text{production capacity per network}$$

element.

d = length of hexagon sides.

78. Conductivity of the staggered line-drive network: 9.26(3) and Fig. 241.

$$Q = \frac{2\pi kh\Delta p}{\mu \log \dfrac{\cosh^4 \pi d/a \, \cosh^3 3\pi d/a}{\sinh^2 \pi r_w/a \, \sinh^4 2\pi d/a \, \sinh 4\pi d/a}} = \text{production}$$

capacity per network element.

d = spacing between the input and output lines.

a = well spacing within the lines.

79. The efficiency of the direct line-drive flood: 9.28(5) and Fig. 246.

$$\text{Efficiency} = \frac{1}{\pi\dfrac{d}{a}\left(\cosh^2\dfrac{\pi d}{a} - 2\right)}\left(\cosh^2\dfrac{\pi d}{a} \log \cosh \dfrac{\pi d}{a} - \right.$$

$$\left. 0.6932 \sinh^2\dfrac{\pi d}{a}\right)$$

= fraction of area of network element flooded by water when it first reaches output wells.

80. The efficiency of the five-spot flooding network: 9.29(3).

Efficiency = 0.723.

81. The efficiency of the seven-spot flooding network: 9.30(3).

Efficiency = 0.740.

82. The efficiency of the staggered line-drive flood: Fig. 246.

FLOW OF COMPRESSIBLE LIQUIDS

83. Steady-state flow of compressible liquids: 10.1(12), 10.1(15).

$Q = Q_0[1 + \beta\bar{p} + 0(\beta^2)]$ = mass flux in compressible-liquid system.

Q_0 = mass flux in same system if the liquid is incompressible.

β = liquid compressibility.

$\bar{p}$ = algebraic mean pressure = $(p_1 + p_2)/2$.

84. Density (and pressure) distribution in a radial system in which the density is preassigned over both boundaries: 10.3(17).

$$\gamma(r,\, t) = \pi \sum_n \frac{\alpha_n{}^2 J_0(\alpha_n r_1) U(\alpha_n r) e^{-\kappa\alpha_n{}^2 t}}{J_0{}^2(\alpha_n r_1) - J_0{}^2(\alpha_n r_2)} \left[\frac{\pi}{2} J_0(\alpha_n r_1) \int_{r_1}^{r_2} r g(r) U(\alpha_n r) dr \right.$$

$$\left. - \kappa J_0(\alpha_n r_2) \int_0^t f_1(\lambda) e^{\kappa\alpha_n{}^2\lambda} d\lambda + \kappa J_0(\alpha_n r_1) \int_0^t f_2(\lambda) e^{\kappa\alpha_n{}^2\lambda} d\lambda \right];$$

γ = liquid density (*cf.* 15).

$U(\alpha_n r) = Y_0(\alpha_n r_2) J_0(\alpha_n r) - J_0(\alpha_n r_2) Y_0(\alpha_n r)$.

$J_0,\ Y_0$ are zero-order Bessel functions of the first and second kinds.

α_n are roots of: $U(\alpha_n r_1) = 0$; $\kappa = \dfrac{k}{f\beta\mu}$.

$f_1(t), f_2(t)$ are values of γ maintained at the boundaries $r = r_1, r_2$.

$g(r)$ = initial distribution of $\gamma = \gamma(r, 0)$.

85. Pressure rise in a well producing a compressible liquid after shutting in: 10.6(3) and Fig. 250.

$$\frac{p_e - p_w}{p_e - p_{wi}} = \frac{2}{\log \rho} \sum_n \frac{e^{-x_n^2 \bar{t}}}{x_n^2 J_1^2(x_n)};$$

p_e = "reservoir" pressure = pressure at r_e.

p_w = bottom-hole pressure at time $\bar{t}$, its initial value at instant of shutting in being p_{wi}.

$\rho = r_e/r_w$. $\bar{t} = \kappa t/r_e^2$.

J_1 = first-order Bessel function of first kind.

x_n = roots of $J_0(x_n) = 0$.

A simpler approximation formula is: (p. 642)

$$\frac{h_e - h_w}{h_e - h_i} = \frac{p_e - p_w}{p_e - p_{wi}} = e^{-ct}; \qquad C = \frac{2\pi k h \gamma_0 g}{\mu A \log r_e/r_w}.$$

h_e, h_w, h_i = fluid heights in well bore corresponding to p_e, p_w, p_{wi}.

h = sand thickness; γ_0 = density of fluid, of viscosity μ.

A = area of flow string.

86. Density (and pressure) distribution in a radial system in which the density is specified over one boundary and the flux over the other: 10.7(10).

$$\gamma = \pi \sum_n \frac{\alpha_n^2 J_0(\alpha_n r_1) U(\alpha_n r) e^{-\kappa \alpha_n^2 t}}{J_0^2(\alpha_n r_1) - J_1^2(\alpha_n r_2)} \left[\frac{\pi}{2} J_0(\alpha_n r_1) \int_{r_1}^{r_2} r g(r) U(\alpha_n r) dr \right.$$

$$\left. - \kappa J_1(\alpha_n r_2) \int_0^t f_1(\lambda) e^{\kappa \alpha_n^2 \lambda} d\lambda - \frac{\kappa J_0(\alpha_n r_1)}{\alpha_n r_2} \int_0^t f_2(\lambda) e^{\kappa \alpha_n^2 \lambda} d\lambda \right].$$

$U(\alpha_n r) = Y_1(\alpha_n r_2) J_0(\alpha_n r) - J_1(\alpha_n r_2) Y_0(\alpha_n r)$.

α_n = roots of $U(\alpha_n r_1) = 0$.

$f_1(t)$ = value of γ maintained at r_1.

$f_2(t)$ = value of $r\dfrac{\partial \gamma}{\partial r}$ maintained at r_2.

87. The theory of the pressure decline in the "East Texas" oil field: Figs. 253 and 254.

88. Density (and pressure) distribution in a radial system in which the flux is specified over both boundaries: 10.11(8).

$$\gamma = \frac{2\kappa}{r_2{}^2 - r_1{}^2} \int_0^t [f_2(\lambda) - f_1(\lambda)]d\lambda + \frac{2}{r_2{}^2 - r_1{}^2} \int_{r_1}^{r_2} rg(r)dr$$

$$+ \pi \sum \frac{\alpha_n{}^2 J_1(\alpha_n r_1) U(\alpha_n r) e^{-\kappa \alpha_n{}^2 t}}{J_1{}^2(\alpha_n r_1) - J_1{}^2(\alpha_n r_2)} \left[\frac{\pi}{2} J_1(\alpha_n r_1) \int_{r_1}^{r_2} rg(r) U(\alpha_n r)dr \right.$$

$$\left. + \frac{\kappa J_1(\alpha_n r_2)}{\alpha_n r_1} \int_0^t f_1(\lambda) e^{\kappa \alpha_n{}^2 \lambda} d\lambda - \frac{\kappa J_1(\alpha_n r_1)}{\alpha_n r_2} \int_0^t f_2(\lambda) e^{\kappa \alpha_n{}^2 \lambda} d\lambda \right].$$

$U(\alpha_n r)$: Same as in 86;
α_n are roots of: $U'(\alpha_n r_1) = 0$.

89. Pressure decline in a closed lens produced by a single well: Figs. 256 and 257.

[FLOW OF GASES

90. Gas flux through a linear sand column: 11.2(4).

$$Q = \frac{k\gamma_0}{\mu(1 + m)L} \{p_2{}^{1+m} - p_1{}^{1+m}\} = \text{mass flux per unit area.}$$

γ_0 = gas density at unit pressure.
m = characteristic of type of flow, and is defined by the equation $\gamma = \gamma_0 p^m$; $m = 1$ for isothermal flow.
p_2, p_1 are the pressures at the ends of the column of length L.

91. Strictly radial (isothermal) gas flow (mass) into a well: 11.3(3).

$$Q = \frac{\pi k h \gamma_0 (p_e{}^2 - p_w{}^2)}{\mu \log r_e/r_w}.$$

92. Gas flow into a well with arbitrary pressures over the boundaries of the system: 11.3(4).

$$Q = \frac{2\pi k h \gamma_0 (\overline{p_e{}^{1+m}} - \overline{p_w{}^{1+m}})}{(1 + m)\mu \log r_e/r_w}.$$

$\overline{p_e{}^{1+m}}$, $\overline{p_w{}^{1+m}}$ = averages of $p_e{}^{1+m}$, $p_w{}^{1+m}$ over the boundaries of radii r_e, r_w (well).

93. General formula for gas flow into a well: 11.3(5) (cf. 35).

$$Q = \frac{2\pi k h \gamma_0 (p_e{}^{1+m} - p_w{}^{1+m})}{(1 + m)\mu \log c/r_w}.$$

94. Production capacities of partially penetrating gas wells: Figs. 262 and 263.

95. Relation of average gas-oil ratios in a field to the depletion of its oil content: 11.8(1).

$$P \geqq \frac{R_i(p_i - \bar{p})}{p_i \bar{R}}.$$

P = fractional oil recovery until the time that the average reservoir pressure has fallen to $\bar{p}$.

p_i = initial reservoir pressure.

R_i = initial formational gas-oil ratio.

$\bar{R}$ = average gas-oil ratio during the pressure drop from p_i to $\bar{p}$. Equality holds when there are no free gas sands; inequality refers to a production accompanied by a free-gas-zone depletion.

96. Gas-oil ratio for communicating gas and oil zones: 11.9(4).

$$R = \frac{\mu_o p_e (1 - y_e/h)}{\mu_g y_e/h} = \text{mass of gas per unit volume of oil}$$
produced.

μ_o, μ_g = viscosities of oil and gas, flowing as homogeneous fluids.

p_e = reservoir pressure.

y_e = thickness of oil zone at large distance from well.

h = total sand thickness.

97. Gas-oil ratios for noncommunicating gas and oil zones: 11.10(3) and Figs. 267, 268, 269.

$$R = R_0 + \frac{k_g \mu_o h_g (p_{eg}^2 - p_w^2)}{2 k_o \mu_g h_o (p_{eo} - p_w)}.$$

R_0 = gas-oil ratio due to the gas originally dissolved in the oil.

k_g, k_o = permeabilities of gas and oil zones.

h_g, h_o = thicknesses of gas and oil zones.

p_{eg}, p_{eo} = reservoir pressures in gas and oil zones.

98. Effect of tubing lowering on the gas-oil ratio from wells exposed to a gas zone overlying the oil zone: 11.11(2), (3) and Figs. 273 and 274.

$h_t \leqslant h_g$:

$$\frac{R'}{R} =$$

$$\frac{(h_g - h_t)\left\{ \bar{p}_e{}^2 - \left[\bar{p}_w - \frac{\gamma g(h_g - h_t)}{2} \right]^2 \right\} + h_t\{ \bar{p}_e{}^2 - [\ \bar{p}_w - \gamma g(h_g - h_t)]^2 \}}{h_g\left\{ \bar{p}_e{}^2 - \left(\bar{p}_w - \frac{\gamma g h_g}{2} \right)^2 \right\}} ;$$

$h_t \geqslant h_g$:

$$\frac{R'}{R} = \frac{h_o(\bar{p}_e - \bar{p}_w)\{ \bar{p}_e{}^2 - [\bar{p}_w + \gamma g(h_t - h_g)]^2 \}}{\left[h_o(\bar{p}_e - \bar{p}_w) - \frac{\gamma g(h_t - h_g)^2}{2} \right]\left[\bar{p}_e{}^2 - \left(\bar{p}_w - \frac{\gamma g h_g}{2} \right)^2 \right]} .$$

R = gas-oil ratio with tubing set above the gas zone.

R' = gas-oil ratio after lowering the tubing to a depth h_t below the top of the gas zone.

h_g, h_o = thicknesses of the gas and oil zones.

$\bar{p}_e, \bar{p}_w$ = reservoir and bottom-hole pressures, as reduced to the level of the gas-oil interface with the tubing set above the gas zone.

γ = density of oil-gas mixtures in well bore opposite the producing horizons.

99. Production decline in a closed gas reservoir drained by a well held at a constant bottom-hole pressure: Fig. 277.

100. Pressure decline in a closed gas reservoir drained by a well producing at a constant rate: Fig. 278.

AUTHOR INDEX

751

SUBJECT INDEX

A

Acid treatment, 404, 420, 428, 448
  of fractured limestones, 425, 427, 451, 741
  of gas reservoirs, 685, 717
  in radial systems, 422, 424, 448, 741
Analogies of porous flow, 139*ff*., 146, 455, 736
Analytic functions, 182
Anisotropic media, 225*ff*., 253, 738
  partially penetrating wells in, 277*ff*., 285, 739
  resultant permeability in, 227
Averaging procedure, 240

B

Bessel functions, 429*ff*., 630*ff*.
  definition of, 431
  differential equation for, 431, 662
  recurrence relations for, 662
Bipolar coordinates, 181
Boundary conditions, 136*ff*., 146
  analytic form of, 139
  mixed, 188
  at surfaces of discontinuity, 401, 448
Boyle's law, 123, 143

C

Calculus of variations, 235
Capillary zone, 30*ff*., 292
  effect on gravity flow, 372, 373
  fluid movements in, 30*ff*.
Classical hydrodynamics, 125*ff*., 144
  derivation of Darcy's law from 131
  discontinuous motion in, 300

Compaction, 14*ff*.
  of clay, 16
  effect on fluid migration, 44
  effect on permeability, 17
  of sand and gravel, 14
Compressible liquids, analytic definition, 132
  equation for density distribution of, 133, 145
  nonsteady state radial flow of, 630*ff*., 632, 642, 655
  sources and sinks for, 666, 676
  steady-state flow of, 627*ff*., 745
  linear flow, 628
  radial flow, 628*ff*.
Conjugate functions, 181, 469
  method of, 181*ff*.
  transformations, 186
  by electrical models, 242
Connate waters, 40*ff*.
  analyses of, 42
  migration of, 43
  occurrence of, 40
Conversion factors, 723, 724
Curvilinear coordinates, 725
Cylindrical coordinates, 141, 726

D

Dams, 3
  coffer, 221*ff*., 252, 748
  pressure distribution at base of, 193, 201, 202, 207, 248, 737
  sand-model experiments on, 205*ff*., 250
  seepage flux under, 208*ff*., 215*ff*., 738
  seepage through, approximate theory of, 338*ff*., 377*ff*., 393, 398
  rigorous theory of, 303*ff*., 314*ff* 388*ff*., 739